CHEMISTRY
IN THE WORLD

THIRD EDITION

BY Dr. Kirstin Hendrickson

ARIZONA STATE UNIVERSITY

cognella® | ACADEMIC PUBLISHING

Bassim Hamadeh, CEO and Publisher
Kassie Graves, Director of Acquisitions
Jamie Giganti, Senior Managing Editor
Jess Estrella, Senior Graphic Designer
Bob Farrell, Senior Field Acquisitions Editor
Natalie Lakosil, Licensing Manager
Kaela Martin, Allie Kiekhofer, and Rachel Singer, Associate Editors
Kat Ragudos, Interior Designer

Cover image copyright © Scott Lefler.
copyright © Scott Lefler.
copyright © Scott Lefler.
copyright © 2011 Depositphotos/Nadezda Razvodovska.

Printed in the United States of America

ISBN: 978-1-63487-540-0 (pbk) / 978-1-63487-541-7 (br)

For my students, from whom I have learned
more than I could ever teach.

DEDICATION

CONTENTS

ACKNOWLEDGMENTS

Developing this book has been a labor of love over the last ten or more years, though the actual writing took (thankfully!) somewhat less than that. There are many without whom this text would never have transitioned from dream to reality—you know who you are, and I thank you. A few individuals, though, require special mention. To Scott for being the catalyst, and providing the gorgeous photographs that bring the ideas to life. To my person…for so many reasons. I need not list, because you know. Finally, to my beautiful daughter: you are a newcomer to this planet since the book was first published, but you inspire me daily to look at the world around me in a different way, and your sense of wonder kindles mine. I look forward to so much more kitchen chemistry with you. Please never stop asking, "*WHY?*" …and I will never stop answering.

TO THE STUDENT AND THE INSTRUCTOR—HOW TO USE THIS BOOK

This text was written to provide a perspective on socially important topics in the field of chemistry, which as the student will soon ascertain, is a broad field indeed! Of course, even an applications-based approach (as this book is) must by necessity introduce sufficient principle and theory to allow for meaningful discussion. It was a primary goal in writing this text to include only those principles absolutely necessary to understanding observable chemical phenomena and relevant chemical issues. At every step of the way, I strove to make readily apparent the answer to the question, *When am I ever going to use this again?* For this reason, I recommend reading this book more like literature than a science text—it tells a story. Granted, the story is occasionally punctuated by equations and symbols, but it's a story nevertheless. Rather than trying to internalize the applications in terms of the chemical principles, use the applications to make sense of the principles—the big picture will help you understand the details.

Each chapter introduces a topic or group of related topics in which we see chemistry on a day-to-day basis, in our lives or in the news. Principles are addressed where needed in the course of discussing the topic. Where relevant, practice problems are provided in boxes labeled *Try This*. It's worth taking a moment to try a problem on your own before moving on—there are many additional practice problems at the end of the chapter. Conceptual questions are occasionally presented in boxes labeled *Concept Check*. It should be fairly easy to answer the question posed in the box after reading a particular section—if not, that's an indication that you may have missed a main point from the previous several paragraphs. Occasionally, *Concept Check* boxes will refer to older concepts that are being brought in to the current discussion, and in such cases, the answers will refer you to the relevant chapter. Key sentences within the reading appear in bold, while glossary terms are in red.

Problems at the end of the chapter are divided into two sections. *Questions and Topics for Discussion* are intended to stimulate thought—in many cases, there are no right or wrong answers. These questions are meant to be suitable for essay or paragraph assignments, stimulation of thinking, or in-class discussions. *Problems* are mathematical or factual in nature—in particular, math problems occur in sets of two (grouped together) for ease of studying and assigning work.

New for the third edition are several case studies, which are meant to give comprehensive examples of real world chemistry in action—often with an eye toward sustainability. These allow the student to examine specific solutions to chemistry problems in the real world; they also highlight some really unique and neat approaches to business, agriculture, and so forth. It is my hope to add more of these case studies in later editions, and I welcome input from either students or faculty with regard to what sorts of case studies you'd find interesting and instructive.

The last section in each chapter is called *The Last Word*, and is an offshoot of something I've done in my classes that has been very popular with my students. In class, I call it "Five Minutes of Really Cool Chemistry," and I'll talk about something at least passingly relevant and always really cool that is based upon the principles we've been discussing. *The Last Word* is its adaptation into written form—these sections are meant to be interesting vignettes that relate to the concepts addressed in the chapter. They're not chapter summaries (which do, incidentally, appear separately at the end of each chapter), but are hopefully a few moments worth of relaxation and wonder—a few minutes of really cool chemistry.

Enjoy the journey!

ABOUT THE COVER

From top to bottom: In addition to making sugar and structural material (cellulose) by using the Sun's energy to combine water and carbon dioxide, this columbine has a lovely color that results from absorbing the lower wavelengths of visible light, while reflecting the shorter (blue and purple) ones (columbine, (c) Scott Lefler, 2009). This active child is deepening and accelerating her respirations in order to provide more oxygen to her hard-working cells. They, in turn, combine the oxygen with sugar in order to generate the chemical energy molecule, ATP, which powers her muscles (running child, (c) Scott Lefler, 2016). Water is one of the most unique and fascinating molecules there is, and is also required for all known life. In addition to making for beautiful scenery, water is the medium of life: in essence, we are all water-based soup (Mooney Falls, (c) Scott Lefler, 2009).

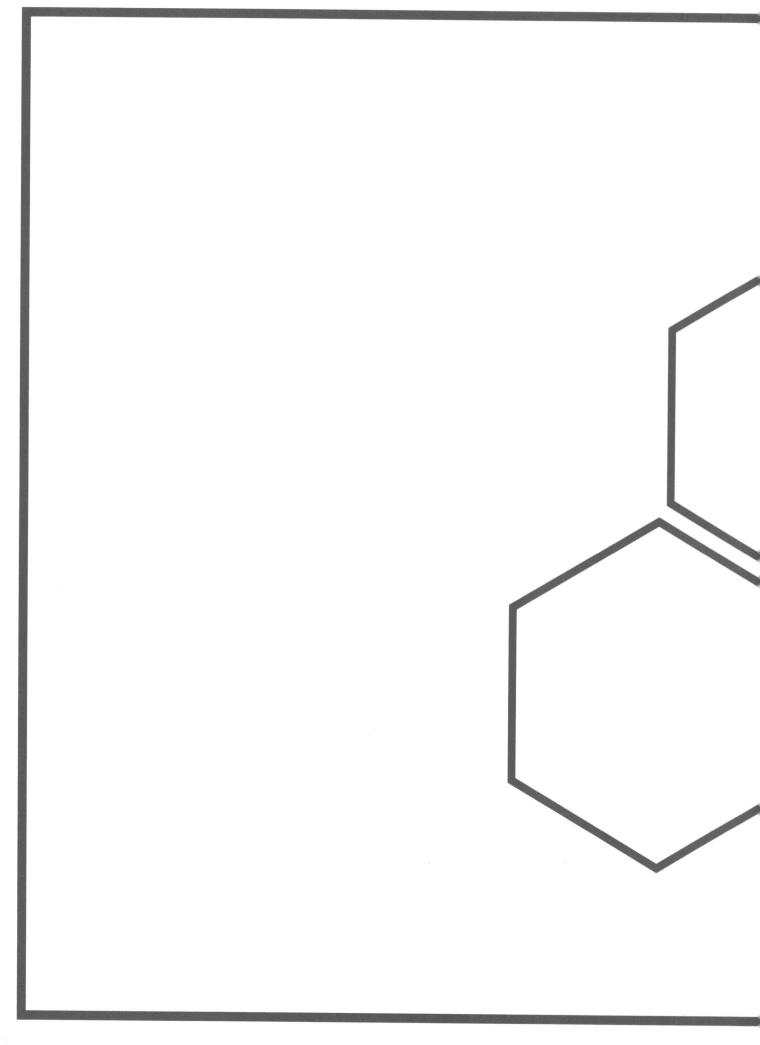

ALL ArOUND US AND INSIDE O F US

Chemistry Is Everywhere

1

The word *chemistry*, especially in the title of a class or book, conjures up images of goggled scientists at their lab benches, surrounded by odd bubbling solutions in beakers and flasks. Depending upon the power of your imagination, this scene might even be punctuated in your mind by the occasional explosion. In any case, as you imagine the white-coated chemist going about her business, mixing and stirring and writing down equations rife with symbols and letters of the Greek alphabet, I practically guarantee the two things you are thinking to yourself are "What in the world does this have to do with real life?" and "When am I ever going to use this information?!" The answers to those (very real, very important) questions are *everything*, and *every day*!

In order to see the applicability of chemistry to real life, we have to stop imagining it as something that takes place in a laboratory and start looking for it in the real world. **Chemists work in laboratories because they are attempting to study what goes on everywhere** else, **not because the lab is the only place chemistry happens!** In this book, we'll look for chemistry everywhere *except* the lab. We'll learn that there's chemistry all around us—in leaves changing colors in the fall, in the atmosphere (which keeps our planet warm and habitable, the same way a greenhouse protects roses from winter frost), in nuclear power plants, and in the food we eat, as image 1.1 demonstrates.

IMAGE 1.1. (a) Leaves change color in the fall as the light-absorbing chemical chlorophyll (which gives them their green color) is no longer produced, leaving behind other light-absorbing chemicals such as carotenes (yellow) and anthocyanin (red). (b) Just as a greenhouse made of glass traps heat to keep plants warm during cooler months, the gases of the atmosphere trap heat from the sun, keeping the planet warm and habitable. (c) Nuclear reactions have the potential to release massive amounts of energy that can be used to generate power. (d) The food we eat is made up of chemicals that provide us with energy and raw material for building muscle and other tissue.

IMAGE 1.2. The complex interplay between chemistry and society is apparent in this picture. We affect the chemistry of the oceans by adding environmental pollutants and toxins that disrupt the ecosystem. As a result, the altered chemistry of the ocean affects us. For example, mercury builds up in high-level predator fish that we use as food (tuna and shark are among these), rendering them toxic to us if consumed more than occasionally.

IMAGE 1.3. Chemistry on the individual level includes such topics as pharmaceuticals (b) and food (e). Cities need to consider the chemistry of local air pollution (c) and water safety (purification plant, a). As a global community, we are all concerned with and affected by climate change. Due to a number of factors, including economic disparity between First- and Third-World nations, international treaties to reduce carbon emissions (such as the Kyoto Protocol, with participation status represented on image d) have been unsuccessful (see Chapter 13).

This isn't just a chemistry textbook; it also addresses ways in which chemistry, chemicals, and chemists affect society. However, this is more complex than it sounds. It's very important to think about the relationship between chemistry—any science, really—and society as a reciprocal one. To be sure, chemistry affects society; pollution, for instance, has a clear and unequivocal negative impact upon our air and water (Image 1.2). However, society also affects chemistry, in that our actions produce and release chemicals into the environment, and we formulate legislation that affects what can and can't be done chemically. **In this text we will be exploring the nature of this relationship and some of the ways in which chemistry and society affect each other.**

It's important to note that society is multifaceted. Individuals, local communities, and the global population represent different levels of society. Any level can affect (and be affected by) any other through chemistry, as illustrated in Image 1.3. For instance, making appropriate choices regarding nutrition and exercise positively impacts personal health, but also reduces the need for medical care, which

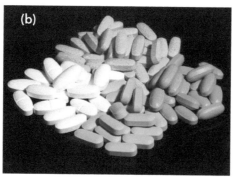

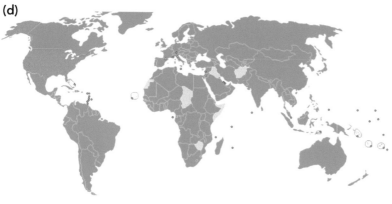

decreases insurance rates for all policyholders. A community that installs a light-rail system benefits from a reduction in local pollution, but there are also global benefits through reduced carbon emissions.

In order to address the reciprocal relationship between chemistry and each level of society, this text is divided into four units. In the first, we'll be discussing some introductory chemical principles in the context of the chemistry we encounter every day. The second unit will address chemical concerns within local communities—things like air pollution and water safety. In Unit 3, we'll shift our focus to the chemistry that takes place within our own bodies. Finally, in Unit 4, we'll expand our thinking outward to encompass issues of global concern. At every step along the way, we will be focusing not simply upon chemical concepts, but on how we affect the chemistry and how the chemistry affects us.

1.1 Classifications of Matter

Stated very simply, chemistry is the study of matter. **Matter is anything that occupies space and has mass, so it is what makes up the physical universe.** Everything we can touch, see, and smell (as well as things we can't see, like the air around us) is made of matter. Because there are several different types of matter, we need to define some general categories.

Let's start with mixtures, because they're actually the easiest kind of matter to understand. **Mixtures are physical combinations of two or more pure substances in variable proportions.** There are two key components to this definition. First, a mixture is a physical combination, which means that each pure substance within the mixture retains its own chemical identity and properties but is interspersed with one or more other pure substances in space (where a pure substance is an element or a compound, both of which will be discussed in greater detail shortly). For instance, making pancake batter involves mixing sugar, baking soda, flour, spices, eggs, and so forth (Image 1.4). Sugar is a pure substance with distinct properties (it tastes sweet, it dissolves in water, and so on). Baking soda is another pure substance with a separate set of distinct properties. As an ingredient like sugar is added to the batter, it mixes in and spreads out through the batter until all parts of the mixture are equally sweet. The sugar, though dispersed through the batter, is still sugar—as a chemical, it remains unchanged. That's the nature of a mixture.

IMAGE 1.4. When table salt (NaCl) is stirred into water (H_2O), neither the salt nor the water changes chemical properties—instead, a mixture forms (a). Beating together the ingredients for pancakes makes a mixture with tasty potential (b, photo © Scott Lefler, 2010).

Saltwater is another example of a mixture. Salt (a common name for the chemical sodium chloride, or NaCl) has distinct properties. Water (a common name for the chemical H_2O) has a different set of properties. They can be stirred together to make saltwater, but the salt is still NaCl, and the water is still H_2O. In other words, physically mixing salt and water means that bits of NaCl are surrounded by bits of H_2O and vice versa, but the individual chemicals retain their identities and properties. If we wanted to, we could "unmix" salt water by boiling it in a pan on the stove, in which case the water would depart as steam (which is still H_2O), and the NaCl would

be left behind as a crust on the bottom of the pan (Image 1.5). At no point in the mixing or separating process does either chemical change identities or properties.

The second key part of the definition of a mixture is that the pure substances are combined in variable proportions. **This means that it doesn't matter how much salt we mix with a given amount of water—whether we use lots of salt or only a bit, we'll still have a mixture of saltwater.**

Chemicals that can be physically combined to produce mixtures are called pure substances, and can be divided into two subcategories: elements and compounds. **Elements are the simplest types of matter and are found listed on the periodic table (Image 1.6); they cannot be broken into simpler substances by any chemical means.** When you look at the periodic table, you'll notice that it's made up of many elements, each with a characteristic atomic symbol, which is a letter or combination of letters that represents the element. C, for instance, is the atomic symbol for the element carbon. Each element has its own unique physical and chemical properties (Image 1.7). Some, like neon (Ne), are gases at room temperature, while others are liquids or solids. Many, like gold (Au), uranium (U), and mercury (Hg) are metals, while others, like sulfur (S), are not. We'll look more closely at properties of specific elements and at the organization of the periodic table later in this chapter.

IMAGE 1.5. A beaker full of a saltwater mixture, if put on a burner until all the water evaporates, is left full of salt crust. The evaporating water vapor could be collected and cooled, in which case it would condense back into ordinary liquid water. The salt crust, if crushed, would be indistinguishable from table salt straight from the shaker.

IMAGE 1.6. The periodic table of the elements.

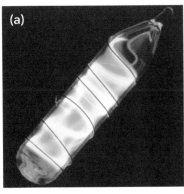
(a)

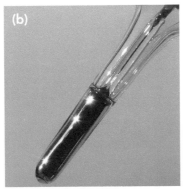
(b)

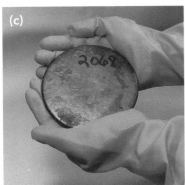
(c)

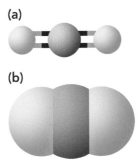
(d)

IMAGE 1.7. Some elements (clockwise from top left): neon (Ne) can be used to make glowing signs(a); mercury (Hg) is the only metal that's liquid at room temperature, and is used in thermometers (b); gold (Au) is a beautiful, butter-yellow metal highly prized for its aesthetic appeal and scarcity (d); uranium (U) is used in nuclear reactors to generate energy (c).

Compounds are chemical combinations of two or more elements in fixed, characteristic proportions. Notice the differences between the definition of a compound and the definition of a mixture. First, where mixtures are physical combinations in which each pure substance retains its identity and properties, compounds are **chemical combinations**, meaning that a **chemical bond** forms between particles of two or more different elements. A chemical bond is a little like glue—it sticks two particles together. **The compound has an identity and properties that are different from the identities and properties of the elements that make it up.** For instance, carbon dioxide (CO_2) is a compound made from a chemical combination of the elements carbon and oxygen, but CO_2 is nothing like either carbon or oxygen—it is its own separate, distinct substance. Another major difference between compounds and mixtures is that compounds must consist of **fixed proportions** of the constituent elements. While we can use any quantity of salt and any quantity of water to make saltwater, CO_2 is only CO_2 if it's made up of one particle of carbon and two particles of oxygen. Image 1.8 provides two ways to visualize the combination of carbon with oxygen to form CO_2. The subscripts in the compound's formula tell us how many particles of each element are chemically combined to form the compound, where the lack of a subscript is always taken to mean *one*. If one particle of carbon and one particle of oxygen combined chemically, the resulting compound would *not* be CO_2. Instead, it would be CO—carbon monoxide—which has very different properties

(a)

(b)

IMAGE 1.8. Two representations of the compound CO2. The figure on the top (a) shows that an atom of carbon is chemically combined with two atoms of oxygen. This representation is often called a "ball and stick" model, because it represents the atoms as "balls" on either end of chemical bonds, which are represented as "sticks." In reality, bonds form when atoms sort of squish together in space, as represented in the image on the bottom. This is called a space-filling model, and while a bit more accurate, it's harder to draw.

IMAGE 1.9. Some commonly encountered compounds include (clockwise from top left): water (H_2O; a), sucrose (table sugar, $C_{12}H_{22}O_{11}$; b); calcite (the major ingredient in marble, $CaCO_3$; e); and ethanol (the alcohol in beer, wine, and liquor, C_2H_6O; c and d). Crater Lake photo © Scott Lefler, 2006.

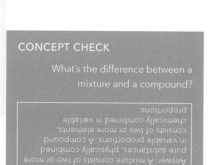

CONCEPT CHECK

What's the difference between a mixture and a compound?

Answer: A mixture consists of two or more pure substances, physically combined in variable proportions. A compound consists of two or more elements, chemically combined in variable proportions.

NOTE[1]

This statement seems paradoxical at first—how can we say that elements are the simplest type of matter, and then say that if we were to divide an atom of an element (gold, for instance), we'd get something that was not gold? We'll discuss this more in Chapter 2, but as it turns out, we can't really "cut" an atom and get pieces that hang around at all. If an atom were (theoretically) divided, the resulting pieces would be what are called subatomic particles, and we don't find them on their own (in a persistent state) in nature.

than CO_2. Carbon dioxide is a waste product of respiration, but it also plays an important role in helping to regulate the acidity of the body's fluids, as we'll see in Chapter 5. Carbon monoxide, on the other hand, is (among other things) a pollutant gas formed through combustion of certain fuels, and it is toxic when inhaled. Much of what we see in the world around us is made up of various compounds (Image 1.9).

CO_2 can be classified further—in addition to being a pure substance, it's also a molecule. **A molecule is a pure substance made from the chemical combination (through formation of bonds) of two or more atoms. An atom is the smallest particle of an element that maintains the identity of that element.** In other words, if I had a chunk of the element gold (Au on the periodic table) and I cut it in half, I'd have a smaller chunk of gold. If I cut that smaller chunk in half, and then cut the half in half, and so forth, eventually I'd get to a very, very tiny particle of gold that couldn't be cut in half (or at least, if it were, it wouldn't be gold anymore[1]). This tiny, indivisible particle of gold is an atom (the word atom comes from the Greek word *atomos*, which means indivisible). All matter is made up of atoms, some of them uncombined and some of them in chemical combination with other atoms. A molecule of CO_2 is made up of one atom of carbon and two atoms of oxygen.

It's important to note that the words compound **and** molecule **are not synonymous.** It's absolutely possible to have a molecule that is not a compound, as long as it's made of two or more atoms *of the same element*. For example, most respiring organisms require oxygen, which is found in our atmosphere as the molecule O_2. However, atoms of the element oxygen can also be found incorporated into compounds (like CO_2). Uncombined atoms of oxygen aren't found in nature, for reasons that will be discussed in later chapters. O_2 is made of two atoms of oxygen, so even though it's a molecule, O_2 is an element rather than a compound. In nature, some elements are found as uncombined atoms; examples include helium (He) and neon (Ne). Other elements occur most frequently as **diatomic molecules** (molecules consisting of only two atoms) such as O_2 and N_2. Still others occur naturally as **polyatomic molecules**, or molecules composed of many (*poly* means "many") atoms. Sulfur, for instance, is most commonly found in nature as S_8 (Image 1.10). Atoms, molecules, elements, compounds, and mixtures are visually summarized in Image 1.11.

One question students often have regarding elements is how to know whether they are found in nature as single atoms, diatomic molecules, or polyatomic molecules. The answer is that, frankly, it's a bit complicated! There are a few general guidelines, though, that are good to know. **The elements that naturally form diatomic molecules are H, N, O, F, Cl, Br, and I** (Image 1.12), and it's quite useful to know this about them, because we'll be talking frequently about several of these elements.

The elements in the far-right column of the periodic table tend to be found as free atoms, and we'll learn why in Chapter 3. One other element whose form in nature is worth mentioning is carbon. Most of the carbon atoms we'll see in this class will be chemically combined with other elements, because carbon is a very common

IMAGE 1.10. The polyatomic molecule S_8 is a crystalline, yellow solid that smells faintly of rotten eggs.

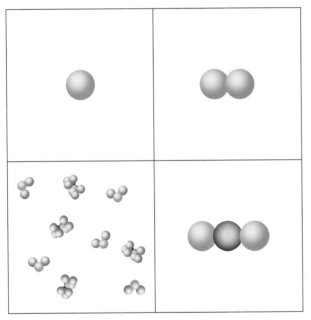

IMAGE 1.11. An atom of oxygen (top left) is a pure substance, an element, and, of course, an atom. A diatomic molecule, oxygen (O_2, top right) is also a pure substance and an element. A molecule of carbon dioxide (CO_2, bottom right) is a pure substance, a compound, and a molecule. At the bottom left, a mixture of ethanol (the larger molecules in the mixture; C_2H_6O) and water (the smaller molecules in the mixture; H_2O) contains molecules of both compounds, physically stirred together but chemically unchanged.

TRY THIS

Which of the following are pure substances, and which are mixtures? Of the pure substances, which are elements, and which are compounds? Which are atoms, and which are molecules?

Fe, $FeCl_2$, sugar water, silver (Ag on the periodic table), maple syrup, Cl_2. Answers: Fe is a pure substance, an element, and an atom. $FeCl_2$ is a pure substance, a compound, and a molecule. Sugar water is a mixture. Ag is a pure substance, an element, and an atom. Maple syrup is a mixture. Cl_2 is a pure substance, an element, and a molecule.

1	2	3	4	5	6	7	8	9	10	11	12	13	14	15	16	17	18
1 H 1.01 Hydrogen																	2 He 4.00 Helium
3 Li 6.94 Lithium	4 Be 9.01 Beryllium											5 B 10.81 Boron	6 C 12.01 Carbon	7 N 14.01 Nitrogen	8 O 16.00 Oxygen	9 F 19.00 Fluorine	10 Ne 20.18 Neon
11 Na 23.0 Sodium	12 Mg 24.3 Magnesium											13 Al 26.98 Aluminum	14 Si 28.09 silicon	15 P 30.97 Phosphorus	16 S 32.07 Sulfur	17 Cl 35.45 Chlorine	18 Ar 39.95 Argon
19 K 39.10 Potassium	20 Ca 40.08 Calcium	21 Sc 44.96 Scandium	22 Ti 47.88 Titanium	23 V 50.94 Vanadium	24 Cr 52.00 Chromium	25 Mn 54.94 Manganese	26 Fe 55.58 Iron	27 Co 58.93 Cobalt	28 Ni 58.69 Nickel	29 Cu 63.55 Copper	30 Zn 65.39 Zinc	31 Ga 69.72 Gallium	32 Ge 72.59 Germanium	33 As 74.92 Arsenic	34 Se 78.96 Selenium	35 Br 79.90 Bromine	36 Kr 83.80 Krypton
37 Rb 85.47 Rubidium	38 Sr 87.62 Strontium	39 Y 88.91 Yttrium	40 Zr 91.22 Zirconium	41 Nb 92.91 Niobium	42 Mo 95.94 Molybdenum	43 Tc (98) Technetium	44 Ru 101.1 Ruthenium	45 Rh 102.9 Rhodium	46 Pd 106.4 Palladium	47 Ag 107.9 Silver	48 Cd 112.4 Cadmium	49 In 114.8 Indium	50 Sn 118.7 Tin	51 Sb 121.8 Antimony	52 Te 127.6 Tellurium	53 I 126.9 Iodine	54 Xe 131.3 Xenon
55 Cs 132.9 Cesium	56 Ba 137.9 Barium	57 La *138.9 Lanthanum	72 Hf 178.5 Hafnium	73 Ta 180.9 Tantalum	74 W 183.9 Tungsten	75 Re 186.2 Rhenium	76 Os 190.2 Osmium	77 Ir 192.2 Iridium	78 Pt 195.1 Platinum	79 Au 197.0 Gold	80 Hg 200.5 Mercury	81 Tl 204.4 Thalium	82 Pb 207.2 Lead	83 Bi 209.0 Bismuth	84 Po (210) Polonium	85 At (210) Astatine	86 Rn (222) Radon
87 Fr (223) Francium	88 Ra (226) Radium	89 Ac *(227) Actinium	104 Rf (257) Rutherfordium	105 Db (260) Dubnium	106 Sg (263) Seaborgium	107 Bh (262) Bohrium	108 Hs (265) Hassium	109 Mt (266) Meitnerium	110 Ds (271) Darmstadtium								

58 Ce 140.1 Cerium	59 Pr 140.9 Praseodymium	60 Nd 144.2 Neodymium	61 Pm (147) Promethium	62 Sm 150.4 Samarium	63 Eu 152.0 Europium	64 Gd 157.3 Gadolinium	65 Tb 158.9 Terbium	66 Dy 162.5 Dyspromium	67 Ho 164.9 Holmium	68 Er 167.3 Erbium	69 Tm 168.9 Thulium	70 Yb 173.0 Ytterbium	71 Lu 175.0 Lutetium
90 Th 232.0 Thorium	91 Pa (231) Protactinium	92 U (238) Uranium	93 Np (237) Neptunium	94 Pu (242) Plutonium	95 Am (243) Americium	96 Cm (247) Curium	97 Bk (247) Berkelium	98 Cf (249) Californium	99 Es (254) Einsteinium	100 Fm (253) Fermium	101 Md (256) Mendelevium	102 No (254) Nobelium	103 Lr (257) Lawrencium

(Top right repeated pair: 1 H 1.01 Hydrogen | 2 He 4.00 Helium)

IMAGE 1.12. The elements that exist in nature as diatoms are indicated on this periodic table.

NOTE²

If graphite and diamond are both made of carbon, why is diamond so much more valuable than graphite? The quick answer is that it's much harder and more durable because of the arrangement of atoms. It's also aesthetically pleasing. These two factors combine to make it a good candidate for jewelry (among other things), and because it's rare, it comes at a high price. From a chemical perspective, diamond is actually less stable than graphite, meaning that very slowly over time, diamond turns into graphite! Don't worry; it doesn't happen on a human timescale, so your diamonds will still be around for your grandchildren. Still, it's interesting to think about!

component of compounds in nature. However, elemental carbon is also common—probably more so than you think—and it can be found in several different forms with very different properties. Two of the most frequently encountered forms of elemental carbon are graphite and diamond. These differ only in the way in which carbon atoms are arranged and chemically bonded to one another.² Looking at the arrangement of atoms in these two forms (Image 1.13) helps us to understand some of the differences in their physical properties and appearances. Graphite's structure is composed of carbon atoms bonded into flat plates. These plates lie on top of each other but are independent and can slide past one another easily. This sliding capability within the structure of graphite makes it soft and easily spread onto a surface (hence its utility as pencil "lead"). Also, it makes graphite a good lubricant in mechanical applications. Diamond, on the other hand, has carbon atoms arranged in a crystal lattice in which each atom is bonded to all nearby atoms. This structure is incredibly rigid, giving diamond its unique hardness.

One of the themes that we will see repeatedly in this class is the pervasiveness of chemistry. Even when we don't think we're talking about chemistry, much of the time we are. For instance, let's ponder jewelry for a moment. This is a topic that seems very far removed from anything we might traditionally associate with chemistry. However, because it occupies space and has mass, jewelry must be made of matter, and chemistry is the study of matter. It becomes a little more complicated

(a)

(b)

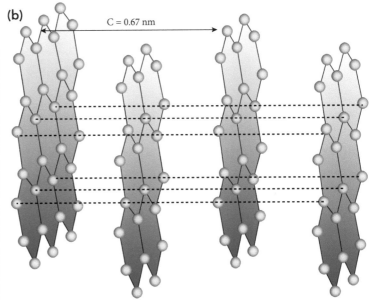
(b)
C = 0.67 nm

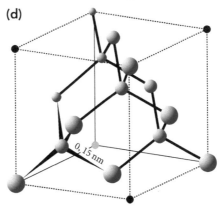
(d)
0.15 nm

IMAGE 1.13. Note the difference in appearance between graphite (top left) and diamond (bottom left), both of which are forms of the element carbon.

when we try to ask ourselves of what *kind* of matter jewelry is made. Take, for example, a gold ring. If you look inside a gold ring, you'll see a stamp that says 14K, 18K, or some other number followed by a "K" to indicate the material of the ring[3] (Image 1.14).

Pure gold is an element (Au on the periodic table), and while aesthetically pleasing and rare, which combine to make it valuable, it's far too soft to be used in jewelry designed for daily wear. As such, elemental gold is mixed with other elements like nickel (Ni) and palladium (Pd). A "gold" ring, therefore, is actually made of a mixture rather than a pure substance.

IMAGE 1.14. The stamp inside a gold ring indicates what fraction (parts out of 24) of the metal mixture is the element Au. Photo © Scott Lefler, 2010.

NOTE[3]

If you're curious, the "K" is for karat, which is just a way of indicating what fraction of the metal mixture is elemental gold. By definition, 24K is elemental gold, in which 24/24 parts of the metal are gold. 18K gold has 18/24 parts elemental gold, and 6/24 parts other metals. Metallurgists blend in various metals to affect the color and properties of the resulting alloy (mixture) in addition to strengthening the gold.

Non-metals

Metals

Metalloids

1 H 1.01 Hydrogen																	2 He 4.00 Helium
3 Li 6.94 Lithium	4 Be 9.01 Beryllium											5 B 10.81 Boron	6 C 12.01 Carbon	7 N 14.01 Nitrogen	8 O 16.00 Oxygen	9 F 19.00 Fluorine	10 Ne 20.18 Neon
11 Na 23.0 Sodium	12 Mg 24.3 Magnesium											13 Al 26.98 Aluminum	14 Si 28.09 silicon	15 P 30.97 Phosphorus	16 S 32.07 Sulfur	17 Cl 35.45 Chlorine	18 Ar 39.95 Argon
19 K 39.10 Potassium	20 Ca 40.08 Calcium	21 Sc 44.96 Scandium	22 Ti 47.88 Titanium	23 V 50.94 Vanadium	24 Cr 52.00 Chromium	25 Mn 54.94 Manganese	26 Fe 55.58 Iron	27 Co 58.93 Cobalt	28 Ni 58.69 Nickel	29 Cu 63.55 Copper	30 Zn 65.39 Zinc	31 Ga 69.72 Gallium	32 Ge 72.59 Germanium	33 As 74.92 Arsenic	34 Se 78.96 Selenium	35 Br 79.90 Bromine	36 Kr 83.80 Krypton
37 Rb 85.47 Rubidium	38 Sr 87.62 Strontium	39 Y 88.91 Yttrium	40 Zr 91.22 Zirconium	41 Nb 92.91 Niobium	42 Mo 95.94 Molybdenum	43 Tc (98) Technetium	44 Ru 101.1 Ruthenium	45 Rh 102.9 Rhodium	46 Pd 106.4 Palladium	47 Ag 107.9 Silver	48 Cd 112.4 Cadmium	49 In 114.8 Indium	50 Sn 118.7 Tin	51 Sb 121.8 Antimony	52 Te 127.6 Tellurium	53 I 126.9 Iodine	54 Xe 131.3 Xenon
55 Cs 132.9 Cesium	56 Ba 137.9 Barium	57 La *138.9 Lanthanum	72 Hf 178.5 Hafnium	73 Ta 180.9 Tantalum	74 W 183.9 Tungsten	75 Re 186.2 Rhenium	76 Os 190.2 Osmium	77 Ir 192.2 Iridium	78 Pt 195.1 Platinum	79 Au 197.0 Gold	80 Hg 200.5 Mercury	81 Tl 204.4 Thalium	82 Pb 207.2 Lead	83 Bi 209.0 Bismuth	84 Po (210) Polonium	85 At (210) Astatine	86 Rn (222) Radon
87 Fr (223) Francium	88 Ra (226) Radium	89 Ac *(227) Actinium	104 Rf (257) Rutherfordium	105 Db (260) Dubnium	106 Sg (263) Seaborgium	107 Bh (262) Bohrium	108 Hs (265) Hassium	109 Mt (266) Meitnerium	110 Ds (271) Darmstadtium								

58 Ce 140.1 Cerium	59 Pr 140.9 Praseodymium	60 Nd 144.2 Neodymium	61 Pm (147) Promethium	62 Sm 150.4 Samarium	63 Eu 152.0 Europium	64 Gd 157.3 Gadolinium	65 Tb 158.9 Terbium	66 Dy 162.5 Dysprosium	67 Ho 164.9 Holmium	68 Er 167.3 Erbium	69 Tm 168.9 Thulium	70 Yb 173.0 Ytterbium	71 Lu 175.0 Lutetium
90 Th 232.0 Thorium	91 Pa (231) Protactinium	92 U (238) Uranium	93 Np (237) Neptunium	94 Pu (242) Plutonium	95 Am (243) Americium	96 Cm (247) Curium	97 Bk (247) Berkelium	98 Cf (249) Californium	99 Es (254) Einsteinium	100 Fm (253) Fermium	101 Md (256) Mendelevium	102 No (254) Nobelium	103 Lr (257) Lawrencium

IMAGE 1.15. This periodic table has metals, nonmetals, and metalloids indicated.

1.2 The Periodic Table

The periodic table of the elements contains a lot of information; we won't go over it all just yet. However, one interesting organizational feature of the table is that it divides elements into three distinct categories based upon their position on the table, as shown in Image 1.15.

Metals, which are shiny, ductile (can be drawn into wires), conductive (can conduct electricity), and malleable (can be hammered into sheets), are located in the two columns on the left-hand side of the table and in the central, recessed portion (these recessed elements are called *transition metals*). Also, the two rows of elements that appear below the table (as though they were kicked out) are metals. The rows classically displayed under the periodic table actually fit into the center of the table as shown in Image 1.16. There are a few explanations commonly given for the fact that these elements, sometimes called the *inner transition elements*, are conventionally placed under the periodic table. One is that they were not yet discovered when the classical periodic table was developed, so room was not made for them. Another explanation is that they make the table absurdly long and difficult to display on wall posters and in text figures!

Periodic Table (Image 1.16):

1 H 1.01 Hydrogen																	2 He 4.00 Helium
3 Li 6.94 Lithium	4 Be 9.01 Beryllium											5 B 10.81 Boron	6 C 12.01 Carbon	7 N 14.01 Nitrogen	8 O 16.00 Oxygen	9 F 19.00 Fluorine	10 Ne 20.18 Neon
11 Na 23.0 Sodium	12 Mg 24.3 Magnesium											13 Al 26.98 Aluminium	14 Si 28.09 silicon	15 P 30.97 Phosphorus	16 S 32.07 Sulfur	17 Cl 35.45 Chlorine	18 Ar 39.95 Argon
19 K 39.10 Potassium	20 Ca 40.08 Calcium	21 Sc 44.96 Scandium	22 Ti 47.88 Titanium	23 V 50.94 Vanadium	24 Cr 52.00 Chromium	25 Mn 54.94 Manganese	26 Fe 55.58 Iron	27 Co 58.93 Cobalt	28 Ni 58.69 Nickel	29 Cu 63.55 Copper	30 Zn 65.39 Zinc	31 Ga 69.72 Gallium	32 Ge 72.59 Germanium	33 As 74.92 Arsenic	34 Se 78.96 Selenium	35 Br 79.90 Bromine	36 Kr 83.80 Krypton
37 Rb 85.47 Rubidium	38 Sr 87.62 Strontium	39 Y 88.91 Yttrium	40 Zr 91.22 Zirconium	41 Nb 92.91 Niobium	42 Mo 95.94 Molybdenum	43 Tc (98) Technetium	44 Ru 101.1 Ruthenium	45 Rh 102.9 Rhodium	46 Pd 106.4 Palladium	47 Ag 107.9 Silver	48 Cd 112.4 Cadmium	49 In 114.8 Indium	50 Sn 118.7 Tin	51 Sb 121.8 Antimony	52 Te 127.6 Tellurium	53 I 126.9 Iodine	54 Xe 131.3 Xenon
55 Cs 132.9 Cesium	56 Ba 137.9 Barium	57 La *138.9 Lanthanum	72 Hf 178.5 Hafnium	73 Ta 180.9 Tantalum	74 W 183.9 Tungsten	75 Re 186.2 Rhenium	76 Os 190.2 Osmium	77 Ir 192.2 Iridium	78 Pt 195.1 Platinum	79 Au 197.0 Gold	80 Hg 200.5 Mercury	81 Tl 204.4 Thallium	82 Pb 207.2 Lead	83 Bi 209.0 Bismuth	84 Po (210) Polonium	85 At (210) Astatine	86 Rn (222) Radon
87 Fr (223) Francium	88 Ra (226) Radium	89 Ac *(227) Actinium	104 Rf (257) Rutherfordium	105 Db (260) Dubnium	106 Sg (263) Seaborgium	107 Bh (262) Bohrium	108 Hs (265) Hassium	109 Mt (266) Meitnerium	110 Ds (271) Darmstadtium								

Lanthanides: 58 Ce 140.1 Cerium | 59 Pr 140.9 Praseodymium | 60 Nd 144.2 Neodymium | 61 Pm (147) Promethium | 62 Sm 150.4 Samarium | 63 Eu 152.0 Europium | 64 Gd 157.3 Gadolinium | 65 Tb 158.9 Terbium | 66 Dy 162.5 Dysprosium | 67 Ho 164.9 Holmium | 68 Er 167.3 Erbium | 69 Tm 168.9 Thulium | 70 Yb 173.0 Ytterbium | 71 Lu 175.0 Lutetium

Actinides: 90 Th 232.0 Thorium | 91 Pa (231) Protactinium | 92 U (238) Uranium | 93 Np (237) Neptunium | 94 Pu (242) Plutonium | 95 Am (243) Americium | 96 Cm (247) Curium | 97 Bk (247) Berkelium | 98 Cf (249) Californium | 99 Es (254) Einsteinium | 100 Fm (253) Fermium | 101 Md (256) Mendelevium | 102 No (254) Nobelium | 103 Lr (257) Lawrencium

IMAGE 1.16. An "extended" periodic table of the elements.

Nonmetals, which are almost always either gases or solids[4] at room temperature, have physical properties opposite those of metals. They are dull rather than shiny, brittle rather than ductile or malleable, and nonconductive. **Nonmetals are located on the right-hand side of the table.** The element hydrogen (H) shows up in two places in many periodic tables. This is because it has some properties in common with metals and some properties in common with nonmetals, as we'll see in more detail in later chapters. **Hydrogen's technical classification, however, is as a nonmetal.**

Along a diagonal line between the metals and nonmetals are the metalloids, whose physical properties are between those of metals and nonmetals, and many of which are also *semiconductors*. Some of the metalloids are useful in making photovoltaic cells for solar-power panels (as we'll see in Chapter 11), as well as in computer applications. Image 1.17 illustrates the visual differences among metals, nonmetals, and metalloids.

TRY THIS

Classify each element as a metal, metalloid, or nonmetal.
Fe, B, I, Kr, Ca, U, Si.
Answers: The metals are Fe (iron), Ca (calcium), and U (uranium). The metalloids are B (boron) and Si (silicon). The nonmetals are I (iodine) and Kr (krypton).

NOTE[4]

There's one liquid nonmetal: bromine (Br_2).

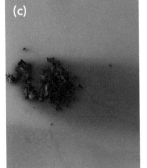

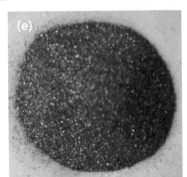

IMAGE 1.17. Copper (a) has all the properties of metal: it's shiny, ductile, conductive, and malleable. Sodium (b) is a metal (shown here in liquid paraffin to keep it from reacting with air), but we don't find it looking traditionally metallic in nature because it's so reactive and combines with other elements in compounds like table salt. Silicon (e) is a metalloid (and a semiconductor) commonly used in computer chips. The nonmetals include gases such as chlorine (a toxic yellowish gas, bottom center), and brittle, nonconductive solids such as iodine (c).

H_2O

IMAGE 1.18. Macroscopic, microscopic or molecular (a), and symbolic (b) representations of the compound water. Havasupai Falls picture © Scott Lefler, 2007.

As you look at the elements on the periodic table, you'll notice a number of names among the metals that belong to substances you may never have thought of as metallic, like sodium (Na) and calcium (Ca). Elemental sodium and calcium are both shiny metals. They have all the properties of more familiar metals like aluminum and silver, but, oddly enough, they are soft. Sodium can actually be cut with a butter knife! The reason we don't find these elements in their metallic form in nature is that they are very reactive and readily combine with other elements to form compounds. For instance, you commonly encounter the element sodium as part of the compound sodium chloride (NaCl), which is table salt.

1.3 Thinking About Chemicals, Chemical Reactions, and Chemical Equations

It's fundamentally a bit difficult to be a student of chemistry, because unlike some of the other sciences, chemistry doesn't really afford us the opportunity to see and manipulate the object of our examination. In biology, we can touch the fetal pig we're dissecting or turn over the leaf we're sketching. Even bacteria, which are invisible to the naked eye, can be viewed with the help of a microscope. Chemistry is a little different. Even though we can go to the lab, mix a few chemicals, and note a color change or bubbling or some other visible evidence of a reaction, we have to take our instructor or lab manual at face value when they tell us what the chemicals are and what reaction is occurring. We can't see the molecules, and we can't watch bonds break and reform. As a result, the study of chemistry involves employing a variety of visualization aids. Let's take the example of water. We can think about water on a macroscopic level, meaning in terms of what we actually see; it's a clear liquid. We can also think about water on a microscopic level, or in terms of how the atoms are arranged; an atom of oxygen is bonded to two separate atoms of hydrogen.

Finally, we can symbolically represent water in a concise way that provides information about its chemical composition; water has the formula H_2O (Image 1.18). Depending upon what it is we're trying to communicate or what we're trying to imagine, we might pick any one or a combination of these strategies to visualize the compound water. For instance, if I wanted to communicate to you that the chemical methane (CH_4) burns in oxygen (O_2), making carbon dioxide (CO_2) and water (H_2O), it would be sufficient to symbolically represent chemicals with their formulas—no further information is needed. In Chapter 13, however, we'll need to think about the shape of a water molecule in order to understand the greenhouse effect. Because of that, we'll spend a lot of time thinking about its microscopic representation, in which we concentrate on the physical arrangement of the atoms in the molecule and its shape in space.

Let's take a closer look at the chemical reaction mentioned in the previous paragraph. Methane (CH_4) is a member of the class of chemicals collectively called **hydrocarbons** because they're composed of carbon and hydrogen. We'll be talking about hydrocarbons frequently in this textbook, in part because they all burn in oxygen (in chemistry, burning is called combustion) to produce carbon dioxide and water. Have you ever barbecued on a gas grill? If so, you've combusted the hydrocarbon propane (C_3H_8). If you're a camper, you've probably used a camp stove that runs on the hydrocarbon butane (C_4H_{10}). Have you fueled your car recently? That was a mixture of, among other things, octane (C_8H_{18}) that you put in the tank, and likely some ethanol (C_2H_5OH) as well.[5] **When we describe the behavior of methane in oxygen, we are describing a chemical reaction, which is a rearrangement of atoms and molecules to form new molecules.** The original chemical species are called the reactants, and the resulting chemical species are called the products. In the example above, methane and oxygen are reactants, while carbon dioxide and water are products.

In addition to describing a reaction in words, we can represent it symbolically. Let's take the example of a very simple chemical reaction with only two reactants and one product. Sulfur (S) combines with oxygen (O_2) to produce sulfur dioxide (SO_2). Represented microscopically, the reaction looks like this:

While this is a convenient way to visualize the rearrangement of atoms and molecules taking place in the reaction, it quickly becomes prohibitive to draw microscopic representations of reactants and products, especially in more complicated reactions. **Chemical equations are symbolic representations of chemical reactions.** The chemical equation for the reaction above would look like this:

$$S + O_2 \rightarrow SO_2$$

The arrow indicates that the reactants sulfur and oxygen[6] (S and O_2) are rearranging to form the product sulfur dioxide (SO_2). Aloud, this would read, "sulfur and oxygen react to form sulfur dioxide."[7]

You may be asking yourself right about now: "Why would I want to know how to write out a chemical equation, since I'm not a chemist?" That's a valid question. The answer is that in order to talk meaningfully about chemistry in everyday life, we'll need to speak a little bit of the language of chemistry, which is written in terms of reactants, products, and reactions.

1.4 Balancing Chemical Equations

Some chemical reactions end up causing us a bit of trouble when we try to write them out as chemical equations. Take, for instance, the true statement that elemental hydrogen (found in nature as a diatomic molecule, remember?) burns in oxygen to form water. When we write out an equation, we get something that looks like this:

NOTE[5]

Compounds like ethanol are called hydrocarbon derivatives—these compounds are mostly carbon and hydrogen, but contain small amounts of other elements, such as N, O, or S. Many of them, like ethanol, also combust in oxygen.

NOTE[6]

Why do we call O_2 *oxygen* when an atom of O is also called *oxygen*? This is a slightly confusing issue, and it's worth addressing. Even though an atom of O is called *oxygen*, atomic oxygen is rarely found on its own in nature—it's almost always combined into molecules. Elemental oxygen in nature is found as the molecule O_2. It's so common, in fact, that rather than saying *molecular oxygen* or *diatomic oxygen*, we simply call it oxygen. Generally, we can figure out what is meant by the word *oxygen* from its context. If I say, "CO_2 consists of one carbon and two oxygens," you pretty well know that I mean atoms of oxygen. If I say, "The atmosphere is made up of 21% oxygen," you should know that elemental oxygen found in nature is always O_2. If the context leaves doubt as to what is meant, we generally reserve *oxygen* for O_2, and say *atomic oxygen* or *an atom of oxygen* for O.

NOTE[7]

Alternately, we could use the words *produce* or, if we want to be very technical, *yield*, instead of *form*.

$$H_2 + O_2 \rightarrow H_2O$$

Looking closely, it seems that we have a problem. There are two atoms of hydrogen (the subscript tells us that one molecule of hydrogen consists of two atoms of hydrogen chemically bonded together) and two atoms of oxygen on the reactant side of the equation. However, there are two atoms of hydrogen and only *one* atom of oxygen on the products side. A microscopic representation of the equation makes this very clear:

Where did the second atom of oxygen go? The answer is *nowhere*—atoms can't disappear (nor can they appear!)—that's a fundamental law of nature. The above is therefore not an accurate representation of the chemistry, because it's not a **balanced equation**. **Balanced equations obey the Law of Conservation of Matter, which means that atoms cannot appear, disappear, or change into atoms of other elements through any chemical reaction.** In order for an equation to be chemically correct, we must ensure that there is the same number of each type of atom on each side of the equation. In order to do this, we can specify that different *amounts* of each reactant or product are involved in the reaction, but under no circumstances can we change the *subscripts* associated with any element in any molecule[8] To specify that there is more than one of any given chemical involved in a reaction, we use a coefficient in front of the species (the lack of a coefficient in front of a species is always taken to mean *one*). In this case, we need two water molecules to balance the oxygen atoms and two hydrogen molecules to account for the hydrogen needed to form the second water molecule. Our equation, correctly balanced, becomes:

$$2\,H_2 + O_2 \rightarrow 2\,H_2O$$

Doing a quick atomic head count, this gives us four hydrogen atoms and two oxygen atoms on each side of the equation:

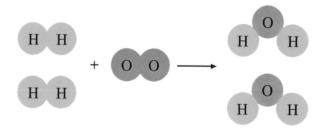

NOTE[8]

Tempting though it might be to balance the oxygen atoms by changing the product to H_2O_2, the chemical truth is that in this reaction, hydrogen and oxygen combine to form water, which is H_2O. H_2O_2 (while appearing to balance the equation nicely) is not water. It's hydrogen peroxide, which is a very different chemical indeed!

Even though we started with three molecules and ended up with two, we've conserved the numbers and identities of the atoms. **In chemical reactions, reactants and products can contain different numbers of molecules, but they must contain exactly the same number and type of atoms, which is why we count atoms rather than molecules when balancing equations.**

Some equations are a bit more difficult to balance. Let's return, for instance, to the reaction between methane and oxygen that produces carbon dioxide and water (Image 1.19):

$$CH_4 + O_2 \rightarrow CO_2 + H_2O$$

An examination of this equation reveals that it's not balanced. Unlike the hydrogen and oxygen reaction, however, this one is difficult to balance simply by looking at it. A good technique for balancing reactions like this follows a series of steps:

1. Identify an element that appears in *one* compound (do not pick elements that appear alone—save these for last!) on either side of the equation. Balance that element by adjusting the coefficients for molecules containing the element.

Oxygen appears in our reaction as a free element (O_2), so we'll save it for last. Carbon and hydrogen are both good Step 1 candidates, as they appear in only one compound on either side. We can therefore start with either carbon or hydrogen.

$$CH_4 + O_2 \rightarrow CO_2 + H_2O$$

Carbon is already in balance; there is one atom of carbon on each side of the equation.

2. Repeat Step 1 for remaining elements other than free elements.

Hydrogen needs to be balanced; we can accomplish this by adding the coefficient 2 in front of the product H_2O.

$$CH_4 + O_2 \rightarrow CO_2 + 2\,H_2O$$

There are now four atoms of hydrogen on each side of the equation.

3. Balance any remaining elements.

Oxygen can now be balanced. There are four total oxygen atoms on the right—two from CO_2, and one from each of the two waters—so we need four total oxygen atoms on the left. We therefore add the coefficient 2 in front of the reactant O_2.

$$CH_4 + 2\,O_2 \rightarrow CO_2 + 2\,H_2O$$

This is a balanced equation.

IMAGE 1.19. A Bunsen burner combusts methane (CH_4) in oxygen to produce carbon dioxide and water.

Shakespeare said in *Romeo and Juliet* that "a rose by any other name would smell as sweet," but as it turns out, naming is critically important in chemistry! You've seen a few chemical formulas coupled with names so far in this chapter, so you may be getting familiar with the way the names tend to sound. For instance, it was mentioned earlier that a compound consisting of sodium (Na) and chlorine (Cl) is called *sodium chloride*, and I'll tell you right now that a compound of rubidium (Rb) and fluorine (F) is called *rubidium fluoride*; there's a pattern there. Just to throw a wrench in the works, however, a compound of carbon and oxygen is *never* called *carbon oxide*. Why? The short answer is that there are several ways in which carbon and oxygen can combine[9] and using the (incorrect) name *carbon oxide* doesn't help us to distinguish them from one another. Clearly, we need to outline some rules for how to name compounds. Knowing a little bit of **nomenclature** (which means a system for naming) helps us to understand one another when we are discussing chemistry. The easiest compounds to name are **binary compounds**, which are made up of two (and only two) different elements. There are lots of compounds made of more than two elements, but their nomenclature is much more complex. Of course, anything made up of only one element is not a compound at all. **Binary compounds may consist of a metal and a nonmetal or a nonmetal and a nonmetal.** Compounds of metals with metals, while they exist, are exceedingly esoteric and atypical, and are far beyond the scope of this class.

TRY THIS

Balance each of the following equations:

$$\_\_\_S_8 + \_\_\_O_2 \rightarrow \_\_\_SO_3$$
$$\_\_\_N_2 + \_\_\_O_2 \rightarrow \_\_\_N_2O$$
$$\_\_\_C_{10}H_{16} + \_\_\_Cl_2 \rightarrow \_\_\_C + \_\_\_HCl$$

Answers:

$$S_8 + 12\,O_2 \rightarrow 8\,SO_3$$
$$2\,N_2 + O_2 \rightarrow 2\,N_2O$$
$$C_{10}H_{16} + 8\,Cl_2 \rightarrow 10\,C + 16\,HCl$$

NOTE[9]

Two examples include CO (carbon monoxide) and CO_2 (carbon dioxide).

Binary Compounds—Metal and Nonmetal

To name a binary compound made of a metal and a nonmetal, we use the name of the metal followed by the name of the nonmetal, adding the suffix *-ide* to the nonmetal. You may need to drop some letters to do this phonetically—sulfur becomes sulfide, and oxygen becomes oxide. These sorts of compounds are often called salts (Image 1.20).

Example: A compound of the metal barium (Ba) and the nonmetal oxygen (O) is called *barium oxide*.

Example: The compound LiF is called *lithium fluoride*.

Note that chemical formulas for binary compounds of metals and nonmetals are written in the same order in which the name is given—metal first.

When there are subscripts in the chemical formula of a compound, we ignore them in naming the compound.

Example: The compound $AlCl_3$ is called *aluminum chloride*.

(a)

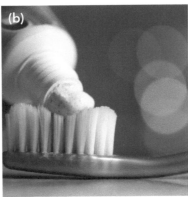
(b)

IMAGE 1.20. Binary compounds of a metal and a nonmetal are also called salts. They are solid compounds with some interesting chemical properties that will be discussed further in Chapter 4. Among the binary compound salts we commonly encounter (besides table salt, of course!) are one form of iron oxide (Fe_2O_3), which is rust, and sodium fluoride (NaF), used in toothpaste.

The reason we don't reference the relative number of atoms of each element when we name the compound is that metals can generally only combine with a given nonmetal in one possible way.[10] In other words, when lithium and fluorine combine, the only possible combination is LiF. When aluminum and chlorine combine, the only possible combination is $AlCl_3$. How would you know this? As it turns out, there are some rules we'll learn in Chapter 4 that we can use to figure out what combination is possible for a given metal and nonmetal. In the meantime, though, we'll have to satisfy ourselves with predicting names from formulas. We can't predict formulas from names just yet because the names don't tell us how many atoms of each element are present.

Binary Compounds—Nonmetal and Nonmetal

Like salts, binary compounds of metals and nonmetals are ubiquitous in the world around us (Image 1.21). To name a binary compound made of two nonmetals (or in the much rarer case of a metalloid and a nonmetal), we name the element closer to the left side of the periodic table first, then the element closer to the right side, adding the suffix -*ide* to the second element. Additionally, we need to make reference to the number of atoms of each element present, using the Greek prefixes listed below:

Example: The compound N_2O is called *dinitrogen monoxide* (nitrogen is closer to the left side of the periodic table, so it is named first and written first in the formula).

If there's only one of the element closer to the left side (the one we'd write or say first), we leave off the prefix. In other words, we never start the name of a compound with "mono."

Example: The compound CO_2 is called *carbon dioxide*.

We may need to drop vowels from the end of the Greek prefix when element names begin with vowels in order to make the name phonetically sensible.

Example: The compound CO is *carbon monoxide*.

TRY THIS

Name each of the following: K_2O, MgS, AlF_3.
Answers: Potassium oxide, magnesium sulfide, aluminum fluoride.

NOTE[10]

As we'll learn in Chapter 4, metals combine with nonmetals by becoming positively charged. Some metals can form more than one possible charge, but the specific charge determines the number of (negatively charged) nonmetal particles with which the metal particle will combine, and for each charged metal particle, there is only one possible combination ratio with a specific nonmetal particle.

IMAGE 1.21. Binary compounds of two nonmetals or of a metalloid and a nonmetal may be gaseous like the volcanic gas sulfur dioxide, liquid like water, or solid like silicon dioxide (sand). Volcano Arenal, Costa Rica © Kirstin Hendrickson, 2002. Costa Rica beach © Kirstin Hendrickson, 2002.

(a)

(b)

Number of atoms of an element	Prefix
1	mono
2	di
3	tri
4	tetra
5	penta
6	hexa

TABLE 1.1. Prefixes used in nomenclature.

TRY THIS

Name each of the following: CF_4, PCl_5,

Answers: Carbon tetrafluoride, phosphorus pentachloride, carbon disulfide.

TRY THIS

Predict formulas for each of these names: silicon tetrafluoride, dinitrogen trisulfide, diboron hexahydride.

Answers: SiF_4, N_2S_3, B_2H_6.

The two examples above demonstrate why we need prefixes to indicate the relative number of each atom; unlike compounds of metals and nonmetals, nonmetals can combine with other nonmetals in many different ways. The prefixes help us know to which of the many possible combinations of two given nonmetals we refer.

1.6 The Last Word—Chemical Free?

One thing you'll notice as you start to look at chemistry in the real world is that chemicals are often treated as something to be avoided. People tend to crave the "natural" and fear or dislike the "unnatural," but the lines drawn to separate the two are often randomly placed and inaccurate. In particular, the word *chemical* is frequently taken to be synonymous with *unnatural* and treated as something to be avoided. Case in point: I was reading the label on a loaf of bread not long ago while making toast. I noticed that, in an attempt to reassure the segment of society interested in "natural" foods, the bread claimed to be "chemical free."

Of course, *chemical* is simply another way of saying *matter*, so no physical material is chemical free any more than it can be matter free. As we've learned in this chapter, atoms and molecules, elements and compounds are *all* chemicals. Just for emphasis, the bread's ingredients included: whole wheat flour (a mixture of elements and compounds, chief among them amylose [a large molecule consisting of lots of C, H, and O] and cellulose [a different large molecule, also made of C, H, and O]); water (the molecule H_2O); honey (a mixture of elements and compounds, including lots of the sugars fructose and glucose, both of which are compounds consisting of C, H, and O); yeast (an organism composed of a variety of chemicals); and sea salt (again, a mixture of chemicals, but comprised primarily of NaCl). The bread contains nothing *but* chemicals!

As you look around, nothing you can see or touch is chemical free; even things you can't see, like the air you breathe, are made of chemicals. The paper in the pages of a book is made from the chemical cellulose (which comes from plants). All the food you eat is composed of various chemicals, just like the bread we've been discussing. The water you drink is a chemical. Air is a mixture of many chemicals, including the O_2 your cells need.

Of course, a devil's advocate would say that the expression *"chemical free"* is nothing more than a concise way to say "free of unnatural chemicals that will cause harm." Touché, but as we'll see later, even the idea that "unnatural" equals "bad" and "natural" equals "good" is fundamentally flawed. Furthermore, it can be quite difficult to draw lines between *natural* and *unnatural*. **More than anything, the point is simply that where chemistry and society intersect, society can respond with a fear of the unknown, and that fear may lead to irrational (and scientifically unfounded) behavior and ideas.** Discussing these departures from logic makes us aware of them, and we may begin to evaluate the reasoning behind our behavior, the logic that informs our reasoning, and the efficacy of our logic.

SUMMARY OF MAIN POINTS

- Chemistry is all around us. Everything that occupies space and has mass is made of chemicals, and chemical reactions take place all around us and inside us all the time.
- When we think about the interaction between chemistry and society, we have to consider the individual, community, and global levels of society.
- Chemistry is the study of matter, which consists of mixtures and pure substances. Pure substances include elements and compounds.
- A pure substance can be atomic or molecular in nature; molecules can either be elements or compounds.
- The elements on the periodic table can be divided into three broad categories by their position on the table: metals, nonmetals, and metalloids. Each of these categories has characteristic properties.
- When we think about chemicals, we can consider them from macroscopic, microscopic, and/or symbolic perspectives, depending upon what we're trying to communicate.
- Chemical equations are symbolic representations of chemical reactions, consisting of reactants and products separated by an arrow.
- Chemical equations must obey the Law of Conservation of Matter, which states that atoms may not appear, disappear, or change into atoms of other elements during a chemical reaction. In other words, equations must be balanced.
- Balancing equations must be done by adjusting coefficients in front of the reactant or product species; subscripts within chemical formulas may not be changed.
- Binary compounds of a metal and a nonmetal (salts) are named without referencing subscripts. The metal is named first, then the nonmetal (with the suffix *-ide*).
- Binary compounds of two nonmetals are named using Greek prefixes to indicate the number of atoms of each element. The element closer to the left side of the periodic table is named first, followed by the one closer to the right side of the periodic table (with the suffix *-ide*).

(a)

HONEY WHOLE WHEAT BREAD

INGREDIENTS: 100% freshly milled whole grain wheat flour (certified chemical free), water, honey, yeast, sea salt.

NET WT. 28 oz
Sell By: 09-01-09

(b) HONEY WHOLE WHEAT BREAD

INGREDIENTS: 100% freshly milled whole grain wheat flour (certified chemical free), water, honey, yeast, sea salt.

IMAGE 1.22. "Certified chemical free" bread. Photos © Scott Lefler, 2010.

1. Provide an example of each of the following relationships with regard to chemistry (do not use examples discussed in the text):

    • Individual behavior affects that individual
    • Individual behavior affects society
    • Behavior of a society affects an individual
    • Behavior of a society affects that society
    • Behavior of a society affects another society.

2. How does chemistry in a lab help us to understand chemistry in the "real world"? What are the limitations of studying chemistry in a lab?

3. With regard to chemistry, do you think that there are any examples of individual behavior that have no effect whatsoever upon society? Justify your answer.

4. What do you think might be some of the challenges associated with the fact that one individual/group/nation can behave in a way that impacts chemistry on a global level?

5. Using only items you could find around yourself in everyday life (i.e., *not* in a lab), give an example of a mixture. Give an example of a compound. Give an example of a pure element.

6. Explain the primary differences between mixtures and pure substances.

7. Explain the primary differences between mixtures and compounds.

8. Are molecules necessarily compounds? Why or why not? Are compounds necessarily molecules? Why or why not?

9. Does matter have to be visible? Think of an example of invisible matter.

10. Provide an example of when the word *nitrogen* would be understood from context to mean an atom of N. Provide an example of when the same word would be understood from context to mean a molecule of N_2. Provide an example of when the meaning might be ambiguous and clarification would be required.

11. Think of a compound that interests you and describe a few of its physical or chemical properties. Look up the physical and chemical properties of its constituent elements (if necessary) and describe them. Are compounds similar to the elements that make them up, or are they different?

12. Provide a "real-life" example of when, in discussing a substance, we'd need to visualize it macroscopically. Microscopically? Symbolically?

13. What is the primary naming difference between binary compounds of metals with nonmetals and binary compounds of nonmetals with nonmetals? What is the reason for this difference?

14. When might knowing how to name a chemical (or knowing the formula from the name) be really important in real life?

15. Why can we predict the formula of a nonmetal–nonmetal compound from its name, but we can't predict the formula of a metal–nonmetal from its name?

16. Why do you think people associate the word *chemical* with something bad? Is there any context in which *chemical* is taken to mean something good? Provide an example.

17. Is it possible to have a "chemical-free" food or cleaning solution? Why or why not? Why do you think companies might want to advertise a product as being "chemical free"?

PROBLEMS

1. Define each of the following and provide an example: matter; mixture; pure substance; element; compound; atom; and molecule.

2. Can an element be a molecule? Give a specific example to justify your answer.

3. Classify each of the following, using as many of the words in Problem 1 as are applicable in each case: CO_2; Ne; N_2; saltwater; glucose ($C_6H_{12}O_6$); HCl.

4. Classify each of the following, using as many of the words in Problem 1 as are applicable in each case: H_2O; S_8; $MgCl_2$; sugar water; C; Fe.

5. What are the primary physical characteristics shared by metals? By nonmetals? By metalloids?

6. Where on the periodic table are the metals? The nonmetals? The metalloids?

7. Classify each of the following elements as metal, nonmetal, or metalloid: Rb; fluorine; Xe; Pb; mercury; Si; germanium; K; Po.

8. Classify each of the following elements as metal, nonmetal, or metalloid: B; Ne; titanium; Cm; tin; At; F; Na; silver.

9. Provide or describe each of the following: the macroscopic representation of mercury; the microscopic representation of carbon dioxide; the symbolic representation of carbon monoxide.

10. Identify the products and reactants in the reaction: $2\ C_2H_2 + 5\ O_2 \rightarrow 4\ CO_2 + 2\ H_2O$. Explain how the coefficients ensure a balanced reaction in terms of atoms.

11. Identify the products and reactants in the reaction: $S_8 + 12\ O_2 \rightarrow 8\ SO_3$. Explain how the coefficients ensure a balanced reaction in terms of atoms.

12. What does it mean for a chemical equation to be balanced?

13. What is the Law of Conservation of Matter, and how does it affect our writing of chemical equations?

14. Can the number of molecules in a balanced reaction be different on the products side as compared with the reactants side? Why or why not?

15. Can the number of atoms in a balanced reaction be different on the products side as compared with the reactants side? Why or why not?

16. What of the following *must* be conserved in a chemical equation: number of atoms, number of molecules, identity of atoms, identity of molecules?

17. Write out a chemical equation for the statement "Elemental nitrogen reacts with elemental hydrogen to produce ammonia (NH_3)." Balance if necessary.

18. Write out a chemical equation for the statement "Elemental nitrogen reacts with oxygen to form nitrogen dioxide." Balance if necessary.

19. Balance each of the following reactions:

    ___C_3H_8 + ___$O_2 \rightarrow$ ___CO_2 + ___H_2O (this is the reaction that takes place when you run a gas grill)

    ___$HgO \rightarrow$ ___Hg + ___O_2 (this is the reaction that helped the early chemist Joseph Priestley discover the element oxygen)

    ___SO_3 + ___$H_2O \rightarrow$ ___H_2SO_4 (this is how acid rain forms)

    ___Fe + ___$O_2 \rightarrow$ ___Fe_2O_3 (this is how rust forms)

20. Balance each of the following reactions:

    ___$O_2 \rightarrow$ ___O_3

    ___$C_6H_{12}O_6$ + ___$O_2 \rightarrow$ ___H_2O + ___CO_2

    ___Al + ___$O_2 \rightarrow$ ___Al_2O_3

    ___I_2 + ___$Na_2S_2O_3 \rightarrow$ ___NaI + ___$Na_2S_4O_6$

21. Name each of the following compounds: $CsCl$; $AgBr$; Ag_2O; Al_2O_3; BaF_2; Na_2S.

22. Name each of the following compounds: $MgCl_2$; NaF; KCl; Na_2O; MgO; K_2O.

23. Name each of the following compounds: NO; N_2O; NO_2; H_2O (name systematically!).

24. Name each of the following compounds: SO_2; SO_3; PBr_3; PCl_3.

25. Name each of the following compounds: $BaBr_2$; CS_2; PCl_5; Al_2S_3; P_2O_3; S_2Cl_2; Na_3N; Mg_3P_2; CaO.

26. Name each of the following compounds: Rb_2O; LiF; Li_2S; N_2O_4; H_2O_2; K_3N; ClO (this one is a little strange; there's an exception to the naming rule that says we always name O *last* if it appears in a formula); CF_4; BaS.

27. Provide formulas for each of the following names: dinitrogen pentoxide; iodine monochloride; carbon tetrachloride; boron trifluoride.

28. Provide formulas for each of the following names: sulfur dioxide; sulfur trioxide; phosphorus pentabromide; silicon dioxide.

ANSWERS TO ODD-NUMBERED PROBLEMS

1. Matter occupies space and has mass; mixtures are composed of two or more pure substances, physically combined, in variable proportions; pure substances are elements and compounds; elements are the simplest type of matter, listed on the periodic table: they cannot be broken into simpler substances by any chemical means; compounds are a combination of two or more elements, chemically combined, in fixed proportions; atoms are the smallest (indivisible) part of matter of a given element; molecules are two or more atoms, chemically combined, in fixed proportions. Examples will vary.

3. CO_2 is matter, a pure substance, a compound, a molecule; Ne is matter, a pure substance, an element, an atom; N_2 is matter, a pure substance, an element, a molecule; saltwater is matter, a mixture; glucose is matter, a pure substance, a compound, a molecule; HCl is matter, a pure substance, a compound, a molecule.

5. Metals are shiny, ductile, conductive, and malleable. Nonmetals are dull, brittle, and nonconductive. Metalloids have properties between those of metals and nonmetals and are semiconductors

7. Rb is a metal; fluorine is a nonmetal; Xe is a nonmetal; Pb is a metal; mercury is a metal; Si is a metalloid; germanium is a metalloid; K is a metal; Po is a metal.

9. Mercury is a shiny silver/grey liquid at room temperature; carbon dioxide is made up of two atoms of oxygen bonded to one (central) atom of carbon; carbon monoxide is CO.

11. S_8 and O_2 are reactants; SO_3 is the product. The coefficients are there to balance the equation (1 C atom on each side, 4 H atoms on each side, 4 O atoms on each side).

13. The Law of Conservation of Matter states that atoms cannot appear, disappear, or change into atoms of any other element by any chemical means. Because of this law, chemical equations must be balanced.

15. The number of atoms, and their identities, must be conserved from reactants to products; reactions are, by definition, rearrangements of atoms into new molecules.

17. $N_2 + 3 H_2 \rightarrow 2 NH_3$.

19. $C_3H_8 + 5 O_2 \rightarrow 3 CO_2 + 4 H_2O$; $2 HgO \rightarrow 2 Hg + O_2$; $SO_3 + H_2O \rightarrow H_2SO_4$; $4 Fe + 3 O_2 \rightarrow 2 Fe_2O_3$.

21. Cesium chloride; silver bromide; silver oxide; aluminum oxide; barium fluoride; and sodium sulfide.

23. Nitrogen monoxide; dinitrogen monoxide; nitrogen dioxide; and dihydrogen monoxide (a funny name for water!).

25. Barium bromide; carbon disulfide; phosphorus pentachloride; aluminum sulfide; diphosphorus trioxide; disulfur dichloride; sodium nitride; magnesium phosphide; and calcium oxide.

27. N_2O_5; ICl; CCl_4; and BF_3.

CREDITS

Chemistry Around Us—Of Atoms and The Atmosphere

2

Our Earth is surrounded by a shell of gases called the atmosphere. We can't see these gases, so we don't tend to think much about them, but our atmosphere is not only critical to life as we know it, it is also the site of some pretty interesting and important chemistry. We'll discuss some of what's going wrong in the atmosphere in Chapters 10 and 13, but for now, let's focus our attention on the chemistry of a healthy atmosphere.

The Earth is 7,926 miles (12,756 km) in diameter. Our atmosphere, by comparison, is a mere 62 miles (100 km) or so in thickness, although it's difficult to pinpoint exactly where the atmosphere stops; since it's gaseous in nature, it just tapers off. An easier way to think about it is that the atmosphere is thickest near Earth's surface and becomes exponentially thinner with increasing altitude. **Ninety percent of the atmosphere by mass is below an altitude of 10 miles, and the remaining 10% is located between 10 and 62 miles above Earth's surface.** For reference, it may be helpful to think about this in terms of the altitude above sea level of some other common things (Image 2.2).

The atmosphere is divided into several major regions by altitude. The two we'll discuss in this text are the troposphere (the region of the atmosphere that we inhabit) and the stratosphere (the location of the ozone layer).

One interesting property of the atmosphere is that it exerts pressure on everything it surrounds. We commonly hear about barometric pressure—the ambient atmospheric pressure in a given location at a given time—in reference to the weather. However, we don't often extend that concept to the realization that the atmosphere truly exerts pressure on us—literally pushes against us—all the time. Pressure is defined in physics as a force (which is generated by a mass) exerted over an area, so if the majority of the atmosphere's mass is located in the first 10 or so miles above Earth's surface, then, by extension, the highest pressures are also closest to sea level. In fact, air pressure, like the density of the atmosphere, falls off exponentially with increasing altitude (Image 2.3).

IMAGE 2.1. The moon through the shell of our atmosphere. Notice that compared to the Earth's radius of curvature, the atmosphere really is quite thin.

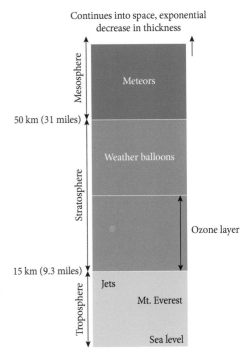

IMAGE 2.2. Regions of the atmosphere and what we find there.

NOTE[1]

If you're curious, the logic here is that most people feel fine going from sea level to 7,000 feet, but many experience symptoms of altitude sickness above that elevation, particularly if they ascend rapidly. Since planes can easily fly at over 35,000 feet above sea level, the cabin must be pressurized. Why not just maintain sea-level pressure? Remember, air pressure comes from particles of gas, and those particles have mass! An airplane pressurized to sea level would weigh more than an airplane pressurized to 7,000 feet, so it improves gas mileage to reduce the pressure inside the plane as much as possible.

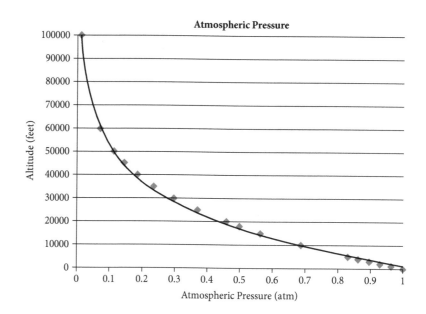

Even small changes in altitude can have dramatic effects upon air pressure. For instance, at 10,000 feet (the approximate elevation of Quito, Ecuador), the air pressure is about two-thirds of that at sea level. On top of Mt. Everest (just over 29,000 feet), the air pressure is about one-third of that at sea level. One common experience that makes this pressure differential quite apparent is the "popping" sensation in the ears or throat that accompanies a change in altitude and air pressure. For instance, passengers on an airplane are subjected to a change in air pressure as the plane takes off and begins to climb. The pressure in the cabin is adjusted to the equivalent of about 7,000 feet above sea level.[1] Assuming the plane took off from an elevation lower than 7,000 feet, this represents a loss of atmospheric pressure, which creates a pressure imbalance between the middle ear and ear canal (the part of the ear open to the outside); these are separated from one another by the impermeable eardrum (Image 2.4). The higher pressure in the middle ear causes the eardrum to bow outward, which is uncomfortable and affects hearing. To return the eardrum to a relaxed, unbowed state, air must escape the middle ear through a normally collapsed tube (called the Eustachian tube) that opens into the back of the throat. The bubbles of air escaping through this tube produce a popping sensation. In preparation for landing, air pressure in the cabin is increased to the equivalent of that at the plane's destination. This, too, can affect the ears; assuming the plane is landing at an elevation below 7,000 feet, pressure in the cabin increases toward the end of the flight. The higher pressure outside the ear bows the eardrum inward, requiring that air enter the middle ear through the Eustachian tube, once again producing the sensation of a "pop."

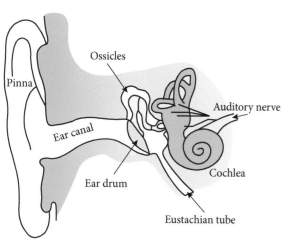

IMAGE 2.4. Air pressure must be the same in the middle ear (the space just to the right of the eardrum in the figure) as in the ear canal to prevent bowing of the eardrum.

2.1　Composition of the Atmosphere

We now know a little about the distribution of the atmosphere, so the next thing we need to think about is its composition. Our atmosphere is composed of a mixture of many different gases, but elemental nitrogen (N_2) is the most prevalent among these. **Nitrogen accounts for 78% of our atmosphere. Most of the rest of the atmosphere is elemental oxygen (O_2), which accounts for 21% of the total.** The remaining 1% is a mixture of other naturally occurring gases, including some water vapor (Image 2.5).

This composition is maintained throughout the troposphere. Other gases are present in the stratosphere, and we'll discuss those more in Chapter 10. For now, though, it's sufficient to remember that **anywhere we go on Earth (even on a high mountaintop), we're breathing an atmospheric mixture of consistent composition.**

Composition of the Atmosphere

IMAGE 2.5. The composition of Earth's atmosphere. Other gases include Ar, CO2, H_2, Ne, Kr, and methane (CH_4).

2.2　Chemistry at High Altitude

If the percentage of O_2 is constant throughout the troposphere, why do you feel short of breath at high altitude and, at sufficient elevation, experience symptoms of altitude sickness?

Where air pressure is high—close to sea level—the atmosphere is more dense, meaning that there are more molecules of gas in a given volume. When we fill our lungs, we fill them with a mix of 78% N_2 and 21% O_2, and there are lots of molecules of each gas. At high elevation, a lungful of air contains fewer total molecules, so even though the percentages of nitrogen and oxygen are the same, 21% of a smaller

IMAGE 2.6. At 22,485 feet in elevation, Aconcagua (in the Andes mountains of South America) is the highest peak in the western hemisphere. It's popular with climbers, but because of its staggering height, they often experience symptoms of altitude sickness. Photos © Kirstin Hendrickson, 2002.

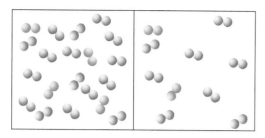

IMAGE 2.7. A representation of the number of molecules of air in a small sample at sea level (left), versus the number in the same size sample at low atmospheric pressure (right). The percentage of oxygen, shown in red, is the same in each case, but there are fewer total oxygen molecules in the low-pressure sample of air.

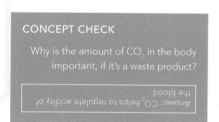

CONCEPT CHECK

Why is the amount of CO_2 in the body important, if it's a waste product?

Answer: CO_2 helps to regulate acidity of the blood.

total means fewer molecules of O_2 per breath (Image 2.7). With less oxygen entering the lungs per breath, less oxygen makes it into the bloodstream, and we begin to experience shortness of breath and other symptoms of oxygen deprivation.

The issue of high-altitude breathing has some interesting chemistry associated with it. At first glance, the solution to the problem of oxygen deprivation seems uncomplicated—just breathe more! Unfortunately, it's not quite that simple. The problem is that every breath we take has two functions: to bring in O_2 and to get rid of CO_2. CO_2 is a metabolic waste product that is produced by cells as they burn sugars and other fuels for energy. Even though CO_2 is a waste product that we exhale, that doesn't mean it is functionless in the body. **CO_2 serves the very important purpose of keeping the blood at a constant level of acidity (pH).** Acidity will be discussed in more detail in Chapter 5, but right now it's sufficient to know that we have to maintain a constant blood acidity in the same way that we have to maintain a constant body temperature (98.6°F) to be comfortable and healthy. As an example, think how awful it feels if your body temperature is off by even a few degrees; a fever of 101°F—just slightly higher than normal temperature—will make you feel terribly ill! Acidity of the blood works the same way, and the body strives to maintain it within strict parameters. **The role of CO_2 in maintaining acidity comes from its reaction with water, of which there's plenty in the bloodstream.** CO_2 released from cells combines with water by the reaction:

$$CO_2 + H_2O \rightarrow H_2CO_3$$

H_2CO_3 is carbonic acid, and we need a constant amount of it in the blood—not too much, not too little—to maintain appropriate acidity. The carbonic acid travels to the lungs, and there it reacts again, regenerating water and CO_2, the latter of which is then exhaled:

$$H_2CO_3 \rightarrow CO_2 + H_2O$$

The cells are always making more CO_2, just as we're always exhaling it, so the level of carbonic acid in the blood stays constant under normal circumstances. **However, if we inhale more (to bring in more O_2 at high altitude), we're also *exhaling* more, which means we quickly get too low on CO_2 and lose too much acidity.** You can try an experiment if you like: if you breathe very rapidly in and out for a while, you'll start to feel a little light-headed. It's not the extra O_2 that's responsible for that feeling; it's the lack of CO_2. In fact, you probably know that if someone is hyperventilating (breathing very hard and fast, perhaps because they're panicking or upset), they are at risk of passing out. It sometimes helps to give them a paper bag to breathe in. Why? The rapid breathing is causing the loss of too much CO_2. By trapping that "lost" CO_2 in a bag, where they can reinhale it with their next breath, they stabilize their CO_2 levels and help to bring their acidity back to normal. Of course, the more a person breathes into a bag, the more oxygen in the bag gets used up, so no one should breathe into a bag for too long.

There's an odd phenomenon at very high altitude that results from the body's competing drives to bring in enough O_2 without losing too much CO_2. When high-altitude travelers are sleeping (and breathing becomes reflexive), they often experience what's called *paradoxical respiration* or *Cheyne-Stokes respiration*. This involves a period of hyperventilation followed by a period of *apnea*, or lack of breathing. The hyperventilation brings in plenty of O_2 but also results in the loss of too much CO_2, which lowers the acidity of the blood. The body senses the change in acidity and responds by shutting off the breathing reflex temporarily, allowing CO_2 levels (and acidity) to increase and normalize. Of course, during this period of apnea, O_2 levels start to fall, and eventually the individual begins to breathe (fast!) to prevent oxygen deprivation. This cycle of hyperventilation and apnea repeats itself all night. It's not dangerous, and it's completely normal in high-altitude sleepers, but it is certainly interesting chemistry!

2.3 Atomic and Molecular Mass

Our discussion of atmospheric pressure raises another interesting issue. Remember that according to physics, pressure is a force exerted over an area, which means that the components of the atmosphere literally push against us and everything else they surround. It stands to reason, then, that the molecules of the atmosphere must have mass, since without mass they wouldn't be able to exert a force. **In fact, all atoms have mass, and we can use the periodic table to determine the mass of an atom of a given element.** On the periodic table, there are two numbers associated with each element. For right now, we'll focus on the number located underneath the atomic symbol. This is called the atomic mass, and it tells us the average mass of an atom of that element. The units of atomic mass are called atomic mass units (amu), where 1 amu = $1.67{*}10^{-24}$ g.[2] Nitrogen, as we see in Image 2.8, has an atomic mass of 14.01 amu. It's important to note that while we can correctly say *the atomic mass of nitrogen is 14.01 amu*, no *single* atom of nitrogen actually weighs 14.01 amu. As we'll learn later in the chapter, individual atoms have whole-number weights; this is the result of their composition. As it turns out, elements exist in several different forms, called isotopes, which have identical chemical properties but can differ from one another with regard to *radioactivity*, a topic that will be explored more in Chapter 12. The atomic mass of an element is a weighted average mass. In the case of nitrogen, approximately 99.64% of all nitrogen in nature weighs 14 amu (this is called nitrogen-14). The remaining 0.36% is slightly heavier, weighing 15 amu (called nitrogen-15). If we were to gather up a random sample of 1,000 nitrogen atoms, they would collectively weigh about 14,010 amu. Dividing by the number of atoms in the sample, we'd find that the average weight of an atom of nitrogen is 14.01 amu, but again, if we were to pick out the individual atoms and weigh them, most would weigh 14 amu, and a few would weigh 15 amu.

7
N
14.01

IMAGE 2.8. The element nitrogen has an atomic mass of 14.01.

NOTE[2]

We generally don't calculate the mass of atoms or molecules in grams, simply because they are so light; it would be like measuring the length of your pinkie fingernail in miles! Atomic mass units are much more convenient (and appropriate) to the scale of atomic mass.

TRY THIS

What is the atomic mass of H? Of C? Of U? What is the mass of the most common form of each of these elements?

Answers: Atomic masses: H = 1.01 amu, C = 12.01 amu, U = 238 amu. Masses of most common forms: H = 1 amu, C = 12 amu, U = 238 amu. The parentheses around the atomic mass of uranium and certain other elements are there because these elements are radioactive; the mass in parentheses is that of the most stable isotope.

We can use an element's atomic mass to determine not only the *average* mass of an atom of an element (14.01 amu for nitrogen, for instance), but also the mass of the *most common isotope of that element* in many instances; in nitrogen's case, because the atomic mass is very close to the whole number 14, this tells us that the preponderance of nitrogen in nature weighs 14 amu.³ For some elements, like chlorine (atomic mass = 35.45 amu), it's not possible to determine the most common isotope from the periodic table, as the atomic mass is not very close to a whole number. Most of the elements we'll see on a regular basis in this text, however—including carbon, nitrogen, hydrogen, and oxygen—have atomic masses very close to whole-number values, and it's therefore easy to determine the most common isotope. We'll routinely use the atomic mass as well as the mass of individual isotopes (including the most common, if applicable) in discussing what atoms weigh. Incidentally, if we want to refer to the mass of a single atom of an element as opposed to the average mass of that element, we are referring to an atom's **mass number**. So, we could say that an atom of nitrogen-14, because it weighs 14 amu, has a mass number of 14. Mass numbers, because they are the masses of individual atoms, are never fractional.

If atoms have mass and atoms combine chemically to make molecules, then molecules must also have mass. We can calculate the mass of a molecule (the **molecular mass**) by adding up the masses of the atoms comprising that molecule. Generally, we're interested in the *average* mass of a molecule of a given species, so we use the atomic masses of the constituent elements, remembering that no individual atoms—and therefore no individual molecules—can have fractional masses. For example, a molecule of water is made up of one atom of oxygen and two atoms of hydrogen. We'd therefore calculate the molecular (average) mass of a water molecule as follows:

Molecular mass of water = mass of oxygen atom + 2(mass of hydrogen atom)
Molecular mass of water = 16.00 amu + 2(1.01 amu) = 18.02 amu

If, for some reason, we were interested in the mass of a specific molecule of water—say, one made up of two atoms of the most common isotope of hydrogen (1 amu each) and one atom of the most common isotope of oxygen (16 amu), we'd use mass numbers instead; this would allow us to determine that the specific water molecule in question would have a mass of 18 amu. This type of calculation is much less common, at least outside of nuclear chemistry, which we'll see in Chapter 12; generally, we're interested in the average mass of a molecule.

Molecular mass of NaCl = atomic mass of sodium + atomic mass of chlorine
Molecular mass of NaCl = 23.0 amu + 35.45 amu + 58.45 amu

2.4 Counting Atoms and Molecules—The Mole

Knowing that the atmosphere is made up of atoms and molecules (and therefore has mass), we can discuss its density in concrete terms. Given that the atmosphere is most dense nearest Earth's surface, then we know that there must be more molecules of gas per unit volume at low altitude. This raises an interesting question, however. How, exactly, does one *count* atoms and molecules? Even a tiny volume of air contains a mind-boggling number of gas molecules; there are nearly a billion molecules of nitrogen and oxygen in 1 mL (1 cm³) of air! Clearly, counting atoms and molecules in a sample of *anything* is going to involve some pretty big numbers.

Sometimes it makes sense to think about certain items in groups rather than as individuals. For instance, eggs tend to come in packages of a dozen (12) at the grocery store. That's because generally, no one needs just *one* egg—so it doesn't make much sense to sell them individually—and most people wouldn't be able to eat 100 eggs before they went bad, so they're not sold in packages of 100. Twelve is a nice size for a package of eggs. If I asked you to count all the eggs at the grocery store, it'd be much easier to count the number of packages (dozens) than to count each individual egg. **In much the same way, it is useful to have a "package size" for atoms and molecules so that, instead of counting them individually, we can think of them in groups.** A fellow named Lorenzo Romano Amadeo Carlo Avogadro di Quaregna e di Cerreto (he lived from 1776 to 1856, and for simplicity's sake, we'll just call him Avogadro) lent his name to a very reasonable package size for calculating the number of atoms or molecules in a sample of air, water, or any substance.

In chemistry, we count atoms and molecules in groups of $6.02*10^{23}$. In the same way that the number 12 is given the nickname *a dozen*, the number $6.02*10^{23}$ is called a **mole** (it is also called **Avogadro's number**).[4] **A mole (mol) of *anything* is $6.02*10^{23}$ of that thing.** We could have $6.02*10^{23}$ atoms of Ne; that would be a mole of Ne. We could have $6.02*10^{23}$ molecules of water; that would be a mole of water. We could even have $6.02*10^{23}$ cats—but we're not likely to, thank goodness; that'd be a lot of kitty litter—and we'd call that a mole of cats. Frankly, because a mole is such a very *big* number, even though we could theoretically have a mole of anything (atoms, molecules, cats, hamburgers, pencils, and so forth), we're not ever likely to find a mole of anything macroscopic, so we generally only use moles to count atoms and molecules.

We can use Avogadro's number to figure out how many atoms or molecules are in a given number of moles of a substance, and vice versa.

Example: How many atoms of Ne are in 2 moles of Ne?

$$2\ mol\ Ne \times \frac{6.02*10^{23}\ atoms}{1\ mol} = 1.20*10^{24}\ atoms\ of\ Ne$$

Example: If we have $5.34*10^{23}$ molecules of H_2, how many moles of H_2 is that?

$$5.34*10^{23}\ molecules\ of\ H_2 \times \frac{1\ mol}{6.02*10^{23}\ molecules} = 0.887\ mol\ H_2$$

Remember that molecules may contain more than one of a given atom, which affects our calculations!

Example: How many molecules of H_2O are in 2.5 mol H_2O? How many *atoms* of O are in that same sample? How many *atoms* of H are in that same sample?

$$2.5\ mol\ H_2O \times \frac{6.02*10^{23}\ molecules}{1mol} = 1.50*10^{24}\ molecules\ H_2O$$

$$2.5\ mol\ H_2O \times \frac{6.02*10^{23}\ molecules}{1mol} \times \frac{1\ atom\ O}{1\ molecule\ H_2O} = 1.50*10^{24}\ atoms\ O$$

$$2.5\ mol\ H_2O \times \frac{6.02*10^{23}\ molecules}{1mol} \times \frac{2\ atom\ H}{1\ molecule\ H_2O} = 3.01*10^{24}\ atoms\ H$$

IMAGE 2.9. Lorenzo Romano Amadeo Carlo Avogadro di Quaregna e di Cerreto, or, more simply, Avogadro.

TRY THIS

How many moles of C_2H_4 are in a sample of $8.56*10^{24}$ molecules? How many moles of C atoms are in the sample? How many H atoms are in the sample?

Answers: 14.2 mol C_2H_4, 28.4 mol of C, $3.42*10^{25}$ atoms of H.
Solution:

$$8.56*10^{24}\ molecules\ C_2H_4 \times \frac{1\ mol}{6.02*10^{23}\ molecules} = 14.2\ mol\ C_2H_4$$

$$14.2\ mol\ C_2H_4 \times \frac{2C}{1C_2H_4} = 28.4\ mol\ C$$

$$8.56*10^{24}\ molecules\ C_2H_4 \times \frac{4H}{1C_2H_4} = 3.42*10^{25}\ atoms\ H$$

IMAGE 2.10. How much is a mole? A mole of O2 would fill just about six one-gallon water bottles. A mole of water is a mere 18 mL. This is because liquids are much more dense than gas, and the same number of molecules takes up less room. Photos © Scott Lefler, 2010.

2.5 Molar Mass

Here's a question: Why on Earth is Avogadro's number such an odd one? Why not a nice round number like $1*10^{24}$, or something to that effect? It turns out that the specific value of a mole makes determining the mass of a mole of atoms or molecules very easy! We know that atoms have mass, so if we wanted to calculate the mass of

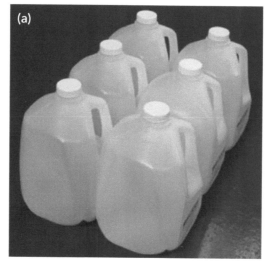

(a)

(b)

IMAGE 2.11. Our thought experiment: if one atom of carbon-12 weighs 12 amu, or 2.00*10-23 g, it would take many, many atoms of carbon (6.02*1023, to be exact) to get a scale to read 12 g.

a mole of atoms, we could multiply the mass of one atom by the number of atoms in a mole. This makes for a lot of math every time we want to determine the mass of a mole of something, but we can do a **thought experiment** to find the beauty in Avogadro's number. **A thought experiment is a way of thinking through a problem or question that is impractical or impossible to test in real life.** This sort of thinking is very important to science because it allows for the development of theories that advance scientific understanding long before technology is available to test those theories.[5] For our thought experiment, let's imagine that we were to pick up, say, one atom of the most common isotope of carbon (atomic mass = 12 amu) and put that atom on a very small scale. The scale would read $2*10^{-23}$ g (12 amu*$1.67*10^{-23}$ g/amu). If we then were to add another atom of carbon-12, the scale would read $4*10^{-23}$ g (24 amu). Let's imagine that we continue to add atom after atom of carbon-12 until the scale reads 12 g. If we then calculated the number of atoms placed on the scale to achieve this mass, we'd find it was $6.02*10^{23}$ (Image 2.11). Put simply, Avogadro's number is the number of atoms of carbon—or any element—that we'd have to put on a scale to get the scale to read in grams the same number it would read in amu if we weighed a single atom of the element in question.

Stated another way, the beauty of Avogadro's number is that it means **the atomic mass of any element is both the mass in amu of an atom of that element *and* the mass in grams of a mole of that element.** We do have to be a bit careful here, though: we know that no single atom of carbon weighs 12.01 amu, though 12.01 is the atomic mass of carbon. In our thought experiment, we put $6.02*10^{23}$ atoms of carbon-12—atoms that each weighed exactly 12 amu—on the scale, and at the end of the experiment, the scale read 12 g. If we had put $6.02*10^{23}$ atoms of carbon on the scale, selecting the atoms from nature—meaning that all naturally occurring isotopes would be represented—we'd see a different picture emerge. Most of the carbon we'd put on the scale would be carbon-12, but we'd have some carbon-13 and a bit of carbon-14 in there as well. After placing $6.02*10^{23}$ atoms, our scale would read 12.01 g. In other words, an element's atomic mass is the mass of an *average* atom of that element in amu and the mass of mole of that element, in its naturally occurring proportions, in grams. The mass of a mole of a chemical is called the chemical's **molar mass**, so for instance, the molar mass of carbon is 12.01 g, or we can say that carbon weighs 12.01 g/mol. If we want to refer to the molar mass of a specific isotope, we can do that as well: the molar mass of carbon-12 is 12 g (or carbon-12 weighs 12 g/mol).

This same principle holds true for molecules; we can also add up the molar masses of elements in a molecule to find the molar mass of the molecule.

NOTE[5]

If a thought experiment isn't tested in "real life," can it really be accurate? The answer is absolutely! For instance, Einstein developed his theory of relativity as a thought experiment. There are examples of accurate theories developed in this manner throughout the history of science.

TRY THIS

What is the molar mass of carbon? Of carbon-13? Of CO_2?

Answers: 12.01 g/mol, 13 g/mol, 44.01 g/mol

Solution: Check periodic table for molar mass of carbon (or any element). Note for the second question that, in specifying that we want the molar mass of carbon-13, we are indicating that each atom of carbon weighs 13 amu (we're isolating a specific isotope), meaning that $6.02*10^{23}$ atoms will weigh 13 g. Molar mass of CO_2 = molar mass of carbon + 2(molar mass of oxygen) Molar mass of CO_2 = 12.01 g/mol + 2(16.00 g/mol) = 44.01 g/mol

TRY THIS

What is the mass of 1.5 mol of nitrogen atoms?

Answer: 21.02 g

Solution:

$$1.5 \, mol \, nitrogen \times \frac{14.01g}{1 \, mol \, nitrogen} = 21.02 \, g \, nitrogen$$

TRY THIS

How many moles of N are in 54.5 g N_2O?

Answer: 2.48 mol N

Solution:

$$54.5 \, g \, N_2O \times \frac{1 \, mol \, N_2O}{44.02g} \times \frac{2N}{1 \, N_2O} = 2.48 \, mol \, N$$

TRY THIS

How many atoms of S are in a 64.0 g sample of SO_2? What is the total mass of the S?

Answers: 6.01*10²³ atoms S, 32.03 g S.

Solution:

$$64.0 \, g \, SO_2 \times \frac{1 \, mol \, SO_2}{64.07g} \times \frac{6.02*10^{23} \, molecules}{1 \, mol} \times \frac{1S}{1SO_2} = 6.01*10^{23} \, atoms \, S$$

$$64.0 \, g \, SO_2 \times \frac{1 \, mol \, SO_2}{64.07g} \times \frac{1S}{1SO_2} \times \frac{32.07g}{1 \, mol \, S} = 32.03 \, g \, S$$

Example: What is the molar mass of water?

Molar mass of water = molar mass of oxygen + 2(molar mass of hydrogen)

Molar mass of water = 16.00 g/mol + 2(1.01 g/mol) = 18.02 g/mol

We can use a chemical's molar mass to convert between the number of moles of a substance and the mass of that substance.

Example: How many moles of water are in 35.6 g of water?

$$35.6 \, g \, H_2O \times \frac{1 \, mol \, H_2O}{18.02g} = 1.98 \, mol \, H_2O$$

Example: What is the mass of 2.5 mol of C_2H_2?

$$2.5 \, mol \, C_2H_2 \times \frac{26.04g}{1 \, mol \, C_2H_2} = 65.1 \, g \, C_2H_2$$

2.6 Interconversions: Grams to Moles to Molecules and Atoms

So far, we've used molar mass to convert between grams and moles and Avogadro's number to convert between moles and molecules or atoms.

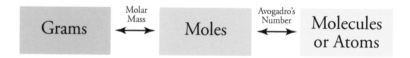

By combining these conversion factors, we can figure out how many molecules or atoms are in a given mass of a substance, or vice versa.

Example: How many atoms of H are in 24.3 g H_2O?

$$24.3 \, g \, H_2O \times \frac{1 \, mol \, H_2O}{18.02g} \times \frac{6.02*10^{23} \, molecules}{1 \, mol} \times \frac{2 \, atoms \, H}{1 \, molecule \, H_2O} = 1.62*10^{24} \, atoms \, H$$

Example: What is the mass of $8.5*10^{20}$ molecules of CH_4?

$$8.5*10^{20} \, molecules \, CH_4 \times \frac{1 \, mol}{6.02*10^{23} \, molecules} \times \frac{16.05g}{1 \, mol \, CH_4} = 2.27*10^{-2} \, g$$

2.7   Stoichiometric Equivalents

Now that we understand the concept of moles, we can look at chemical equations in a slightly different way. Formerly, we defined the coefficients in a chemical equation as the number of each species involved in the reaction, but they really just represent a ratio. For example, let's look at the familiar reaction below:

$$2\,H_2 + O_2 \rightarrow 2\,H_2O$$

The coefficients tell us that two molecules of hydrogen react with one molecule of oxygen to form two molecules of water. However, we could just as easily imagine four molecules of hydrogen reacting with two molecules of oxygen (this would be like having the reaction written above happen twice). The products in this case would be four molecules of water. The coefficients in a chemical equation are **stoichiometric equivalents**, which simply means they provide us with the ratios in which reactants are required and products appear in a chemical reaction. In the case of the reaction above, it is therefore possible to react *any* number of oxygen molecules, as long as there are twice as many hydrogen molecules present. Also from the balanced reaction, the number of water molecules formed would be the same as the number of hydrogen molecules reacted. For instance, for three oxygen molecules to react, the coefficients tell us that six molecules of hydrogen would be required and that six molecules of water would form. Because a mole is just a number, we could react a mole of oxygen molecules ($6.02*10^{23}$ molecules) if we wanted to. We'd need *two* moles of hydrogen molecules, though, to have a stoichiometric equivalent of hydrogen ($1.20*10^{24}$ hydrogen molecules), and we'd make two moles of water. A variety of stoichiometric equivalents for this reaction is listed in Table 2.1.

Looking at the last line of Table 2.1, we notice something odd. We're not used to seeing fractional numbers in chemical reactions; it makes no sense to think about half an atom or half a molecule. However, a mole is a large number that can easily be split in half (just as half of a dozen is six), so it's not inappropriate at all to talk about fractions of *moles* when we refer to stoichiometric equivalents. If only one mole of hydrogen were to react, it would react with *half* a mole of oxygen (that's still a lot of oxygen—it's $3.01*10^{23}$ molecules), to produce a mole of water. If you're wondering how or why we'd ever need the concept of stoichiometry in the real world, rest assured that the applications are numerous. For example, if a pharmaceutical manufacturer is making a drug, its scientists will want to understand the reaction's stoichiometry so that they avoid wasting expensive chemicals (by adding too much of one reactant relative to another) and maximize their product (by making sure that they have enough of each reactant to ensure the amount of product desired).

From now on, whenever we see a chemical equation, we should remember that the coefficients merely represent ratios; reactions can involve just a few molecules or many moles of molecules, as long as the ratios are satisfied. Because we can use molar mass to relate moles to grams, we can even think about stoichiometric

Amount of H^2	Amount of O^2	Amount of H_2O
2 molecules	1 molecule	2 molecules
4 molecules	2 molecules	4 molecules
6 molecules	3 molecules	6 molecules
$1.20*10^{24}$ molecules	$6.02*10^{23}$ molecules	$1.20*10^{24}$ molecules
2 moles	1 mole	2 moles
1 mole	0.5 mole	1 mole

TABLE 2.1. Stoichiometric equivalents in the reaction of H^2 and O^2 to form water.

CONCEPT CHECK

Why can we react half a mole of O_2, even though we can't have the coefficient 0.5 in a chemical equation?

Answer: The coefficients in chemical equations must represent the ratio of each species that reacts. These numbers must be integer values so they are valid for individual atoms and molecules as well as for moles.

TRY THIS

For the equation $2\,S + 3\,O_2 \rightarrow 2\,SO_3$, fill in the table of stoichiometric equivalencies:

Amount of S	Amount of O_2	Amount of SO_3
4 atoms		
	12 molecules	
		2 moles
	1.5 moles	
		44 g

Answer:

Amount of S	Amount of O_2	Amount of SO_3
4 atoms	6 molecules	4 molecules
8 atoms	12 molecules	8 molecules
2 moles	3 moles	2 moles
1 mole	1.5 moles	1 mole
18 g (0.55 moles)	26 g (0.83 moles)	44 g (0.55 moles)

equivalents in terms of masses. However, we have to bear in mind—and this is absolutely critical—that while we can do some math to determine equivalent masses by using the ratio of coefficients in a reaction, **masses of species in a reaction are not present *in* the ratio indicated by the coefficients.** In other words, when elemental hydrogen and elemental oxygen react to make water, we need twice as many molecules (or moles) of hydrogen as oxygen to have stoichiometric equivalency. This does *not* mean, however, that we will need twice the mass of hydrogen that we have of oxygen, because a molecule of oxygen is so much heavier than a molecule of hydrogen. To determine the stoichiometrically equivalent mass of one chemical relative to another, we use the mass of the first substance to determine the number of moles of that substance, then use stoichiometric coefficients to find the equivalent number of moles of the second substance. Finally, we find the mass of the second substance using its molar mass:

Example: What mass of H_2 is required to react with 65 g O_2 in the reaction $2\,H_2 + O_2 \rightarrow 2\,H_2O$, and what mass of water is formed?

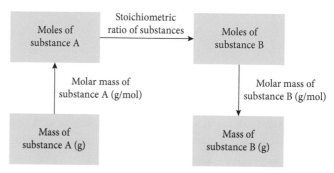

$$65g\,O_2 \times \frac{1\,mol\,O_2}{32.0g} \times \frac{2\,mol\,H_2}{1\,mol\,O_2} \times \frac{2.02\,g\,H_2}{1\,mol} = 8.2\,g\,H_2$$

$$65g\,O_2 \times \frac{1\,mol\,O_2}{32.0g} \times \frac{2\,mol\,H_2O}{1\,mol\,O_2} \times \frac{18.02\,g\,H_2O}{1\,mol} = 73.2\,g\,H_2O$$

2.8 Subatomic Particles

The last puzzle piece we need in order to make sense of atomic mass (and learn why the mass of an individual atom is always a whole number) is to know where that mass comes from. Atoms are made of subatomic particles, which were mentioned briefly in Chapter 1 as the bits you'd (theoretically) divide an atom into if you were able to cut it into pieces, which you can't actually do by any chemical means.

Particle	Mass (g)	Mass (amu)	Charge (C)	Charge (relative)
Proton	$1.67*10^{-24}$ g	1 amu	$+1.6*10^{-19}$ C	+1
Neutron	$1.67*10^{-24}$ g	1 amu	0	0
Electron	$9.11*10^{-28}$ g	~0 amu	$-1.6*10^{-19}$ C	-1

TABLE 2.2. The subatomic particles. Masses are given in grams and amu, and charge is given in Coulombs (C), as well as in relative terms.

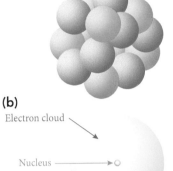

(a)

(b)

Electron cloud

Nucleus

IMAGE 2.12. The nucleus consists of densely packed protons and neutrons (left) and is located at the center of an atom, surrounded by an electron cloud.

Subatomic particles are responsible for making an atom what it is (chemically and physically), but they are not really found on their own, separate and apart from atoms, so we are very correct to say atoms are indivisible.

There are three subatomic particles that go into making up atoms: **protons**, **neutrons**, and **electrons**. Each particle can be defined in terms of its mass and its charge (Table 2.2).

There are a few points to note with regard to these particles. First, as per our discussion of the masses of atoms and molecules, it doesn't really make sense to measure the mass of these very light particles in grams. This is why we use atomic mass units (amu). Note that protons and neutrons have identical mass; they weigh 1 amu each. Electrons do technically have mass, but because they are so light relative to protons and neutrons (nearly 2,000 times lighter), their mass contributes virtually nothing to the overall mass of an atom. An atom would have to have well over 1,000 electrons for them to come anywhere close to contributing significant mass, and no atom has that many! For this reason, we consider the electron essentially massless in our discussion of subatomic particles. A second point is that the charge (measured in Coulombs) of a proton is equal in magnitude but opposite in sign to that of an electron. In other words, they have the same amount of charge, but protons are positively charged and electrons are negatively charged. Because of this, it's easier to refer to them as having "units" of charge, where a proton has one unit of positive charge and an electron has one unit of negative charge. This simplifies discussions of charge significantly. **A brief and relevant bit of information at this juncture is that opposite charges attract, and like charges repel each other.** As such, protons are repelled by other protons, and are attracted to (and by) electrons. These attractions and repulsions are called **electrostatic forces**.

The distribution of subatomic particles within an atom is very organized. Atoms have a very tiny, central portion called a **nucleus**, which is made up of protons and neutrons. Because only protons and neutrons contribute significantly to an atom's mass, and because all an atom's protons and neutrons are located in the nucleus, the mass of an atom is localized in its nucleus. The nucleus is surrounded by a large, virtually massless "cloud" of electrons (Image 2.12).[6]

Armed with this information, we can now begin to make sense of the subatomic structure of an atom. Since mass comes from protons and neutrons (each of which weighs 1 amu), we can use the mass of an atom to figure out how many total protons and neutrons are in the atom's nucleus:

Mass of atom (amu) = (number of protons*1 amu/proton) + (number of neutrons*1 amu/neutron)

Or more simply, mass of atom = number of protons + number of neutrons

| 7 |
| N |
| 14.01 |

IMAGE 2.13. Nitrogen's atomic number is 7.

NOTE[6]

Because electrons move around a lot, we say the cloud is large even if we're talking about an atom with very few electrons.

What is the number of protons,
neutrons, and electrons in atomic O?
In atomic F?

Answer:
O has eight protons, eight neutrons,
eight electrons. F has nine protons, 10
neutrons, nine electrons.

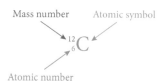

IMAGE 2.14. A shorthand
representation of carbon-12.

Without referring to a periodic table,
how many protons, neutrons, and
electrons are in $^{31}_{15}P$? In $^{200}_{80}Hg$?

Answer:
P has 15 protons, 16 neutrons, and
15 electrons. Hg has 80 protons, 120
neutrons, and 80 electrons.

As we saw in a "Try This" earlier in the
chapter, the reason that some mass
numbers on the periodic table are in
parentheses is that these elements have
multiple isotopes but are radioactive,
with some isotopes being very unstable.
The mass number in parentheses is the
one that corresponds to the most stable
isotope of the element.

For instance, the most common isotope of nitrogen—nitrogen-14—weighs 14 amu. We therefore know that there are 14 particles (protons and/or neutrons) in nitrogen's nucleus. With a bit of additional information, we can even determine how many of each species are present. Recall that each element has two numbers associated with it: the larger of the two (always underneath the atomic symbol) is the atomic mass. The smaller of the two (positioned above the atomic symbol) is called the **atomic number**. **The atomic number indicates the number of protons in the nucleus of an atom of any element.** Nitrogen, for instance, with an atomic number of seven, has seven protons. Using the example of nitrogen-14, which has a mass of 14 amu, we can calculate the number of neutrons in the nucleus of a nitrogen atom:

Mass number = number of protons + number of neutrons

14 = 7 + number of neutrons

Number of neutrons = 7

Nitrogen-15 is still nitrogen, albeit a different isotope. As a form of nitrogen, it has the atomic number 7, indicating that there are seven protons in its nucleus. **All isotopes of a given element have the same atomic number, and therefore the same number of protons in their nuclei.** However, when we subtract the atomic number (7) from nitrogen-15's mass number (15), we find that it has eight neutrons in its nucleus. **Isotopes of an element differ from one another in the number of neutrons in their nuclei.**

To determine how many electrons an atom has, we use the charges of subatomic particles. Since both protons and electrons have charge, we can determine the total charge on an atom by adding up the sum of positive charges (from protons) and the sum of negative charges (from electrons). It is important to know that any atom that has not reacted chemically will be **neutral**, which is to say that the sum of its positive and negative charges will be zero. Remember that protons have a charge of +1 and electrons have a charge of −1. **Therefore, the number of protons and the number of electrons in a neutral atom must be the same; all atoms in their elemental form are neutral.** For example, atomic nitrogen has seven protons (and therefore the sum of positive charge is +7), so it must have seven electrons (and therefore the sum of negative charge is −7) for the net charge to be zero.

We can also convey information regarding atomic number and mass number via shorthand representations of elements (Image 2.14). These representations allow us to quickly represent or determine (without needing to refer to a periodic table) how many of each subatomic particle make up an atom of a given element. Note that these representations are specific for a given isotope of an element, so we always use mass number rather than atomic mass. In other words, the shorthand representation in image 2.14 is of carbon–12, not carbon in general.

There are a few things worth noting at this point. First, as you look at a periodic table, notice that atomic masses are not unique: Po and At, for instance, both have atomic masses of 210.[7] Further, you'll notice that while atomic masses generally increase left to right and top to bottom on the periodic table, this trend is not absolute. I (number 53), for instance, has a smaller atomic mass than Te (number 52). These inconsistencies with regard to atomic mass relate to the fact that while the number of neutrons *generally* increases as the number of protons increases, there are exceptions to this trend. Atomic numbers, however, follow much stricter rules. Across the periodic table, atomic numbers increase sequentially. They are always integer values (because they indicate the number of protons, which cannot be fractional) and they are always unique to a given element; there are no two elements with the same atomic number. Because of this, **the atomic number is the most important identifying feature of an atom**. For instance, carbon is the only element with the atomic number 6; an atom whose nucleus contains 10 protons must be an atom of Ne.

It's worth noting the relationship between number of protons and number of neutrons in atoms of various elements. The lighter elements tend to have the same number of protons as neutrons (carbon, for instance, has six of each), while the heavier elements have more (sometimes *many* more) neutrons than protons. Why? To answer this question, we need to know a little about what holds an atom's nucleus together. The nucleus is a tight mass of protons and neutrons, which is a bit counterintuitive given what we know about protons. If protons are all positively charged and like charges repel, how is it possible to pack many positively charged particles into a very tiny space? It turns out that **nucleons** (a collective term for protons and neutrons) are strongly attracted to one another over *very* short distances by a very strong force (which is cleverly named the **strong force**). As a result, even though protons are electrostatically repelled by other protons, if they're packed in very tightly, they're also attracted by and to other protons, due to the strong force. Neutrons are not electrostatically attracted to or repelled by protons, but they are held strongly to each other and to protons over very small distances by the strong force (Image 2.15).

This explains two nuclear phenomena. First, the nucleus is tiny because it has to be very tightly packed in order for strong forces to be able to hold the protons together; if the nucleus were less dense than it is, protons would feel only repulsion. Second, the more protons in a nucleus, the harder it is to hold all those like-charged particles together, since there's more electrostatic repulsion and the radius of the nucleus is greater, which weakens the strong forces. Neutrons act like nuclear "glue": they contribute additional strong force without increasing electrostatic repulsion, so bigger nuclei have more neutrons relative to the number of protons. If a nucleus

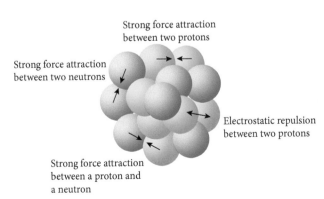

Strong force attraction between two protons

Strong force attraction between two neutrons

Electrostatic repulsion between two protons

Strong force attraction between a proton and a neutron

IMAGE 2.15. Neutrons attract each other and protons via the *strong force*. Protons attract other protons via the *strong force* but also repel them electrostatically. The more neutrons in a nucleus, the more attraction between nucleons without additional repulsion.

IMAGE 2.16. Radioactive elements include (clockwise): plutonium (Pu; a), polonium (Po; b), and uranium (U; c). We'll discuss radioactivity further in Chapter 12.

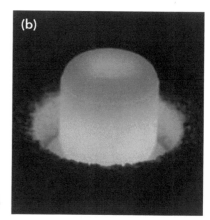
(a)

(b)

(c)

(d)

is big enough, no number of neutrons is really sufficient to hold all those protons together. This makes very large nuclei fundamentally unstable, and they tend to fall apart. This phenomenon is called radioactive decay, and while it can be observed in small elements, it's ubiquitous in the large ones. In fact, **every element with an atomic number equal to or greater than that of bismuth (83) is radioactive!**

A few final words about atomic structure: we all know atoms are small (so by extension, subatomic particles must be smaller), but it's really hard to conceive of exactly *how* small. One interesting way to think about the size of atoms is to imagine an apple. If we were to expand an apple to the size of the earth, each atom in the apple would end up approximately the size of the original apple.

Further, we tend to think about matter (and therefore atoms, since matter is made up of atoms) as being solid, but it turns out that since almost all the mass in an

IMAGE 2.17. The size difference between an atom in an apple and the apple itself is the same as the size difference between an apple and the planet Earth!

(a)

(b)

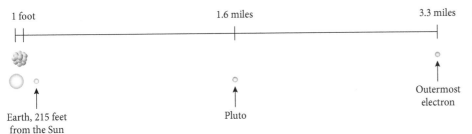

1 foot   1.6 miles   3.3 miles

Earth, 215 feet from the Sun

Pluto

Outermost electron

IMAGE 2.18. A comparison of the amount of empty space in the solar system and in an atom.

atom is clustered into a tiny central nucleus, most of the atom is actually empty space! In order to conceive of this, imagine scaling down the solar system until the Sun was a mere one foot in diameter. Earth would be 215 feet away from the Sun and the (dwarf) planet Pluto would be 1.6 miles from the Sun. That's a lot of emptiness, which fits with our notion of the solar system as mostly empty space.[8] However, if we imagined at an atom of gold on the same scale, expanding the nucleus to a diameter of 1 foot, the outermost electron would be 3.3 miles away—almost double the radius of the solar system! In that sphere with a total diameter of 6.6 miles, we'd find only 79 electrons. Clearly, atoms contain a *lot* of empty space!

One more interesting bit of information: the nucleus is *so* dense that a matchbox full of nuclei (separated from their electron clouds so that they'd take up less space) would weigh 2.5 billion tons (that's 5 trillion pounds!).

NOTE[8]

In this scaled model of the solar system, the nearest star other than our Sun—which in this model is 215 feet away from Earth—would be Alpha Centauri, at over 10,000 miles away! In the real world, Alpha Centauri is about 4.4 light years from Earth, or $2.6*10^{13}$ miles. This is a truly mind-boggling distance.

2.9   The Last Word—Land Animals Breathe Air, Aquatic Animals Breathe Water?

A whole chapter's worth of discussing atoms and the atmosphere leads to an interesting question: what do humans (and other land animals) breathe? Most people would answer "air." Well, yes, we *do* breathe air, but as it turns out, we don't *need* to breathe air. We *need* O_2, and we breathe air because it's a readily available mixture that contains the right amount of oxygen. As it turns out, we're perfectly capable of breathing a substance other than air and staying alive and well, as long as that substance has an appropriate percentage of O_2 mixed in. Let's think about fish for a moment: like land animals, fish also require O_2. They do not, however, breathe air. Fish get their oxygen from water, which has a small amount of atmospheric O_2 dissolved (mixed) into it. Unlike our atmosphere, however, which is 21% oxygen, the percentage of O_2 dissolved in water is in the range of 0.5–3%. Fish handle this just fine; their gills are adapted to allow them to pull O_2 from the water extremely efficiently. We are not nearly as efficient at removing O_2 from mixtures, and as such would not be able to survive by breathing water. Oddly enough, though, we are able to breathe liquid as long as that liquid dissolves O_2 at a higher percentage than water does. The liquid *perfluorodecalin* ($C_{10}F_{18}$, Image 2.19) is much better at dissolving O_2 than water is; perfluorodecalin can dissolve up to 40% O_2, making it a perfectly acceptable medium for humans (or other land animals) to breathe. In Image 2.20, we see a mouse which has been submerged in perfluorodecalin; he is alive and

(a)

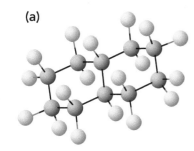

(b)

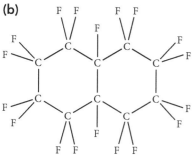

IMAGE 2.19. The compound perfluorodecalin is composed of the elements carbon and fluorine.

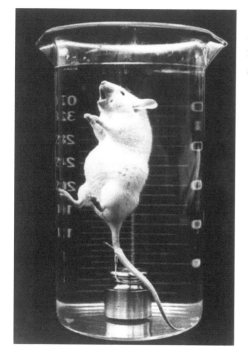

IMAGE 2.20. This mouse is alive and well (if not necessarily happy), submerged in perfluorodecalin. The weight tied to its tail keeps it under the surface of the fluid, which is quite dense and in which the mouse would otherwise float.

well and will be able to return to getting his O_2 from air once he's released from his beaker. Humans have also experimented with breathing liquid media; while it's reported to be psychologically uncomfortable to submerge one's head in liquid and take a deep breath—an observation that is quite understandable—it's apparently not painful. Anecdotal reports do indicate that the process of clearing the lungs of liquid in order to return to breathing air is a bit uncomfortable, however.

Aside from the fact that it's a scientific curiosity, what is the function of breathing a liquid? There are actually several medical applications being explored. First, breathing a liquid medium could reduce the incidence of infant respiratory distress syndrome (IRDS), a condition resulting from the tendency of premature infants' underdeveloped lungs to collapse. Infants suspended in perfluorodecalin would get the oxygen they need without the risk of lung collapse associated with breathing air. Further, liquids do not expand and compress significantly (like gases do) when exposed to changes in pressure. As such, liquid breathing media are being investigated for deep-sea-diving applications; they could potentially prevent complications associated with extended periods of breathing high-pressure air, which include the painful and potentially fatal condition of decompression sickness, also known as "the bends." Additionally, since perfluorodecalin is a liquid, it has the ability to "wash" the lungs as it delivers oxygen, and might find applications in medicine for removing debris and foreign matter. Finally, because of its ability to dissolve high concentrations of oxygen, perfluorodecalin has the potential to function as a component of artificial blood (the job of the blood is to carry oxygen from the lungs to the tissues using an O_2-carrying molecule called *hemoglobin*). All around, it's pretty interesting stuff!

SUMMARY OF MAIN POINTS

- Earth is surrounded by a thin shell of gases. This is the atmosphere.
- The atmosphere exerts pressure, and pressure varies inversely with altitude.
- The atmosphere is composed primarily of nitrogen and oxygen and its composition is nearly uniform throughout the troposphere.
- Atoms of elements have mass indicated by their atomic mass on the periodic table. Atomic masses are measured in units of amu.
- The mass of a molecule is determined by adding the masses of its constituent atoms.
- Atoms and molecules are counted in moles, where a mole is $6.02*10^{23}$ (Avogadro's number) of *anything*.
- The mass of a mole of an element is indicated by its atomic mass on the periodic table, where molar masses are measured in units of grams.
- The molar mass of a molecule is determined by adding the molar masses of its constituent atoms.
- Avogadro's number can be used to convert from atoms or molecules to moles and vice versa.
- Molar mass can be used to convert from mass in grams to moles and vice versa.

- Stoichiometric equivalents are quantities of reactant or product that satisfy the ratios indicated by coefficients in chemical equations in atoms/molecules, moles, or grams.
- The atom is made of protons and neutrons (in a dense, low-volume nucleus) and electrons (in a nearly massless, high-volume cloud around the nucleus).
- The mass number is the number of protons plus neutrons in an atom's nucleus.
- The atomic number is an atom's identifying feature and is equal to the number of protons.
- The number of electrons in an elemental atom is equal to the number of protons in that atom.

QUESTIONS AND TOPICS FOR DISCUSSION

1. Why is the air on mountaintops sometimes described as "thin"? How is the air there similar to the air at sea level? How is it different?

2. What kind of relationship do altitude and atmospheric pressure have? Is it a linear correlation (doubling altitude cuts pressure in half), or is it exponential? At what altitude is pressure approximately 50% of its sea level value? At what altitude is it approximately 10% of its sea-level value?

3. Why do ears pop with changes in altitude?

4. What causes mountain climbers to experience paradoxical respirations when sleeping at altitude?

5. How would hyperventilation affect the acidity of the blood, and why? What about apnea (temporary cessation of respiration)?

6. Why can we add the masses of the atoms in a molecule to determine the mass of the molecule? What does this show us about chemical reactions (making molecules) and conservation of mass?

7. What is the difference between atomic mass and mass number? When might we talk about an atom's atomic mass, and when might we be more interested in its mass number?

8. Why are atomic and molecular masses measured in amu rather than grams? What is an amu?

9. How is the concept of a mole similar to the concept of a dozen? How are they different? What sorts of things do we measure in moles (practically speaking)?

10. What is a thought experiment? Do a little Internet research to find another famous thought experiment (there are many). Why do scientists sometimes do thought experiments? Are they dependable?

11. Why is the atomic mass for any element equal to both the mass of one atom of that element in amu and its molar mass in grams? Is that merely coincidence?

12. Why can we add the molar masses of the elements in a molecule to determine the molar mass of the molecule?

13. Why does a mole of carbon weigh 12.01 g, but a mole of carbon-12 weighs 12.0 g? Under what circumstances would a mole of carbon weigh 13 g?

14. The atomic mass of nitrogen is 14.01 amu; are there any nitrogen atoms that have this mass? Explain.

15. The molecular mass of water is 18.02 amu; are there any molecules of water that have this mass? Explain.

16. What does "stoichiometric equivalent" mean? Do these values refer to individual atoms/molecules, moles, or both?

17. Are the stoichiometric coefficients in a chemical equation representative of the ratio of masses in the equation? Why or why not? How can we use these coefficients to determine stoichiometrically equivalent masses?

18. Why is an electron considered massless?

19. Where is most of the mass in an atom? Where is most of its volume? What does this mean with regard to how "full" or "empty" the space in an atom is?

20. Explain how to use the periodic table to find the number of protons, neutrons, and electrons in an atom.

21. What do two isotopes of an element have in common? How do they differ?

22. Why do smaller atoms have approximately equal numbers of protons and neutrons, while larger atoms have many more neutrons than protons?

23. Why are all the large atoms radioactive?

24. Why can't humans breathe water? Under what circumstances could a human breathe liquid?

1. Using Image 2.3 in the chapter, what is the atmospheric pressure at 10,000 feet? At 65,000 feet?

2. Using Image 2.3 in the chapter, at what altitude is the atmospheric pressure 0.8 atm? 0.25 atm?

3. What percent of our atmosphere is made of N_2? What percent is O_2?

4. Where on the periodic table can you find an element's atomic mass?

5. What is the atomic mass of hydrogen? Of helium? What is the mass of an atom of the most common isotope of hydrogen in amu? In grams?

6. What is the atomic mass of polonium (Po)? Of oxygen? What is the mass of an atom of the most stable isotope of radium in amu? In grams?

7. Calculate the molecular mass of the following: $CaCO_3$ (calcite); Fe_2O_3 (rust); H_2O_2 (hydrogen peroxide, a sterilizer).

8. Calculate the molecular mass of the following: C_2H_6O (alcohol); $C_6H_{12}O_6$ (glucose, a sugar); NO_2 (nitrogen dioxide, an atmospheric pollutant).

9. Calculate how many atoms are in each of the following samples: 1.5 moles C; 10.2 moles He; 0.75 moles H.

10. Calculate how many atoms are in each of the following samples: 7.2 mol N; 1.3 mol Rn; 0.24 mol Ca.

11. Comparing a mole of elemental oxygen and a mole of water, which contains more total molecules (or are they the same)? Which contains more total atoms (or are they the same)?

12. Calculate how many molecules are in each of the following samples: 0.60 mol H_2O; 1.2 mol NaCl; 23.5 mol $C_4H_{10}O$ (ether, an anesthetic).

13. Calculate how many molecules are in each of the following samples: 12.6 moles NaF (an ingredient in toothpaste); 1.5 moles N_2O (nitrous oxide or laughing gas); and 0.65 moles CO.

14. Determine the number of moles in each of the following: $5.4*10^{22}$ atoms Xe; $1.22*10^{23}$ molecules CO_2; $3.5*10^{25}$ eggs (OK, we're not likely to have this many eggs, but the point remains that you *could* count eggs in moles).

15. Determine the number of moles in each of the following: $3.2*10^{25}$ atoms Ti; $7.2*10^{24}$ molecules H_2O; $3.5*10^{20}$ cats (again, not likely to find this many cats, but the point remains).

16. How many atoms of C are in each of the following samples: 1.2 moles CH_2O (formaldehyde, a preservative); 5.7 moles CH_4O (wood alcohol); 10.4 moles $C_2H_4O_2$ (vinegar)?

17. How many atoms of O are in each of the following samples: 4.3 moles CH_2O_2 (formic acid, one of the irritants in ant bites); 0.50 moles H_2SO_4 (sulfuric acid, a component of acid rain); 1.3 moles $NaHCO_3$ (baking soda)?

18. What is the molar mass of S? Of SO_3 (volcanic gas)?

19. What is the molar mass of Cl? Of CCl_2F_2 (Freon, responsible for ozone depletion)?

20. Calculate the mass of each of the following: $1.2*10^{23}$ molecules of CO_2; $5.4*10^{25}$ atoms of K; 3.5 moles CO_2; 0.65 moles K.

21. Calculate the mass of each of the following: $6.5*10^{22}$ molecules of HNO_3 (nitric acid, a component of acid rain); $4.4*10^{25}$ atoms of B; 0.23 moles HNO_3; 12.5 moles B.

22. Calculate the number of moles in each of the following samples: 118 g H_2O; 75 g $C_6H_{12}O_6$ (glucose, a sugar); 0.3 g NaCl.

23. Calculate the number of moles in each of the following samples: 212 g NO_2 (nitrogen dioxide, an atmospheric pollutant); 61 g O_3 (ozone); 6.2 g O_2.

24. Calculate the number of molecules in each of the following samples: 118 g H_2O; 75 g $C_6H_{12}O_6$ (glucose, a sugar); 0.30 g NaCl.

25. Calculate the number of molecules in each of the following samples: 212 g NO_2 (nitrogen dioxide, an atmospheric pollutant); 61 g O_3 (ozone); 6.2 g O_2.

26. Calculate the number of atoms (total) in each of the following samples: 118 g H_2O; 75 g $C_6H_{12}O_6$ (glucose, a sugar); 0.30 g NaCl.

27. Calculate the number of atoms (total) in each of the following samples: 212 g NO_2 (nitrogen dioxide, an atmospheric pollutant); 61 g O_3 (ozone); 6.2 g O_2.

28. For the reaction $3 H_2 + N_2 \rightarrow 2 NH_3$, fill out the following table with stoichiometrically equivalent quantities:

Amount of H_2	Amount of N_2	Amount of NH_3
6 molecules		
	$5.25*10^{23}$ molecules	
		2 moles
0.75 mole		

29. For the reaction $C_2H_4 + 3 O_2 \rightarrow 2 CO_2 + 2 H_2O$, fill out the following table with stoichiometrically equivalent quantities:

Amount of C_2H_4	Amount of O_2	Amount of CO_2	Amount of H_2O
	9 molecules		
$2.9*10^{24}$ molecules			
			2 moles
		0.33 mole	

30. For the reaction $3 H_2 + N_2 \rightarrow 2 NH_3$, fill out the following table with stoichiometrically equivalent quantities:

Amount of H_2	Amount of N_2	Amount of NH_3
6.5 g		
		32.4 g
	125 g	

31. For the reaction $C_2H_4 + 3 O_2 \rightarrow 2 CO_2 + 2 H_2O$, fill out the following table with stoichiometrically equivalent quantities:

Amount of C_2H_4	Amount of O_2	Amount of CO_2	Amount of H_2O
		35.4 g	
	1.2 g		
			518.2 g

32. What is the mass (in amu) and relative charge of a proton? Of a neutron? Of an electron?

33. What is the atomic mass of sulfur (S)? What is the mass number of the most common isotope of sulfur?

34. What is the atomic mass of calcium (Ca)? What is the mass number of the most common isotope of calcium?

35. How many protons, neutrons, and electrons are in an atom of the most common isotope of each of the following (Hint: find the most common isotope by rounding the atomic mass to the nearest whole number value): K; Cm; Pt; Kr?

36. How many protons, neutrons, and electrons are in an atom of the most common isotope of each of the following (Hint: find the most common isotope by rounding the atomic mass to the nearest whole number value): Li; Mn; Br; Pb?

37. Without using the periodic table, how many protons, neutrons, and electrons are in $^{65}_{30}Zn$? In $^{98}_{43}Tc$? In $^{24}_{12}Mg$?

38. Without using the periodic table, how many protons, neutrons, and electrons are in $^{210}_{84}Po$? In $^{184}_{74}W$? In $^{32}_{16}S$?

39. Write out shorthand representations $\left(^{mass\ number}_{atomic\ number}symbol\right)$ of each of the following: Nickel (Ni); selenium (Se); astatine (At).

40. Write out shorthand representations $\left(^{mass\ number}_{atomic\ number}symbol\right)$ of each of the following: Barium (Ba); tin (Sn); zinc (Zn).

ANSWERS TO ODD-NUMBERED PROBLEMS

1. 0.69 atm; ~0.06 atm.
3. 78%; 21%.
5. 1.01 amu; 4.00 amu; 1 amu; $1.69*10^{24}$ grams.
7. 100.09 amu; 159.7 amu; 34.02 amu.
9. $9.0*10^{23}$ atoms C; $6.14*10^{24}$ atoms He; $4.5*10^{23}$ atoms H.
11. A mole of O_2 and a mole of H_2O contain identical numbers of molecules $(6.02*10^{23})$, but the mole of water contains more total atoms.
13. $7.59*10^{24}$ molecules NaF ; $9.0*10^{23}$ molecules N_2O ; $3.9*10^{23}$ molecules CO.
15. 53 moles Ti ; 12 moles H_2O ; $5.8*10^{-4}$ moles cats.
17. $5.2*10^{24}$ atoms O; $1.2*10^{24}$ atoms O; $2.3*10^{24}$ atoms O.
19. 35.45 g/mol; 120.91 g/mol.
21. 6.8 g; 790 g; 14.5 g; 135 g.
23. 4.61 moles NO_2; 1.3 moles O_3; 0.19 moles O_2.
25. $2.78*10^{24}$ molecules NO_2; $7.8*10^{23}$ molecules O_3; $1.1*10^{23}$ molecules O_2.
27. $8.34*10^{24}$ atoms; $2.3*10^{24}$ atoms; $2.2*10^{23}$ atoms.
29. For the reaction $C_2H_4 + 3\ O_2 \rightarrow 2\ CO_2 + 2\ H_2O$, fill out the following table with stoichiometrically equivalent quantities:

Amount of C_2H_4	Amount of O_2	Amount of CO_2	Amount of H_2O
3 molecules	9 molecules	6 molecules	6 molecules
$2.9*10^{24}$ molecules	$8.7*10^{24}$ molecules	$5.8*10^{24}$ molecules	$5.8*10^{24}$ molecules
1 mole	3 moles	2 moles	2 moles
0.165 mole	0.495 mole	0.33 mole	0.33 mole

31. For the reaction $C_2H_4 + 3\,O_2 \rightarrow 2\,CO_2 + 2\,H_2O$, fill out the following table with stoichiometrically equivalent quantities:

Amount of C_2H_4	Amount of O_2	Amount of CO_2	Amount of H_2O
11.3 g	38.6 g	35.4 g	14.5 g
0.35 g	1.2 g	1.1 g	0.45 g
403.5 g	1380 g	1257 g	518.2 g

33. Atomic mass of S = 32.07 amu. Mass number of most common form of S is 32.

35. K has 19p, 20n, 19e. Cm has 96p, 151n, 96e. Pt has 78p, 117n, 78e. Kr has 36p, 48n, 36e.

37. Zn has 30p, 35n, 30e. Tc has 43p, 55n, 43e. Mg has 12p, 12n, 12e.

39. $^{59}_{28}Ni$, $^{79}_{34}Se$, $^{210}_{85}At$.

SOURCES AND FURTHER READING

Guo, Z. L., Lu, G. P., Ren, T., *et al.* (2009). Partial liquid ventilation confers protection against acute lung injury induced by endotoxin in juvenile piglets. *Respiratory Physiology and Neurobiology, 167*(3), 221–26.

Shaffer, T. H., Wolfwon, M. R., Greenspan, J. S., *et al.* (1996). Liquid ventilation in premature lambs: Uptake, biodistribution and elimination of perfluorodecalin liquid. *Reproduction, Fertility and Development, 8*(3), 409–16.

Spence, R. K., Norcross, E. D., Costabile, J., *et al.* (1994). Perfluorocarbons as blood substitutes: The early years. Experience with Fluosol DA-20% in the 1980s. *Artificial Cells, Blood Substitutes, and Immobilization Biotechnology, 22*(4), 955–63.

Suy, G. Y., Chung, M. P., Parks, S. J., *et al.* (1999). Partial liquid ventilation with perfluorocarbon improves gas exchange and decreases inflammatory response in oleic acid-induced lung injury in beagles. *Journal of Korean Medical Science, 14*(6), 613–22.

CREDITS

Chemistry Inside Us—The Elements of Life

3

Of all the elements on the periodic table, only a handful of them are represented to any significant extent in living organisms. **Whether an organism is composed of one cell (such as a bacterium) or many (like us), cells are made mostly of just four elements: C, H, O, and N.** A few additional elements show up in trace amounts, collectively comprising a mere 4% of the human body by mass (Image 3.1). We sometimes refer to the elements of life as falling into two categories: "major" (C, H, O, and N) and "trace" (important elements that are represented in smaller amounts), as per Image 3.2.

Even some of the elements not highlighted in Image 3.2 have applicability to life, albeit in very small quantity or in specific forms of life only. Dr. Charles Cockell of the University of Edinburgh has assembled a really fabulous "astrobiological periodic table" that includes not only the source of each element, but also its role, if any, in

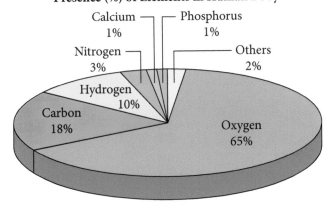

IMAGE 3.1. The elements of life, percent by mass.

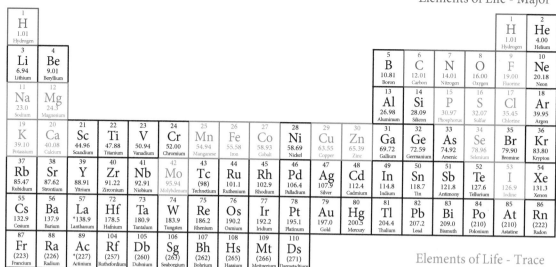

Elements of Life - Major

Elements of Life - Trace

The Astrobiological Periodic Table
© Charles S Cockell, v. 1.0 [June 2015]: The Astrobiological Periodic Table

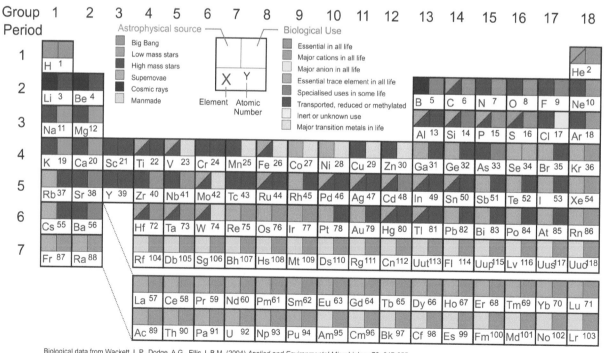

Biological data from Wackett, L.P., Dodge, A.G., Ellis, L.B.M. (2004) *Applied and Environmental Microbiology* **70**, 647-655.

biology (Image 3.3). Note that because Dr. Cockell uses slightly different criteria for assigning categories to elements (phosphorus, for instance, is not a major element of life, as it is not represented in large quantities, but it is essential for life), there are some minor differences between this table and the table presented in Image 3.2. The latter assigns elements to "major" and "trace" on the basis of proportion in the majority of life forms.

3.1   Carbon-Based Life Forms

You've probably heard before that we are "carbon-based life forms." If this is true, then why does oxygen account for 65% of our elemental breakdown by mass? The reason is that we—like all living organisms—are composed largely of water, and water contains oxygen. We'll look at the elemental makeup of the body in three different ways: by mass, by number of atoms, and by number of atoms with water removed (Table 3.1).

Discounting water, we are primarily carbon (by mass) and hydrogen (by sheer quantity). The large molecules that make up cells, which in turn make up organisms, have carbon "backbones" (Image 3.4).

So why carbon? What's so special about this element, that it is the foundation upon which we are all built? In order to answer this question, we need to begin to look at the nature of chemical bonding. And in order to do that, we need to first go deeper into the structure of the atom itself.

3.2   Electron Shells

In the last chapter, we discussed the "electron cloud" that surrounds the nucleus of an atom, accounting for most of the volume, but only the tiniest fraction (so tiny, in fact, that we ignore it) of the mass. In reality, describing electrons as forming a nebulous "cloud" around the nucleus does nothing to help us understand what is, in actuality, a form of organization of electrons within an atom.[1] **Electrons are distributed in shells around the nucleus, where shells are numbered according to their radius (shell 1 is closest to the nucleus), as Image 3.5 illustrates.**

Element	Percent by Mass	Percent by Number of Atoms	Percent by Number of Atoms (no H_2O)
Carbon	18%	9%	39%
Hydrogen	10%	63%	52%
Oxygen	65%	26%	1.6%
Nitrogen	3%	1.25%	5.6%

TABLE 3.1. Elemental composition of the human body.

NOTE[1]

Actually, there's no one *best* way to describe the arrangement of electrons in an atom—subatomic particles are strange beasts, and it's nearly impossible to use macroscopic, familiar images as metaphors for describing them with any rigor. For this reason, it's pretty common to think about atoms in many ways, picking analogies as relevant to emphasize a particular point. In considering the size of the nucleus relative to the volume occupied by electrons, we do well to imagine the electrons as a cloud. In conceptualizing the arrangement of electrons within shells in an atom, we tend to picture a little "solar system." It's probably best to remember that these are all analogies, and while they each add to our understanding of atomic structure, no one of them is *better* or *more correct* than any other.

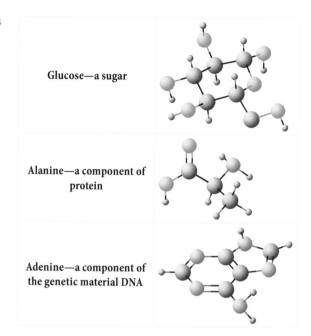

Glucose—a sugar

Alanine—a component of protein

Adenine—a component of the genetic material DNA

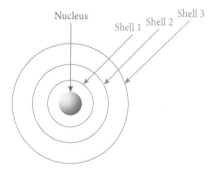

IMAGE 3.5. The first three electron shells (diagram not to scale). Radius increases with increasing shell number.

Because electrons are negatively charged, they're attracted to (and by) protons. For this reason, even though each electron in an atom must be located within a shell, **electrons are most stable in shell 1 because of its proximity to the nucleus.** You can think about the attractive force a bit like gravity; if you were to hold a ball (electron) away from the earth (nucleus of an atom), the ball would be attracted to the earth due to the force of gravity (electrostatic force between protons and electrons). If you let go of the ball, it would fall toward Earth's surface, getting as close as it could, unless something (like a table) got in its way. The shells are a bit like tables; electrons "fall" as close to the nucleus as they can get, but they can't get closer than shell 1, so they stop there. Departing a bit from our ball, Earth, and table analogy (as with all atomic structure conceptualizations, each analogy only gets us so far), it turns out that **each shell can only hold a certain number of electrons.** After that, they are full, and no more electrons can "fall" into them. Shell 1, for instance, can only hold two electrons. As such, if we consider an atom with one electron—that would be hydrogen, which has one proton and therefore one electron—we expect to find that electron in shell 1. For an atom with two electrons (helium), we'd expect to find them *both* in shell 1. However, an atom with three electrons would have two of them in shell 1, and the third electron would be located in shell 2. Electrons always "fall" as close as they can get to the nucleus, but they can't enter full shells!

How can we tell when a shell is full? Conveniently, the periodic table provides us with a guide. **Rows on the periodic table correspond to shells, so we can count the number of elements in any row and determine how many electrons it takes to fill the correspondingly numbered shell.** Row 1 on the periodic table contains two elements, which reminds us that shell 1 holds two electrons. Row 2 contains eight elements, which reminds us that shell 2 holds eight electrons. Row 3 also contains eight elements, so shell 3, like shell 2, holds eight electrons—but then things get a little weird. *Technically*, shell 3 holds 18 electrons, but it acts full with eight, as do shells 4 and up. Very briefly, shell 3 finishes filling partway through the filling of shell 4—the first row of transition metals represent the remainder of shell 3, the second row of transition metals are the remainder of shell 4, and so on.

To avoid the confusion associated with some of these higher-numbered shells, we use three simple rules for the purposes of this text:

1. Aside from shell 1 (which is full with two electrons), each shell is (or acts as if it is) full when it contains eight electrons.
2. Because rows on the periodic table correspond to shells, elements at the right end of each row have a filled outer electron shell.
3. For elements in row 3 and the first two elements of row 4 (K and Ca), shell 3 holds up to eight electrons. For all elements after the transition metals in row 4 (from Ga onward), shell 3 is completely full with 18 electrons.[2]
4. Determining how many electrons are in shells 4 and up is beyond the scope of this text, and we'll simply use the inviolate rule that regardless of the shell number, elements at the end of each row have a full outermost shell.

Let's look more closely at the second rule. Helium (He) is at the end of row 1. We know that shell 1 holds two electrons, and remembering that He has two electrons, He must have a full shell 1. Neon (Ne) is at the end of row 2, and has a total of ten electrons. The first two electrons must be in shell 1, and the remaining eight that can't fit in shell 1 are left to fill shell 2, which is consequently full. We see the same thing with argon (Ar), which has 18 electrons. The first two are in shell 1, there are eight in shell 2, and the remaining eight go into shell 3, filling it. Even though things get a bit confusing in terms of our electron count after row 3 on the periodic table, the pattern regarding the elements at the end of each row holds true. **Elements at the end of each row *always* have full outermost shells, containing (at that point, regardless of how many they might eventually hold) eight electrons; helium is the only exception, as its full outermost shell holds only two electrons.** Krypton, for instance, has 36 total electrons. The first two are in shell 1. The next eight are in shell 2. Krypton is at the end of row 4, so shell 3 is filled with a total of 18 electrons. We've accounted for 28 of krypton's electrons so far, so the last eight must be in shell 4, filling it.

We can use the information about rows and their corresponding shells to figure out where the electrons are in any element, not just those at the ends of rows. Elements in row 2 will have their first two electrons in shell 1 and any remaining electrons in shell 2. Notice that of the elements in row 2, however, only Ne has a filled shell 2 (Image 3.6).

NOTE[2]

Shell 3 finishes filling up (from eight to 18 electrons) over the course of the transition metals in row 4. However, within the transition metals, there are some weird exceptions to normal filling patterns, so a discussion of their electron configurations is beyond the scope of this text. Suffice it to say, K and Ca have eight electrons in shell 3, while anything beyond Ga on the periodic table has 18 electrons in shell 3. Frankly, the transition metals break a lot of the rules we learn that are valid for the other elements.

Li   Be   B   C   N   O   F   Ne

IMAGE 3.6. Electron configurations of the row 2 elements.

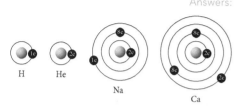

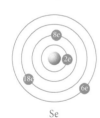
3.3 Valence

Aside from helping us to know where specifically the electrons are within an atom, what can we do with our understanding of configuration? It turns out that **only the electrons in an atom's outermost shell participate in bonding and chemical reactions, and the number of electrons in this outermost shell gives us information about an element's chemical behavior.** In terms of understanding the chemistry of carbon, for instance, it's more important to know that it has four electrons in its outer shell—these are called valence electrons—than that it has six electrons total. Furthermore, a closer look at the periodic table provides another revelation: **all elements within a given column on the periodic table have the same number of valence electrons, even though their total electron counts differ!** We already saw this trend in the column on the far right of the periodic table: neon, argon, and the rest of the elements in this column all have full outer shells of eight electrons. Two columns to the left, oxygen, sulfur, selenium, and the rest of that column's elements have six valence electrons (Image 3.7).

Many periodic tables number the columns; this provides a convenient way of determining an atom's valence without having to sketch a full electron configuration. Elements in the first column (called group 1) have one valence electron, elements in the second column (group 2) have two valence electrons, and so forth (Image 3.8). Even if a periodic table's columns aren't numbered, it's still possible to count the columns from left to right (skipping over the transition metals) to determine an atom's valence.

If valence electrons are the only ones that participate in bonding and reactions, and the number of valence electrons is, as stated earlier, more important than the total number of electrons in predicting how an atom acts chemically, then it should come as no surprise that **elements in the same group have similar chemical properties.** Lithium, sodium, potassium, and the other members of group 1 show striking chemical similarities: they are all, for instance, very reactive soft metals that catch fire or explode when exposed to water (Image 3.9).

The same is true for groups 2 through 8. Group 8, in particular, is interesting: these elements behave similarly to one another in that they do *not* tend to react or form bonds. In fact, their tendency not to form chemical bonds with other elements has resulted in the group's name: they're called the *noble gases*, for their "refusal" to associate with "lesser" elements (Image 3.10).[3]

The noble gases were discovered much later than some of the other elements, simply because they don't tend to react or form compounds, which made them hard to detect. This inertia on their part is even reflected in their elemental names. Aside from helium (so named because it's made in the sun, which is *helios* in Greek), the noble gases are neon ("new one"), argon ("lazy one"), krypton ("hidden one"), xenon ("strange one"), and radon ("radioactive one"). It's no coincidence that the noble gases are both unreactive *and* have full electron shells: chemical reactivity is a product of valence. Full outer shells are very stabilizing for atoms: those that have them

Group numbers correspond to valence

I	II											III	IV	V	VI	VII	VIII
1 H 1.01 Hydrogen																1 H 1.01 Hydrogen	2 He 4.00 Helium
3 Li 6.94 Lithium	4 Be 9.01 Beryllium											5 B 10.81 Boron	6 C 12.01 Carbon	7 N 14.01 Nitrogen	8 O 16.00 Oxygen	9 F 19.00 Fluorine	10 Ne 20.18 Neon
11 Na 23.0 Sodium	12 Mg 24.3 Magnesium											13 Al 26.98 Aluminum	14 Si 28.09 Silicon	15 P 30.97 Phosphorus	16 S 32.07 Sulfur	17 Cl 35.45 Chlorine	18 Ar 39.95 Argon
19 K 39.10 Potassium	20 Ca 40.08 Calcium	21 Sc 44.96 Scandium	22 Ti 47.88 Titanium	23 V 50.94 Vanadium	24 Cr 52.00 Chromium	25 Mn 54.94 Manganese	26 Fe 55.58 Iron	27 Co 58.93 Cobalt	28 Ni 58.69 Nickel	29 Cu 63.55 Copper	30 Zn 65.39 Zinc	31 Ga 69.72 Gallium	32 Ge 72.59 Germanium	33 As 74.92 Arsenic	34 Se 78.96 Selenium	35 Br 79.90 Bromine	36 Kr 83.80 Krypton
37 Rb 85.47 Rubidium	38 Sr 87.62 Strontium	39 Y 88.91 Yttrium	40 Zr 91.22 Zirconium	41 Nb 92.91 Niobium	42 Mo 95.94 Molybdenum	43 Tc (98) Technetium	44 Ru 101.1 Ruthenium	45 Rh 102.9 Rhodium	46 Pd 106.4 Palladium	47 Ag 107.9 Silver	48 Cd 112.4 Cadmium	49 In 114.8 Indium	50 Sn 118.7 Tin	51 Sb 121.8 Antimony	52 Te 127.6 Tellurium	53 I 126.9 Iodine	54 Xe 131.1 Xenon
55 Cs 132.9 Cesium	56 Ba 137.9 Barium	57 La *138.9 Lanthanum	72 Hf 178.5 Hafnium	73 Ta 180.9 Tantalum	74 W 183.9 Tungsten	75 Re 186.2 Rhenium	76 Os 190.2 Osmium	77 Ir 192.2 Iridium	78 Pt 195.1 Platinum	79 Au 197.0 Gold	80 Hg 200.5 Mercury	81 Tl 204.4 Thalium	82 Pb 207.2 Lead	83 Bi 209.0 Bismuth	84 Po (210) Polonium	85 At (210) Astatine	86 Rn (222) Radon
87 Fr (223) Francium	88 Ra (226) Radium	89 Ac *(227) Actinium	104 Rf (257) Rutherfordium	105 Db (260) Dubnium	106 Sg (263) Seaborgium	107 Bh (262) Bohrium	108 Hs (265) Hassium	109 Mt (266) Meitnerium	110 Ds (271) Darmstadtium								

58 Ce 140.1 Cerium	59 Pr 140.9 Praseodymium	60 Nd 144.2 Neodymium	61 Pm (147) Promethium	62 Sm 150.4 Samarium	63 Eu 152.0 Europium	64 Gd 157.3 Gadolinium	65 Tb 158.9 Terbium	66 Dy 162.5 Dysprosium	67 Ho 164.9 Holmium	68 Er 167.3 Erbium	69 Tm 168.9 Thulium	70 Yb 173.0 Ytterbium	71 Lu 175.0 Lutetium
90 Th 232.0 Thorium	91 Pa (231) Protactinium	92 U (238) Uranium	93 Np (237) Neptunium	94 Pu (242) Plutonium	95 Am (243) Americium	96 Cm (247) Curium	97 Bk (247) Berkelium	98 Cf (249) Californium	99 Es (254) Einsteinium	100 Fm (253) Fermium	101 Md (256) Mendelevium	102 No (254) Nobelium	103 Lr (257) Lawrencium

IMAGE 3.8. Group numbers and valence.

(the noble gases) have no reason to react, and so they don't. **Atoms that do not have full valence shells react chemically in order to achieve them.** The chemical reactions we'll see in this text are driven by the atoms' energetic need to obtain full valence shells. Because all shells (except shell 1) are full when they contain eight electrons, the tendency of an atom to react so as to fill its valence is called the octet rule.

3.4 Lewis Dot Structures

At this point, it's hopefully become clear that in discussing chemical reactivity, the number of valence electrons matter more than the total electron count. A chemist named Gilbert Lewis developed—when fresh out of graduate school—a very elegant way to keep track of and portray an atom's valence: the element's atomic symbol is written circumscribed by dots representing the valence electrons. These shorthand representations are called **Lewis dot structures** (Image 3.11).

While there are not strict rules regarding the placement of electrons in Lewis dot structures—lithium's single electron, for instance, could be placed to the left of the symbol or to the right, above or below—it is nevertheless common practice

NOTE[3]

Other groups have names as well. For instance, group 1 elements are called *alkali metals*, group 2 elements are called *alkaline earth metals*, and group 6 elements are called *chalcogens*. Aside from group 8's name, though, the only one you'll probably encounter commonly is the name for the group 7 elements, which are called the *halogens*.

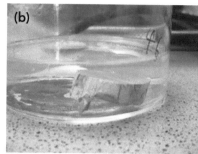
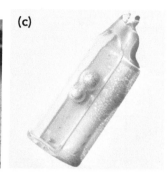

IMAGE 3.9. Lithium (a) , sodium (b) and potassium (c) are all soft, reactive metals. They're generally stored in ether or paraffin to keep them away from air and water, with which they react.

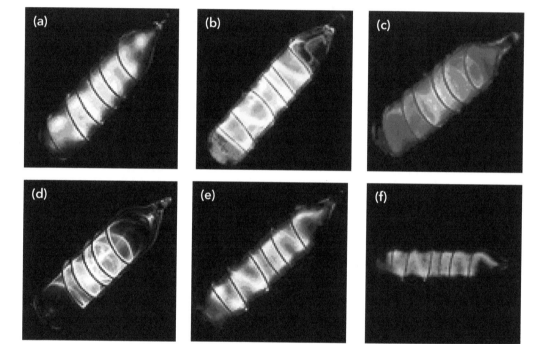

IMAGE 3.10. Glowing tubes of the noble gases (counterclockwise from top left): helium, neon, argon, krypton, xenon, radon. Each of these gases glows when subjected to electricity, which is why neon is used to make signs.

to imagine an invisible square surrounding the atomic symbol, with electrons distributed symmetrically around the square, as in Image 3.11. Electrons typically aren't paired on any one side of the square until all sides have at least one electron (note the Lewis dot structure for carbon versus that of nitrogen in Image 3.11). This convention has *nothing whatsoever* to do with the electrons' actual location within the valence shell; it's merely shorthand. However, the convention is useful because rapidly determining an atom's valence from a Lewis dot structure is much easier if electrons are placed sensibly rather than scattered. Compare the two structures in Image 3.12; it's much easier to glance at the correctly drawn Lewis dot structure on

Li· ·Be· ·Ḃ· ·Ċ· ·Ṅ· ·Ö: :Ḟ· :Ne:

IMAGE 3.11. Lewis dot structures of several elements.

the left and note that fluorine has seven valence electrons than it is to try to count the dots in the improperly drawn structure on the right.

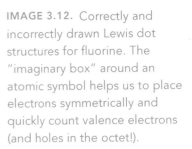

IMAGE 3.12. Correctly and incorrectly drawn Lewis dot structures for fluorine. The "imaginary box" around an atomic symbol helps us to place electrons symmetrically and quickly count valence electrons (and holes in the octet!).

3.5   Covalent Bonding

One very convenient characteristic of Lewis dot structures is that aside from allowing us to deduce the number of valence electrons rapidly, they also make "holes" in the valence readily apparent. In the Lewis dot structures of fluorine, there is one electron "hole," precluding a full octet. Fluorine would be more stable if it had a full outer shell like its group 8 neighbor, neon. Unfortunately for fluorine, electrons can't be conjured out of empty space. However, by reacting chemically, fluorine can gain an electron and complete its octet. Let's imagine two atoms of fluorine. They each have a "hole" in their valence, and they each need one more electron.

As it happens, electrons belonging to one atom can be shared with another as long as the two are in close enough proximity that their valence shells overlap. As a result, the fluorine on the left (we'll call it "blue fluorine" for the sake of this example) can let the fluorine on the right ("red fluorine") share one of blue fluorine's electrons. If that were to happen, red fluorine would have a full octet—seven electrons of its own, and one that technically belongs to blue fluorine. This is great for red fluorine, but what about blue fluorine? In a lovely *quid pro quo* (a Latin expression meaning a fair trade) exchange, red fluorine shares one of its electrons with blue fluorine.

TRY THIS

Draw Lewis dot structures for: H, Si, S, Cl.

Answers:

red shares an electron with blue

blue shares an electron with red

As long as the two fluorine atoms stay close to one another in space, they each get the benefit of an additional electron in their valence, so they each have complete octets.

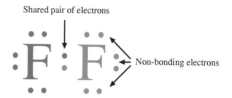

Shared pair of electrons

Non-bonding electrons

Now you know why F is one of
the elements found as a diatomic
molecule, or F_2. As a lone atom,
fluorine has an unfilled octet, but in a
molecule, its octet is filled!

The shared pair of electrons is a physical connection between the two atoms—almost like they're holding hands—and it keeps them stuck together. This is one type of **chemical bond**, which, because it's made up of electrons contributing to *both* fluorines' valences, is also called a **covalent bond** (co-valent = "both valences"). Note that neither fluorine atom has truly *gained* an electron (which would unbalance the protons and electrons in the fluorine atom and result in it no longer being neutral in charge). Instead, each atom maintains ownership of its individual electrons; the shared electrons are more *lent/borrowed* than *given*. If we wanted to do a quick electron head count at this point, we'd say that each fluorine atom still has seven "personal" electrons, but that as a result of the shared electron pair, each fluorine also has a total of eight valence electrons—a full octet. **The diagram of the bonded fluorine atoms above is a Lewis dot structure of a molecule; the molecule in this case is F_2.**[4] When drawing the Lewis dot structure for a molecule, we have the option of showing a pair of dots between the atoms to represent the shared electrons. Alternately, we can use a line to indicate the bond.

A line (bond) is always interpreted to mean a shared pair of electrons. Also, even if we're showing the bond as a line rather than a pair of dots, we always show the remainder of each atom's electrons (the nonbonding electrons, also called **lone pairs**). This helps us keep track of the electrons and each atom's valence.

3.6   Multiple Covalent Bonds

Sometimes atoms need to fill more than one hole in their valence. Oxygen, for instance, has a valence of six electrons and therefore needs to fill two holes. As you might guess, oxygen needs to form *two* bonds. One possibility is for oxygen to bond to two different atoms, as in a molecule of water:

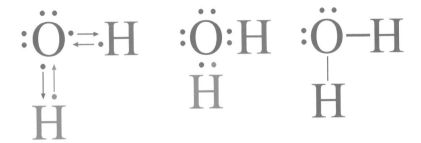

Note that in the example above, oxygen is filling its valence, and the two hydrogen atoms are also filling theirs. Another possibility is for oxygen to form two bonds to one other atom, as it might with another atom of oxygen to form the molecule O_2. Because two pairs of electrons are shared by the atoms in the O_2 molecule, this is called a **double bond**, and can be portrayed either as four dots between the atoms or two parallel lines.

It's also possible for atoms to form **triple bonds**, sharing three pairs of electrons. Take the example of nitrogen: because each atom of N has three holes in its valence, it needs to form three bonds. This can be achieved by forming a single bond to each of three different atoms (as in NH_3), by forming a double bond to one atom and a single bond to another, or by forming a triple bond to one other atom (as in N_2).

In predicting Lewis structures of molecules, there are a few things worth remembering. **First, if a molecule consists of only two atoms, we make as many bonds as necessary (one, two, or three—there's no such thing as a quadruple bond) between the atoms to fill their valences.** If a molecule is made up of more than two atoms, Lewis dot structures can be harder to predict without additional information about the order in which the atoms are bonded to one another. There is one large category, though, for which we can still predict connectivity (and Lewis structures): cases in which a molecule consists of two elements, with one atom of one element, and more than one atom of the second element (as in CO_2, for instance, or CH_4). **In this case,**

CONCEPT CHECK

How do we know how many bonds an atom will need to make?

Answer: An atom will make enough bonds to fill the hole(s) in its valence—one hole means one bond, two holes means two bonds, and so forth.

How are you to know when a molecule is complicated enough to warrant following the steps? It's good practice to use them every time, at least until you get used to drawing molecules.

the correct Lewis structure will have the element of which there is only one atom at the center of the structure, surrounded by atoms of the other element (C is at the center of CH_4, surrounded by and bonded to four hydrogen atoms). Predicting Lewis structures for molecules more complicated than this is largely beyond the scope of this text.

With a molecule like NH_3 or N_2, it's pretty easy to put together a Lewis structure intuitively. However, as molecules get bigger, the sharing can be quite complicated and counterintuitive, and it makes sense to have a systematic approach. Let's look at the steps necessary to drawing a correct Lewis structure for a more complicated molecule, using the example of SO_2:[5]

1. Count up the total number of valence electrons contributed by each atom (the total in the molecule *must* be the sum of the individual valences).
   Sulfur has a valence of six electrons; oxygen has a valence of six electrons.
   Total electrons in SO_2 = 6 + 6 + 6 = 18 electrons.

2. Put together a skeleton structure of the molecule, following the hints for connectivity given above. Don't worry about double and triple bonds or nonbonding pairs for right now; just connect everything with single bonds.
   SO_2 consists of one sulfur atom and two oxygen atoms, so sulfur should go in the middle.

$$O-S-O$$

3. Subtract the number of electrons used to make bonds in Step 2 from the total to determine the number of electrons that still need to be placed.
   Two single bonds so far = 2 × 2 = 4 electrons
   18 − 4 = 14 electrons remaining

4. Place remaining electrons in pairs around OUTER atoms; do not overfill octets. If outer octets fill before electrons run out, place leftover electrons on central atom. Be as symmetrical as possible
   Twelve of the remaining electrons can be placed around the oxygen atoms. Two electrons are left over, so they are placed on the central sulfur.

5. Count electrons around each atom to determine whether octets are full, remembering that each bond consists of a pair of electrons that counts toward the valence of both atoms. Where octets are left unfilled, share pairs from outer atoms with inner atom to fill all octets. Be as symmetrical as possible

Each oxygen has a full octet, but sulfur has only six electrons. One of the pairs currently on an oxygen atom will need to be shared with sulfur. There's no way to do this symmetrically, so we simply pick an oxygen atom; it doesn't matter which one.

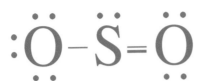

6. Verify that each atom has a filled octet.

Sulfur has one non-bonding pair, one pair shared with the oxygen on the left, and two pairs shared with the oxygen on the right for a total of 8 electrons.

Oxygen has three non-bonding pairs and one pair shared with sulfur for a total of 8 electrons.

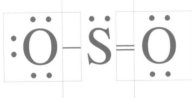

Oxygen has two non-bonding pairs and two pairs shared with sulfur for a total of 8 electrons.

Our SO₂ molecule is now complete.

Notice that the correct Lewis structure of SO_2 is not what we might have anticipated: based upon their valence holes, we anticipate atoms of sulfur and oxygen to make two bonds.[6] The sharing is quite a bit more complex here. This sort of thing happens sometimes (especially in some of the pollution and greenhouse-gas molecules we'll see in later chapters), and rather than try to figure out when it's going to happen and when it's not, we're better off simply using the steps described above to generate the correct Lewis structure every time. The most important thing this example teaches us is that even though the bonding didn't come out exactly the way we might have expected, all the octets are filled. Furthermore, the case above is definitely less common than cases in which the number of bonds is exactly what we expect, so it's fine to anticipate that normally, we'll see oxygen forming two bonds, and so forth.

One more helpful hint: correct Lewis dot structures tend to be as symmetric as possible. In determining the structure of CO_2, for instance, we can fill valences of all atoms by drawing double bonds between the carbon and each oxygen atom; *or* we can draw a single bond between carbon and one oxygen and a triple bond between carbon and the other oxygen. The correct Lewis structure is the symmetrical one, in which CO_2 has two double bonds.

NOTE[6]

If you've had more advanced chemistry classes, you may know that there is another way to draw a Lewis structure for SO_2 that *does* have two bonds coming from each oxygen. This results in a violation of the octet rule, however, and as such, it's far beyond the scope of this text. The structure of SO_2 as drawn is perfectly valid and much easier to explain without having to enter into some tricky chemical explanations, so it is the one we'll use throughout this book.

TRY THIS

Draw Lewis dot structures for: HCl, H₂, HCN, CH₄.

Answers:

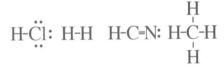

IMAGE 3.13. A *literal* "carbon backbone"—a molecule specially designed in a lab (see Sources and Further Reading) to look like a running person with oxygen (red) for eyes! Hydrogen atoms have been left out of this picture of the structure for ease of viewing. Researchers sometimes build "specialty" molecules like this to engender interest in chemistry while simultaneously refining synthetic (molecule-building) techniques that can be used in drug design and other applications.

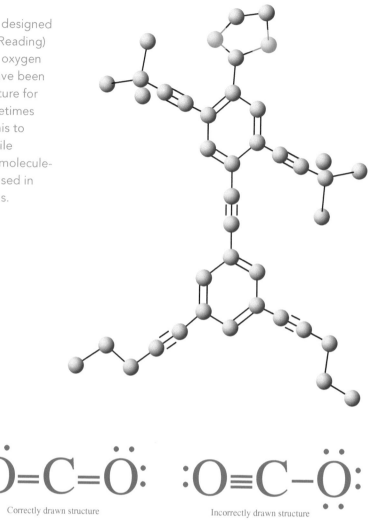

Correctly drawn structure

Incorrectly drawn structure

3.7 So Why Are We Carbon Based?

Let's return to the questions posed at the beginning of this chapter. First off, given the incredible variety of elements in the universe, why are so few of them major components of living systems?

The answer to this question requires us to look a bit more closely at the nature of covalent bonding. When fluorine, for instance, needs to fill the hole in its valence, it enters into a relationship with another atom (of fluorine or some other element). The second atom shares an electron with fluorine, and in return, is able to share one of fluorine's electrons. **The nature of a covalent bond is that it requires an even exchange.** Oxygen, in order to fill its valence by sharing two electrons belonging to other atoms, must allow those other atoms to share two of oxygen's own electrons. Nitrogen shares three electrons belonging to other atoms, and in return allows them to share in three of its own. In the case of carbon, all electrons end up being shared; carbon gets four shared with it, and in exchange must share its own four electrons.

Continuing along this line of reasoning, we encounter a problem. Boron has five holes in its valence, so it would need to have five electrons shared with it to fill its octet, which would require forming five bonds. However, boron *can't* form five bonds, because it only has three electrons of its own to offer; the largest number of normal covalent bonds it can form is three.[7] Beryllium would need to form six bonds to fill its valence but can't, because it has only two electrons of its own to offer up in exchange. Lithium is in a real pickle; it needs to form seven bonds, but has only one electron of its own to share. Many metals have problems when it comes to filling their octets exclusively by forming covalent bonds; they simply don't have enough electrons of their own to share. **As a result, many metals—particularly those in groups 1 and 2—do not use covalent bonds to fill their octets.**[8] They still react chemically; they just do so slightly differently. We'll discuss the chemical reactivity of some metals in Chapter 4. Suffice it to say that in our look at real-world chemistry in this text, we'll only see covalent bonds forming between nonmetals (or metalloids) and nonmetals. Remember, though, that hydrogen has a sort of "split personality": it acts a bit like a group 1 element and a bit like a group 7 element. On that note, now we know why: it has one valence electron (like the group 1 elements), and one valence hole (like the group 7 elements). In any case, while it technically belongs to group 1, it is formally classified as a nonmetal. In its ability to form covalent bonds, we see hydrogen displaying its group 7 character.

Even though we won't talk about the chemical reactivity of metals just yet, it's worth mentioning that the covalent bond is the most stable, strongest kind of bond there is. As a result, compounds made up of nonmetallic elements are much longer-lived. A biological system depends upon stability; cells can't have their structural components falling apart! Thus, you'll notice that the major elements of life are all nonmetals, and the reason is that these elements form stable, long-lasting bonds with each other.

The second question posed at the beginning of the chapter was: of all the elements that are major components of living systems, why does carbon form the structural "backbone" of the molecules of life? Given that elements form bonds based upon the number of valence holes they have to fill (group 7 elements form one bond, group 6 elements form two bonds, and so forth), the group 4 elements are of particular interest. Carbon (and the other group 4 elements) atoms form more bonds than any of the other nonmetals. As a result, they display great versatility and variety of bonding patterns (Table 3.2). Furthermore, their prodigious bonding ability allows for the development of extensive networks of bonds and the formation of very large molecules. **In light of this reasoning, it makes sense that the backbone of a biomolecule is made of a group 4 element: versatility of bonding patterns and strength of bonds are both characteristics very desirable in the molecules of life.**

There's just one question left: why carbon rather than another group 4 element? Silicon is metalloid, as is germanium, and the others in the column are metallic. Even

NOTE[7]

There are cases in which boron forms a fourth bond, but the electrons in the fourth bond both come from the other atom. As such, this is not a normal covalent bond (and is beyond the scope of what we'll be covering here).

NOTE[8]

Metals can, under some circumstances, form covalent bonds. All sorts of "rule breaking" happens in chemistry: the octet rule gets violated, quadruple bonds—which aren't expected to form by bond theory—can in some cases be made to form, and noble gases occasionally react chemically. There are very few rules, in fact, that are completely hard and fast in this field (particularly in the artificial conditions of the laboratory). Happily, though, the molecules we'll see here are well behaved and don't require us to learn esoterica of chemistry. For this reason, while I often indicate by footnote that a rule I've just presented isn't absolute, this is solely for the purpose of being extremely precise (perhaps obsessively so). The real-world chemistry we talk about in this text is relaxingly predictable, no matter what sort of chemical aerobics intrepid scientists are able to accomplish in the lab.

Bonding Patterns Available to Group IV, V, VI, and VII Elements (based upon the number of covalent bonds they can each form)

TABLE 3.2. The elements of life have a number of possible bonding patterns in which they can engage.

though metalloids can form covalent bonds, the very strongest bonds are formed between nonmetals and nonmetals. Silicon actually forms strong bonds, however, and has every bit of the bonding versatility of carbon, making it the most popular "backbone element" for fiction writers attempting to create an alternative life scenario. Further, silicon is incredibly abundant on Earth: nontropical sand, for instance, is silicon dioxide (SiO_2); it's overall the second most abundant element in Earth's crust. Clearly a lack of availability of silicon is not the reason the molecules of life are carbon based. Living organisms don't just require stable molecules, however. They also need a way to move the backbone element (carbon or, in our alternative scenario, silicon) in and out of the organism for metabolic purposes. We, for instance, get carbon into our cells through the food we consume (more on that in Chapter 7), and we eliminate it through exhaled CO_2. Assuming we were silicon based, one imagines that we would consume silicon-rich foods to get more of the element. Getting rid of it, though, presents a problem. The presumed metabolic waste product in a silicon life form would be the aforementioned SiO_2, which is a solid at livable temperatures and which therefore cannot be exhaled. Logistically speaking, silicon-based life would be problematic. **Ultimately, carbon forms stable bonds with great versatility of bonding patterns and is simple to shuttle in and out of living organisms.**

3.8   Molecular Shapes

Actually, the story of carbon's special stability and bond-forming ability goes even deeper than simply that it forms four strong bonds. We've been representing molecules so far in this chapter as nothing more than a collection of atoms, bonded to one another in a certain order. But there are two very important things we need to remember about chemical bonds:

1. Bonds are made up of pairs of electrons.
2. Electrons repel other electrons.

This takes us pretty naturally to the realization that bonds repel each other, and therefore separate from one another as much as possible in space. At first glance, it seems as though that's what we've been portraying. After all, we haven't been showing methane (CH_4) with the bonds all smashed together like the picture below on the left; we've been showing the bonds spaced out as shown on the right:

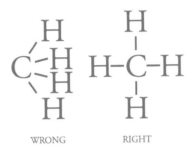

WRONG   RIGHT

However, even a reasonable Lewis dot structure of methane (like the one above) doesn't take advantage of an important fact about molecules: they exist in three-dimensional space. Our Lewis dot structure portrays the bonds as separated as possible on a two-dimensional page, but in three dimensions, bonds can get further apart than 90°. In fact, four objects are maximally spaced when placed at the corners of a four-sided pyramid. When its bonds are separated as much as possible, the methane molecule looks like this:

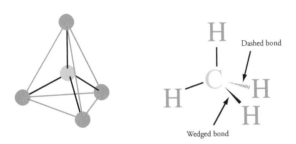

The "wedged" and "dashed" bonds are there to remind us that not all of the hydrogen atoms attached to the central carbon are in the plane of the paper; any two of them can be (just as any two corners and the center of a four-sided pyramid are in the same plane), but of the other two, one must be in front of and the other behind the plane. In this perspective representation, the wedged bond is taken to mean "in front of the plane of the page," and the dashed bond is taken to mean "behind the plane of the page." Just as the center of a four-sided pyramid is in the same plane as any two of its corners, the center of the methane molecule (the carbon atom) is in the same plane as any two of its hydrogens. **If you were to calculate the angle from one corner of the pyramid to the center (where the carbon is) back to another corner, you'd find that the bond angle is 109.5°; this arrangement clearly**

affords the electrons more space than they'd have in only two dimensions. Using perspective to portray shape is just one more way of communicating information about molecules. Our Lewis dot structure of methane told us everything we needed to know about connectivity, but didn't accurately represent shape, since methane isn't flat. **This is actually very important; the pyramid formed by a carbon bonded to four other atoms (this geometry is called tetrahedral) is very versatile and stable.** If you need reminding of just *how* stable this shape can be, go back and look at Image 1.13 in Chapter 1: the structure of diamond is nothing more than a tetrahedral arrangement of carbon atoms, each of which is bonded to four other carbons.

Let's revisit some of the molecules for which we drew Lewis dot structures earlier in the chapter. The molecule ammonia (NH_3) has a central nitrogen atom with three bonds to hydrogen and a lone pair of electrons. **Just like the carbon in methane, the nitrogen in ammonia has four groups of electrons around the central atom (a lone pair is an electron group, because in the same way that a pair of electrons sticks together in a bond, a nonbonding pair sticks together).** Thus, ammonia will have the same tetrahedral arrangement of electron groups that we saw with methane. The only difference will be that instead of an atom at one of the corners of the pyramid, there will be a lone pair of nonbonding electrons there:

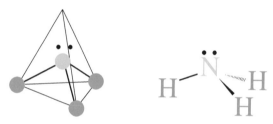

We can portray ammonia using wedge-dash notation just as we did with methane. **Remember, molecules are three-dimensional objects that can be turned over and upside down.** As such, there's no reason the lone pair of electrons has to be shown on top of the nitrogen in ammonia; the molecule has the freedom to turn any which way. It's often easiest to get the perspective drawing to "look right" if we draw the molecule this way, but it's completely legitimate to turn it around or over and draw it from any perspective you like!

One more thing about ammonia: notice that the electrons in each bond are pulled slightly out and away from the central atom—because they're shared between two atoms that are somewhat separated in space—but nitrogen's lone pair is held closer because nitrogen isn't sharing it. This means that the lone pair is exerting a little extra repulsive force—electrostatic forces are strongest at close range—so the electrons in the N-H bonds are repelled more by the lone pair than they would be by a bond. **The result is that the bond angle from one of the Hs to the central N to a second H is slightly less than that of a perfect tetrahedron; ammonia has bond angles of about 107°.** We still consider this tetrahedral geometry, though, despite the fact that the bond angles are a bit constricted.

Water also has four electron groups around its central atom: two lone pairs, and two bonds. Like methane and ammonia, the groups will arrange themselves naturally into a tetrahedron, giving the water molecule a "bent" shape:

CONCEPT CHECK

Why don't we typically use wedged and dashed lines to draw water?

Answer: In any tetrahedral arrangement, two of the bonds are in the plane of the paper. Since water only has two bonds, we can show them both in the plane.

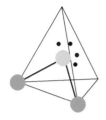

For simplicity's sake, we generally don't use wedged and dashed lines when portraying water; it's easier to show the two O-H bonds in the plane of the paper, remembering that the lone pairs (which we never put on the ends of wedged or dashed lines, because there are no "bonds" to lone pairs) are in front of and behind the plane. Just as in ammonia, the lone pairs in water condense the H-O-H bond angle. **With two lone pairs, water's bond angle is even more affected than that of ammonia, condensed to around 104.5°.**

It's possible for molecules to have fewer than four electron groups around a central atom. Consider the molecule COH_2—this is formaldehyde, if you're curious—which has a central C attached to an O and two H's. To fill octets, the C and O must share a double bond:

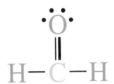

In this case, there are three electron groups around the central C. **The two electron pairs that make up the C-O double bond must stay together between the C and O, and because they can't separate from each other, they are treated as one group.** The furthest from one another that three electron groups can get in three-dimensional space is 120°. The outer atoms therefore lie at the corners of a triangle:

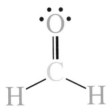

We don't need wedges and dashes to portray a molecule with geometry like this; the trigonal planar shape created by three electron groups around a central atom is in a single plane. Incidentally, the same bond condensation effect that we saw in ammonia and water takes place in trigonal planar molecules. SO_2, with two bonds and a lone pair, has a bond angle slightly less than 120°.

There's another important shape family that we see in the molecules of life, which is observed in molecules with two electron groups around a central atom. **These molecules, including carbon dioxide, have a linear shape with 180° bond angles.** Portraying a linear molecule does not require wedge-dash perspective, as the molecule exists in a single plane:

$$\ddot{O} = C = \ddot{O}$$

Carbon, as we've seen, can form bonds with a variety of geometries: tetrahedral, trigonal planar, and linear. As Image 3.14 shows, this results in an incredible variety of shapes within a single large molecule, and many of the molecules of life are large. As we'll learn more in Chapter 7, the shape of a molecule of life contributes tremendously to its function. Considering molecular shape is just one more way to think about molecules and helps to explain some of the chemistry we observe in the world, such as the ubiquity of carbon.

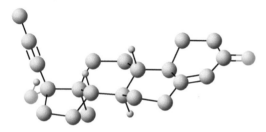

IMAGE 3.14. Carbon can produce molecular backbones with an incredible variety of shapes. This is norethisterone, a synthetic progestin (pregnancy hormone) used in oral contraceptives. Most hydrogen atoms have been left off this image of the molecule for ease of viewing the carbon backbone. Note areas of tetrahedral shape, trigonal planar shape (along double bonds) and linear shape (near the C-C triple bond).

3.9   The Paradox of Bond Formation

Our discussion of molecular shape arose from the principle that like charges repel, so electrons in bonds or lone pairs repel other electrons in bonds or lone pairs. One might wonder why bonds can form at all, given that they (by necessity) bring two electrons (for a single bond) or four (for a double bond) or six (for a triple bond) into close proximity. Don't the electrons in a bond repel each other, and if so, how and why does the bond form? The answer has to do with stability, but can be explained using a simple cost–benefit analysis. Let's imagine that you're trying to make a decision—for example, perhaps you're considering buying a new car. This action has costs associated with it: a higher payment, higher insurance premiums, and maybe a greater risk of getting the car stolen if it's something really nice. There are also definite benefits associated with buying a new car. Just to name a few, you'd have a nicer vehicle to drive around in, your gas mileage would probably improve, and maybe you'd impress someone special. In any case, while considering your purchase, you'd probably spend some time thinking about whether the benefits were worth the costs, or whether, in the end, it simply wasn't a good exchange. In the same way, most of what happens in chemistry has stability costs (the system becomes less stable) and stability benefits (the system becomes more stable). If the sum of the stability lost is greater than the sum of stability gained in a process, then the system becomes less stable overall, which is not good. Conversely, if the stability benefits outweigh the stability costs, the system will become *more* stable, which is favorable. **Chemical reactions aren't magic. They happen for one reason, and** one reason only: **because the result is more stable than the starting point (just as a dropped ball falls toward Earth because it's more stable on the ground than in the air).** We'll discuss this concept a little more when we talk about energy in Chapter 8.

What does this have to do with bond formation? When a bond forms, the atoms involved gain something: they get their valence shells filled. That's a definite stability benefit to each atom. The electrons involved in forming the bond *also* get something: even though they're brought close together (a stability cost), they get sandwiched between two nuclei, to which they're attracted. Since proximity to a nucleus stabilizes electrons, being close to two nuclei is more stabilizing than being close to one. Imagine two hydrogen atoms getting ready to bond. Before bonding, each atom would have an unfilled valence shell, and each electron would be stabilized by one nucleus:

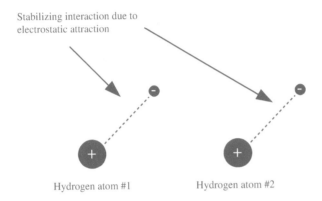

After bonding, each atom has a full valence shell, and each electron is stabilized by two nuclei:

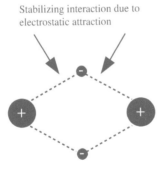

It turns out that the benefits outweigh the cost of bringing two electrons close together. The only other problem with this arrangement is that in bringing the atoms close enough to bond, the positively charged nuclei are also brought close together. However, this is not as big a problem as one might anticipate. The electrons sandwiched between the nuclei "shield" the nuclei from each other, and mitigate the repulsive force. **All in all, forming a bond is stabilizing for atoms** (it's possible to do the math, but for the purposes of this text, we'll simply take it on faith and on the basis of the observation that bonds do, in fact, form); that's why it happens!

The entire Unit 1 of this text has been focused on one main point: chemistry is everywhere. It's a much bigger part of our lives than we sometimes think it is. Still, though, humans are high-order animals who think, feel, reason, and dream. Surely not *everything* that goes on inside us can be reduced to chemistry, right? Well, let's examine that.

It turns out that the science of attraction—why we are drawn to or develop crushes on certain individuals—has a lot of chemistry to it! Researchers have been trying for some time to determine whether humans have *pheromones*, as many other animals do. Pheromones are chemical signals that help animals to communicate with each other. They can indicate pregnancy, availability to mate, health status, and so forth. These chemicals are not detected with the nose in most animals, but are instead sensed by specialized cells that make up the *vomeronasal organ*. While the presence of a vomeronasal organ in humans has been debated, there is some solid evidence for its existence. Further, there is considerable evidence that—vomeronasal organ or no—we use our regular sense of smell to make decisions. **For instance, several studies that show we use smell to help us select the partner with whom we are most likely to have healthy children.** How does this work?

It's been known for quite some time that one function of the immune system (which protects us from infection) is to produce proteins called *major histocompatibility complexes (MHCs)*. These proteins are located on the surface of every body cell and help that cell recognize threats from bacterial and viral invaders. The ability to recognize a wide variety of potential invaders is critical to fighting off infection, and is therefore directly linked to survival. Because of this, individuals are healthiest and most likely to survive if they have the greatest possible variety of MHC proteins. Since the genes for these proteins are passed from parent to child, the healthiest child will result from parents who have very different MHC genes. Because MHCs are microscopic proteins on the surface of cells, though, it's impossible to look at another person and determine their MHC profile. How, then, do we find a partner with whom we will have children with the most robust possible immune systems?

Researchers have recently determined two interesting bits of information that help answer this question. First, it appears that our MHC profile is reflected to some extent in our natural body odor; people with similar MHC profiles smell alike. Second, we instinctively respond to another person's scent, and research has shown we are most attracted to individuals who smell differently than we do. During one experiment, researchers asked a group of men to wear the same t-shirt to bed for several nights in a row (they were asked to use no scented hygiene products during this time, to ensure natural body odor would be the only thing on the shirt). A group of women was then asked to smell several shirts: each woman smelled three shirts from men who had been genetically tested and determined to have similar MHC profiles to hers, and three from men whose MHC profiles were different. **Women** strongly preferred the shirts worn by men whose MHC profiles were different from their own. Turns out, our initial sense of attraction to another person is strongly influenced by chemicals, and these chemicals provide us with valuable information.

OK, so initial attraction involves chemicals, but what about love? Is this complex emotion really just chemistry at work? As a matter of fact, much of the emotion of love is dependent upon chemicals, including the "butterflies in the stomach" feeling of a great first date and the deep sense of contentment and security we feel when cuddling with a partner or child. The "butterflies" feeling comes primarily from dopamine (Image 3.15, left), a chemical released in the pleasure centers of the brain as a sort of "reward" molecule. Dopamine interacts with neurons in these pleasure centers to produce feelings of excitement, focused attention, and intense energy. As a result, a person who is falling in love feels great, thinks about their sweetheart constantly, and can easily stay up all night on a date or on the phone.

A different chemical is involved in deep, abiding love. **This compound, called oxytocin (Image 3.15, right), is produced and released in response to cuddling and physical affection, and results in a feeling of closeness and security.** It's also the chemical, incidentally, that floods the brain of a new mother, helping her to fall in love with her baby (which helps to ensure that she'll care for the baby, and therefore promotes human survival). Knowing the chemistry behind these emotions doesn't make them any less magical or real, but it's interesting to know where they come from. Furthermore, next time you refer to a great romantic connection by saying, "we've got good chemistry," you'll know you're absolutely right—literally!

SUMMARY OF MAIN POINTS

- Living organisms are composed primarily of the elements C, H, O, and N.

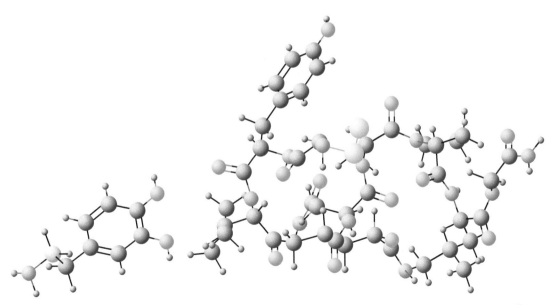

IMAGE 3.15. Dopamine (left) and oxytocin (right). The two greenish atoms in oxytocin are sulfur.

- Electrons are organized into shells around the nucleus; the shell closest to the nucleus is shell 1.
- Radii of shells increase with shell number; electrons in shell 1 are most stable.
- Shells fill with electrons from the inside out; the number of electrons in a shell is the same as the number of atoms in the corresponding row in the periodic table.
- Most shells are (or act) full with eight electrons.
- Only electrons in the outermost shell are available for chemical reactions; these are the valence electrons.
- Elements with the same number of valence electrons (same periodic group) react similarly.
- Atoms react chemically to fill their valence—this is the octet rule.
- Lewis dot structures portray just an atom's valence electrons.
- Atoms form covalent bonds to fill their octets.
- Lewis dot structures of molecules show shared electron pairs.
- The major elements of life are nonmetals because of their ability to form strong, covalent bonds with each other.
- Electrons in bonds and nonbonding pairs spread out as much as possible in three-dimensional space—this results in the characteristic shapes of molecules.
- Shapes of molecules can be portrayed using perspective bonds (wedge-dash).
- The backbone element of life is carbon because it has the greatest variety of bond shapes and structures available to it.
- Bond formation is stabilizing to the atom in many ways—it fills both atoms' valences, stabilizing electrons with additional nuclei, and nuclei are shielded from one another by electrons.

QUESTIONS AND TOPICS FOR DISCUSSION

1. Given that most of the elements on the periodic table don't show up in living cells to any significant degree, where on Earth are they found? Pick three elements, do a little Internet research and find out where they are commonly found.

2. C, H, O, and N make up the bulk of the atoms in the human body. However, there are many trace elements present as well. Pick three of these trace elements and do a little Internet research to find out where they're found and what they're used for in the body.

3. Why is the percent of hydrogen in the human body is much higher than the percent of carbon by number of atoms, but by mass, the percent of carbon is much higher than that of hydrogen?

4. Why is shell 1 the first electron shell that fills in any atom?

5. How can you use the periodic table to help determine how many electrons fit in each shell?

6. Explain what the octet rule means.

7. Why do atoms in the same group on the periodic table act similar chemically to one another?

8. How does the number of valence electrons in an element change across a row? Down a column?

9. What do all the elements in the rightmost column of the periodic table have in common (with regard to their electrons)?

10. Explain what it means to say that atoms react chemically to achieve full valence shells.

11. Why do we draw Lewis dot structures of atoms with the electrons spaced out around four sides of an imaginary box?

12. When a covalent bond forms, atoms add additional electrons to their valence. However, their charges do not become unbalanced. Why not?

13. Why does the number of valence electrons affect an atom's chemical behavior more than the total number of electrons?

14. What holds the atoms together in a covalently bonded pair?

15. Based upon everything you've learned in this chapter, why do you think hydrogen often appears in two different places on the periodic table? What about it is similar to the group 1 metals? What about it is similar to the group 7 nonmetals? Which is it more like, in your opinion?

16. Explain why the major elements of life are nonmetals, and why of these, carbon is the "backbone" element.

17. Explain why even though there are more hydrogen atoms (by number) in a living cell than carbon atoms, we still say carbon is the "backbone" element of life.

18. Why do electron groups (bonds and nonbonding pairs) spread out as much as possible in space?

19. Why does a molecule like COH_2 (a central carbon with a double bond to oxygen and a bond to each of two hydrogens) have a different bond angle than

a molecule like CH_4, despite the fact that, in each case, the central carbon is surrounded by a full valence octet?

20. Imagine making molecular models using balloons. If you were to tie two balloons together, what sort of "bond angle" would they form? Do they bunch up, or spread out?

21. What would happen to the shape of the balloon model in question 13 if you tied three balloons together? Four? What does this tell us about molecular shapes and where they come from?

22. Draw a perspective (wedge-dash) NH_3 so that it looks like the N is pointed up, resting on a tripod of the three H's. Now mentally turn it 180° along a vertical axis, and redraw. Now flip it over (upside down) and redraw. Remember that *every* molecule is a three-dimensional object that can be turned and flipped! We can draw these from any perspective we like.

23. Why are covalent bonds able to form, despite the fact that forming bonds involves bringing multiple electrons (and multiple positively charged nuclei) into close proximity?

PROBLEMS

1. List the four major elements of life.

2. List the trace elements of life.

3. Collectively, what percent of the atoms in the human body do C, H, O, and N comprise when water is included?

4. Which element is represented in the highest percentage (by mass) in the human body when water is included?

5. Which element is represented in the highest percentage (by number of atoms) in the human body when water is included?

6. Which element is represented in the highest percentage (by number of atoms) in the human body when water is excluded?

7. Which element is represented in the highest percentage (by mass) in the human body when water is excluded? (Hint: you'll need to use the data for number of atoms, and then take mass into account).

8. What information do we get about an atom from knowing which row of the periodic table it's in?

9. What information do we get about an atom from knowing which column (main groups only) of the periodic table it's in?

10. How many electrons does shell 1 hold? Shell 2? Shell 3?

11. Using the periodic table as your guide and without drawing electron configuration diagrams, in what shell are the outermost electrons in each of the following: Sr; As; Xe; Na; B; He.

12. Using the periodic table as your guide and without drawing electron configuration diagrams, in what shell are the outermost electrons in each of the following: Ba; I; Rn; Ar; Al; Si.

13. Why do we consider shell 3 to hold eight electrons for elements up through Ca, and to hold 18 electrons for elements from Ga on?

14. Draw electron configurations for the row 3 elements.

15. Draw electron configurations for the row 4 elements (do not include transition metals).

16. Draw electron configurations for the first three group 2 elements. What do they all have in common?

17. Draw electron configurations for the first three group 7 elements. What do they all have in common?

18. What feature of an atom is singly most responsible for its chemical behavior?

19. List elements from the periodic table that are most chemically similar to Li.

20. List elements from the periodic table that are most chemically similar to N.

21. List elements from the periodic table that are most chemically similar to H (careful!)

22. Why do Lewis dot structures include only valence electrons?

23. Draw Lewis dot structures for the row 3 elements.

24. Draw Lewis dot structures for the row 4 elements (do not include transition metals).

25. Draw Lewis dot structures for the first three group 2 elements. What do they have in common?

26. Draw Lewis dot structures for the first three group 7 elements. What do they have in common?

27. Draw Lewis structures for each of the following molecules: Br_2; H_2O_2 (the order of atoms is HOOH); HF; H_2S; O_3 (the order of atoms is OOO).

28. Draw Lewis structures for each of the following molecules: I_2; PCl_3; CH_2Cl_2 (put the C in the middle); CO; PH_3.

29. A single bond is considered an electron "group" for purposes of determining molecular shape. What are the other electron "groups"?

30. What general shape is formed by four electron groups around a central atom?

31. What general shape is formed by three electron groups around a central atom?

32. What general shape is formed by two electron groups around a central atom?

33. What bond angles are found in the shape formed by four electron groups around a central atom?

34. What bond angles are found in the shape formed by four electron groups, one of which is a lone pair of electrons, around a central atom?

35. What bond angles are found in the shape formed by four electron groups, two of which are lone pairs of electrons, around a central atom?

36. What bond angles are found in the shape formed by three electron groups around a central atom?

37. What bond angles are found in the shape formed by three electron groups, one of which is a lone pair of electrons, around a central atom?

38. What bond angles are found in the shape formed by two electron groups around a central atom?

39. Why do lone pairs condense bond angles?

40. Draw each of the following molecules in perspective, showing shape: NCl_3; $COCl_2$ (put C in the middle with a double bond to O); H_2S; SO_2; CO_2.

41. Draw each of the following molecules in perspective, showing shape: CH_2Cl_2; O_3; CCl_4; $COBr_2$ (put C in the middle with a double bond to O); NI_3.

42. Provide expected bond angles for each of the following: NCl_3; $COCl_2$; H_2S; SO_2; CO_2.

43. Provide expected bond angles for each of the following: CH_2Cl_2; O_3; CCl_4; $COBr_2$; NI_3.

44. Compare the shapes of CO_2 and SO_2. Despite the similarity of the chemical formulas, the shapes are quite different. Why?

ANSWERS TO ODD-NUMBERED PROBLEMS

1. C, H, O, N.
3. 96%.
5. Hydrogen.
7. Carbon.
9. The column number tells us how many valence electrons the atom has.
11. Sr—shell 5; As—shell 4; Xe—shell 5; Na—shell 3; B—shell 2; He—shell 1.
13. Shell 3 fills to a total of eight electrons, then stops filling while shell 4 starts to fill, then resumes filling. By the end of the row 4 transition metals, shell 3 has 18 electrons.

15.

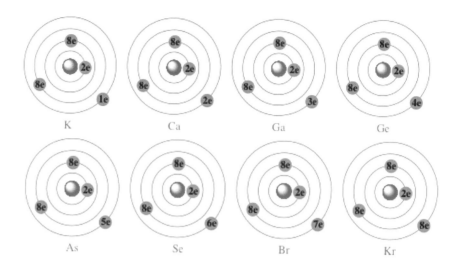

17. Each of the configurations has seven electrons in the outer shell.

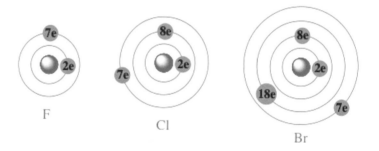

19. H, Na, K, Rb, Cs, Fr.

21. Answers may vary. H has some similarities (one valence electron) to group 1 elements, but some similarities (one valence hole) to group 7 elements. It's actually quite unique.

23.

Na·   ·Mg·   ·Al·   ·Si·   ·P·   ·S·   :Cl·   :Ar·

25. Each of the structures has two valence electrons.

·Be·   ·Mg·   ·Ca·

27.

:Br—Br: H—O—O—H   H—F: H—S—H   :O—O=O

29. Electron groups include (in addition to single bonds) double bonds, triple bonds, and lone pairs of electrons.

31. Trigonal planar.

33. 109.5°.

35. ~104.5°.

37. <120°.

39. Lone pairs are closer to the central atom than electrons in bonds are, so their repulsive force is greater.

41.

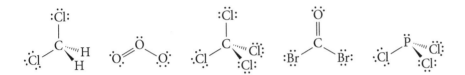

43. 109.5°; <120°; 109.5°; 120°; ~107°.

SOURCES AND FURTHER READING

Anthropomorphic Molecules

Chanteau, S. H., Tour, J. M. (2003). Synthesis of anthropomorphic molecules: The NanoPutians. *Journal of Organic Chemistry, 68*(23), 8750–66.

Pheromones

Lawton, G. (1997). Change of mood about pheromones. *Chemistry and Industry, 16*, 635.

Wright, K. (1994). The sniff of legend: Human pheromones? Chemical sex attractants? And a sixth sense organ in the nose? What are we, animals? *Discover, 15*(4), 60–67.

MHCs and Attraction

Herz, R. S. (2002). Sex differences in response to physical and social factors involved in human mate selection: The importance of smell for women. *Evolution and Human Behavior, 23*(5), 359–64.

Jacob, S., McClintock, M. K., Zelano, B., *et al.* (2002). Paternally inherited HLA alleles are associated with women's choice of male odor. *Nature Genetics, 30*, 175–79.

Santos, P. S. C., Schinemann, J. A., *et al.* (2005). New evidence that the MCH influences odor perception in humans: A study with 58 Southern Brazilian students. *Hormones and Behavior, 47*(4), 384–88.

Dopamine, Oxytocin, and Love

Bales, K. L., Mason, W. A., Catana, C., *et al.* (2007). Neural correlates of pair-bonding in a monogamous primate. *Brain Research, 1184*, 245–53.

Stein, D. J., Vythilingum, B. (2009). Love and attachment: The psychobiology of social bonding. *CNS Spectrums, 14*(5), 239–42.

Zeki, S. (2007). The neurobiology of love. *FEBS Letters, 581*(14), 2575–79.

CREDITS

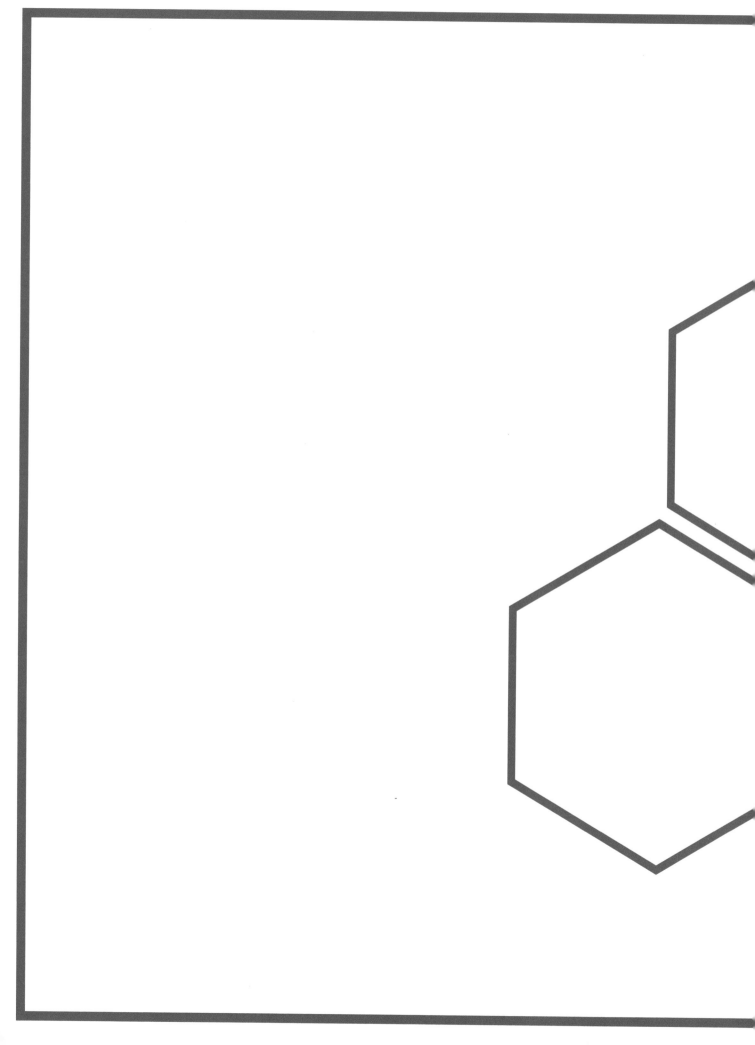

Unit 2

COMMUNITY CHEMISTRY

Water—the Solvent of Life

4

If a chemist or biologist were asked what single chemical is most critical to life, the answer would surely be water. Water forms the internal environment of cells, it is the medium through which cells communicate with one another, it is combined chemically with carbon dioxide to produce new structural material as plants grow, and it serves as the environment for countless aquatic species. There is not a single known living thing that can function without water. In addition to its importance to life, water is also a very strange little molecule in many ways, both chemically and physically. Happily, water's oddities make it uniquely suited to supporting life. An exploration of the nature of water will help us learn how and why it serves the function it does and will also help us understand some properties of the chemicals that dissolve in and interact with water.

4.1   Solutions and Solubility

Water is sometimes (via a bit of hyperbole) referred to as the "universal solvent." **Solvents** are chemicals that can dissolve other chemicals, and water is good at dissolving many things (but not everything! See Image 4.2).

When a solvent dissolves other chemicals (which are called **solutes**), the resulting mixture is a **solution**. For instance, in a solution of salt water, water is the solvent, and salt is the solute. **Like all mixtures, solutions are physical combinations of chemicals, meaning that neither the solute nor the solvent is chemically changed in the process of mixing.** Further, like all mixtures, solutions can be made in varying proportions. The proportion of solute to a given volume of solution is called the solution's **concentration**. It's worth knowing that a solution made using water as the solvent is called an **aqueous solution** (aqueous comes from the Latin word for water). Due to water's ubiquity and its prodigious ability to function as a solvent, the most common kind of solution is aqueous.

IMAGE 4.1. A beautiful waterfall like this one in the Eagle Cap Wilderness is just one of the ways in which we experience the interesting and unique chemical that is water. Eagle Cap, Oregon © Scott Lefler, 2009.

IMAGE 4.2. Cartoon © Kirstin Hendrickson, 2010.

IMAGE 4.3. Dark drops of balsamic vinegar bead up and sink to the bottom of a bowl of olive oil.

We routinely encounter solutes that dissolve well in water; these include salt, sugar, many coloring and flavoring agents used in cooking, and liquids like vinegar and ethanol, just to name a few. Other things, however, do not dissolve well in water (Image 4.3). Oil, for instance, floats on top of the aqueous vinegar solution in a bottle of salad dressing. Shaking the dressing temporarily combines the oil and vinegar-water, but they separate again in fairly short order.

Why is it that some molecules dissolve well in water and others do not? The answer to this question has to do with the chemical structures of the solutes and solvents.

4.2 Dipole Moments

We learned in the last chapter that atoms can form covalent bonds, which are electron-sharing interactions, to fill their octets. It turns out, though, that two covalently bonded atoms don't always share electrons equally. In many cases, one atom "hogs" the electrons, pulling them closer to its nucleus and away from the other atom's nucleus. This electron-pulling behavior comes from the atomic property of **electronegativity**. An atom's electronegativity is a measure of its ability to pull electrons in a bond—even those that originally came from another atom—toward itself. Electronegativity is a **periodic trend**, meaning that it can be predicted from looking at the periodic table (Image 4.4). **More electronegative atoms are found at the right end of a row, and at the top of a column.** Remembering that hydrogen *technically* belongs on the left side of the table in group 1, and that the group 8 elements are not really relevant to a discussion of electronegativity— since they don't generally form bonds—fluorine emerges as the element furthest to the right and closest to the top of the table, indicating that fluorine is the most electronegative element.

While a periodic table can often be used to determine which of a pair of atoms is more electronegative, it doesn't help us if the pair is situated diagonally with one element closer to the top and one closer to the right. For cases like this, there's a mnemonic device[1] that is useful in remembering how the elements we most

NOTE[1]

The word *mnemonic* comes from the Greek word for memory. A mnemonic device is a trick that makes things easy to remember

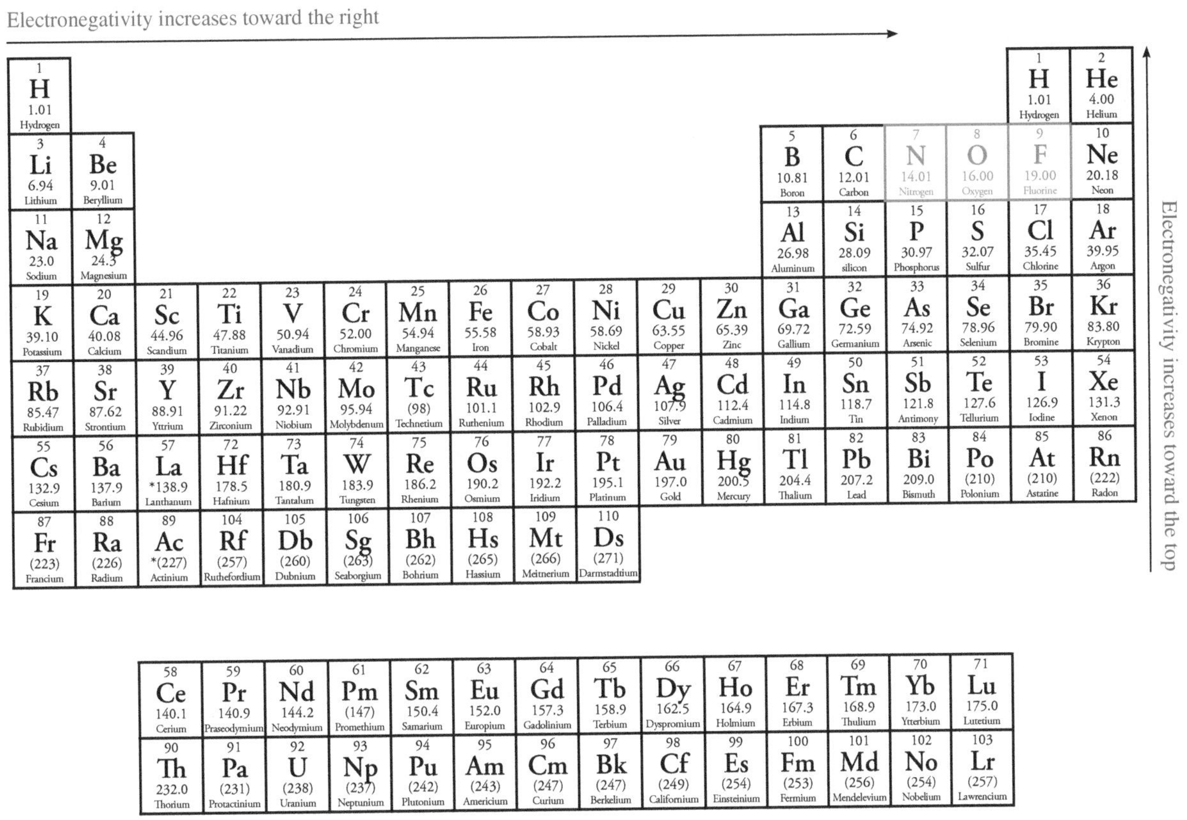

1 **H** 1.01 Hydrogen																	
3 **Li** 6.94 Lithium	4 **Be** 9.01 Beryllium											5 **B** 10.81 Boron	6 **C** 12.01 Carbon	7 **N** 14.01 Nitrogen	8 **O** 16.00 Oxygen	9 **F** 19.00 Fluorine	10 **Ne** 20.18 Neon
11 **Na** 23.0 Sodium	12 **Mg** 24.3 Magnesium											13 **Al** 26.98 Aluminum	14 **Si** 28.09 silicon	15 **P** 30.97 Phosphorus	16 **S** 32.07 Sulfur	17 **Cl** 35.45 Chlorine	18 **Ar** 39.95 Argon
19 **K** 39.10 Potassium	20 **Ca** 40.08 Calcium	21 **Sc** 44.96 Scandium	22 **Ti** 47.88 Titanium	23 **V** 50.94 Vanadium	24 **Cr** 52.00 Chromium	25 **Mn** 54.94 Manganese	26 **Fe** 55.58 Iron	27 **Co** 58.93 Cobalt	28 **Ni** 58.69 Nickel	29 **Cu** 63.55 Copper	30 **Zn** 65.39 Zinc	31 **Ga** 69.72 Gallium	32 **Ge** 72.59 Germanium	33 **As** 74.92 Arsenic	34 **Se** 78.96 Selenium	35 **Br** 79.90 Bromine	36 **Kr** 83.80 Krypton
37 **Rb** 85.47 Rubidium	38 **Sr** 87.62 Strontium	39 **Y** 88.91 Yttrium	40 **Zr** 91.22 Zirconium	41 **Nb** 92.91 Niobium	42 **Mo** 95.94 Molybdenum	43 **Tc** (98) Technetium	44 **Ru** 101.1 Ruthenium	45 **Rh** 102.9 Rhodium	46 **Pd** 106.4 Palladium	47 **Ag** 107.9 Silver	48 **Cd** 112.4 Cadmium	49 **In** 114.8 Indium	50 **Sn** 118.7 Tin	51 **Sb** 121.8 Antimony	52 **Te** 127.6 Tellurium	53 **I** 126.9 Iodine	54 **Xe** 131.3 Xenon
55 **Cs** 132.9 Cesium	56 **Ba** 137.9 Barium	57 **La** *138.9 Lanthanum	72 **Hf** 178.5 Hafnium	73 **Ta** 180.9 Tantalum	74 **W** 183.9 Tungsten	75 **Re** 186.2 Rhenium	76 **Os** 190.2 Osmium	77 **Ir** 192.2 Iridium	78 **Pt** 195.1 Platinum	79 **Au** 197.0 Gold	80 **Hg** 200.5 Mercury	81 **Tl** 204.4 Thalium	82 **Pb** 207.2 Lead	83 **Bi** 209.0 Bismuth	84 **Po** (210) Polonium	85 **At** (210) Astatine	86 **Rn** (222) Radon
87 **Fr** (223) Francium	88 **Ra** (226) Radium	89 **Ac** *(227) Actinium	104 **Rf** (257) Rutherfordium	105 **Db** (260) Dubnium	106 **Sg** (263) Seaborgium	107 **Bh** (262) Bohrium	108 **Hs** (265) Hassium	109 **Mt** (266) Meitnerium	110 **Ds** (271) Darmstadtium								

Top-right inset:

1 **H** 1.01 Hydrogen	2 **He** 4.00 Helium

58 **Ce** 140.1 Cerium	59 **Pr** 140.9 Praseodymium	60 **Nd** 144.2 Neodymium	61 **Pm** (147) Promethium	62 **Sm** 150.4 Samarium	63 **Eu** 152.0 Europium	64 **Gd** 157.3 Gadolinium	65 **Tb** 158.9 Terbium	66 **Dy** 162.5 Dysprosium	67 **Ho** 164.9 Holmium	68 **Er** 167.3 Erbium	69 **Tm** 168.9 Thulium	70 **Yb** 173.0 Ytterbium	71 **Lu** 175.0 Lutetium
90 **Th** 232.0 Thorium	91 **Pa** (231) Protactinium	92 **U** (238) Uranium	93 **Np** (237) Neptunium	94 **Pu** (242) Plutonium	95 **Am** (243) Americium	96 **Cm** (247) Curium	97 **Bk** (247) Berkelium	98 **Cf** (249) Californium	99 **Es** (254) Einsteinium	100 **Fm** (253) Fermium	101 **Md** (256) Mendelevium	102 **No** (254) Nobelium	103 **Lr** (257) Lawrencium

commonly see in real-world chemistry rank with regard to electronegativity. In order of decreasing electronegativity, the common elements fall into the order:

FONClBrISCH—pronounce as a word: "fonclibrish"; it sounds like something right out of Lewis Carroll's "Jabberwocky!"

Even though "fonclibrish" isn't a real word, it's silly-sounding enough that perhaps you can remember it. If you can, you can use it to quickly determine which of two elements is more electronegative in difficult cases. Take, for instance, nitrogen and sulfur. Nitrogen is closer to the top of the periodic table, but sulfur is further to the right. The mnemonic helps us remember that nitrogen is more electronegative. It also reminds us of the important fact that fluorine, oxygen, and nitrogen are the three most electronegative elements. **It's also worth knowing that carbon and hydrogen are exceedingly similar to one another—almost identical—in electronegativity.** There is unequal electron distribution (carbon pulls electrons from hydrogen, as the mnemonic reminds us), but the inequality is negligible in most—but not all—applications. To keep things simple, up through the end of Chapter 12, we'll treat carbon and hydrogen as though they have identical electronegativities. In Chapter 13, the very minor difference between the two becomes important, and we'll see it come into play.

What does all of this have to do with bonding? When two atoms of the same element (such as two hydrogen atoms) bond to one another, they each pull electrons in

IMAGE 4.4. The three most electronegative elements are F, O, and N.

TRY THIS

Pick the element from each pair that is more electronegative: I and Cl; C and H; C and O; H and O; S and C.

Answers:
Cl, C, O, O, S.

CONCEPT CHECK

Why would an atom of nitrogen pull electrons toward it when bonded to carbon, but not when bonded to oxygen?

Answer: Nitrogen is more electronegative than carbon but less electronegative than oxygen. The more electronegative atom always pulls electrons toward itself.

the bond toward their own nuclei. However, since two atoms of the same element have the same electronegativity, they pull electrons with equal strength. Just as in a game of tug-of-war between two equal-strength contestants, the electrons don't go anywhere:

Conversely, when two atoms of different elements with different electronegativities (such as hydrogen and fluorine) bond to one another, the more electronegative atom pulls the electrons in the bond toward its nucleus, and away from the nucleus of the less electronegative atom. The electrons are still being shared, but they're not exactly halfway between the nuclei. Instead, they're being "hogged" by one of the atoms:

While shared electron pairs are always called covalent bonds (regardless of whether one atom is pulling the electrons with greater strength), if there is an unequal distribution of electrons, we say the bond is **polarized**. We can also call a bond between atoms of different electronegativities, like the bond between hydrogen and fluorine, a **polar covalent bond.**

Another ramification of the H-F bond polarization is that since fluorine is not only holding on to its own shared electron, but hogging hydrogen's shared electron as well, fluorine's atomic charges become slightly off balance. From the periodic table, we know that fluorine has nine protons, or a charge of +9 in its nucleus. It also has nine electrons, contributing a charge of -9, and making the atom neutral overall. Bonded to hydrogen, though, it's as though fluorine has partially acquired a tenth electron, since it's pulling both shared electrons closer to its own nucleus. This creates a slight charge imbalance on the fluorine atom, giving it a **partial negative charge**. The polarized bond affects hydrogen as well. Hydrogen has one proton, for a nuclear charge of +1. It also has one electron, and is normally charge-neutral. However, with fluorine partially taking hydrogen's electron, hydrogen becomes slightly charge imbalanced, and has a **partial positive charge**. We can indicate these charge imbalances in two ways. A **bond dipole** arrow (a dipole is a separation of charge) can be used to indicate the direction of polarization. **The dipole arrow has a plus sign at the less electronegative atom, and points in the direction in which the electrons are being pulled—toward the more electronegative atom:**

A nice mnemonic for remembering which way dipole arrows are oriented is the short rhyme *arrows show where electrons go*. Alternately, atoms can be marked with a plus or minus sign preceded by the Greek letter *delta* (d), which is taken to mean *partial charge*:

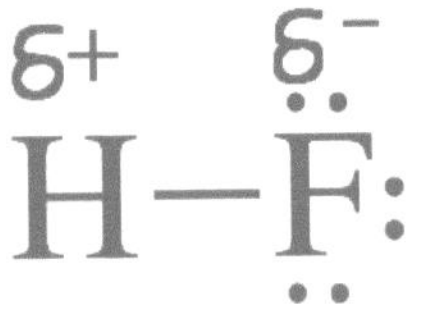

Any time two atoms of different elements are bonded together, they'll have different electronegativities, and their bond will be polarized. While the electrons in the bond still contribute to both atoms' valences, the atoms will be partially charged instead of being neutral.

In a molecule consisting of a single polar covalent bond like HF, the bond dipole also represents the net dipole, or the molecule's overall polarization. For a molecule consisting of more than one polar covalent bond, the bond dipoles must be summed to determine the net dipole. This process requires taking the bond dipoles *and* the molecular shape into account. For instance, the molecule CO_2 has two polar covalent bonds, with dipoles that completely oppose each other:

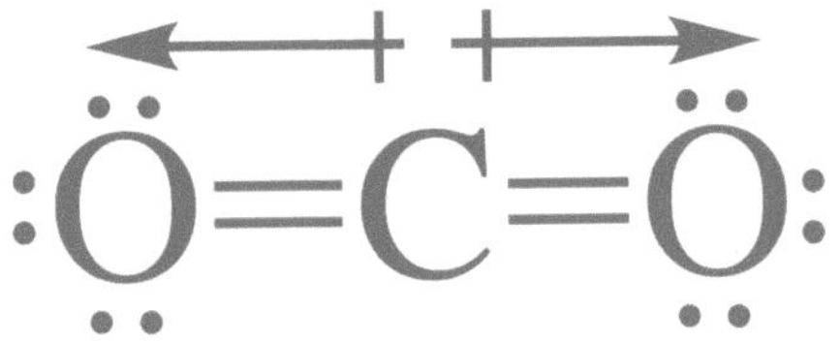

As a result of the molecule's linear shape, the bond dipoles completely cancel one another out on a molecular level, and CO_2 has no net dipole. In other words, CO_2 is completely symmetric with regard to its charge distribution. Regardless of symmetry, CO_2 still has bond dipoles, so it still has partial charges:

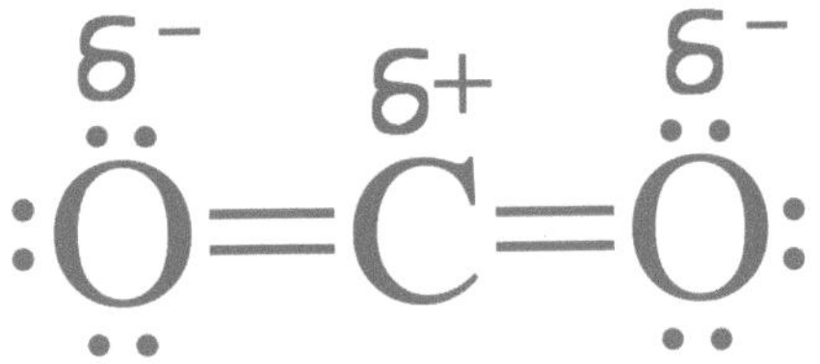

A molecule like CO_2 that has no net dipole is called a **nonpolar** molecule. Some molecules, however, are asymmetric with regard to their charge distribution. This means the dipoles don't completely cancel each other out, and instead sum to form a net dipole. These are called **polar** molecules. HF, because it has only one bond dipole (which therefore *is* its net dipole), is polar. H_2O is also polar, because while it has two bond dipoles, they do not oppose one another completely:

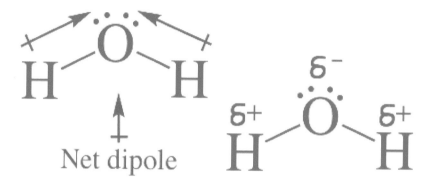

In the image above, the two O-H dipoles oppose each other on an imaginary x-axis (as the molecule is currently drawn), but both point (along an imaginary y-axis, as the molecule is currently drawn) toward oxygen.[2] **Notice that to sum dipoles correctly, both bonds *and* shape must be taken into account.** We have to know that water has a bent shape to know that its dipoles don't completely oppose each other like CO_2's do. As a result of the orientation of water's bond dipoles, they sum to produce a net dipole toward the oxygen atom. It's tempting to say there's a net dipole *upward*, but remember that water (like all molecules) is a three-dimensional object that can be turned around and flipped over. If we were to draw water with the oxygen toward the bottom of the page and the hydrogen atoms toward the top, the dipole would point down. Regardless of the orientation, however, the dipole in water always points toward oxygen. A net dipole indicates that with regard to the molecule as a whole, electrons are clustered at one particular end or side—in this case, the side with the oxygen.

When we consider molecules with tetrahedral geometry, determining net dipoles requires some three-dimensional visualization. For example, the molecule CCl_4 has four bond dipoles:

NOTE[2]

For those with a background in simple physics, this is just addition of vectors.

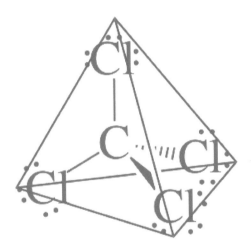

In a two-dimensional image, it's very difficult to picture the sum of the dipoles. However, if we remember that the perspective drawing is nothing more than a way to portray a three-dimensional object, it gets a little easier to imagine what's going on. In three dimensions, our molecule occupies a tetrahedral space, with each chlorine atom at the corner of a four-sided pyramid (the bond from C to each Cl has been removed for the sake of simplicity, and only bond dipole arrows appear)[3]:

Note that each bond dipole points out symmetrically toward one corner of the pyramid from the center of the molecule. This means the dipoles cancel each other out completely, and CCl_4 is nonpolar.

With regard to our discussion of solubility, remember that charge attracts opposite charge. **As a result, the partially negative end or side of one polar molecule is attracted to the partially positive end or side of another polar molecule. For this reason, polar compounds tend to be attracted to, and therefore dissolve in, other polar compounds. Nonpolar compounds, lacking distinct positive and negative ends/sides, are not similarly attracted to polar compounds and therefore do not tend to dissolve in them.** Fats, for instance, which consist primarily of carbon and hydrogen (remember that the C-H dipole is so small as to be virtually nonexistent on account of the atoms' similar electronegativities), do not dissolve well in water, which is extremely polar.

CONCEPT CHECK

Why do the dipoles cancel each other out in CO_2, but not in H_2O?

Answer: The shape of CO_2 means the dipoles completely oppose each other, but in H_2O they do not.

TRY THIS

Draw bond and net dipoles, and indicate partial charge, for: HCl, F_2, SO_2.

Answers:

NOTE[3]

In the CH_2Cl_2 molecule, bond dipoles along the C-H bonds are shown here in the interest of absolute precision. Many chemists (and chemistry teachers) do not show dipoles along these bonds, as they are so minute as to be inconsequential in determination of the net dipole. Similarly, in the CH_4 molecule, it is also considered correct to portray neither bond dipoles nor a net dipole. In the vast majority of applications, the C-H dipole is so negligible as to be disregarded completely. However, these very small dipoles are shown here as a reminder that they will come into play--likely for the only time you'll ever see them do so--in Chapter 13, when we talk about the greenhouse effect.

Draw bond and net dipoles, and indicate partial charge, for: NH_3, CH_2Cl_2, CH_4.

Answers:

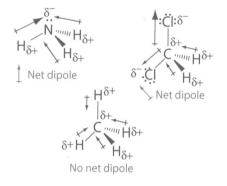

NOTE[4]

Remember from Chapter 3 that the metals are not able to fill their valences by forming covalent bonds; they bond in a different way that will be discussed later in this chapter.

NOTE[5]

How do we indicate larger versus smaller dipoles? The size of a dipole arrow is sometimes used as a visual cue in molecules with more than one type of bond dipole. The symbol δ, however, contains no quantitative information whatsoever. In general, we just keep track mentally—a larger difference in electronegativity means a bond dipole will be large.

Ammonia (NH_3), however, is very polar, and dissolves well in water. A maxim by which to remember this principle is *like dissolves like*, meaning that polar molecules dissolve other polar molecules. The maxim can also be used in reference to nonpolar molecules, which dissolve other nonpolar molecules. This attraction between nonpolar molecules is a bit more complex, since nonpolar molecules don't have permanent attractions to each other. Instead, it has to do with a number of factors, including the exclusion of these molecules by water. Because nonpolar molecules don't dissolve well in water (their lack of partial charge means water is not at all attracted to them), the water molecules tend to exclude (push out) nonpolar molecules. The excluded molecules therefore cluster together, dissolving in and mixing with each other. This solution behavior is referred to as a **hydrophobic** effect (hydrophobic means water-fearing). Water is an excellent solvent because, while it can't dissolve nonpolar molecules, it's very adept at solvating anything polar, as we'll discuss further.

4.3 Hydrogen Bonds

One of water's molecular oddities is that it has really big dipoles compared to most compounds. The most electronegative elements are fluorine, oxygen, and nitrogen. The least electronegative element that still forms covalent bonds is hydrogen.[4] As a result, the biggest bond dipoles (those with the greatest polarization and largest resulting partial charges) are along F-H, O-H, and N-H bonds. The only compound with an F-H bond is HF, hydrofluoric acid. In practical terms then, the O-H and N-H bonds have the largest dipoles we find routinely in a variety of compounds. The magnitude of the bond dipoles and partial charges is so large that the attraction between molecules is very strong.[5] In fact, the electrostatic attractions are so strong as to be almost bond-like. Though they are not true bonds, these interactions are called **hydrogen bonds**:

Note that the term "hydrogen bond" doesn't refer to the covalent bond between oxygen or nitrogen and hydrogen in a given molecule; instead, it refers to the attraction between the partially negative oxygen or nitrogen of one molecule and the partially positive hydrogen of another molecule. Additionally, while water and ammonia are two of the simplest molecules that form hydrogen bonds, *any* compound with one or more O-H or N-H bonds is capable of the same interaction; a small selection of hydrogen-bonding molecules appears in Image 4.5.

In addition to molecules forming hydrogen bonds to other molecules of the same compound, it's also possible for hydrogen bonds to form between molecules of different compounds, as long as both are capable of hydrogen bonding. Water, for instance, forms hydrogen bonds with ammonia (and any other compound with O-H or N-H bonds):

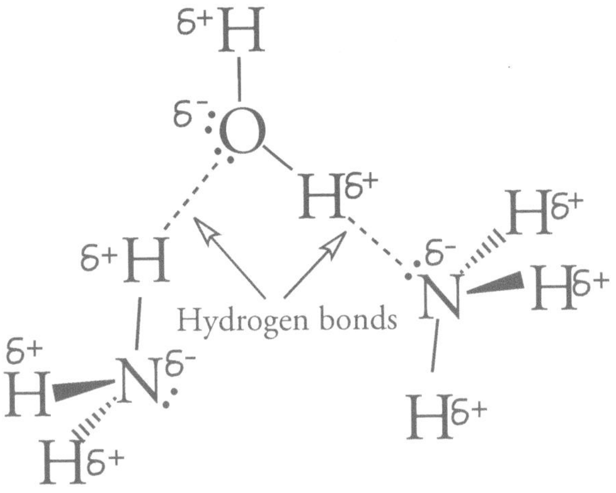

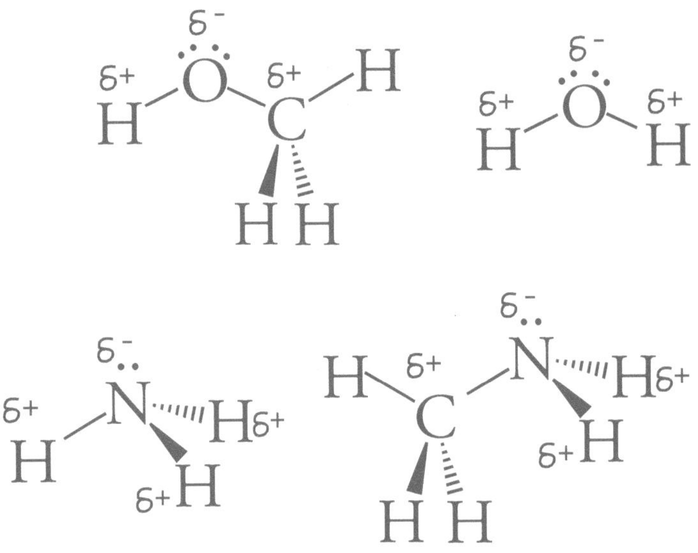

IMAGE 4.5. A few of the many compounds capable of hydrogen bonding.

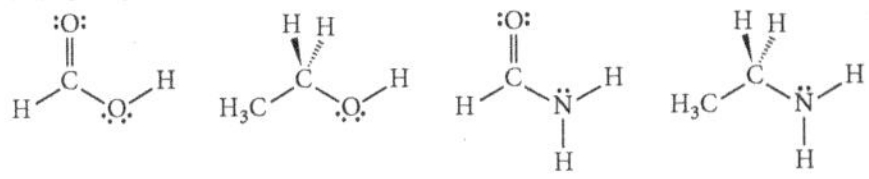

4.4 Unique Properties of Water

Hydrogen bonding has several effects on the physical behavior of water. First, as already discussed, water is very good at dissolving polar compounds. We can now add that in particular, water is an excellent solvent for compounds capable of hydrogen bonding. Secondly, though, of all the hydrogen-bonding compounds, water is the one with the strongest and most robust hydrogen-bonding network. It's composed of nothing but O-H bonds, and O-H bonds produce slightly stronger hydrogen bonds than N-H bonds, since O and H are more disparate in electronegativity. As a result of water's extensive hydrogen-bonding network, molecules of water are very strongly attracted to each other. This results in some interesting physical properties.

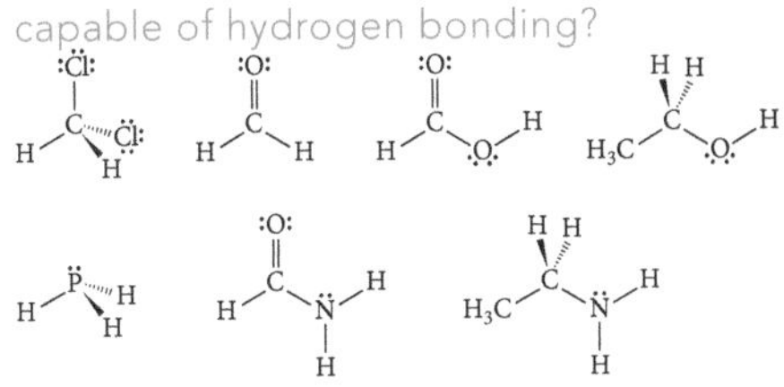

TRY THIS

Which of the molecules shown is capable of hydrogen bonding?

Answers:

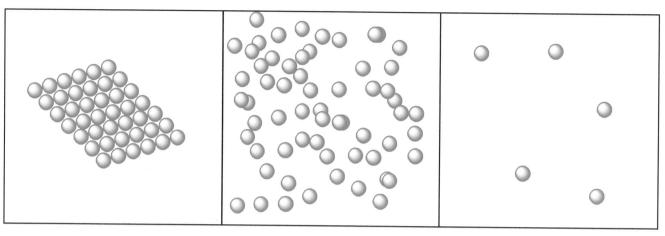

IMAGE 4.6. Solids are densely packed (left), liquids have less attraction and therefore slightly more space between particles, giving them more freedom of motion (center). Gases have a great deal of space between particles with minimal attraction, leading to complete freedom of motion (right).

NOTE⁶

The relationship between molecular size and strength of the van der Waals forces explains why methane (CH$_4$) and ethane (C$_2$H$_6$) are gases, while hexane (C$_6$H$_{14}$) and octane (C$_8$H$_{18}$) are liquids.

Compound	Melting Point	Boiling Point
H$_2$O	0°C	100°C
NH$_3$	-77.7°C	-33.3°C
CH$_2$Cl$_2$	-96.7°C	40°C

To understand why, we need to think about the three phases of matter—solid, liquid, and gas (Image 4.6). Solids form when particles that are very attracted to one another pack tightly with essentially no space between them. The lack of empty space means solids are incompressible (they can't be stuffed into a smaller volume), and the strong attractive forces between particles means they don't slide past each other, so they don't take the shape of their container. If particles are slightly less attracted to one another, a liquid forms. In this phase of matter, the particles are still quite close in space, so liquids are also incompressible. However, the weaker attractive forces allow particles to slide past one another, so liquids can take the shape of their container. Gases form when very weakly attracted particles spread out with a lot of empty space between them. Gases can therefore be compressed, and the particles are free to move around and take the shape of their container. The process of melting involves breaking down the very strong attractions between particles in a solid just a bit, in order to make a liquid. This takes energy, which is why we have to heat a solid to melt it. Boiling breaks down those attractions even more; again, this requires energy. Why is energy required to break down the attractive forces between particles? Think about water and other polar compounds: the molecules are attracted to one another because of opposite partial charges, so pulling the charges apart to decrease the attraction requires energy, much as it requires energy to pull apart two magnets. It also takes energy to separate nonpolar compounds. This is because they're also attracted to one another, though via a different mechanism. The attractions between nonpolar molecules are called van der Waals forces. While a complete discussion of the nature of van der Waals forces is beyond the scope of this text, it's sufficient to know that they are the weak attractions between nonpolar molecules and that they are stronger in larger molecules than they are in smaller ones.⁶

As a result of water's extensive hydrogen bond network, it takes a lot of energy to break down the attractive forces between water molecules. As a result, water has a very high melting point and boiling point for its size. Note the vast difference in melting points and boiling points between water and the other similarly sized molecules, NH$_3$ and CH$_2$Cl$_2$:

Additionally, water is much more dense and cohesive than many other liquids due to its attraction to itself. Density is a measure of how much matter is packed into a given volume of space. Cohesiveness is a measure of the extent to which a chemical tends to stick to itself rather than spreading out on a surface.

Further, water's extensive hydrogen bonding network makes it resistant to temperature change, a property very important to living creatures. Life depends upon cells maintaining a fairly stable temperature, and water's enhanced ability to maintain its temperature despite losing or absorbing small amounts of heat contribute to making it the perfect cellular medium. For a demonstration of water's ability to maintain its temperature, you could place a cup of cool water and a similarly sized block of cool metal on the sidewalk in full sunlight on a very hot July day. After an hour in the sun, you could still put a finger in the water without getting burned, but you'd dare not touch the metal! Despite absorbing the same amount of the sun's heat, the water's temperature would not have changed nearly as much as that of the metal.

Another of water's interesting properties is that, again due to its extensive hydrogen-bond network and high cohesiveness, it's very lubricating. This helps reduce friction between cell membranes in multicellular organisms and protects delicate organ tissue from damage due to rubbing against other body organs.

Perhaps water's most interesting and unique property of all, however, is that unlike almost any other substance, solid water floats on top of liquid water. At first glance, this is unremarkable: we've all enjoyed a tall glass of iced tea on a hot day, and we know the ice floats.

The experience of floating ice is so common to us that we often don't stop to consider just how much it's at odds with what we know about states of matter and density. **If solids are closely packed and liquids are slightly less so, while gases contain a lot of empty space, then a solid should be the most dense physical state of any given substance, and a gas the least dense.** As a result, a solid should not float, but *sink* in its own liquid! We don't generally observe this since there are very

CONCEPT CHECK

Why does water's hydrogen-bonding network contribute to its melting and boiling points, density, and cohesiveness?

Answer: Hydrogen bonding means molecules of water are very attracted to each other and pull each other into a smaller volume of space (density and cohesiveness). They also take lots of energy to break (MP and BP).

IMAGE 4.7. Dewdrops demonstrate water's tendency to bead up due to its high cohesiveness.

(a)

(b)

IMAGE 4.8. Solid water floats in a glass of water, soda, or tea. Icebergs float on the ocean.

NOTE[7]

You can't try this at home; the vinegar in your kitchen or bathroom is not pure acetic acid. In fact, it's mostly water and will therefore act like water, so the experiment won't work at all. It's quite difficult to obtain pure acetic acid outside the lab.

few substances we have the opportunity to experience in multiple states. However, pure acetic acid (the substance that, when diluted with water, is called vinegar) solid sinks like a rock in liquid acetic acid.[7] The same goes for ammonia, ethanol, and almost every other substance. So what's different about water that its solid is evidently less dense than its liquid? Again, the answer to this question has to do with hydrogen bonding. The hydrogen-bonding network formed by water when it freezes into ice takes the shape of a series of tetrahedrons, with an atom of oxygen at the center of each (Image 4.9). This lattice is very stable—actually, it looks just like the structure of diamond that we saw in Chapter 1—but it contains a lot of empty space compared to most solids. In order to maximize hydrogen bonding, the water molecules need to space themselves out a bit, which reduces the density of solid water.

IMAGE 4.9. The structure of liquid water (a) shows considerably closer packing of molecules than that of solid water (b).

(a)

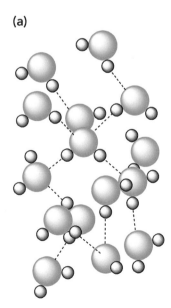

(b)

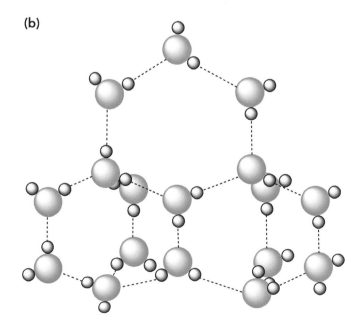

IMAGE 4.10. The amazing complexity and symmetry of snowflakes arises from the hydrogen bonding properties of water.

By comparison, as solid water melts, the crystal lattice begins to break down. There is still a network of hydrogen bonds, but it's not nearly as extensive. As molecules rotate and slide past one another, the network becomes much less orderly than that of ice. **As a result, the molecules aren't forced into crystal lattice positions and, as a result of both their freedom of movement and their strong attraction to one another, they cluster closer together than in solid water. This makes water more dense as a liquid than as a solid.** It's a good thing, too; in addition to the convenience of being able to float ice in a cold drink, this phenomenon also allows for turnover of lake water. In winter, surface water cools, and as it becomes more dense,[8] it descends to the lake bottom. This carries fresh oxygen from the surface into the depths. Incidentally, the crystal lattice of solid water is also directly responsible for the beautiful and symmetric patterns of snowflakes (Image 4.10). Incidentally, due to the structure of crystalline water, snowflakes will always exhibit sixfold symmetry (regardless of what all the winter decorations depicting five-, seven-, or eight-armed snowflakes might have you believe)!

NOTE[8]

Water is most dense at about 4°C.

Water's large bond dipoles give it a number of interesting properties and explain its ability to dissolve other polar compounds. There's another class of compounds that dissolves readily in water; these are the salts. **Salts form when there is such a large difference in electronegativity between two atoms that when they try to bond, the more electronegative atom actually completely takes one or more of the less electronegative atom's electrons.** The result isn't a covalent bond, since there are no shared electrons. Instead, the more electronegative atom now has extra electrons, and is negatively charged. A negatively charged atom is called an anion. The less electronegative atom is now missing electrons and is positively charged; this is a cation.[9] Collectively, cations and anions are called ions. **Even though a covalent bond does not form via this electron transfer, the resulting cation and anion are attracted to one another because of their opposite charges.** They therefore stick together, held not by a physical bond, but by electrostatic forces. We call this interaction between atoms an ionic bond.

Consider a very electronegative element like fluorine trying to form a bond with a minimally electronegative element like sodium. If they tried to share a pair of electrons (one contributed by sodium, one by fluorine), the dipole would be tremendous. In fact, the dipole would be *so* large as to pull the pair of electrons entirely into fluorine's electron cloud, away from sodium. When this happens, the fluorine atom becomes an ion of fluoride; we add the suffix *-ide* to the name of an atom when it becomes an anion. Further, fluoride is electrostatically attracted to the sodium cation (names of atoms don't change when they become cations), and the attraction, called an ionic bond, results in the ions sticking together to form the compound *sodium fluoride*:

$$Na \cdot \rightharpoonup \cdot \ddot{F} \colon \longrightarrow Na^+ \quad \colon \ddot{F} \colon^- \longrightarrow Na^+ \colon \ddot{F} \colon^-$$

Fluorine literally pulls sodium's electron right out of the electron cloud

The resulting ions are attracted to one another

The ionic compound NaF results—the bond is not physical (covalent), but electrostatic (ionic)

NOTE[9]

To help you remember which ion type has which charge, you can either imagine the word "ca+ion" spelled with a plus sign instead of a "t," or you can use the mnemonic "CATions are PAWSitive."

Note that while we can use Lewis dot structures to portray ions just as we can use them for neutral atoms, there's an important difference. When an electron is lost from sodium's outer shell of electrons—shell 3—the new structure for sodium continues to depict the *original* valence shell, which is now empty. Charges for ions are always included in Lewis dot structures.

When will two atoms form an ionic rather than a covalent bond? **As a rule, ionic bonds will form when metals (which have very low electronegativities) react with nonmetals.** We saw in Chapter 3 that metals do not form covalent compounds (also called *molecular compounds*); we now know that they do form ionic compounds.

Don't forget that hydrogen, despite its position on the far left of the periodic table, has an electronegativity virtually identical to carbon's and can form covalent bonds.

Let's spend a little more time on the ionization reaction that takes place between sodium and fluorine, because it explains more than simply how metals can be incorporated into compounds. Ionization introduces a new strategy for an atom to obtain a complete valence shell of electrons. We discussed in Chapter 3 that we don't expect to see group 1, 2, or 3 elements forming covalent bonds because they don't have enough electrons to share. Sodium, for instance, with one electron in its valence, would need to gain seven electrons in order to complete its octet through covalent bonding. With only a single electron to share, though, forming this many bonds is simply not possible for sodium. However, obtaining enough electrons to fill shell 3 is not the only way for sodium to achieve a full octet. The other possibility is to lose one electron, emptying shell 3. This leaves sodium with only its shell 1 and shell 2 electrons, and its new outermost shell (shell 2) is full. As such, not only does ionization result naturally from the differences in electronegativity between sodium and fluorine, it also leaves them both with complete octets!

What happens if an atom has more than one electron that it needs to lose? Magnesium, for instance, has two valence electrons. If it tries to bond to fluorine, the result is ionization, because of the difference in electronegativity. To complete fluorine's octet, fluorine takes one electron from magnesium. However, magnesium still has an electron left in its valence:

Fluorine can take an electron from magnesium

The resulting magnesium, however, still has a shell 3 electron

In order to lose that remaining electron, magnesium needs to react a second time. Because we know that atoms are typically found in sizeable samples, we can assume there's plenty of fluorine around. The magnesium can therefore lose a second electron to another, as yet unreacted, fluorine atom:

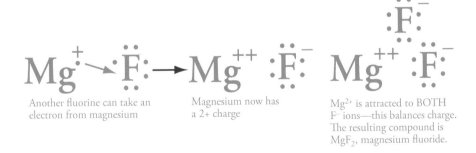

Another fluorine can take an electron from magnesium

Magnesium now has a 2+ charge

Mg^{2+} is attracted to BOTH F^- ions—this balances charge. The resulting compound is MgF_2, magnesium fluoride.

Using the naming conventions we learned in Chapter 1, the product is named without reference to the number of each species involved, and is called *magnesium fluoride*. **In any combination of a metal and a nonmetal, atoms continue to ionize as necessary until *all* atoms involved have gained or lost sufficient electrons to achieve full valence shells.** Keeping track of valence allows us to predict reactions without having to rely on Lewis structures. For instance, sodium needs to lose one electron to have a full shell and oxygen needs to gain two. As a result, two sodium atoms would need to react with one oxygen atom:

$$2\,Na + O \rightarrow 2\,Na^+ + O^{2-} \rightarrow Na_2O \text{ (sodium oxide)}$$

We don't really need to include the intermediate step, showing the charges of the individual ions, but it can be helpful when learning to put together ionic compounds. It is perfectly acceptable to write this reaction in a more simplified form:

$$2\,Na + O \rightarrow Na_2O$$

In another example, since magnesium has two electrons to lose, magnesium and oxygen react on a one-to-one ratio:

$$Mg + O \rightarrow Mg^{2+} + O^{2-} \rightarrow MgO \text{ (magnesium oxide)}$$

Note that each of the compounds above is neutral overall: the charges sum to zero. This is true of all ionic compounds. **As a result, a quick and easy way of determining how many of the metal or nonmetal atoms are required to fill all octets is simply to balance the charges on the resulting ions.** This technique is convenient simply because charges are so predictable: group 1 metals always have one valence electron, so when they ionize, they'll form +1 cations. Similarly, group 2 metals always have two valence electrons, so they'll ionize to form +2 cations. Group 3 elements form +3 cations, but it's difficult to build up that much charge in one place, so the only group 3 element we'll commonly see ionizing (within the scope of this text) is aluminum. We won't see any group 4 elements ionizing; it's simply too difficult to gain or lose four electrons. Instead, we'll see these elements form covalent bonds. From group 5 on, the elements (all nonmetals) can either ionize or form covalent bonds to fill their octets. Fluorine, for instance, can form one covalent bond (if bonding to another nonmetal), or it can take one electron from a metal and form the fluoride anion. Just as with the metals, though, the charges on these elements (if they ionize) are predictable. Group 5 elements have three valence holes, but don't often ionize for the same reason group 3 elements don't; that's an awful lot of charge. It's generally more favorable for group 5 elements to simply form three covalent bonds. Occasionally, though not in this text, one might run across N^{3-}, the nitride anion. Group 6 elements can ionize to form -2 anions, and group 7 elements, with only one valence hole, can ionize to form -1 anions.[10]

NOTE[10]

Unfortunately, it's exceedingly difficult to predict the charges on ionized transition metals. Like all metals, they form cations, but many can form more than one possible cation. Copper, for instance, can ionize to Cu^{+1} or Cu^{+2}. Though we will occasionally discuss these ions, charge will be provided. Predicting the charges on transition metal cations is far beyond the scope of this text.

Let's use this information to predict the formula of an ionic compound made from aluminum and oxygen. To make our prediction, we need to approach the problem methodically:

1. Determine what the charges on each element will be after ionization.

   Al is in group 3—it will form Al^{3+}. O is in group 6—it will form O^{2-}.

2. Determine how many of each ion is required to produce a neutral ionic compound. Mathematically, we're looking for the least common multiple of the charges.

   The least common multiple of 2 and 3 is 6. This means we expect the total positive charge in the compound to be +6 (from $Al^{3}+$), and the total negative charge to be -6 (from O^{2-})—note that these will sum to zero, making our compound neutral. It takes 2 Al^{3+} and 3 O^{2-} to achieve these quantities, so our compound is Al_2O_3.

This shortcut technique gets us to the same answer at which we would have arrived had we used Lewis dot structures.

Another interesting bit of information to come out of our discussion of ionic compounds is that we now know why it's unnecessary to use Greek prefixes in the names of binary compounds of metals and nonmetals. **Because the charges on ions determine how many of each will be present in the neutral compound, we don't need prefixes to help us predict the chemical formula.** For example, lithium chloride is composed of lithium (a metal) and chlorine (a nonmetal). Because we're dealing with a metal and a nonmetal, we know this will be an ionic compound. Further, lithium is a group 1 metal, so it will form Li^+. Chlorine is a group 7 nonmetal, so it will form Cl^-. To balance charge, only one of each ion is necessary. The compound, then, is LiCl, and LiCl is the only possible combination of lithium and chlorine. As such, when we say "lithium chloride," there is no ambiguity possible.

It's worth mentioning that it's not really accurate to envision a single "molecule" of an ionic compound. Rather, if a compound forms (as, say, in the case of LiCl), we can assume that there are *many* Li^+ and *many* Cl^- ions present, and each cation is attracted to multiple anions, and vice versa. As such, each Li^+ cation is actually surrounded on all sides by Cl^- anions, and each Cl^- anion is surrounded on all sides by Li^+ cations. While we refer to "molecules" of a salt, in reality, the formula simply represents the simplest possible "unit" of that salt's composition. LiCl, then, is a shorter (and much more convenient) way of saying *a salt composed of alternating units of lithium cation and chloride anion, on a 1:1 ratio* (Image 4.11).

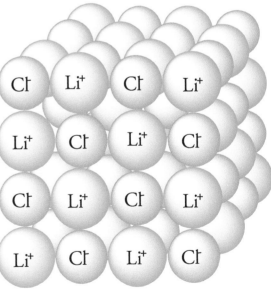

IMAGE 4.11. LiCl is a network of lithium cations and chloride anions.

Given that polar compounds dissolve well in water and hydrogen bonding compounds (which are *very* polar compounds) dissolve well in water, you might guess that ionic compounds also dissolve well in water. You'd be right. Opposite charges attract, and the more charge there is, the greater the attraction. Thus, water (with its partial charges) is very attracted to ionic compounds, because of their full units of charge. In fact, water is *so* attracted to ionic compounds that when stirred into water, the cations and anions in an ionic compound separate from one another and become surrounded by water molecules. This doesn't mean the compound is chemically reacting, nor does it mean the compound doesn't exist anymore. Remember, the nature of an ionic bond is different than that of a covalent bond. There's nothing physically holding the cation and anion together; they simply stick together under normal circumstances because of attraction. Water can also satisfy that attraction of charge. When we make salt water, for instance, we stir NaCl into H$_2$O. The Na$^+$ cations separate in space from the Cl$^-$ anions, and both become surrounded by water molecules (Image 4.12).

The reason NaCl seems to "disappear" when we dissolve it in water is that the crystal lattice falls apart, and the ions spread out in solution. Why would the crystal lattice fall apart, allowing the sodium and chloride ions to separate from each other? **From an atomic perspective, there's no such thing as "better" or "worse"**

(a)

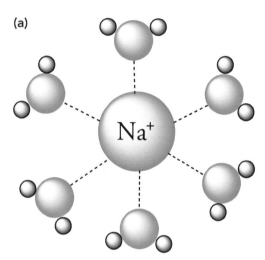

(b)

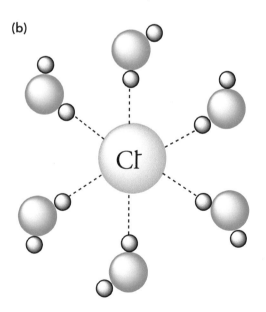

(c)

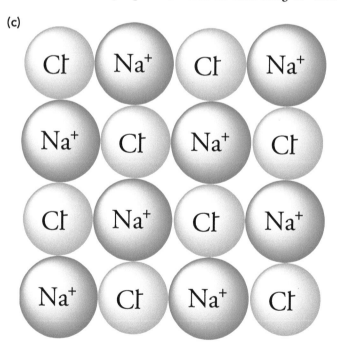

IMAGE 4.12. Solid NaCl (c). In water, the sodium cations separate from the chloride anions. Each cation is surrounded by the partially negatively charged oxygen atoms of water, and each anion is surrounded by the partially positively charged hydrogen atoms of water.

charge for the purposes of attraction; Na^+ is electrostatically attracted to Cl^- when Cl^- is the most prevalent negatively charged substance around. **In aqueous solution, however, there's far more water (which has partial negative charges on oxygen) than Cl^-, so Na^+ is more likely to become surrounded by water.** By the same token, in aqueous solution, it's more likely that chloride will be drawn to the partially positive hydrogen atoms of water than to the sodium cation, by sheer number. However, if we were to boil off the water, the Na^+ and Cl^- would resume their ionic attraction to each other (since there would longer be any water to be attracted to), and solid NaCl would recrystallize.

The separation of cations and anions from one another in aqueous solution is referred to as **dissociation**. Compounds that dissociate in water—ionic compounds—are called **electrolytes**, which are solutes that conduct **electricity** in aqueous solution. We can refer to an aqueous solution of such a compound as an *electrolyte solution*. Electricity is a current of charged particles moving through a wire or other medium, so only materials with mobile charged particles can conduct electricity. Metals make good conductors because they hold their valence electrons so loosely that the electrons form something of a "communal electron cloud" around the entire sample of metal. For this reason, electrons can easily be pulled to one end or another of a sample of metal, and this movement of electrons is an electrical current. **Similarly, ions in aqueous solution are free to move throughout the solvent, and can therefore transmit charge.** Molecular compounds (those consisting of purely covalent bonds) do not contain ions, and therefore cannot dissociate in solution. This means solutions containing molecular solutes are not capable of transmitting electricity. Sucrose (table sugar, $C_{12}H_{22}O_{11}$), for instance, is composed entirely of nonmetals. It is therefore molecular, and while it dissolves well in water (it happens to have lots of O-H bonds), it does not dissociate. Such solutions are not electrolytic (we could also say sugar is one example of a **nonelectrolyte**).

As an extension of this discussion, it turns out that pure water is *not* a good conductor of electricity; after all, it's molecular! Still, we all think of water as a good conductor—electrical appliances have labels warning us not to use them in the tub, for instance—so what's going on? As it happens, water is such a good solvent that, while it's theoretically possible to purify water completely (generally in the lab), it's simply not practical. Our tap water is far from pure; it contains lots of salts and other dissolved particles. Essentially, any water you run across is going to have many contaminants (that is to say, things other than water) dissolved in it. **For this reason, then, almost all water we encounter is actually an aqueous solution of many things, and is electrolytic.**

IMAGE 4.13. You've probably heard that sports drinks contain electrolytes. This is really just a way of saying they contain salts (though doubtless the beverage companies would like you to believe that an electrolyte is something exotic and hard to come by, outside of their beverage). A sweating body loses lots of salts (especially NaCl), and these beverages replenish those salts in order to keep the body functioning normally.

 TRY THIS

Identify the electrolytes among the following: glucose ($C_6H_{12}O_6$), $CaCl_2$, LiF, NH_3.

Answers:
$CaCl_2$, LiF. The others are compounds of nonmetals and are molecular.

4.7 Water Distribution and Safety

The surface of the planet Earth is approximately 71% water, but most of this water is unfit for human consumption. In fact, 97% of the planet's water is saline (saltwater), contained in the oceans. Ocean water requires so much processing to be made **potable** (fit for drinking)—which, in the case of ocean water, requires desalination to remove salt as well in addition to purification—that its use is very energy inefficient.

The remaining 3% of the water on Earth is fresh, but even much of this is unusable or inaccessible. Of the freshwater supply, nearly 69% is tied up in glaciers and icecaps, which are mainly localized at the poles and unavailable to large portions of the population. Thirty percent of freshwater is **groundwater**; it's found below the surface of the earth in large underground pools called **aquifers**. This water can be accessed via wells, and is generally potable without extensive treatment. A very small percentage of water (0.03% of fresh, or 0.009% of the total) is **surface water**, which includes lakes, rivers, swamps, and the like. While surface water can be made drinkable, it requires extensive treatment to remove impurities and unsafe contaminants.

It is worth mentioning that safe drinking water does not need to be 100% pure H_2O. As mentioned earlier, it is incredibly difficult to purify water to the point of having no contamination whatsoever. In addition to being impractical to fully purify water for drinking purposes, it's not necessary. Small amounts of dissolved chemicals, including salts of sodium, magnesium, and calcium, do no harm. **In any case, a colloquial reference to "pure water" should always be taken to mean potable water, not water that is chemically pure.**

In order to render surface water potable, the water is filtered in a treatment plant to remove dirt, leaves, and other bits of organic matter. Subsequent filtration through a smaller-pore filtering system then removes bacteria, viruses, and other organisms. Disinfectant chemicals such as chlorine are often added as well, and depending upon the source of the water, chemical contaminants may be removed using activated carbon, which *adsorbs* (sticks to and removes) the chemicals. Groundwater is filtered naturally due to the fact that it has passed through various subterranean layers on its way down to the aquifers, and it typically does not need as extensive a treatment process as surface water. Both surface and groundwater typically have fluoride levels adjusted to optimized concentrations: sufficiently high to produce the benefit to dental health without being concentrated enough to cause staining of the teeth.[11]

Residents of the United States are lucky to have convenient, unrestricted access to safe drinking water. **Per the United Nations, 18–20% of the world's population**

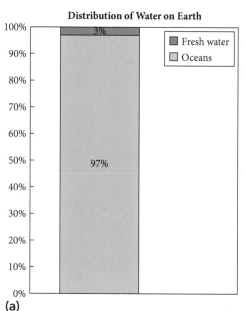

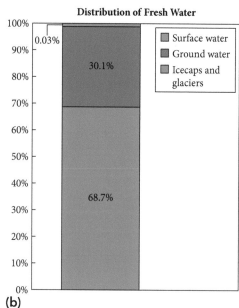

IMAGE 4.14. Global water distribution by category. Note that despite its visibility, surface water forms only the tiniest percentage of water on the planet.

(a)

(b)

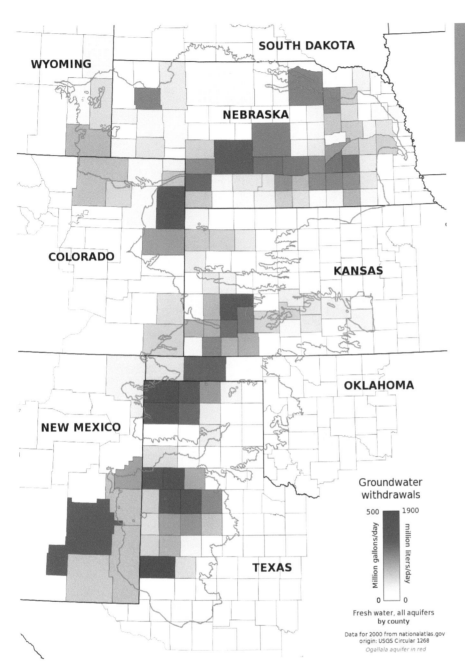

Groundwater
withdrawals

500 ▬ 1900

Million gallons/day

million liters/day

0   0

Fresh water, all aquifers
by county

Data for 2000 from nationalatlas.gov
origin: USGS Circular 1268

Ogallala aquifer in red

IMAGE 4.15. The Ogallala Aquifer provides for the water needs of much of the Midwest. Shading indicates withdrawals from the aquifer by region. Map from U.S. Geological Survey.

lack access to a sanitary water supply (Image 4.16). This leads to millions of deaths each year from diseases associated with unsafe water and unsanitary living conditions, including cholera, giardia, and malaria. Many global organizations, including the United Nations (UN), UNICEF, and the World Health Organization (WHO), have initiatives focused on increasing global access to safe water, particularly in underdeveloped and developing nations.

Total Drinking Water Coverage 2006

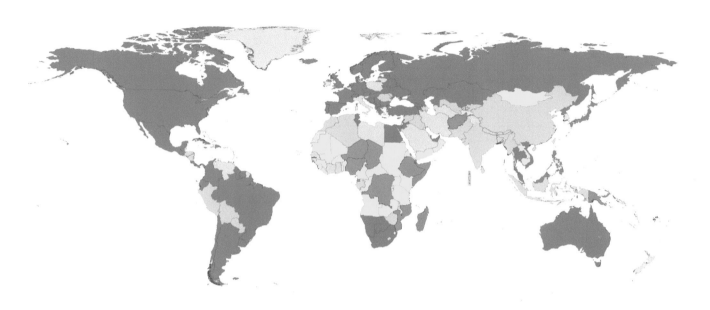

Total drinking water coverage, 2006

- Less than 50%
- 50 – 75%
- 76 - 90%
- 91 - 100%
- No or Insufficient data

Note: The boundaries and names shown and the designations used on this map do not imply official endorsement or acceptance by the United Nations.

IMAGE 4.16. Percentage of the population with access to safe drinking water (rural and urban areas) by country. Map from United Nations Environmental Programs (UNEP).

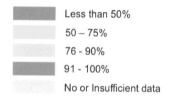

NOTE[12]

Minerals will be discussed further in Chapter 8, but in short, include the metal cations and the halides (halogen anions).

4.8   Hard Water

The Environmental Protection Agency (EPA), a United States governmental regulatory organization, maintains strict water safety standards, such that tap water in the US is quite safe to drink. Because potable does not mean free from solutes, however, tap water may contain various dissolved compounds. Of particular interest are the cations Ca^{2+} and Mg^{2+}, since these are responsible for the phenomenon of **hard water**. Ground and surface water dissolve minerals from the rock and soil. Some of these minerals[12]—namely magnesium and calcium—have the ability to complex with anions in the water. The resulting ionic compounds form difficult-to-remove salt crusts around water fixtures and in sinks and tubs (Image 4.17). This **calcification** (formation of these crusts is called *calcification* regardless of whether the cation responsible is calcium or magnesium) can damage fixtures and impair water flow. There's plenty of Na^+ (and other group 1 cations) in tap water as well, but calcium and magnesium salts display the interesting property of having low water solubility. Despite the fact that ionic compounds should theoretically be water soluble, some of them are held together so tightly by electrostatic forces that they don't tend to dissolve. This phenomenon is commonly observed in salts containing group 2 and group 3 cations. As a result, sodium and other group 1 cations in water don't produce faucet calcification and aren't a factor in water hardness, but the group 2 minerals do, and are thus a factor.

Soap introduces another issue. Soap molecules are anionic,[13] meaning that they can also complex with calcium and magnesium. Sodium and soap form soluble salts, but calcium and magnesium result in formation of insoluble soap salts. These calcium and magnesium soap salts don't behave the way soap generally does, which is why it's difficult to form a lather in hard water, and why soap scum tends to precipitate— meaning collect in solid form—on tub walls in the familiar tub ring. **Hard water is not a health hazard; it's merely a nuisance.** However, it's possible to "soften" water by exchanging the undesirable calcium and magnesium ions for sodium ions (Image 4.18). Water passes through a softening device that performs an ion exchange. Ca^{2+} and Mg^{2+} ions displace Na^+ ions from the exchange resin (a resin is a bit like a wax), and the sodium ions end up in the water, while calcium and magnesium ions stick to the resin. Since sodium salts are soluble in water, neither calcification of faucets nor precipitation of soap occurs.

Not all municipal water is hard; regions (often coastal) of the United States that utilize treated rainwater, which contains no dissolved minerals, have very soft water. Soft water doesn't result in calcification of fixtures and soap lathers effectively. However, soft water doesn't rinse soap from the skin as easily as hard water, which can leave skin feeling a little slimy when bathing or washing hands.

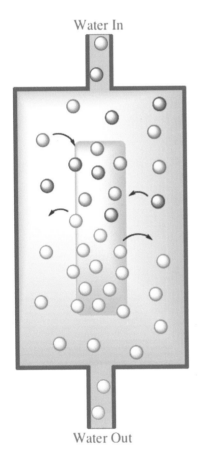

IMAGE 4.18. An ion-exchange resin forms the heart of a water softener. Water full of Ca2+ and Mg2+ ions (green and red) enters the softener and the cations stick to the resin, displacing Na+ ions (purple). The water exits the softener enriched in sodium, and without its calcium and magnesium.

Water In

Water Out

IMAGE 4.17. A calcified faucet.
© Scott Lefler, 2010.

 NOTE[13]

Soaps will also be discussed further in Chapter 8, but it is sufficient for now to know that they contain large, negatively charged molecules.

There are places in the world where clean, safe drinking water is hard to come by. The United States is not one of these places. Still, bottled water manufacturers have managed to turn consumers' concern with the safety of their drinking water into a multibillion-dollar business (soda is the only beverage whose sales outstrip those of bottled water; beer, juice, and milk lag far behind). However, Americans' love affair with bottled water comes at a high social price: the energy associated with producing bottles, processing and bottling water, and transporting the finished product is immense.

Several studies have evaluated and reported the energy cost associated with bottling water, and have determined that the energy input is multifaceted. In the first place, petroleum and petroleum products are used in the manufacture of bottles; plastic bottles are literally made of oil. Second, energy is required to treat and process the water that is to be bottled. Bottled water falls into one of two categories, per United States Food and Drug Administration (FDA) regulations. *Springwater* is derived from an underground source, while *purified water*, which represents a huge share of the bottled water market, is merely municipal tap water that has been further distilled, deionized, or otherwise treated—and unnecessarily so, as EPA regulations

IMAGE 4.19. The image that bottled water manufacturers try to evoke (a) is at odds with the source of most bottled water (b). Waterfall © Scott Lefler, 2009.

strictly maintain the safety of municipal tap water. This processing represents an increased energy cost above that of tap water production. Energy is then required to package bottled water and still more energy to transport it (actually, transportation represents one of the greatest energy inputs due to the weight of a truckload of water). **Overall, the energy cost per bottle of water in terms of oil consumed has been estimated as equivalent to one quarter of the volume of the bottle itself! This represents an energy expenditure of 2,000 times that associated with producing municipal tap water.** Given the issues of dwindling fossil-fuel (oil) supplies and the environmental ramifications of unnecessary energy expenditure, both of which we'll discuss in Chapter 13, bottled water represents a huge environmental burden.

Another cost associated with the consumption of bottled water is the price. The United States Department of Agriculture (USDA) advises us to drink eight 8-oz. glasses of water per day. If that consumption were filled entirely by bottled water, the cost could be upwards of $1,400 annually.[14] By comparison, the annual price tag on that same quantity of tap water would be about 50 cents.

Presumably, given that the cost of water is astronomically high (both energetically and financially), consumers are enamored of it for good reason. Perhaps it's purer, healthier, or tastier than tap water? Unfortunately, none of these rationales for bottled water consumption stand up to scrutiny. In terms of purity and health-fulness, municipal water is regulated by the EPA, which has strict processing and purity standards. Tap water also contains appropriate amounts of fluoride, routinely demonstrated to have a positive effect on dental health. **Bottled water, on the other hand, falls outside the regulatory control of the EPA. As a product, it's meant to be regulated by the FDA, though most of it falls outside their purview since very little crosses state lines, exempting it from federal scrutiny.** Further, independent studies show no improved purity of bottled water over tap, and less than 5% of bottled water contains the concentration of fluoride recommended by the American Dental Association. In the end, most bottled water is nothing but low-fluoride tap water that's been subjected to additional processing, producing an infinitesimal improvement upon the impressive purity of municipal water, if any improvement at all.

Regarding taste, while no controlled scientific studies have addressed this issue (taste being a subjective matter), there is overwhelming nonscientific evidence that in blind taste tests, bottled water can't be distinguished from tap. In fact, many cities routinely hold water "taste-offs" to raise awareness of municipal water quality, with consistent success. Regardless, even those for whom tap water has an undesirable taste have the option of a home tap or pitcher filter.

A final issue to consider is that plastic water bottles, while recyclable, are often used once and discarded. This increases community (and global) waste. Further, even those bottles that are recycled are reprocessed at an additional energy cost. It's somewhat interesting (and distressing) to consider the costs we're willing to accept to obtain bottled water, despite having instant access to incredibly pure, minimal-cost water available from every tap in our homes. **Overall, the decision to avoid bottled water saves money, saves a tremendous amount of energy, and comes at no cost to health or wellness.**

NOTE[14]

Bottled water costs, in general, about 2–2.5 times the cost of gasoline by volume, and ironically, we happily pay for the former while complaining about the price of the latter.

IMAGE 4.20. A polluted beach in Malaysia (a) and a river in France (b) are stark examples of where a great many water bottles end up.

4.10 The Last Word—Superwaters

Some bottled waters go above and beyond simply evoking the image of a mountain spring: market shelves are now filled with bottles upon bottles of "superwaters" that contain flavorings, vitamins, and so forth. These superwaters promise to do more than just quench thirst: claims range from improving sports performance to promoting weight loss. In a particularly bizarre misconception of both chemistry and basic physiology, some waters are *oxygenated*: O_2 has been bubbled into the water, resulting in greater oxygen saturation than is normally found in drinking water. These products promise enhanced cellular oxygenation and suggest that oxygenated water boosts performance, cognition, energy level, and a host of other functions. Unfortunately, this claim simply doesn't, well, hold water. Humans, like other land animals, absorb oxygen through the lungs. We have no mechanism whatsoever for absorption of oxygen through other body surfaces or membranes. Supplementing the gut with oxygen from overly fancy water does nothing but increase the amount of gas in the stomach, which has no effect whatsoever except perhaps to contribute to eructation (an overly fancy word for belching).

SUMMARY OF MAIN POINTS

- Water is critical to life.
- Solutions are mixtures of solvents and solutes—water is an excellent solvent.
- Electronegativity is a measure of an atom's ability to pull electrons in bonds.
- Pure covalent bonds form between atoms with identical electronegativity.
- Polar covalent bonds form between atoms with different electronegativity, and result in dipole moments.
- A bond dipole is unequal distribution of electrons in a bond, and results in partial charges on the atoms in the bond.
- A net dipole is the sum of a molecule's bond dipoles.
- Atoms with net dipoles are polar; atoms with no net dipole are nonpolar.

- Water is extremely polar.
- Polar things dissolve well in other polar things; nonpolar things dissolve well in other nonpolar things.
- Compounds with very large bond dipoles (along F-H, O-H, or N-H bonds) interact strongly with each other via hydrogen bonding—water is an excellent hydrogen bonder.
- Due to water's formation of hydrogen bonds, it has a high boiling point, melting point, and density. It is resistant to temperature change, is very cohesive, and forms a solid that is less dense than its liquid.
- Ionic bonds form between atoms with very different electronegativity—metals and nonmetals—and are the result of complete electron transfer.
- Cations and anions—the result of electron transfer reactions—are electrostatically attracted to one another and form ionic salts.
- The charge on an atom when it ionizes can be predicted from its group number on the periodic table.
- Ionic compounds always form so that the charges balance.
- Ionic compounds dissolve well in water.
- Ionic compounds dissociate in water, forming electrolytic solutions that can conduct electricity (molecular compounds are not electrolytes).
- Earth's water is primarily unavailable for drinking—only 1% of the planet's water is fresh *and* accessible.
- To make water potable, it must be treated—surface water requires more extensive treatment than groundwater.
- The availability of safe drinking water is an important global health issue.
- Water with large amounts of dissolved Ca^{2+} and Mg^{2+} is called hard water and is a nuisance; it can be softened via ion exchange.
- Bottled water is incredibly expensive financially and energetically, and affords no real advantage over municipal water.

QUESTIONS AND TOPICS FOR DISCUSSION

1. Why might a chemist or biologist call water the single most critical chemical to life?

2. Why is the process of making a solution not a chemical reaction?

3. Explain the concept of electronegativity. How does it affect chemical bonding?

4. To which element is hydrogen's electronegativity considered to be nearly identical? What is the ramification of this upon bonds between hydrogen and this element?

5. What information about electronegativity can we get from the periodic table?

6. Why do bond dipoles result in formation of partial charges?

7. What kinds of molecules have bond dipoles? Net dipoles? What kinds of molecules have bond dipoles but *no* net dipole?

8. Why is being able to determine a molecule's shape critical to finding whether it's polar or nonpolar?

9. Consider the molecules CH_4, CH_3Cl, CH_2Cl_2, $CHCl_3$, and CCl_4. How are they similar? How are they different? Which have net dipoles?

10. What is the primary predictor of solubility? What sorts of compounds dissolve well in water?

11. What does *like dissolves like* mean?

12. What is hydrogen bonding?

13. How does water's hydrogen bonding affect its physical properties, making it unusual as a molecule?

14. How are particles arranged in solids? In liquids? In gases? What generalizations regarding attraction between particles, mobility of particles, compressibility, and density can be made for each?

15. Why does ice float on liquid water? Why is this odd behavior for a chemical?

16. Why do some elements form covalent compounds, while some form ionic compounds? Which types of element pairs will form covalent compounds? Which types of pairs will form ionic compounds?

17. From the perspective of trying to obtain a full octet, why do metals react by ionizing rather than forming covalent bonds? Why can many nonmetals do either?

18. Why is it unnecessary to use Greek prefixes in the naming of ionic compounds?

19. Why is it not strictly correct to think of an ionic compound as a "molecule"? What is different about a sample of an ionic compound in terms of delineating a single molecule?

20. How do we know what charge an ion will have when it forms?

21. What is an electrolyte? What can electrolytes do that nonelectrolytes can't?

22. What are some examples of electrolytes? Of nonelectrolytes?

23. Describe the behavior of an ionic compound when dissolved in water. How does it interact with water?

24. What is hard water? Is it dangerous?

25. What are the effects of hard water?

26. How is rainwater similar to chemically softened water? How is it different?

27. How does hard water form? How can it be softened?

28. What types of water are found on Earth, and what percentage of total water falls into each category?

29. Is most of Earth's water drinkable? Is most of Earth's drinkable water accessible?

30. What is the difference between groundwater and surface water? Which generally requires less processing to be potable, and why?

31. Which parts of the world have the least access to safe drinking water?

32. How is drinking water processed in the United States? Is the U.S. tap water safe and drinkable?

33. What is the difference between potable water and 100% H_2O?

34. What are the major social, economic, and financial ramifications of drinking bottled water? Do you think bottled water is good for a community? For the world?

35. What are the differences between governmental oversight and regulation of tap water versus bottled water?

PROBLEMS

1. Define each of the following: solute; solvent; solution; aqueous solution; concentration.

2. For each of the following, identify the solute and the solvent: sugar water; vitamin A dissolved in fat; salt water.

3. Put the following in order of increasing electronegativity: F; H; C; Cl; Br; Li.

4. Put the following in order of increasing electronegativity: S; N; Mg; I; H; O.

5. What are the three most electronegative elements?

6. Of the elements that routinely form covalent bonds, which is the least electronegative?

7. Define each of the following: dipole; electronegativity; partial charge; periodic trend; pure covalent bond; polar covalent bond.

8. Sketch each of the following molecules, indicating bond dipoles: HBr; Cl_2; COH_2 (the C is in the middle); CH_2I_2.

9. Sketch each of the following molecules, indicating bond dipoles: HCN; O_2; H_2S; CF_2Cl_2.

10. Sketch each of the following molecules, indicating partial charges: HBr; Cl_2; COH_2 (the C is in the middle); CH_2I_2 (indicate partial charge on outer atoms only).

11. Sketch each of the following molecules, indicating partial charges: HCN (indicate partial charge on H and N only; O_2; H_2S; CF_2Cl_2.

12. Sketch each of the following molecules, indicating net dipole: HBr; Cl_2; COH_2 (the C is in the middle); CH_2I_2.

13. Sketch each of the following molecules, indicating net dipole: HCN; O_2; H_2S; and CF_2Cl_2.

14. Classify each of the following molecules as polar or nonpolar: HBr; Cl_2; COH_2 (the C is in the middle); CH_2I_2.

15. Classify each of the following molecules as polar or nonpolar: HCN; O_2; H_2S; and CF_2Cl_2.

16. Which of the molecules shown is capable of hydrogen bonding? Sketch the hydrogen bonding interaction.

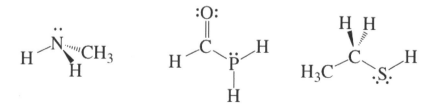

17. Which of the molecules shown is capable of hydrogen bonding? Sketch the hydrogen-bonding interaction.

18. Predict the charges on ions of each of the following elements: K; Al; P; I.

19. Predict the charges on ions of each of the following elements: Cs; Ba; Se; Cl.

20. Draw Lewis dot structures showing the formation of an ionic compound from each of the following combinations of elements (show electron transfer and resulting charges): Cs and F; Cs and O; Ca and O; Ca and Cl.

21. Draw Lewis dot structures showing the formation of an ionic compound from each of the following combinations of elements (show electron transfer and resulting charges): Rb and S; Rb and N; Sr and F; Sr and S.

22. Draw Lewis dot structures showing the formation of an ionic compound from each of the following combinations of elements (show electron transfer and resulting charges): Mg and N; Al and S.

23. Determine the chemical formula of each of the following: lithium oxide; potassium selenide; magnesium oxide; calcium fluoride.

24. Determine the chemical formula of each of the following: rubidium fluoride; sodium oxide; sodium nitride; beryllium chloride.

25. Indicate which of the following are electrolytes: H_2S; PF_3; Li_2O; CH_4; NaF.

26. Indicate which of the following are electrolytes: C_2H_4; LiF; KCl; CH_2O; $CHCl_3$.

27. Indicate which of the following will dissociate in water: CH_2O; $COCl_2$; $MgCl_2$; Na_2S; and KCl.

28. Indicate which of the following will dissociate in water: $MgCl_2$; H_2O_2; Na_2O; COI_2; RbCl.

29. What type of water represents the largest percentage of Earth's water? The largest percentage of Earth's freshwater?

30. Define the following: aquifer; groundwater; surface water; potable water.

31. What ions are responsible for hard water? What ions are used to soften water?

32. Why does the "hardness" of water impact soap function?

33. What is soap scum, and where does it come from?

34. What is calcification, and where does it come from?

35. What are the four ways in which energy is required as an input to producing bottled water?

ANSWERS TO ODD-NUMBERED PROBLEMS

1. A solute is the chemical being dissolved; a solvent is the chemical doing the dissolving; a solution is a mixture made up of a solute and solvent; an aqueous solution has water as the solvent; concentration is a measure of the relative proportion of solute to solvent.

3. Li; H; C; Br; Cl; F.

5. F, O, N.

7. A dipole is an unequal distribution of electrons in a bond (or a separation of charge); electronegativity is a measure of how tightly electrons are drawn toward an atom's nucleus; partial charge results from partial electron loss or gain through a dipole; a periodic trend is one that can be predicted from looking across rows or up/down columns on the periodic table; a pure covalent bond is one in which there is equity of electron sharing; a polar covalent bond is one in which there are differences in electronegativities of atoms in a bond, resulting in unequal electron sharing.

9.

11.

13.

15. HCN is polar; O_2 is nonpolar; H_2S is polar; CF_2Cl_2 is polar.

17.

19. Cs^+; Ba^{2+}; Se^{2-}; Cl^-.

21.

23. Li_2O; K_2Se; MgO; CaF_2.

25. Li_2O and NaF are electrolytes.

27. $MgCl_2$, Na_2S, and KCl dissociate in water.

29. Ocean water is the largest percentage of Earth's water; glaciers/icebergs are the largest percentage of fresh water.

31. Ca^{2+} and Mg^{2+} are responsible for hard water; Na^+ is used to soften water.

33. The insoluble salt formed from the combination of Ca^{2+} and Mg^{2+} with anionic soap molecules precipitates out on fixtures and tub/shower walls; this is soap scum.

35. Energy (oil) to provide raw materials, energy to produce bottles, energy to process water and package, energy to ship water.

Role of Water in Life

Alpert, P. (2005). Sharing the secrets of life without water. *Integrative and Comparative Biology, 45*(5), 683–84.

Rothschild, L.J., Mancinelli, R.L. (2001). Life in extreme environments. *Nature, 409,* 1092–101.

Water Distribution and Safety

Azoulay, A., Garzon, P., Eisenberg, M. (2001). Comparison of the mineral content of tap water and bottled waters. *Journal of General Internal Medicine, 16*(3), 168–75.

www.epa.gov/ow

www.un.org/waterforlifedecade

www.usgs.gov

www.who.int/topics/drinking_water/en

Bottled Water

Gleick, P. H., Cooley, H. S. (2009). Energy implications of bottled water. *Environmental Research Letters, 4*(1), 014009.

Lalumandier, J. A., Ayers, L. W. (2000). Fluoride and bacterial content of bottled water vs. tap water. *Archives of Family Medicine, 9*(3), 246–50.

CREDITS

The Chemistry of Acids and Bases

5

We encounter acids every day of our lives. Vinegar in salad dressing is acetic acid; vitamin C is ascorbic acid; we put muriatic acid (also called hydrochloric acid) in our pools; the burning sensation in hardworking muscles comes from lactic acid, a metabolic waste product (Image 5.1). We even take advantage of acids in food preparation: carbonic acid causes bread to rise, while lactic acid turns milk into cheese. Some common acids have negative effects on us and on the environment. For instance, pollution results in acid rain, which kills aquatic life and destroys architecture. Oxalic acid is found in many foods, including leafy green vegetables, and can build up enough to cause kidney stones in some individuals. Spilled industrial or laboratory acid can cause burns and act as an environmental pollutant. Like all substances, acids can be either useful or detrimental, either safe or harmful. What is an acid, though, from a chemical standpoint? What gives it the physical and chemical properties that distinguish it from a nonacid, or from its chemical opposite, a base? Most importantly, in what context are acids helpful and safe, as opposed to dangerous and toxic?

5.1   Acids

An acid is a chemical that behaves in a particular way in aqueous solution. The simplest definition of an acid is that it dissociates in water (just like an ionic compound) to produce the cation H^+ and an anion. Let's examine this statement further. First of all, it may be surprising to see hydrogen ionizing to form H^+; after all, thus far we've only seen it form covalent bonds. Remember, though, the nature of hydrogen is dichotomous, hence its common representation in two different places on the

IMAGE 5.1. The vitamin C in a grapefruit and other citrus fruit, the chemical making this hardworking runner's legs burn, and the substance helping to prevent contamination of a swimming pool are all acids.

periodic table. As a nonmetal (which it is, since as an element it does not have the physical properties of a metal), we expect to see hydrogen forming a single covalent bond to another nonmetal to fill its one valence shell hole. Of course, given hydrogen's very low electronegativity, this bond is generally highly polarized (unless it is bonding to another atom of hydrogen or to carbon). As a group 1 element, though, hydrogen has properties in common with the other elements in its group. Like lithium, sodium, and so on, hydrogen has a single electron in its outermost shell. The group 1 metals ionize in the presence of sufficiently electronegative elements (nonmetals), emptying their valence shells and leaving the atoms stable (and cationic). Hydrogen, it turns out, can do the same under the right circumstances. **While hydrogen's bonds to very electronegative elements act covalent (molecular) *most* of the time, the bonds can polarize completely in aqueous solution.** This leaves hydrogen cationic, and the atom (or group of atoms) to which it was bonded (which is now in possession of both of the electrons from the former covalent bond) becomes anionic. It is this ionization of hydrogen in aqueous solution that defines an acid, and as such, it becomes a bit easier to predict what sorts of chemicals will be acids—by necessity, they must contain hydrogen bonded to an electronegative element. For instance, all of the compounds of hydrogen with group 7 nonmetals are acids, and there are many examples of acids in which hydrogen is bonded to oxygen. Unfortunately, this definition only works in one direction: while acids always contain hydrogen bonded to an electronegative element, not all chemicals in which hydrogen is bonded to an electronegative element are acids—water is an obvious example.[1] Conveniently, the acids we'll see in this text all have the word *acid* in their name, as in *hydrochloric acid, sulfuric acid,* and *lactic acid.* For now, it's sufficient to know that the acids we'll be discussing fall into one of three groups: binary acids, oxyacids, and carboxylic acids. **Binary acids** are molecules made up of hydrogen and only one other element; in the examples we'll see in this text, the second element will be a halogen, as in HCl. **Oxyacids** are combinations of one or more atoms of oxygen with another nonmetal, plus one or more atoms of hydrogen. Examples of oxyacids include nitric acid (HNO_3) and sulfuric acid (H_2SO_4). **Carboxylic acids** are an incredibly large and varied class of compounds, but all include the characteristic combination of atoms shown here:

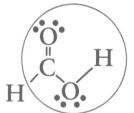

The circled area represents the arrangement of atoms typical of carboxylic acids. This is formic acid.

In a chemical formula, the combination shown above is written *–COOH*. Formic acid's chemical formula, for instance, would be HCOOH. Acetic acid, shown in Image 5.2, has the chemical formula CH_3COOH. These formulas look a bit different than the ones we've seen thus far in the text. Why, for instance, is formic

NOTE[1]

Electronegativity is not the sole determinant of an atom's ability to ionize its bond with hydrogen in aqueous solution. For instance, despite oxygen's high electronegativity, water is not acidic. However, despite the fact that sulfur is not as electronegative as oxygen, H_2S is acidic. It turns out that a combination of factors dictate acidity (including atomic size and properties of the resulting anion after loss of hydrogen), and electronegativity is only one of these factors. It's quite safe to say, though, that molecules in which hydrogen is bonded to an element with low electronegativity (such as carbon) do not result in the release of H^+ in aqueous solution.

acid written HCOOH rather than CH_2O_2? The bigger molecules get, the harder it becomes to draw Lewis structures from formulas—just imagine trying to figure out how atoms are connected to one another in $C_2H_4O_2$, which is a small molecule in the grand scheme. By writing the formulas in ways that give information about structure (the presence of *–COOH*, for instance, always indicates a particular atomic arrangement), it makes it easier for us to determine the structure of a molecule from its formula. Technically, CH_2O_2 is a perfectly accurate chemical formula for formic acid, but looking at that formula, you'd not be able to tell right away that the compound is acidic. HCOOH, however, makes the presence of the *–COOH* arrangement of atoms very apparent. Since this arrangement always indicates that a compound is an acid, you can instantly tell that the formula HCOOH refers to an acidic compound without having to draw the structure or hear the name.

To visualize an acid's behavior in water, we simply imagine *both* electrons from the bond between hydrogen and whatever it's attached to being pulled so tightly toward the more electronegative element that ionization occurs. This leaves us with an H^+ cation, and an anion made up of the rest of the molecule:

When acid dissolves in water, the electronegative element pulls both electrons from the covalent bond into its valence

The result is an H^+ cation, and an anion made up of the remaining parts of the acid molecule

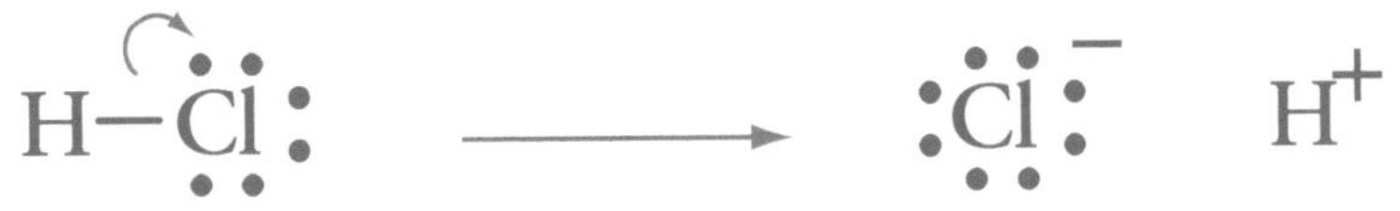

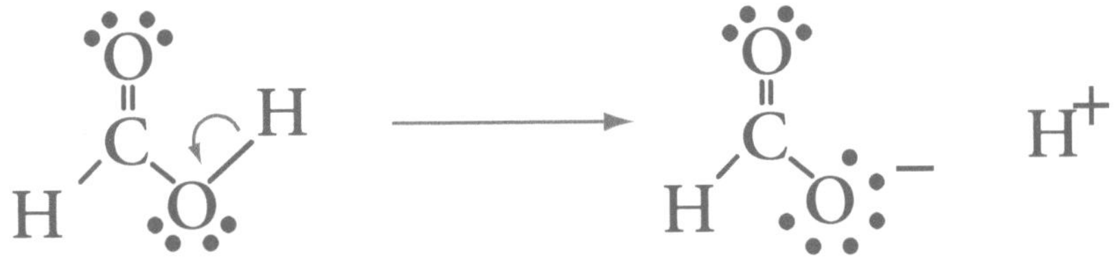

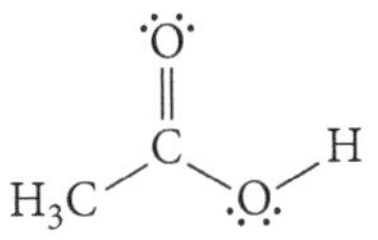

IMAGE 5.2. Acetic acid (vinegar) is a carboxylic acid with the chemical formula CH_3COOH.

Acid Formula	Acid Name
HCl	Hydrochloric acid
H_2SO_4	Sulfuric acid
HNO_3	Nitric acid
H_2CO_3	Carbonic acid
HCOOH	Formic acid
CH_3COOH	Acetic acid (vinegar)

TABLE 5.1. Names worth knowing: some of the most commonly encountered acids.

5.2 Polyatomic Ions

In the dissociation of formic acid above, we see the formation of an interesting species: $HCOO^-$, also called the *formate anion*. It's fairly elementary to see why hydrogen becomes cationic when dissociation occurs—it's lost its electron. Why, though, does the oxygen end up with a negative charge? Both in formic acid and in formate, oxygen has a full octet of eight electrons, so there's been no change in the valence. We can verify that by counting electrons: there are two bonds and two

CONCEPT CHECK

In the context of an acid, how does hydrogen act like a nonmetal? How does it act like a metal?

Answer: It forms a covalent bond to another atom (like a nonmetal) that polarizes and leads to dissociation in water (like a metal).

Formula	Name
NH_4^+	Ammonium
OH^-	Hydroxide
NO_3^-	Nitrate
SO_4^{2-}	Sulfate
PO_4^{3-}	Phosphate
CO_3^{2-}	Carbonate
HCO_3^-	Bicarbonate

TABLE 5.2. Names and chemical formulas of common polyatomic ions.

lone pairs around oxygen in formic acid, and there is one bond and three lone pairs around oxygen in formate. That's a full octet in both cases. Normally, ionization reactions involve a change in an atom's electron count, so what's going on? In reality, oxygen *has* experienced a change in its electron count! Remember that in formic acid, the electrons in the O-H bond don't *both* belong to oxygen; one of them is oxygen's and one is hydrogen's. Oxygen can count both of the electrons toward its valence (that's the nature of a covalent bond), but it doesn't have personal "ownership" of hydrogen's electron, so that electron doesn't throw off oxygen's proton/electron balance, and oxygen is therefore still neutral. **When the electrons in the O-H bond are pulled entirely away from hydrogen by oxygen, oxygen takes ownership of hydrogen's electron.** This leaves hydrogen cationic, because it has lost its one "personal" valence electron, and makes oxygen anionic, because it has acquired a seventh "personal" valence electron, as compared to the six it started with.

Ions such as formate, consisting of two or more atoms covalently bonded together, are called polyatomic ions, and they are relatively common in chemistry (especially in the chemistry of acids and bases). Names, charges, and formulas of the most common polyatomic ions appear in Table 5.2.

When we discussed formation of ionic compounds in Chapter 4, we imagined that the reaction took place in two stages. First, the more electronegative atom(s) took electrons from the less electronegative atom(s), forming anions and cations. Next, the resulting ions, because of their electrostatic attraction to each other, formed an ionic bond. For now, we will not address the formation of polyatomic ions—we'll just assume they already exist. We'll see some instances of how they form in the real world later in the text (sulfate, for instance, is a by-product of volcanic eruption and combustion of fuel), but for now, we don't need to concern ourselves with these details. Suffice it to say that **ions are not permanently "stuck" with the atoms with which they originally reacted**—even if a sodium cation, for instance, formed when sodium lost an electron to fluorine, it's possible for the NaF to break apart (as we saw in our discussion of electrolytes). When this happens, new ionic bonds can form. This is exceedingly common in aqueous solution and allows for the rearrangement of cations and anions into different compounds. Given all of this, assuming for now that the polyatomic ions have formed through some chemical means, we can figure out how they might combine with other ions to form compounds.

Polyatomic ions can combine with other anions or cations to form ionic compounds the same way that monoatomic ions do. **Just as in binary compounds consisting of a metal and a nonmetal, salts containing polyatomic ions form such that the resulting compound is neutral.** Hydroxide, for instance, with its -1 charge, can combine with one sodium cation to form NaOH. A magnesium cation, however, with a +2 charge, will need to combine with two hydroxide anions to create a neutral compound. Magnesium hydroxide therefore has the formula $Mg(OH)_2$. Note the use of parentheses around the OH; this is to indicate that the subscript refers to the entire OH group, and not simply the H. This compound, per the subscript, consists of an ion of magnesium and two units of the OH^- anion. $MgOH_2$, written without parentheses, would refer to a compound made up of one magnesium atom, one oxygen atom, and two hydrogen atoms (and, incidentally, isn't a real chemical).

Compounds containing polyatomic ions are named using the same scheme as binary ionic compounds. **Just as in compounds of metals and nonmetals, when naming ionic compounds with polyatomic ions, we name the cation first (without changing its name), then the anion.** Greek prefixes are not used.

Example: NH_4Cl is called *ammonium chloride*.

Example: K_2SO_4 is called *potassium sulfate*.

TRY THIS

Provide the name and formula of a compound consisting of Ca^+ and CO_3^-.

Answer: $CaCO_3$, calcium carbonate.

5.3   Bases

Unlike acids, bases do not ionize in water to produce hydrogen cations. Instead, they "force" water to ionize by taking one of its hydrogen atoms:

NaOH is ionic; it's composed of an Na^+ cation, electrostatically attracted to the OH^- (hydroxide) anion. It dissociates in water.

Hydroxide uses one of its lone pairs of electrons to form a bond to one of water's hydrogens. The hydrogen leaves its shared pair of electrons behind, on water's oxygen.

The molecule that was OH^- is now H_2O, and the molecule that was H_2O is now OH^-.

There are two important points to note regarding the reaction shown above. First, the compound NaOH is ionic, as it consists of sodium (which never forms covalent bonds) and what we now recognize as the hydroxide anion. NaOH, like all ionic compounds, will therefore dissociate in water. From that point on, the interesting chemistry is perpetrated entirely by OH^- and other species present—Na^+ is not an active participant in the chemical reaction. This is actually quite typical of Na^+ and other metal cations; once they've formed, they're quite stable (and therefore unreactive), and while we'll see them electrostatically attracted to different anions—essentially, they're attracted to whatever negatively charged particles are present in the solution mixture—they don't participate in reactions. **Ions such as the Na^+ in the reaction above are called spectator ions because they "watch" the reaction rather than taking part in it.** A second important point has to do with charge and why neutral atoms become charged (and vice versa) during the course of the reaction. For clarity, let's name the species in the reaction above "blue hydroxide" and

TRY THIS

Provide the name and formula of a compound consisting of NH_4^+ and SO_4^{2-}.

Answer: $(NH_4)_2SO_4$, ammonium sulfate.

"green water." When blue hydroxide "steals" a hydrogen from green water, the hydrogen leaves behind both of the electrons it shared with the oxygen in green water. Because green water's oxygen has now acquired an additional "personal" electron, the oxygen becomes negatively charged. Note that its valence hasn't changed; the difference is merely that what was a shared electron has become an unshared electron. Further, because any chemical bond must consist of a pair of electrons, blue hydroxide's oxygen must provide *both* electrons that will go toward making the bond with green water's hydrogen (the hydrogen contributes no electrons toward the bond, because it left them behind). As such, the blue hydroxide's oxygen is essentially "giving" an electron to green water's hydrogen—it's taking a nonbonding pair and turning it into a shared pair. This means green hydrogen can now claim one of blue oxygen's electrons. Blue oxygen has in effect lost an electron—not from its valence, because it still has an octet, but from its "personal" electron count—and it is now neutral rather than negatively charged.

Not all bases are ionic salts of hydroxide, however, as we see in the case of ammonia (NH_3), which is also basic:

Ammonia uses its lone pair of electrons to form a bond to one of water's hydrogens. The hydrogen leaves its shared pair of electrons with oxygen.

Because ammonia has acquired an ionized hydrogen (an H^+) without acquiring any additional electrons, it is now positively charged (NH_4^+). What was water is now OH^-.

In this case, we are not starting with a dissociation reaction—NH_3 is purely covalent, and the only way to predict its behavior in water is to know it's a base. Bases are quite a bit harder to recognize than acids, but it's worth knowing that the most common bases are salts of the OH^- anion and ammonia. Like any base, ammonia reacts with water by using a nonbonding pair of electrons[2] to steal hydrogen away from water. Hydrogen releases both electrons it shared with oxygen, resulting in the formation of the OH^- anion. Because nitrogen from NH_3 has used a nonbonding pair of electrons to make a bond to hydrogen, it has essentially "lost" an electron—it is now positively charged—and becomes the NH_4^+ cation. **Note that charge is always conserved in these reactions; if neutral starting materials react (as in the case above), the products might be charged, but the sum of the charges will be zero.** Also note that when a base takes a hydrogen from water, the base will *always* lose a unit of negative charge. If it was negative before the reaction (like OH^-), it will become neutral. If it was neutral before the reaction (like NH_3), it will become positive.

NOTE[2]

There's a pattern here. Bases all react with water by taking one of water's hydrogens, so they must all have a lone pair of electrons available to form a bond to that hydrogen. Be careful, though; all bases have lone pairs, but not all molecules or atoms with lone pairs are bases!

NOTE³

If you're trying to imagine what "meaty" tastes like, think of soy sauce or a portobello mushroom: the meaty flavor doesn't solely come from meat.

While acids and bases are chemically relevant to many applications, one of the places we experience them most frequently is in our sensation of flavor. What we think of as the *taste* of food is mostly smell. Nuances of flavor, such as the vanilla and cinnamon flavors in sugar cookies or the complex combination of vegetables, meat, and cheese in lasagna, come from molecules in the food interacting with the *nasal epithelium*, which contains the smell-detecting receptors in our noses. We don't have the ability to literally taste many different flavors, as is well evidenced by the experience of eating food while suffering a bad head cold. When the nose is out of commission, we're reduced to five basic taste groups: sweet, sour, salty, bitter, and meaty.³ The tongue contains taste buds specialized for different flavors; these are roughly, but not completely, distributed into different tongue regions. We taste sweet best with the tips of our tongues, sour is most noticeable along the sides, salty is strongest on the sides and toward the front, and the back of the tongue is best at sensing bitter flavors. Our ability to recognize a meaty flavor is relatively newly discovered, and this sensation appears to be generalized across the entire tongue. Given that we use the nose for the majority of the flavor experience we get from food, what is the purpose of the tongue's ability to recognize these five basic flavors? It turns out that each of them provides us (unconsciously) with important information about what we're putting in our mouths!

IMAGE 5.3. The sweet flavor of cookies, the sour of oranges, saltiness of chips, bitter taste of broccoli, and meaty flavor of a nicely done steak are all chemical signals that help us instinctively eat the foods we need to, and avoid the foods that might harm us.

Sweet foods, as we'll learn in Chapter 8, contain sugars, which provide energy. **A sweet taste, therefore, lets us know that we're eating something capable of fueling our activities.** In fact, we respond favorably to sweet foods on a deep and instinctive level—cross-culturally, babies react positively to bottles of sugar water from the moment they're born. **Salty flavors in food alert us to the presence of NaCl, which provides the important mineral, sodium.** A typical modern Western diet is not lacking in sodium, but the mineral is somewhat difficult to come by in a hunter-gatherer's or forager's diet,[4] so the ability to sense the flavor of salt (and the corresponding positive response to this flavor in low concentration) helped to guide early humans toward foods that would provide necessary sodium. **Meaty flavors direct us toward consuming plenty of protein**, where proteins are large molecules used to build muscles and structural components of cells. **The remaining flavors—sour and bitter—are lent to foods by acid and base respectively.** The mildly sour flavor of an orange is due to the presence of citric acid and vitamin C (ascorbic acid). The much less tolerable, more intense tartness of a lemon comes from higher concentrations of those acids. Our ability to sense acid (via the sensation of sour) directs us toward potential sources of important nutrients such as vitamin C, while directing us away from the unpleasantly sour-tasting acids strong enough to do damage to our teeth and digestive tracts. Bitter flavors, such as those of some vegetables, coffee, and many herbs, are lent to foods by basic compounds (sometimes called *alkaloids*). While not all alkaloids are toxic, many toxic compounds are alkaloids. As a result, our ability to sense them through flavor helps us avoid ingesting toxins. Studies of newborns show that in addition to responding very favorably to sweet flavors, they find mildly sour and salty flavors to be interesting. They respond very unfavorably, however, to bitter flavors, indicating an instinctive avoidance of potential toxins. The ability to tolerate and even enjoy the bitter flavors of some foods, including broccoli, tea, coffee, and beer, is developed over time—often late in childhood or into early adulthood—as we become capable of overruling our instincts.

NOTE[4]

Our word "salary" comes from the Latin for salt (sal). Important to the diet and pleasurable in food but difficult to come by away from the sea coast, salt was once the currency in which Roman soldiers were paid.

5.5 Molarity and Concentration

We saw in Chapter 4 that the proportion of solute to volume of solution is referred to as a solution's concentration. **The most common quantification of concentration in chemistry is molarity (M), which is defined as moles of solute per liter of solution:**

$$molarity = \frac{moles\ of\ solute}{liters\ of\ solution}$$

TRY THIS

What is [CaCl$_2$] if 2.5L of solution contain 1.3 mol CaCl$_2$?

Answer: [CaCl$_2$] = 0.52M.
Solution:

$molarity = \dfrac{1.3\ mol\ CaCl_2}{2.5\ L\ solution} = 0.52\ \dfrac{mol}{L} = 0.52\,M$

Example: What is the molarity of 4.0 L of aqueous solution containing 1.75 moles of NaCl?

$$molarity = \frac{1.75 \ mol \ NaCl}{4.0 \ L \ solution} = 0.44 \ \frac{mol}{L} = 0.44 \ M$$

We can also use brackets to indicate a reference to a solution's concentration. For the solution above, we'd write: [NaCl] = 0.44M.

We can use molarity as a conversion factor to determine how many moles of a solute are in a given volume of solution, or what volume of solution would contain a given amount of solute.

Example: How many moles of NH_3 are in 2.0L of a 0.50M solution?

$$2.0 \ L \ solution \ x \ \frac{0.50 \ mol \ NH_3}{1 \ L \ solution} = 1.0 \ mol \ NH_3$$

Example: What volume of a 0.85M NaCl solution contains 2.0 mol NaCl?

$$2.0 \ L \ mol \ NaCl \ x \ \frac{1 \ L \ solution}{0.85 \ mol \ NaCl} = 2.35 \ L \ solution$$

TRY THIS

1.5L of solution contains how many moles of KF, if [KF] = 0.75M?

Answer: 1.1 mol KF.
Solution:

$$1.5 \ L \ solution \ x \ \frac{0.75 \ mol \ KF}{1 \ L \ solution} = 1.1 \ mol \ KF$$

TRY THIS

What volume of solution contains 0.50 mol NH_3, if [NH_3] = 1.25M?

Answer: 0.40 L solution.
Solution:

$$0.50 \ mol \ NH_3 \ x \ \frac{1 \ L \ solution}{1.25 \ mol \ NH_3} = 0.40 \ L \ solution$$

5.6   The pH of a Solution

Concentration and molarity are very relevant to our discussion of acid. We can use the concentration of H^+ in an acidic solution to calculate the **pH** of the solution. Most of us have heard of pH in reference to a solution's acidity; pH values fall between 0 and 14 for most routinely encountered solutions (Image 5.4). If a solution has a pH of 7, the solution is referred to as **neutral**; that is, it's neither acidic nor basic. Solutions with a pH greater than 7 are **basic**, and solutions with a pH less than 7 are **acidic**.

Mathematically, pH is calculated as:

$$pH = -\log[H^+]$$

Example: What is the pH of a solution for which [H^+] is 0.050M?

$$pH = -\log[H^+] = -\log(0.05.) = 1.3$$

Example: What is [H^+] in a solution of pH = 2?

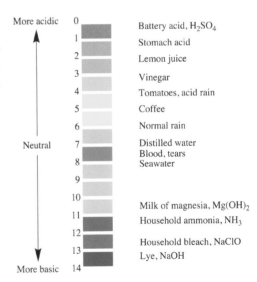

IMAGE 5.4. The pH scale. Common household items have pHs ranging from 0–14.

$$ph = 2 = -\log[H^+], 10^{-2} = [H^+] = 0.01M$$

There are two important points to notice regarding the pH scale. First, **a lower pH corresponds to a higher [H⁺], and a higher pH corresponds to a lower [H⁺]**. Second, because pH is a logarithmic function, a decrease of one pH point corresponds to a tenfold increase in [H⁺]. A two-point decrease in pH corresponds to a hundredfold increase in [H⁺], and so forth.

Example: How many times more [H⁺] is in a solution of pH 4 vs. in a solution of pH 7?

For pH 4: $\ ph = 4 = -\log[H^+], 10^{-4} = [H^+] = 0.0001M$

For pH 7: $\ ph = 7 = -\log[H^+], 10^{-7} = [H^+] = 0.0000001M$

[H⁺] is 1000 times higher in the solution of pH 4.

5.7   Neutralization

One important bit of chemistry involving acids and bases is the way they interact. When acids and bases react in solution, they **neutralize** each other. To understand this, remember that acids dissociate in water to produce H⁺, while bases "steal" H⁺ ions from water. Bases don't *need* to take H⁺ from water, in particular, but do so if water is all that's around. Put together with an acid, from which an H⁺ is easy to abstract due to the acid's tendency to dissociate, the base will react directly with the acid. Note in the reactions below that, left to its own devices, HCl dissociates (top reaction), and NaOH abstracts an H⁺ from water (middle). Together in solution, HCl and NaOH do not involve water in their reaction. Instead, NaOH abstracts a hydrogen directly from HCl (bottom):

Acid dissociates to produce H^+

Base dissociates to produce OH^-

When HCl reacts with NaOH, the OH^- from NaOH bonds with the H^+ from HCl, producing water. The Na^+ and Cl^- are left over, and form NaCl in solution.

Neutralization reactions involving bases that contain the OH^- anion produce water and an ionic salt. The salt is always formed by spectator ions—the reaction above, for instance, forms NaCl. An interesting application of acid neutralization is the use of antacids to relieve the symptoms of acid stomach. Our stomachs produce an acidic secretion called *gastric acid*, which consists largely of HCl in solution at a pH of approximately 2, and which is used to chemically break down the food we eat. Occasionally, insults to the stomach (including eating greasy or very spicy foods, or excessive alcohol consumption) can result in overproduction of gastric acid, leading to the sensation of heartburn or indigestion. This is treated by neutralizing the excess acid with a base, often magnesium hydroxide, $Mg(OH)_2$, commonly called *milk of magnesia*. When $Mg(OH)_2$ neutralizes HCl in stomach acid, water and an ionic salt are formed. In this case, because the spectator ions are Mg^{2+} and Cl^-, the ionic salt will be $MgCl_2$. The balanced chemical reaction will look like this:

$$2\ HCl + Mg(OH)_2 \rightarrow 2\ H_2O + MgCl_2$$

Note that since $Mg(OH)_2$ has *two* hydroxide units (each of which needs to react with hydrogen), *two* HCl molecules per magnesium hydroxide molecule are required to balance the reaction. This results in the formation of two molecules of water. Conveniently, it also provides the two chloride anions necessary to balance the charge on the magnesium cation so that a neutral salt can form.

Another common household remedy for acid indigestion is a solution of bak-ing soda and water. **Baking soda is the ionic salt $NaHCO_3$, sodium bicarbonate.** Bicarbonate (HCO_3^-) is a polyatomic anion. The neutralization reaction of HCl and $NaHCO_3$ is very similar to the reaction of HCl with a hydroxide-containing base. Just as it always does, HCl dissociates in aqueous solution to produce H^+ and Cl^-. $NaHCO_3$, like all ionic salts, dissociates in solution to produce Na^+ and HCO_3^-. HCO_3^- acts as a base, taking the H^+ liberated from HCl. Na^+ and Cl^- are spectator ions:

The products of this reaction are an ionic salt (NaCl), and the compound H_2CO_3, carbonic acid, which would be acidic except that it is quite unstable, and quickly reacts to form carbon dioxide and water:

$$H_2CO_3 \rightarrow CO_2 + H_2O$$

We saw this reaction going in the opposite direction in Chapter 2—carbon dioxide and water can react to produce carbonic acid. In fact, the direction in which the reaction proceeds is determined by the relative concentrations of the reactants and products. In a system in which there is lots of CO_2, the CO_2 will tend to combine with water to form carbonic acid. This is the reason that an individual who holds their breath and accumulates CO_2 in the bloodstream will produce carbonic acid, lowering the pH of the blood. If there is a lot of carbonic acid relative to the amount of CO_2, the carbonic acid will tend to fall apart. When an individual takes a bicarbonate antacid, even though CO_2 is formed, it's relatively promptly expelled via burping. As a result, CO_2 concentrations in the stomach stay low, and carbonic acid keeps reacting to form carbon dioxide and water. Since carbon dioxide from the antacid is expelled, the neutralization products left in the stomach are simply NaCl and water. As an antacid, baking soda is just as effective as milk of magnesia, but they each have their own unique side effects. The large amount of CO_2 generated by baking soda can lead to burping, and milk of magnesia is not only an antacid, but also a laxative.

CONCEPT CHECK

Why will a base react with an acid instead of with water in aqueous solution?

Answer: It's easier to take a hydrogen ion from acid than from water.

5.8   Chemistry of Baking

The production of carbon dioxide through the reaction of sodium bicarbonate and acid makes baking soda useful for much more than neutralizing stomach acid. An important part of baking is the ability to **leaven** bread, or cause the bread to rise. **This is accomplished by chemically generating gas within the dough matrix, resulting in air bubbles in the finished baked goods (Image 5.5).** Leavened baked goods—including cakes, cookies, and many breads—have a lighter texture than unleavened baked goods such as tortillas and pita bread.

One technique used to produce leavening is to incorporate baking soda into the dough mixture. In the presence of an acidic ingredient such as buttermilk or (a)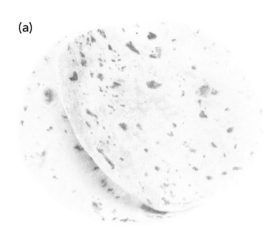

(b)

IMAGE 5.5. The tortillas on the left are unleavened and don't contain air bubbles in the dough. The French bread on the right, however, has been allowed to rise, resulting in its light texture. Gaps in the dough matrix formed by trapped carbon dioxide are clearly visible in the slice of bread.

yogurt (both of which contain lactic acid), the baking soda will neutralize the acid, producing carbonic acid:

$$CH_3CHOHCOOH \text{ (lactic acid)} + NaHCO_3 \rightarrow CH_3CHOHCOO^- \text{ (lactate)} + Na^+ + H_2CO_3$$

The carbonic acid then reacts further to produce water and carbon dioxide. As the dough bakes, the carbon dioxide gets trapped in the dough, expands, and causes the dough to rise:

$$H_2CO_3 \rightarrow CO_2 + H_2O$$

This reaction is quite rapid, and the dough can be baked immediately (in fact, breads made with baking soda are often called *quick breads*). Because of the rapid generation of carbon dioxide upon contact of baking soda with an acidic ingredient, however, it's important not to over-mix quick breads or cookies—to do so would prematurely release the CO_2, and it wouldn't get trapped in the dough.

A drawback to leavening baked goods with baking soda is that the leavening action requires an acidic ingredient. Since a common source of acid for cooking is lactic acid from partially fermented dairy products (which have a distinct sour flavor), this impacts the flavor of the final product. **An alternative to baking soda is** baking powder, **which is baking soda with added starch and an acidic salt, often cream of tartar or calcium hydrogen phosphate.** The acidic salt precludes the need to use a wet acid ingredient—the salt is activated (and becomes acidic) as soon as any moisture is added to the dry ingredient mixture.[5] The starch in baking powder helps to slow the reaction. It does not affect overall CO_2 production, but slowing the rate at which the gas is formed helps to space out the air bubbles more evenly and prevents flattening of bread due to over-mixing.

Understanding the chemistry of leavening allows us to make substitutions in recipes. **Where a recipe calls for baking soda, it's ok to use baking powder instead** (but because the baking powder consists of things other than $NaHCO_3$, it's generally necessary to add about three times as much baking powder as the recipe calls for

NOTE[5]

Kitchen tip: This is the reason many recipes instruct you to mix dry ingredients in one bowl, wet ingredients in a second bowl, and then combine the dry and wet right before baking—mixing separately helps to avoid activating the dry acid salt too soon.

soda). **Where a recipe calls for baking powder, though, it won't work to substitute baking soda, since this substitution reduces the amount of acid in the mixture.** Some chefs will add a bit of vinegar to a recipe—particularly if the taste will be covered up by other strong flavors or complemented by bitter flavors such as chocolate—to serve as an acid when cooking with baking soda rather than baking powder. Others will use lemon juice, and still others will replace part of the milk or water in a recipe with yogurt or buttermilk in a recipe that contains no other acidic ingredients.

Incidentally, baking soda and baking powder are not the only ways to leaven baked goods. A common alternative used more frequently in breads than in cookies or cakes is yeast. **Yeasts are fungal microorganisms** (saccharomyces cerevisiae) **that consume sugar for energy and release CO_2 as a waste product.** If yeast is given a source of nutrition and sufficient time, it can cause bread to rise through natural metabolic action. If you look at a bread recipe that's leavened with yeast, you'll notice a few key elements. First, the yeast (purchased dry) is reconstituted in water that has a bit of sugar stirred in. The sugar provides the yeast with a food source. Second, once the dough is mixed up, the yeast is given time to metabolize (digest) the sugar, and produce CO_2. Yeast breads often need to rise for an hour or more, and recipes recommend that the dough be kept covered in a warm location (in order to keep the yeast active) during the rising period.

IMAGE 5.6. Freshly mixed yeast dough (a), and the same dough after having been allowed to rise for nearly an hour.

5.9 Acid and Microbes

While yeasts metabolize sugar to produce carbon dioxide, other microbes, including the bacterial species *Lactobacillus acidophilus*, metabolize sugar to produce lactic acid. **A favorite sugar of the *L. acidophilus* bacterium is lactose, the sugar in milk.** While any sample of milk has a small bacterial load, the population of *L. acidophilus* and other bacteria steadily increases as milk sits in a container. The bacteria metabolize the milk sugar and produce lactic acid, which eventually causes

IMAGE 5.7. The holes in Swiss cheese (a) come from CO_2 produced by the bacteria of the family *Propionibacter*. Cheddar (b) is an aged, heavily salted cheese. The blue veins in Stilton (c) and other blue cheeses are due to the fungus *Penicillium roqueforti*.

the milk to curdle, or spoil. If this spoiling process is carefully controlled (and if only certain bacterial species are allowed to participate), the result is any one of a number of desirable products; yogurt, buttermilk, and cheese are all products of milk fermentation by *L. acidophilus*. Acid causes one of the proteins in milk, called casein, to collect into soft, semisolid clumps called *curds*. These can be eaten as they are (cottage cheese is a lightly fermented cheese curd mixture, rinsed of acid to reduce the sour taste) or further processed (sometimes with the help of additional bacteria or fungi) into other types of cheese (Image 5.7).

While some microbes are incredibly beneficial to humans, we are susceptible to infection by others. **A function of gastric acid in the stomach, beyond digesting food, is to help protect us from infection by ingested bacteria, many of which are sensitive to acid.** We harbor native bacterial populations in our digestive tracts, including several strains of *Escherichia coli*. Other strains of *E. coli*, however, are pathogenic (harmful to us), and if ingested via consumption of contaminated food, can cause infection. Some of the harmful strains (including *E. coli* O157:H7, responsible for numerous human deaths in recent years) produce toxins that can severely damage human organs. Thankfully, most *E. coli* are unable to survive in the stomach's low pH environment, and are killed shortly after they're swallowed.

In an interesting interaction between chemistry and society, however, our desire to provide ourselves with a ready supply of meat has resulted in the selective breeding of acid-tolerant strains of *E. coli* and an increased incidence of human infection by the bacteria. Cattle are one of a number of animals with multiple digestive compartments, in contrast to our single stomach. Unlike the very acidic stomachs of humans, the digestive tract of the cow is relatively neutral in pH when the cow is being fed its native diet of grasses and hay. However, high-calorie grains such as corn cause cattle to put on weight very quickly, which brings them up to appropriate slaughter size quickly. This has made corn and other grain popular as cattle feed on industrial farms and in concentrated animal feeding operations (CAFOs). While the corn feeding of cattle has ramifications far beyond its effect upon their weight and digestive tracts, it is nevertheless the digestive effect of the corn that is most relevant here. **Grain, it turns out, significantly acidifies the cows' digestive systems.** Not only is this harmful to the cow (resulting in ulceration of the digestive organs, among other things), it also provides the perfect environment for acid-tolerant bacteria. A particularly unappetizing bit of relevant information is that when cattle are slaughtered, meat often becomes contaminated with colonic bacteria. Under ordinary circumstances (in the case of a grass-fed cow) this is not particularly dangerous to consumers of the meat. The bacteria that live comfortably in a cow's neutral digestive tract are quickly killed by the acidic environment of the human stomach. **Corn-fed beef, however, becomes contaminated at slaughter with acid-tolerant bacteria that are *not* killed upon contact with stomach acid, and if these bacteria are of a toxic strain (such as *E. coli* O157:H7) they will produce potentially lethal infections in humans.**

5.10 Acid Spills and Industrial Acid Disasters

We've seen so far that acids are important and useful to us in many ways, including food preparation and chemical protection from microbe infection. Unfortunately, however, the very property of acid that makes them dangerous to microbial invaders also makes them damaging to structures and nonmicrobial living creatures. Acids, particularly if concentrated, have the ability to erode solid surfaces and burn living tissue. As a result, acid spills and accidents can have disastrous consequences on humans and on the environment.

HF, hydrofluoric acid, is a common industrial acid used in oil refining (Image 5.8). It is also used in the production of stainless steel and to clean silicon for use in semiconductors. Due to some of its unique properties as an acid, however, HF also represents a unique toxicity liability in the case of an accidental spill.

Many acids, including HCl (hydrochloric acid) and H_2SO_4 (sulfuric acid), react at the surface of the skin, causing a painful burn on contact. Unlike most acids, however, HF is absorbed into deep tissues, where it interferes with nervous system

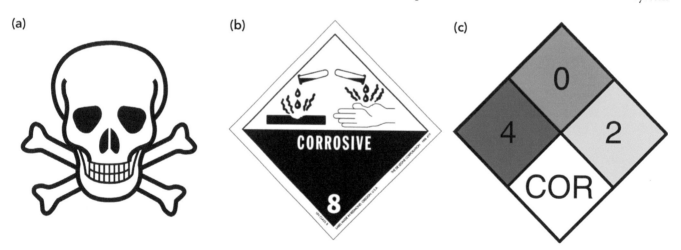

IMAGE 5.8. Symbols commonly associated with hydrofluoric acid. The skull and crossbones represent toxicity, while the "corrosive" symbol indicates the acid's ability to damage surfaces and tissue. Four-colored "fire diamonds," like the diamond for HF shown here, are standard mechanisms for providing information about chemicals and are found on chemical bottles and transport vehicles (as well as anywhere else industrial chemicals are routinely encountered). The number in red is an indication of flammability, blue represents health hazard, and yellow indicates reactivity. Additional important information is abbreviated in the white area. Chemicals are ranked on a scale of 0–4 in each of these categories. HF is not flammable, represents a maximum health hazard, is moderately reactive, and is corrosive.

function and often prevents the burn victim from feeling the extent of their injuries. This may delay burn treatment, which exacerbates tissue damage. Further, upon dissociation, the F⁻ ion has the ability to bind to Mg^{2+} and Ca^{2+} ions, which are found in the bloodstream. Ca^{2+}, in particular, is vital to normal biological function, and formation of CaF_2 (which has very low solubility in the blood, and precipitates out of solution) leads to lowered blood concentrations of Ca^{2+}. This can result in extreme tissue damage and pain, and can even lead to complications such as cardiac arrhythmias (irregular heartbeat) in severe cases. To prevent the formation of CaF_2 from blood calcium, HF burns are generally treated with the chemical calcium gluconate, which is an ionic salt of Ca^{2+}. F⁻ therefore forms ionic bonds with the Ca^{2+} from the calcium gluconate (simply because there's a lot of it), which prevents it from bonding to blood calcium. Of course, large-scale spills of hydrofluoric acid represent a tremendous danger to the surrounding community, as evidenced by the 2009 evacuation of 5,000 local residents following the spill of 33,000 pounds of hydrofluoric acid when a truck overturned near Plainfield, Pennsylvania.

IMAGE 5.9. Acid for industrial applications is transported in containment vehicles like this one.

5.11 The Last Word—Fermentation

(a)

Earlier in this chapter, we discussed that yeast consume sugar and can be used to produce carbon dioxide for leavening baked goods. The other by-product of sugar metabolism by yeast is ethanol (CH_3CH_2OH), which is the alcohol in beer and wine. While ethanol concentrations in yeast bread are so low as to be inconsequential (the alcohol bakes out, as it has a low boiling point), the very same chemical reaction responsible for leavening bread also produces alcohol and carbonation (CO_2 dissolved in liquid) in alcoholic beverages. As in baking, the yeast requires a source of sugar. Commonly used sources include malted barley (to make beer) and pressed grapes (to make wine). There are many different strains of yeast; specifically, the organism *Saccharomyces cerevisiae* is used in many food applications.

Why do yeast produce carbon dioxide and ethanol? With regard to carbon dioxide, yeast makes it for the same reason we do—chemically breaking down sugars[6] for energy results in the production of CO_2. In fact, most organisms that use sugar (or starch) for energy produce carbon dioxide waste. Though yeast are not the only organism that produces CO_2, they are used for leavening bread because they are noninfectious to us, and because they are quite small and can easily be added to a dough mixture. **The fact that they produce ethanol (which animals do not) is a function of one of their particular metabolic mechanisms, called fermentation.** Early cultures discovered fermentation—and the interesting effects of consuming fermented beverages—when airborne yeast got into stored containers of food. Consumption of these yeast-fermented foods and beverages resulted in the feelings we associate with consumption of beer and wine. Just as many individuals now find these effects enjoyable, so did early peoples, and they worked to determine ways of purposely fermenting various substances. Any starch-containing substance can be

IMAGE 5.10. Beer and wine are both produced through the same reaction that leavens bread: yeast consume sugar and produce carbon dioxide and ethanol (alcohol) as by-products.

(b)

NOTE[6]

While the chemical nature of sugars and starches will be discussed in more detail in Chapter 8, it's worth knowing that starches are chemically very similar to sugars and are processed by organisms in much the same way.

fermented. While wine and beer are made by many different cultures, grapes and grain are by no means the only fermentable foods. Geographical and cultural regions have characteristic alcoholic drinks, representative of their most ready source of fermentable sugar or starch. For instance, vodka can be made from potatoes, rum from sugar cane, and tequila from the agave plant. Some Asian cultures even ferment mare's milk.

Cultures throughout the world find ethanol to have pleasant side effects, including social relaxation, euphoria, and a feeling of well-being. **However enjoyable its effects may be, ethanol is technically a toxin; for this reason, individuals under the influence of alcohol are referred to as *intoxicated*.** Ethanol has a short-term effect on the nervous system, affects consciousness and thought, and changes the activity of certain brain chemicals. In high doses, it can cause acute alcohol toxicity, which includes decreased blood flow to the brain, depressed breathing, unconsciousness, and even death. Because ethanol is toxic, the liver (our major detoxification organ) works hard to remove it from the body. As a result, excessive alcohol use over time severely damages the liver, and can lead to *cirrhosis* (scarring of the liver). Interestingly, ethanol isn't just toxic to us; it's also toxic to yeast. While low concentrations of ethanol are tolerable to yeast, they die at higher concentrations. For this reason, it used to be rare to find wine containing more than 12–13% alcohol; that was the maximum concentration the yeast could survive. Now, however, wines commonly contain up to 17% alcohol, thanks to the efforts of intrepid biochemists and microbiologists who have been selectively producing strains of yeast with higher ethanol tolerance!

SUMMARY OF MAIN POINTS

- Acids dissociate in water to produce an anion and the H^+ cation.
- Most acids are binary (usually hydrogen with a halogen), oxyacids, or carboxylic acids (-COOH).
- When an acid dissociates, the electrons from the bond to hydrogen remain with whatever atom hydrogen was bonded to. This produces an anion.
- Ions composed of two or more atoms covalently bonded to one another are called polyatomic ions.
- Polyatomic ions form ionic compounds with other ions—compounds can be predicted from the charges of the ions involved, and nomenclature is comparable to that of binary ionic compounds.
- Bases react with water, taking a hydrogen from it to produce the OH^- anion.
- Our perception of flavor is primarily based on smell, but there are five main tastes; each one provides important information: sweet (sugar), sour (acid), salty (sodium), bitter (potential toxin), and meaty (protein).
- The concentration of a solute in a solvent is often measured in moles of solute per liter of solution and is called the solution's molarity (mol/L, or M).
- The acidity of a solution is related to the molarity of H^+, and is measured as pH, where $pH = -\log[H^+]$.

- Typical pH values range from 0–14, where lower pH corresponds to a more acidic solution, and a pH of 7 is neutral (distilled water).
- Acids and bases react with each other through neutralization reactions.
- Common examples of neutralizations include antacids used for heartburn and the leavening action of baking soda (base) with an acid.
- Baked goods can also be leavened with yeast, which metabolize sugar to produce CO_2.
- Lactic acid is produced by some microorganisms and is used in the food industry to make cheese.
- Acid in the stomach protects us from microorganism infection, but conditions that select for acid-tolerant bacteria (such as the acidic digestive tract of corn-fed cattle) produce bacteria that can't be killed by our acidic stomach, which can cause epidemic infection.
- Acids have the potential to be very damaging, particularly in the case of large-scale spills or industrial accidents.
- Hydrofluoric acid is a commonly used industrial acid that, if spilled, has the potential to be highly toxic or deadly.

QUESTIONS AND TOPICS FOR DISCUSSION

1. In what ways does hydrogen act like a group 1 element? Like a group 7 element? How should it be grouped on the periodic table?

2. Do you think the average person thinks of acid as something good? Something bad? What do you think *most* people think of?

3. In general, how do acids behave in water?

4. Some acids are stronger than others. For instance, HCl is a stronger acid than H_2S. Based upon what you know about the periodic table, propose a reason.

5. Is the bond between H and Cl in HCl more properly characterized as covalent or ionic? Justify your answer.

6. Why do you think it becomes increasingly important to write out chemical formulas in atomic groupings (i.e., CH_3COOH as opposed to $C_2H_4O_2$) as molecules get larger?

7. What are some of the acids you've encountered in everyday life, and where have you come across them?

8. Why do polyatomic ions have charges? Use the example of nitrate, below, to compare the number of protons to the number of electrons (total!) in the molecule:

9. Why does the polyatomic ion ammonium (NH_4^+) form when ammonia reacts with water?

10. In general, how do bases behave in water?

11. Do you think the average person thinks of base as something good? Something bad? What do you think *most* people think of?

12. In the reaction of KOH with water, is there a spectator ion? If so, what is it, and what makes it a spectator?

13. Why is it important for us to be able to taste the basic flavors of sweet, sour, salty, bitter, and meaty?

14. What information do we get about our food from the basic flavors (sweet, sour, salty, bitter, and meaty)?

15. Under what circumstances are sweet, sour, salty, and bitter pleasant tastes? Under what circumstances (if any) are they unpleasant?

16. Why do you think it's important to use smell to get information from our food above and beyond the five basic flavors?

17. Describe the sensation of eating with a stuffy nose. If you can't remember the last time you did so, try eating while holding your nose. How does this compare to a typical eating experience?

18. Think of three bitter foods that children typically don't like, but that adults tend to enjoy. Do a little Internet research—are these foods alkaloid containing?

19. Why do adults train themselves to enjoy bitter flavors? (You may need to identify some common foods that children don't enjoy, but adults do, in order to determine the motivation).

20. Why is it necessary to have a measure of concentration for solutions?

21. Molarity is defined as moles of solute per liter of solution. Would a solution made by adding 1 mol of NaCl to 1 liter of water be 1 M? Why or why not?

22. What is the difference in hydrogen ion concentration between a solution of pH = 1 and pH = 2? Explain.

23. Which is more acidic: a solution of pH = 1 or a solution of pH = 2? Explain.

24. Why does pH decrease as acidity increases?

25. Why do you think the pH scale is logarithmic rather than linear? What purpose does this serve?

26. Explain the general mechanism through which a base reacts with an acid. What is this kind of reaction called?

27. How are acid-base neutralizations useful in everyday life? Think of an example besides those mentioned in the text.

28. Find a recipe that calls for baking soda but not baking powder (you may need to use the Internet as a reference). Is there a source of acid in the recipe? If so, what is it? How could you modify the recipe if you had no baking soda?

29. Find a recipe that calls for baking powder but not baking soda (you may need to use the Internet as a reference). Is there a source of acid in the recipe? If so, what is it? How could you modify the recipe if you had no baking powder?

30. How can yeast be used to make bread dough rise?

31. How can microorganisms be used to make cheese?

32. Why is the human stomach acidic; what function(s) does this serve?

33. Why does the food cattle eat affect the likelihood of human infection upon eating meat from the cattle?

34. Using the Internet as a resource, what are the negative ramifications of feeding cattle grain other than favoring growth of acid-tolerant bacteria?

35. What are some of the applications for which hydrofluoric acid is useful?

36. Why does applying calcium gluconate cream to a hydrofluoric acid burn help to mitigate the damage?

37. How does hydrofluoric acid act similarly to other acids in the case of a burn? How does it act differently?

PROBLEMS

1. Which is the acidic hydrogen in CH_3CH_2COOH? How do you know?

2. Using arrows to indicate where electrons in bonds go, show the behavior of each of the following acids in water: CH_3COOH; HBr; and HI.

3. Using arrows to indicate where electrons in bonds go, show the behavior of each of the following acids in water: CH_3CH_2COOH; HF; and H_2S.

4. Define each of the following: acid; polyatomic ion; binary acid; oxyacid; and carboxylic acid.

5. Categorize each of the following as a binary acid, oxyacid, or carboxylic acid: HF; H_3PO_4; CH_3CH_2COOH; and H_2S.

6. Categorize each of the following as a binary acid, oxyacid, or carboxylic acid: HI; HBr; $HClO_4$; and HCOOH.

7. Provide the chemical formula and name for a compound made of each of the following pairs: ammonium and bromine; ammonium and carbonate; sulfate and cesium; and carbonate and magnesium.

8. Provide the chemical formula and name for a compound made of each of the following pairs: lithium and hydroxide; phosphate and sodium; phosphate and ammonium; and nitrate and magnesium.

9. Provide the chemical formula for each compound: rubidium phosphate; ammonium bicarbonate; calcium sulfate; and calcium phosphate.

10. Provide the chemical formula for each compound: barium carbonate; barium nitrate; strontium phosphate; and sodium hydroxide.

11. Provide the name for each of the following: NH_4NO_3; $LiHCO_3$; Li_3PO_4, and $(NH_4)_3N$.

12. 12. Provide the name for each of the following: NH_4Cl; $Ca(NO_3)_2$; $SrSO_4$; and $Sr(OH)_2$.

13. Using arrows to indicate where electrons in bonds go, show how LiOH behaves in water.

14. When KOH reacts with water, the products are the same as the reactants. Why is this considered a chemical reaction, rather than a "no reaction"?

15. Using arrows to indicate where electrons in bonds go, show how CH_3NH_2 behaves in water (Hint: the N acts just like the N on ammonia).

16. Calculate the molarity of a solution containing 1.5 mol KF in 2.5 L.

17. Calculate [HCl] in a solution containing 0.25 mol HCl in 5.2 L.

18. Calculate [NaCl] in 5.0 L of a solution containing 3.5 mol NaCl.

19. Calculate $[NH_3]$ in 3.5 L of a solution containing 1.25 mol NH_3.

20. How many moles of $Ca(OH)_2$ are in 0.75 L of a 2.5 M solution of $Ca(OH)_2$?

21. How many moles of $NaHCO_3$ are in 1.4 L of solution, if $[NaHCO_3] = 1.2$ M?

22. What volume of a 2.2 M solution of LiOH contains 0.65 mol LiOH?

23. What volume of a 0.45 M solution of $NaNO_3$ contains 1.5 mol $NaNO_3$?

24. What is the $[H^+]$ in a solution of pH = 2.25?

25. What is the hydrogen ion concentration in a solution of pH = 12.4?

26. What is the pH of a solution in which $[H^+] = 2.2 \times 10^{-2}$ M?

27. What is the pH of a solution with a hydrogen ion concentration of 5.7×10^{-13} M?

28. What is the pH of a 1 M solution of HCl (assume all the HCl dissociates)?

29. What is the pH of a 0.01 M solution of HCl (assume all the HCl dissociates)?

30. How does $[H^+]$ in a solution of pH = 6 compare to that of pH = 9?

31. How does $[H^+]$ in a solution of pH = 2 compare to that of pH = 4?

32. Which pH values are acidic? Basic? Neutral?

33. What does it mean (in terms of the chemicals present in solution) to say a solution is acidic?

34. What does it mean (in terms of the chemicals present in solution) to say a solution is basic?

35. What does it mean (in terms of the chemicals present in solution) to say a solution is neutral?

36. Using arrows to show where electrons in bonds go, show the behavior of each of the following pairs in water: LiOH and HI; HBr and NH_3; and HCOOH and $NaHCO_3$.

37. Using arrows to show where electrons in bonds go, show the behavior of each of the following pairs in water: HCl and KOH; NH_3 and CH_3COOH; and HBr and $NaHCO_3$.

38. Write out a balanced chemical equation for the reaction of $Ca(OH)_2$ and HCl.

39. Write out a balanced chemical equation for the reaction of HNO_3 and NaOH.

40. Write out a balanced chemical equation for the reaction of NH_3 and HI.

41. Write out a balanced chemical equation for the reaction of H_2SO_4 and LiOH (Hint: both Hs will dissociate from the sulfuric acid).

42. Write out a chemical reaction showing how baking soda ($NaHCO_3$) produces carbonation for leavening bread. Use CH_3COOH as your acid.

1. The last H (from the COOH); this is always the acidic H in the case of a carboxylic (-COOH) acid.

3.

5. HF is a binary acid; H_3PO_4 is an oxyacid; CH_3CH_2COOH is a carboxylic acid; H_2S is a binary acid.

7. NH_4Br, ammonium bromide; $(NH_4)_2CO_3$, ammonium carbonate; Cs_2SO_4, cesium sulfate; $MgCO_3$, magnesium carbonate.

9. Rb_3PO_4; NH_4HCO_3; $CaSO_4$; $Ca_3(PO_4)_2$.

11. Ammonium nitrate; lithium bicarbonate; lithium phosphate; ammonium nitride.

13.

15.

17. 0.048 M.
19. 0.36 M.
21. 1.7 mol.
23. 3.3 L.
25. 3.98×10^{-13} M.
27. pH = 12.2.
29. pH = 2.
31. pH = 2 has 100x higher $[H^+]$ than pH = 4.

33. Acidic solutions have concentrations of H⁺ greater than 10^{-7}.

35. Neutral solutions have concentrations of H⁺ equal to 10^{-7}.

37.

39. $HNO_3 + NaOH \rightarrow H_2O + NaNO_3$

41. $H_2SO_4 + 2\ LiOH \rightarrow 2\ H_2O + Li_2SO_4$

SOURCES AND FURTHER READING

Taste

Steiner, J. E. (1974). Innate, discriminative human facial expressions to taste and smell stimulation. *Annals of the New York Academy of Sciences, 237,* 229–33.

Pathogens and Acidity

Callaway, T. R., Carr, M. A., Edrington, T. S., *et al.* (2009). Diet, *Escherichia coli* O157:H7, and cattle: A review after 10 years. *Current Issues in Molecular Biology, 11,* 67–79.

Diez-Gonzales, F., Callaway, T. R., Kizoulis, M. G., *et al.* (1998). Grain feeding and the dissemination of acid-resistant Escherichia coli from cattle. *Science, 281*(5383), 1666–68.

Pennsylvania Acid Spill

CNN. (2009, March 21). Acid spill evacuation ends for 5,000 Pennsylvania residents. Retrieved from http://www.cnn.com/2009/US/03/21/pennsylvania.spill/index.html.

Yeast and Alcohol

Moneke, A. N., Okolo, B. N., Nweke, A. I., *et al.* (2008). Selection and characterization of high ethanol tolerant Saccharomyces yeasts from orchard soil. *African Journal of Biotechnology, 7*(24), 4567–75.

Nevoight, E. (2008). Progress in metabolic engineering of Saccharomyces cerevisiae. *Microbiology and Molecular Biology Reviews, 72*(3), 379–412.

Pretorius, I. S., (2000). Tailoring wine yeast for the new millennium: Novel approaches to the ancient art of winemaking. *Yeast, 16*(8), 675–729.

CREDITS

The Chemistry of Pollution

6

If you live in a major metropolitan area, no doubt you've looked out over the city and seen a dense, smoggy cloud obstructing your view. This cloud (which can vary in color from white to blue through gray to brown) is composed of various atmospheric pollutions that are produced as the result of human activity.

In addition to being aesthetically unappealing, pollution is harmful. Studies show that high levels of air pollution are associated with greater numbers of hospitalizations for headache, cardiovascular events, respiratory difficulty and respiratory distress—particularly among asthmatics and sufferers of chronic obstructive pulmonary disease (COPD). Pollutants produced through the combustion of coal and oil have also been identified as carcinogens. In addition to direct health effects, pollution results in the formation of acid rain, leads to climate change and alteration of the environment, and depletes the ozone layer.[1] A discussion of pollution raises several important questions. From a chemical perspective, what is pollution? Where does it come from, and of what is it composed? How is it harmful to individuals and the environment? Most importantly, what can we do as a society to reduce the amount of pollution produced?

IMAGE 6.1. Santiago, the capital of Chile, lies in a valley between the Chilean coastal mountain range on the west and the towering Andes on the east. Thermal inversions in the winter trap pollutants in the valley, making Santiago one of the most polluted cities in South America.

NOTE[1]

There are many pollutant compounds, each with a different effect upon the atmosphere and environment. In this chapter, we'll focus only upon the toxicity of these substances. Climate change and ozone depletion will be addressed in later chapters.

6.1 Common Pollutants—Carbon Monoxide

One theme we'll see as we begin our discussion of pollutants is that the vast majority of them are products of combustion reactions, either directly or indirectly. Recall from Chapter 1 that combustion is any chemical reaction in which something (often a hydrocarbon or hydrocarbon derivative) burns in the presence of oxygen. We've already seen that methane (a gaseous hydrocarbon that is the major component of natural gas) burns in oxygen by the reaction:

$$CH_4 + 2\,O_2 \rightarrow CO_2 + 2\,H_2O$$

In an analogous reaction, octane, a hydrocarbon constituent of gasoline, also combusts:

$$2\,C_8H_{18} + 25\,O_2 \rightarrow 16\,CO_2 + 18\,H_2O$$

The combustion reactions shown above, however, are somewhat idealized. Focusing on the combustion of octane for a moment, a significant amount of oxygen (25 molecules for every two molecules of octane) is required to completely transform octane to CO_2 and H_2O. Because combustion is an exceedingly rapid reaction, fuels of all sorts (including coal, natural gas, and gasoline) also burn via an alternate reaction that requires less O_2. Octane, for instance, can react as follows:

$$2\,C_8H_{18} + 17\,O_2 \rightarrow 16\,CO + 18\,H_2O$$

Here, the major product aside from water is CO—carbon monoxide. In reality, the combustion of octane results in the production of both carbon dioxide and carbon monoxide, as the two reactions shown above above occur simultaneously. Carbon monoxide is a colorless gas with no discernible odor. As an environmental pollutant, however, it's quite toxic. **Carbon monoxide's toxicity arises in part from its structural similarity to oxygen:**

Body cells require oxygen in order to function, and hemoglobin (a protein[2] in red blood cells, Image 6.2) is responsible for carrying oxygen to the cells. In order to transport oxygen effectively, hemoglobin must chemically bind O_2 at the lungs, and release O_2 at the tissues. As a result of, among other things, its structural similarity to O_2, CO can also bind to hemoglobin. However, CO binds so strongly that, unlike oxygen, it is not released by the hemoglobin for a long period of time (many hours). Hemoglobin with CO bound is unavailable to transport O_2. **If sufficient CO is inhaled into the lungs, a large percentage of the hemoglobin in the body will be saturated with CO and will therefore be incapable of delivering oxygen to the tissues; the result of this is carbon monoxide poisoning, which is a form of chemical suffocation.** Mild carbon monoxide poisoning produces headache, nausea, and weakness, as well as exacerbation of existing cardiovascular or respiratory disease. More severe cases cause confusion, seizures, and death. The treatment for carbon monoxide poisoning is the administration of high-concentration oxygen; this helps to saturate the lungs with O_2, and dislodges CO from hemoglobin approximately four times faster than it would otherwise be dislodged.

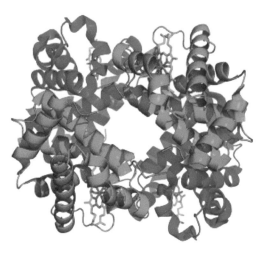

IMAGE 6.2. Hemoglobin, a large molecule, transports oxygen from the lungs to the tissues. It's made of protein (shown as red and blue ribbons) and four molecules called porphyrins (shown in green). Each porphyrin has an Fe^{2+} ion at the center that binds a molecule of O^2.

Carbon monoxide is both an indoor and an outdoor pollutant. Indoors, unvented heaters, leaking chimneys, wood-burning stoves and fireplaces, and gas-powered equipment can result in high levels of carbon monoxide. The EPA sets limits for acceptable levels of environmental pollution due to various contaminants. While there are no regulatory standards for indoor carbon monoxide, the EPA nevertheless recommends that measures be taken (particularly during the winter months and in colder climates) to reduce potential indoor exposure. Outdoors, carbon monoxide sources include smoke from fires, coal-fired power plants, and automobile exhaust. In particular, carbon monoxide builds up as a result of heavy and congested traffic. In order to reduce CO emissions from vehicles, the catalytic converter[3] was invented in the 1950s and refined in the 1970s. A catalytic converter functions by combusting carbon monoxide in more oxygen to produce CO_2:

$$2\,CO + O_2 \rightarrow 2\,CO_2$$

Catalytic converters serve the further purpose of reacting any uncombusted hydrocarbon fuel that makes its way into the exhaust system. It's worth mentioning that while catalytic converters are fairly efficient at converting carbon monoxide to carbon dioxide, we mustn't assume that CO_2 is a harmless solution to the problem: as we'll discuss in Chapter 13, CO_2 is a potent greenhouse gas and has been identified by scientists as a major cause of global warming.

NOTE[3]

The word *catalyst*, as in catalytic converter, refers to a substance that speeds up a chemical reaction without being consumed in the reaction. In the case of an automobile catalytic converter, the catalyst in question is generally a metal—often platinum, palladium, or rhodium. We'll see catalysts again in Chapter 10.

6.2 Common Pollutants—NOx

NOx (pronounced "nocks") is the collective term for mononitrogen oxides—chemicals consisting of one nitrogen atom and one or more oxygen atoms. NOx are a subclass of the nitrogen oxides, which are binary compounds of nitrogen and oxygen, and include NO (nitrogen monoxide), NO_2 (nitrogen dioxide), N_2O[4] (dinitrogen monoxide), and several others. NOx are produced when atmospheric nitrogen combines with oxygen in various proportions. Despite the abundance of elemental nitrogen (78% of the atmosphere), its triple bond makes N_2 quite inert, or chemically unreactive. Under very high temperature conditions, however, O_2 decomposes to form highly reactive atomic oxygen, which in turn can react with elemental nitrogen to form NOx. The required conditions for this reaction exist in a lightning storm (a natural source of NOx) or in the combustion chamber of an engine or coal-fired power plant. In order to provide oxygen for the fuel reaction, combustion chambers have an air intake. Since air is a mixture of nitrogen and oxygen (as well as traces of other substances), and since the temperature of an internal combustion engine can easily exceed 2000K, all the required pieces for formation of NOx are in place:

NOTE[4]

N_2O is also commonly called *nitrous oxide*, or, more colloquially, *laughing gas*. It's sometimes used by dentists during routine procedures, in part because it is an anesthetic and reduces the sensation of pain. Because many dental procedures are not so much painful as they are psychologically uncomfortable, however, nitrous oxide's popularity may be due mostly to its powerful relaxant effect. The chemical has the ability to produce feelings of euphoria and dissociation—essentially, a sense of not particularly caring what a medical professional is doing to one's teeth!

$$O_2 + heat \rightarrow 2\,O \text{ (this reaction begins the process)}$$

$$O + N_2 \rightarrow NO + N \text{ (this reaction couples with the next to form a continuous loop)}$$

$$N + O_2 \rightarrow NO + O$$

NO and NO_2 can be interconverted by some relatively complex chemistry. Suffice it to say, while most NOx are emitted as NO, the rate of conversion to NO_2 is rapid, and within minutes of emission, both NOx species are well represented in exhaust. **NO_2 is quite toxic: exposure causes respiratory irritation, particularly in individuals with underlying respiratory conditions.** Long-term exposure can result in permanent damage to the lungs. It also has an acrid odor and a yellowish-brown color that can be observed in many urban pollution clouds (Image 6.3).

In an attempt to reduce the environmental impact of the NOx pollutants (which as we'll see later in this chapter are also implicated in acid rain formation), modern catalytic converters decompose NO and NO_2 back to N_2 and O_2:

$$2\,NO \rightarrow N_2 + O_2$$

$$2\,NO_2 \rightarrow N_2 + 2\,O_2$$

Unfortunately, the catalytic material used to effect the reactions above can also result in the formation of several undesirable products, including NH_3, which is toxic upon inhalation, and N_2O, a potent greenhouse gas.

6.3 Common Pollutants—Volatile Organic Compounds and Polycyclic Aromatic Hydrocarbons

Volatile organic compounds (VOCs) are small carbon-based molecules (organic means carbon based) that vaporize easily and enter the atmosphere. Polycyclic aromatic hydrocarbons (PAHs) are large carbon-based molecules. **VOCs and PAHs are large classes of pollutant compounds, as opposed to pure substances, but are classified as groups by the EPA and other regulatory agencies due to the similarities in their sources and in the hazards they present.**

VOCs are both indoor and outdoor pollutants. Indoors, common pollutant culprits include formaldehyde (CH_2O), an industrial solvent that leeches into the air from building materials and irritates mucous membranes.[5] Outdoors, VOCs come largely from uncombusted hydrocarbon fuel released in the exhaust of engines and coal-fired power plants. The EPA has found that outdoor VOC concentrations rarely reach levels of concern with regard to human exposure. Indoor VOC concentrations, by contrast, are often many times higher than outdoor concentrations, and activities that result in increased shedding of organic compounds from building materials—such as paint stripping—can take the indoor VOC concentration to 1,000 times those found outdoors. Regardless of the low outdoor concentration of these compounds, they nevertheless lead to formation of another atmospheric pollutant: ozone (see below).

As a class, PAHs are much larger molecules than VOCs. Like VOCs, they are products of incomplete fuel combustion, the primary sources of which are motor vehicle engines, coal-fired power plants, and residential wood and coal burning. **PAHs are of concern as airborne pollutants because of their documented capacity as mutagens (substances that cause genetic mutation) and carcinogens (substances that cause cancer).**

6.4 Common Pollutants—Ozone

In Chapter 10, we'll see that ozone plays a vital role in protecting our planet and its living occupants from overexposure to damaging radiation from the sun. The ozone we depend upon to protect us, however, is not at ground level. **Instead, our protective ozone layer is sequestered in the stratosphere, well away from Earth's occupants. Tropospheric ozone, on the other hand, is an environmental pollutant.**

Ozone (O_3) is an allotrope of oxygen, meaning that it's an alternate form of the element. It's a bluish-gray gas (the color of which can sometimes be observed in summer smog clouds over sunny, hot metropolitan areas) with a sharp, almost metallic odor.[6] Ozone is far less stable than O_2, and in addition to being highly reactive, it's a toxic respiratory irritant. Its effects are particularly hazardous to asthmatics and others with chronic respiratory disorders. In sufficient concentrations, however, respiratory irritation from ozone is felt even by those with completely

healthy lungs. **Increased risk of infection, shortness of breath (particularly during exertion), and coughing or sore throat are common symptoms of ozone exposure.**

Ozone, unlike NOx and CO, is not a direct product of combustion. Instead, it forms when VOCs and NOx react with the sun's ultraviolet radiation (indicated in chemical reactions by the shorthand hv[7]). NO_2 reacts as below:

$$NO_2 + hv \rightarrow NO + O$$

$$O + O_2 \rightarrow O_3$$

The reactions by which VOCs produce ozone are quite complex and involve chemicals called **free radicals**,[8] which are highly reactive chemicals with unpaired electrons. In short, VOCs accelerate conversion of NO (a pollutant in its own right, but not directly implicated in ozone production) to NO_2, thus increasing concentrations of NO_2 available to react with sunlight and produce ozone.

NOTE[7]

Why is hn used to represent ultraviolet radiation in chemical equations? Actually, it's used to represent *any electromagnetic radiation*. It is interpreted to mean *radiation of the appropriate wavelength*, and it's assumed that the individual reading the chemical equation will know (or find out) what that radiation is. This concept will be discussed further in Chapter 10, but for now, hn in a chemical equation can be taken to mean *light energy*.

6.5 Common Pollutants—Particulate Matter (PM_{10} and $PM_{2.5}$)

Particulate matter (PM) is the collective term for solid particles and liquid droplets that may become suspended in air. A variety of materials may be categorized as particulate matter: soot, dust, aerosolized metal, smoke, organic chemicals, and even small molecules.

NOTE[8]

You've likely come across the term *free radical* in the context of nutrition and wellness; many products advertise that they reduce the activity of free radicals in the body and thus diminish the effects of aging, risk of cancer, and so forth. We'll discuss this concept further in later chapters.

IMAGE 6.4. PM^{10} and $PM^{2.5}$ contribute to the opaque nature of this pollution from a processing plant in China.

(a)

(b)

Particulate matter is divided into two classes. *Inhalable coarse particles*, also called PM_{10}, are particles of diameters greater than 2.5 μm (μm is the symbol for a *micrometer*, which is 10^{-6} m, or 1/1000 of a millimeter) but less than 10 μm. These particles are able to pass into the lungs, where they can severely affect health. There are also *fine particles*, or $PM_{2.5}$, which have a diameter less than 2.5 μm. As these particles are smaller than PM_{10}, they penetrate more deeply into the lungs. **Health effects of both PM classes include exacerbation of respiratory symptoms in those with preexisting conditions, development of respiratory symptoms (including difficulty breathing, coughing, and susceptibility to infection) in those without preexisting conditions, and irregular heartbeat or other cardiovascular symptoms.** In addition, $PM_{2.5}$ pollution is the major cause of haze and visibility reduction in cities and the surrounding countryside (Image 6.5).

IMAGE 6.5. The dramatic difference in visibility in Bryce Canyon (southern Utah) on a clear day versus a hazy day. PM is the major cause of visibility reduction. The EPA tracks visibility in the national parks as part of an effort to control and reduce air pollution.

6.6 Common Pollutants—Sulfur Dioxide

All the common fuels combusted for energy are hydrocarbons—natural gas is mainly CH_4, while petroleum and coal are mixtures of much larger and more complex hydrocarbons. As we'll see in Chapter 13, however, the source of petroleum and coal means they also contain organic compounds that include the elements oxygen, nitrogen,[9] and sulfur. **Coal, in particular, can be quite high in sulfur** (up to 7%, with sulfur content varying greatly by region). This sulfur can combine with oxygen during combustion, resulting in the formation of SO_2 (**sulfur dioxide**):

$$S + O_2 \rightarrow SO_2$$

Sulfur dioxide gas is a respiratory irritant, and causes symptoms including coughing and difficulty breathing. It's released naturally (in small amounts, relatively speaking) by volcanic activity. The primary **anthropogenic**—meaning derived from human activities—source of SO_2 is coal combustion. The gas is colorless (though the high water content of volcanic plumes gives them their characteristic white, cloudy look, as in Image 6.6), and has the odor of burnt matches.

NOTE[9]

The nitrogen in petroleum and coal can react with oxygen when burned, providing yet another mechanism for NOx formation.

IMAGE 6.6. SO₂ is released in a plume of volcanic gas (a) from Halema'uma'u Crater in Hawaii. Pingyao, China (b), is often covered with a dense haze of smog due to its many coal-fired power plants.

NOTE¹⁰

The EPA's recommendations regarding the mercury in certain fish species are not the only recommended limitations on seafood consumption. Several organizations (notably the Monterey Bay Aquarium, which has a free Seafood Watch app online and available for mobile devices) have published lists of seafood that should be avoided due to endangerment of the species, nonsustainable harvesting practices, and nonsustainable farming (see Sources and Further Reading).

6.7 Common Pollutants—Mercury and Lead

In addition to containing sulfur, coal also contains trace amounts of the element mercury. When coal is burned, the mercury is aerosolized, or released into the atmosphere, and eventually settles into the ground and water. Aerosolized mercury can travel quite some distance before settling; the EPA has estimated that upwards of 75% of mercury emissions from any given source enters global atmospheric circulation and becomes widely distributed. **Mercury is quite toxic—it can cause birth defects, has been linked to cancer, and is a potent neurotoxin (a compound that damages the brain and nervous system).** Exposure is particularly dangerous to developing embryos.

Environmental mercury exposure occurs primarily through breathing contaminated air and eating contaminated food products. **Mercury does not pass through an organism the way some toxins do. Instead, it accumulates in the tissues and skeletal system. This means it also bioaccumulates in the food chain.** That is to say, when mercury pollution enters an aquatic system, it is taken up first by microorganisms such as bacteria and plankton. These organisms are consumed by small invertebrates, which are in turn eaten by fish. Predatory fish prey on smaller fish, and humans consume predatory fish. Through each step in the chain, the higher organism acquires the mercury from all the prey species it consumes, meaning that mercury concentrations in predatory fish can reach startling and dangerous levels. The EPA currently recommends that consumption of shark, king mackerel, tilefish, and swordfish be avoided, as mercury levels are too high for safety (Image 6.7). Other EPA recommendations limit amounts of various fish and seafood that should be consumed within a given period of time—these recommendations are most stringent for pregnant and nursing women, and for children.¹⁰ While fish and seafood pose a fairly well recognized threat with regard to mercury exposure, there are lesser-known sources of mercury in food. **Researchers have recently determined that measurable levels of mercury can be found in many samples of high-fructose corn syrup, a ubiquitous sweetener in processed foods and beverages.** The significance of high-fructose corn syrup as a source of mercury in the diet depends upon the manufacturer of the sweetener as well as the dietary habits of the individual; those who consume large amounts of soda, fast food, and processed foods are at greater risk.

IMAGE 6.7. The EPA recommends avoiding king mackerel and other large predator fish due to their high levels of mercury.

Lead, unlike mercury, is no longer a major airborne pollutant in the United States. Until the 1980s, lead was a common gasoline additive. In recent decades, however, leaded gasoline is no longer marketed to the public, and airborne lead pollution is restricted mainly to the areas surrounding lead processing plants (including manufacturers of lead batteries and other lead-containing products). **These days, the most common mechanism of exposure to lead in the United States is either through paint on the walls of older homes or through toys and some ceramics manufactured outside the United States which may be made of lead-containing material or coated with lead-based paint.** Lead-based paint was commonly used on both interior and exterior walls of houses until 1978 in the US, at which time the government banned its use. Paint on walls of older houses, however, commonly contains lead and represents a hazard (even if the original paint has been covered over). Old lead paint frequently peels (Image 6.8), and the lead imparts a sweet taste to the paint chips. Children, particularly very young ones, have a tendency to pick up the paint chips and consume them, owing to a child's natural inclination to put things in their mouths and the pleasant taste of the chips. Further, if paint is scraped or sanded down, lead contaminates the soil around the house. This provides further potential for exposure to lead if soil is aerosolized and inhaled (as it would be by children playing in a dusty yard), or if soil is consumed (owing again to the predilection of young children to explore their environments orally). **Lead, like mercury, is a potent neurotoxin.**

IMAGE 6.8. Lead paint on the walls of older homes (such as the paint shown in this home flooded by Hurricane Katrina) presents a health hazard to occupants, particularly children.

6.8   Pollutant Ramifications—Acid Rain

Apart from their intrinsic toxicities, airborne pollutants are implicated in two additional environmental concerns. The first, that most are potent greenhouse gases, will be addressed in Chapter 13. The second concern is that they are directly involved in the production of acid rain, a phenomenon that presents a danger to aquatic and terrestrial environments (Image 6.9), as well as to man-made structures.

It turns out that while distilled water is neutral, with a pH of 7, normal rain is actually slightly acidic. This is due to the slight solubility of carbon dioxide (which occurs naturally in the atmosphere) in water, and the subsequent reaction of the two to form carbonic acid:[11]

IMAGE 6.9. A forest in the "Black Triangle" area of the Czech Republic near the borders with Germany and Poland. Heavy post–World War II industrialization led to pollution and significant precipitation of acid rain. Spruce trees in the former forest have been decimated.

NOTE[11]

Recall that we've seen this reaction several times already. The concentration of CO_2 in the bloodstream affects the pH of the blood, and the production of carbonic acid in baked goods (from the reaction of baking soda with acid) results in formation of CO_2 bubbles.

NOTE[12]

In the same way that NOx is read aloud as "nocks," SOx is read as "socks." Reading a sentence containing both NOx and SOx together makes one feel as if one is reciting the works of Dr. Seuss!

$$CO_2 + H_2O \rightarrow H_2CO_3 \rightarrow H^+ + HCO_3^-$$

The hydrogen ion concentration, therefore, is higher in rainwater than in pure water, giving rainwater a pH of around 5.6. **Other natural processes may slightly affect the pH of rain, but it's generally accepted that normal rain has a pH between 5 and 7.**

Over-acidification of rain (pH < 5) cannot be accomplished by CO_2. **Instead, SOx[12] (monosulfur oxides including SO_2) and NOx emissions are responsible for acid rain formation.** The chemistry of this process revolves around formation of the highly reactive free radical species OHx, the hydroxyl radical. We're used to seeing OH^-, the hydroxide anion. In this case, the species in question is not an ion, and has no charge. Instead, it is an oxygen atom covalently bonded to a hydrogen atom, with an unpaired electron represented as a dot. This is an example of a free radical species. The chemistry of free radicals is somewhat complex, but all have a radical, or unpaired electron—that is to say, an electron that is neither part of a bond *nor* part of a nonbonding pair. As you might imagine, this unpaired electron means there is a hole in the atom's valence (an octet can only be filled if all electrons are paired), and since atoms are most stable with full octets, free radicals are highly reactive. Formation of the hydroxyl radical occurs in the troposphere as a by-product of the reaction of pollutant ozone with sunlight:

$$O_3 + h\nu \rightarrow O_2 + O$$

$$O + H_2O \rightarrow 2\ OH\cdot$$

The hydroxyl radical goes on to react with NO_2, resulting in the production of the highly acidic species nitric acid (HNO_3) which, when formed in the troposphere, can dissolve in the water of forming raindrops. Once dissolved, nitric acid (like all acids) dissociates in water to produce hydrogen ions:

$$OH\cdot + NO_2 \rightarrow HNO_3 \rightarrow H^+ + NO_3^-$$

SO_2 also reacts in the atmosphere with water and hydroxyl radicals to form the acidic species H_2SO_3 (sulfurous acid) and H_2SO_4 (sulfuric acid):

$$SO_2 + H_2O \rightarrow H_2SO_3 \rightarrow 2\ H^+ + SO_3^{2-}$$

$$H_2SO_3 + 2\ OH\cdot \rightarrow H_2SO_4 + H_2O$$

$$H_2SO_4 \rightarrow 2\ H^+ + SO_4^{2-}$$

Note that in the cases above, each acid (H_2SO_4 and H_2SO_3) has two ionizable hydrogens, both of which contribute to the pH of the resulting rain. The other common SOx species, SO_3, is formed from SO_2, and reacts with water to form H_2SO_4:

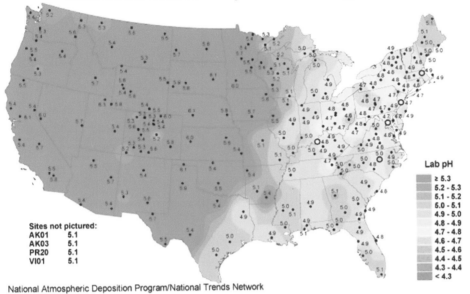

Hydrogen ion concentration as pH from measurements made at the Central Analytical Laboratory, 2009

Sites not pictured:
AK01   5.1
AK03   5.1
PR20   5.1
VI01   5.1

Lab pH

≥ 5.3
5.2 - 5.3
5.1 - 5.2
5.0 - 5.1
4.9 - 5.0
4.8 - 4.9
4.7 - 4.8
4.6 - 4.7
4.5 - 4.6
4.4 - 4.5
4.3 - 4.4
< 4.3

National Atmospheric Deposition Program/National Trends Network

IMAGE 6.10. Acidity of rain in the United States as measured by the National Atmospheric Deposition Program, supported by the EPA.

$$2\,SO_2 + O_2 \rightarrow 2\,SO_3$$

$$SO_3 + H_2O \rightarrow H_2SO_4$$

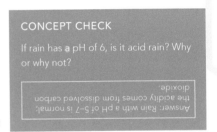

CONCEPT CHECK

If rain has a pH of 6, is it acid rain? Why or why not?

Answer: Rain with a pH of 5–7 is normal; the acidity comes from dissolved carbon dioxide.

While there are natural sources of SOx and NOx, anthropogenic sources are far more significant, and are the major contributors to global acid rain. The EPA estimates that up to 67% of SOx emissions in the United States are the result of burning coal for electricity.[13] Burning of biomass, such as in slash-and-burn tropical farming, is a second major global anthropogenic source of SOx and NOx emissions.

The ramifications of acid rain are significant and far-reaching. Because of global atmospheric cycles, pollutant emissions have the ability to travel hundreds of miles or more, meaning that acid rain is not limited to heavily industrialized areas (though it's more common and more concentrated in these areas). When it falls on fields or wooded areas, acid rain damages plant leaves. This limits the ability of the plant to obtain nutrients (leaves are a tree's sunlight-gathering and food-producing unit), and can weaken or kill them. The decreased pH of soil upon which acid rain has fallen further affects trees and smaller plants by reacting with nutrients and leeching them from the soil. Acidification of surface water is a serious concern: fish gills can be eroded by acidified lakes and rivers, and acid damages fish skeletons. Shelled invertebrates and other aquatic species are also adversely affected. The compound $CaCO_3$, calcium carbonate, is a major component of fish skeletons and aquatic invertebrate shells. While not water-soluble ($CaCO_3$ is also the major component of classroom chalk, which does not dissolve in water), the compound reacts with acid to form a familiar product—carbonic acid:

NOTE[13]

It's worth noting briefly that electricity is sometimes promoted as a "clean" energy source—electric cars, for instance, are advertised as nonpolluting. However, this sort of advertising fails to acknowledge the source of the electricity used to power the vehicles. In the United States, most electricity is generated by coal combustion, which, as we've seen, is highly polluting

IMAGE 6.11. A gargoyle from a European cathedral. The limestone (largely $CaCO_3$) from which this carving was made is not water soluble, but dissolves in acidic solution, hence the "melting" of the gargoyle's once-distinctive features.

$$CaCO_3 + 2\ H^+ \rightarrow Ca^{2+} + H_2CO_3$$

$$H_2CO_3 \rightarrow CO_2 + H_2O$$

In general, aquatic species begin to suffer the effects of acidity below a pH of 6, and most fish eggs can't hatch below a pH of 5. Low-pH lakes eventually become incapable of supporting life.

In addition to the effect of acid rain upon the natural world, human-made structures are adversely affected. Automobile paint and clear coating can be etched by acid rain. Metals, such as the bronze used in many statues, are corroded. Stone, including the limestone and marble used in many older buildings and statues, literally dissolves under the assault of acid rain (Image 6.11). **$CaCO_3$ is a major component of both marble and limestone, rendering these substances susceptible to acid, just as aquatic shells are.**

While the destruction of man-made structures by acid rain does not affect the health of humans or other organisms, structures, statues, and art are nevertheless culturally important and represent a further impact of acid rain on society. Further, these visual manifestations of acid rainfall act as reminders of its presence and the prevalence of pollutant emissions.

The effects of acid rain upon the ecosystem and upon human establishments are made particularly clear in areas such as the "Black Triangle" in Eastern Europe, an area of the northern Czech Republic that industrialized rapidly in the wake of World War II. The industrialization resulted in a dramatic increase in SOx and NOx emissions and heavy acidification of the local rain. Forests died, lakes and streams became too acidic to support aquatic life, and structures began to deteriorate.

6.9 The Industrial Revolution

In addressing the question of when pollution began to affect society, we must look to the Industrial Revolution. This period of history, which began in the mid- to late-eighteenth century and was in full swing by the mid-1800s, represents a major transition in human society in many ways. Foremost among these, the period marked the changeover from human- and animal-powered labor and manufacturing to machine powered. **The hallmark of the Industrial Revolution was the development of coal-fired steam generation, which was used to power machines and engines. This technology was followed closely by the invention of the internal combustion engine. In short, this period of history marked the beginning of humankind's major dependence upon fossil fuels.**[14] Prior to this time, the environmental problem of air pollution simply didn't exist; any combustion that took place, such as burning of wood to heat buildings, was on too small a scale to result in the production of significant quantities of pollutant material. As the Industrial Revolution began to change the shape of society in Europe and worldwide, however, its effects became apparent.

NOTE[14]

We'll discuss fossil fuels in more detail in Chapter 13, but for now, these include natural gas, petroleum, and coal. Early in the Industrial Revolution, coal was the major accessible fossil fuel, purely for logistical reasons.

NOTE[15]

Despite the fact that the leading causes of death in Cincinnati at that time were tuberculosis, pneumonia, and bronchitis—all respiratory in nature.

Early industrialized societies, including England and the United States, quickly began to recognize the beginnings of a pollution problem. At first, this awareness was limited to the observation of the unsightly black smoke that poured from factory and steam engine smokestacks (Image 6.12); the soot tended to blacken the sides of buildings and was observed to worsen the number of respiratory problems in the city. Several unsuccessful acts and orders were passed both in the United States and in Europe: the city of Cincinnati, for instance, ordered smoke-reduction measures in 1881, though the order was not enforced.[15] An attempt by the city of St. Louis to pass an ordinance outlawing the emission of thick black or gray smoke was overturned by the Missouri State Supreme Court, which called the legislation *unreasonable*.

Pollution awareness came to a head in 1952, when a smog—a combination of smoke and fog—settled over the city of London, England (Image 6.14). While this event wasn't the first of its kind (many similar events were observed in the years

IMAGE 6.13. This historic photograph shows the English town of Widnes under a pall of heavy coal smoke in the late nineteenth century.

IMAGE 6.14. The Great Smog of 1952 in London, England.

NOTE[16]

The sulfur dioxide concentration during the London Fog is estimated to have reached 10 times the EPA's current acceptable standard for U.S. air, while the particulate matter concentration is estimated to have topped out at nearly 30 times the current acceptable U.S. standard.

NOTE[17]

The EPA had determined in 2003 that Congress had not given them the authority to regulate greenhouse gases and that even if they had the authority to do so, they would not use it. It was determined by the court that the regulation of CO_2 does indeed fall under the auspices of the Clean Air Act, and as such, under the EPA. In 2015, the EPA and President Barack Obama announced the Clean Power Plan, which establishes target carbon dioxide emissions standards in an effort to reduce carbon pollution. As with other pollutants, however, the potential for carbon-emissions trading weakens the overall positive impact of the plan.

between the beginning of the Industrial Revolution and the London event), it was well publicized and had a shocking death toll. The *London Fog*, also called the *Great Smog of 1952*, was the result of a temperature inversion that held fuel emissions over the city rather than allowing them to dissipate into the higher atmosphere and around the countryside. Pollutants, including sulfur dioxide, nitrogen dioxide, particulate matter, and other toxic emissions reached deadly concentrations[16] within a matter of hours, with concentrations remaining high for four days. Reportedly, London became so darkened and obscured by the smog that even in the middle of the day, men had to walk in front of city buses with lanterns to light the street for the driver. Conservative estimates hold that 4,000 individuals lost their lives as a result of the London Fog, but recent reexamination of the event suggests the total number of deaths may have been closer to 12,000.

6.10 Pollution Regulation

In response to increasing awareness of the dangers of pollution, the U.S. Congress passed a series of Clean Air Acts in the 1960s and 1970s (also creating the EPA in 1970), with a further Clean Air Act amendment in 1990. **One of the major provisions of the Clean Air Act was that the EPA would develop programs to regulate and reduce the common air pollutants (identified as PM, SO₂, NOx, ozone, CO, and lead).** A second major requirement was for a phaseout of leaded gasoline (further discussion on this topic later in the chapter). Further provisions dealt with emissions from motor vehicles (responsible for over half of NOx emissions, and approximately half of VOC and other pollutant emissions) and emissions from "nonroad" motorized equipment such as recreational vehicles and lawnmowers.

For 20 years, the Clean Air Act was arguably one of the most successful pieces of environmental legislation in U.S. history, with impressive reductions in the six common pollutants identified for regulation and reduction by the EPA (Image 6.15). Unfortunately, the initial legislation failed to address another key airborne emission and potent greenhouse gas: carbon dioxide. This led to a 2007 U.S. Supreme Court case in which a number of environmental organizations, twelve states and several cities sued the EPA for failure to regulate carbon dioxide and other known greenhouse gases.[17] A further weakness of the act is that the 1990 amendment allowed for emissions trading, a practice whereby heavy polluters can buy unused pollution allowances from light polluters, rather than coming into compliance of their own accord. This, combined with recent lax enforcement of the act by the EPA, has somewhat reduced the effectiveness of the legislation in recent years.

As a result of Clean Air Act stipulations regulating SOx emissions, coal plants began to preferentially process lower-sulfur coal. This reduced, but did not eliminate, SOx emissions. Furthermore, coal-fired power plants now employ one or more coal scrubbing (chemical washing) or filtering techniques to reduce SOx emissions. The most common technique relies upon wet limestone slurry (largely

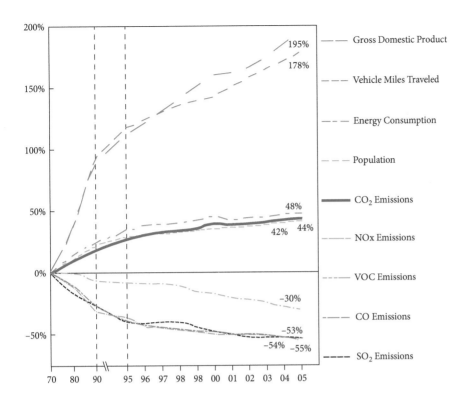

IMAGE 6.15. The Clean Air Act's effect on major airborne pollutant emissions per EPA data, 1970–2005. Pollution values are given as percent change above or below 1970 values.

composed of $CaCO_3$), which absorbs some of the sulfur from combustion gases. While nearly 90% efficient (when operating correctly), smaller sulfur-containing particles are still able to pass through scrubbers and coal pollution still contains measurable SOx. Further, the waste absorbed by the scrubber is neither eliminated nor rendered chemically inert—rather, it is isolated within the scrubbing sludge instead of being aerosolized, rendering the sludge into a land and water pollutant. **All in all, scrubbing is a tremendously resource-intensive process.** It is estimated that a typical 500-megawatt coal plant (the size necessary to power a city of about 140,000 people) burns approximately 1,430,000 tons[18] of coal in a year, and uses 2.2 billion gallons of water and 146,000 tons of limestone for scrubbing. Despite the scrubbing, 10,000 tons of SOx are still emitted from each plant on an annual basis, along with 10,200 tons of NOx, 700 tons of carbon monoxide, and 500 tons of PM_{10} and $PM_{2.5}$. Further, each plant produces nearly 200,000 tons of scrubber sludge annually. The sludge is disposed of in landfills or is converted into concrete and drywall, but is high in mercury and lead. There is concern that these heavy metal pollutants leach out of the sludge, ending up in the ground and in water.

CONCEPT CHECK

Why is coal considered such a dirty fuel?

Answer: It contains nitrogen (source of NOx) and sulfur (source of SOx) in higher proportions than other fuels. It also contains mercury and lead; sludge from scrubbing is an environmental pollutant.

6.11   Air Quality Index

One problem faced by the EPA in quantifying the quality of air on a given day is that each airborne pollutant is dangerous in different concentrations. **The EPA uses two factors to determine acceptable concentrations of any given pollutant: toxicity**

NOTE[18]

One ton is 2,000 pounds, or 907 kg.

IMAGE 6.16. AQI values, health concerns, and corresponding colors.

Air Quality Index (AQI) Values	Levels of Health Concern	Colors
When the AQI is in this range:	*...air quality conditions are:*	*...as symbolized by this color:*
0–50	Good	Green
51–100	Moderate	Yellow
101–200	Unhealthy for sensitive groups	Orange
151–200	Unhealthy	Red
201–300	Very unhealthy	Purple
301–500	Hazardous	Maroon

IMAGE 6.16. AQI values, health concerns, and corresponding colors.

and exposure. Toxicity relates to the inherent harmful potential of a given pollutant—that is, how likely the pollutant is to do damage. **Exposure** is a measure of the amount of substance likely to be encountered: it is a combination of the concentration of a pollutant and the amount of time the pollutant is in the atmosphere. As a means of providing citizens with a simple schema for judging air quality, the EPA developed the **Air Quality Index (AQI)**. Pollutant concentrations are normalized on a scale of 0–500, which is divided into subcategories corresponding to the level of health concern (Image 6.16).

Each range of AQI values is symbolized by a particular color. Additionally, each value range and color correspond to a level of health concern. *Good* air quality conditions are not reasonably expected by the EPA to cause health problems in most individuals, though *moderate* conditions may cause extremely sensitive members of the population some difficulty. Air quality conditions falling into the category of *unhealthy for sensitive groups* may place certain groups at risk, including the very young and elderly and those with underlying cardiovascular and respiratory conditions. *Unhealthy* air is likely to adversely affect nearly everyone, while *very unhealthy* air would cause relatively severe symptoms in all members of a population. *Hazardous* air represents an air quality emergency. **It is important to note that a separate AQI value is assigned to each airborne pollutant being monitored, but that unhealthy levels of even one pollutant create a health concern; not all pollutants are necessarily at the same AQI value at all times.**

Using the normalized AQI (available for cities with air-monotoring programs through the EPA-run Web site www.airnow.gov), residents of the United States can easily assess air quality without having to keep track of the disparate concentrations of each pollutant associated with each air quality category. For instance, in order to fall into the "good" range as defined by the EPA, eight-hour average ozone concentrations must be no higher than 0.064 **ppm** (meaning parts per million, or ozone molecules per million molecules that make up the troposphere), while eight-hour average carbon monoxide concentrations can rise as high as 4.4 ppm (Image 6.17).

Even without committing these concentrations to heart, though, a concerned individual need only glance at the AirNow website to determine that if, for instance,

CONCEPT CHECK

Why is the AQI a helpful tool for getting information about air quality?

Answer: It normalizes the acceptable concentrations of various pollutants to a single, easy-to-read scale.

Air Quality Index (AQI) Values	PM$_{2.5}$ 24-hour Average (micrograms/m^3)	8-hour Ozone Concentration (ppm)	Carbon Monoxide 8-hour average (ppm)
0–50	0–15.4	0–0.064	0.0–4.4
51–100	15.5–40.4	0.065–0.084	4.5–9.4
101–200	40.5–65.4	0.085–0.104	9.5–12.4
151–200	65.5–150.4	0.105–0.124	12.5–15.4
201–300	150.5–250.4	0.125–0.155	15.5–30.4
301–500	>250.4	>0.155	>30.4

IMAGE 6.17. The relationship between AQI value and concentration for three separate pollutants.

their city is "green" for carbon monoxide, but "orange" for PM$_{2.5}$, carbon monoxide is in the acceptable range, but particulate matter levels may cause problems for sensitive groups. This helps individuals choose outdoor activities appropriate to the air quality and their personal health—an asthmatic might not engage in strenuous outdoor activity on an "orange" pollutant day, for instance.[19]

6.12 Changing Perceptions of Risk

One fascinating area of interplay between chemistry and society is in the public perception of risk associated with an activity or substance and how that perception varies with time and circumstance. The advent (and subsequent phaseout) of leaded gasoline in the United States is an excellent example.

Early in the development and popularization of the automobile, the four-stroke internal combustion engine (Image 6.18) used to power the vehicle was subject to a phenomenon known as *knocking*, which occurred when fuel combusted prematurely.

Knocking resulted in an engine that, rather than supplying power continuously, did so in a series of power bursts (similar to a set of explosions). This reduced vehicle power and overall efficiency. Thomas Midgley,[20] an inventor and mechanical engineer, was working for General Motors when he determined that the knocking phenomenon was due to a problem with the fuel used in engines rather than the engines themselves. In 1921 he added tetraethyl lead, (CH$_3$CH$_2$)$_4$Pb, to gasoline in order to smooth combustion and prevent engine knocking. For a variety of reasons, including the patentability and profit potential of the additive, leaded gasoline was heavily promoted and became pervasive in the marketplace, despite concern expressed by several scientists and health experts.[21] The dangers of lead were not in the forefront of public awareness at that time, though they became more apparent as the manufacture, distribution, and sale of leaded gasoline began to take a toll on individuals and the environment. Midgley himself, after working with lead in the lab, found he was developing lead poisoning and took time off to recover. Plant workers were among the first to succumb to lead poisoning, experiencing a variety

NOTE[19]

Every city is different, depending upon the population, time of year, local industries, and so forth. That said, a resident of Phoenix, AZ would often find that the city is "yellow zone" for ozone in the summer—when high levels of sunlight catalyze the production of ozone from NOx—and in the "yellow zone" for carbon monoxide in the winter, when temperature inversions tend to keep pollution in the city center. During a summer dust storm or a particularly severe winter inversion, Phoenix will often rise into the "orange" range for particulate matter. Check AirNow at different times of year to see where your city typically stands!

NOTE[20]

Remember his name—we'll see him again in Chapter 10!

NOTE[21]

While the dangers of lead poisoning were known to the industry at this time, they were not yet widely recognized by the public. Still, GM took the marketing precaution of referring to the fuel additive as *ethyl* to avoid any mention of lead during its introduction to the marketplace.

IMAGE 6.18. The four parts of a four-stroke engine cycle: Intake (fuel is brought into combustion chamber), compression (fuel is compressed for increased generation of power from combustion), ignition (fuel is ignited with a spark), and exhaust (products of combustion are released).

of symptoms culminating in insanity and death, in numbers that raised the concerns of authorities. A conference was convened in 1925 by the Surgeon General to address these concerns and to make a determination on the wisdom of proceeding with production and sale of lead additives. Several of the scientists present at the conference presented strong arguments against the use of lead. Of particular note, the director of a large metropolitan hospital made a statement well in line with what is now known as the **precautionary principle**, which has its roots in maxims such as "better safe than sorry." The idea is that in the absence of proof and with the possibility of harm, it's far better to take a more conservative approach than wait for evidence of danger before initiating change or regulation. While this principle has a multitude of origins and no "conception" as such, it is generally accepted that formalization of the principle took place in 1998 during a meeting of scientists and policy makers, and states:

> When an activity raises threats of harm to human health or the environment, precautionary measures should be taken even if some cause and effect relationships are not fully established scientifically. In this context the proponent of an activity, rather than the public, should bear the burden of proof. The process of applying the precautionary principle must be open, informed and democratic and must include potentially affected parties. It must also involve an examination of the full range of alternatives, including no action (Wingspread Conference on the Precautionary Principle, 1998).

The hospital director suggested that the use of lead additives ought to be suspended until research demonstrating its safety was available. In contrast, however, a doctor named Robert Kehoe suggested that the ruling on lead use should be made only upon the basis of research available at that time. Since there was little research available, he suggested that lead should not be barred from manufacture or use. **In the end, the surgeon general adopted Kehoe's perspective, stating that until there was evidence of harm, tetraethyl lead would be allowed.** It's well worth noting that Dr. Kehoe was initially a consultant of the lead industry and was later director of the Kettering Laboratory—a research group funded by the lead industry to produce lead safety studies. It is easy to conjecture that the economic and professional pressure on Dr. Kehoe to avoid producing any studies demonstrating the toxicity of lead was considerable.

In a further effort to demonstrate the innocuous nature of the additive (and with questionable wisdom, given his previous bout with lead poisoning), Midgley famously washed his hands with tetraethyl lead during a press conference. Despite the best efforts of the industry, however, public health concerns over exposure to the lead additive through misuse, mishandling, spills, and accidents increased. Further, lead oxides from fuel exhaust contaminated the air, ground, and water, leading to widespread and rampant lead pollution. One study in particular examined lead concentrations in New York City dust, finding concentrations 50% higher in 1934 than ten years prior. This study and others like it initially received little attention, and U.S. consumption of leaded gasoline continued to increase. **By the 1970s, though, growing concern over lead pollution and toxicity resulted in the aforementioned EPA-mandated phaseout of leaded gasoline.** While industries vigorously opposed the notion that lead from the gasoline was leaching into the environment and argued that the phaseout was unwarranted and unnecessary, the second U.S. National Health and Nutrition Examination Survey, published in 1980, extinguished any doubt of a connection. The survey showed that as use of leaded gasoline decreased from 1976 to 1980, blood lead levels closely paralleled that decrease. This led the EPA to more closely examine the dangers of leaded gasoline, and eventually resulted in the much more aggressive phaseout of leaded gasoline mandated in 1982. Lead is no longer a major airborne pollutant in the United States, thanks to these measures, and lead pollution and risk of exposure are now isolated to manufacturing sites, sources of old paint, and a small number of household items manufactured outside the US. Countries in which lead is still used in gasoline, however, continue to have problems with lead toxicity and pollution.

Of note in this discussion is that despite concerns expressed by environmental researchers and experts, production and marketing of tetraethyl lead as a gasoline additive was allowed to proceed. Only after many preventable deaths and 60 years' worth of unnecessary lead pollution of the environment was legislation finally put in place to curtail the use of lead fuel additives. During this period of time, public opinion regarding the dangers of lead changed markedly. **It's worth mentioning that in a battle for control of public opinion between independent researchers and corporations or corporation-sponsored researchers, those with access to the advertising dollars will most likely win public favor.** Lessons from lead can be applied as we face modern environmental challenges and questions: it's important to remember that manufacturers of products that may be environmentally harmful are highly motivated to produce results that support their claims of safety and to encourage suppression of results that don't. In the end, there are two questions worth asking whenever evaluating scientific information. First, is there data supporting the claim of risk or safety? Second, who funded the study?

IMAGE 6.19. Cigarette companies strive to conjure images of nature and health (such as that of this tobacco leaf) rather than images of soot, tar, and products of chemical combustion.

6.13   The Last Word—Smoking

Very few psychologically healthy individuals, particularly if armed with information about the health dangers of pollutants, would purposely concentrate and inhale the smoke from a coal-fired power plant or the tailpipe of a car. However, to compare these clearly off-putting inhalants to the smoke from a cigarette is not a scientific stretch—in fact, in many ways, cigarette smoke is even worse. While the cardiovascular and respiratory consequences of smoking are well known at this point and while smoking is acknowledged to cause cancer, it's interesting to consider cigarettes from another, less articulated angle: that of their chemistry.

From a chemical standpoint, cigarette addiction is due in large part to the compound nicotine. There are additional (significant) psychosocial influences on smoking behavior and addiction, but a discussion of these is beyond the scope of this text. Nicotine is a powerful **neurochemical**,[22] inducing a feeling of alertness that is followed some time later by relaxation. It also suppresses appetite, which proves to be a motivating factor for smoking—and a deterrent to smoking cessation—for some weight-conscious individuals. Regardless of the myriad chemical effects of nicotine, however, the molecule itself does not cause cancer. The carcinogenic properties of cigarettes are due to the hundreds of other chemicals present in both the raw materials and the products of combustion reactions taking place within a lit cigarette.

Cigarettes consist of tobacco leaf (which contains the nicotine) and as many as 599 additives, a list of which was finally made available to the Centers for Disease Control and Prevention (CDC) in 1994. While the additives are all federally approved for use in food, they have largely not been tested under combustion conditions, which in many cases result in the formation of carcinogenic products. Among disturbing additives from the list are vanilla, honey, and chocolate, which are included in cigarettes for their sweetness and appealing aroma and suspected by anti-tobacco lobbies to be a mechanism for luring children into smoking. Ammonia is added to enhance nicotine absorption and increase addiction. Other additives serve various purposes, including preserving the tobacco, influencing taste, and masking symptoms of smoke inhalation and illness. These include benzene, formaldehyde, methanol, and acetylene torch fuel (all VOCs), as well as cyanide (a highly toxic chemical used in gas chamber executions in some states), arsenic (a potent inhibitor of cellular energy production), and lead.

The list above addresses only the raw ingredients in cigarettes; a further litany of compounds is produced during combustion. These include all the major airborne pollutants—NOx, SOx, CO, PAHs, VOCs, and so forth—as well as a number of carcinogens not generally found in exhaust. In addition, pollutants are found in cigarette smoke at concentrations much higher than those routinely encountered even in highly polluted air. The CO concentration of undiluted cigarette smoke, for instance, is about 4.5 times higher than that of undiluted, concentrated car exhaust.

Clearly the chemistry of an unlit cigarette is complex enough, to say nothing of the chemistry that takes place upon lighting. This is further complicated by the chemistry (much of it still unknown or incompletely understood) of these compounds upon absorption into the body. In many ways, smoking can be compared to lead: for a long time, large corporations either prevented public dissemination of studies or funded their own in order to control what knowledge was available to the public. Public perception of the risk of cigarette smoking was very different in the early- and mid-1900s than it is today; even now, the transition toward perceiving cigarettes as something that should be regulated is incomplete. This is due jointly to the desire of individuals to maintain freedom of choice and the desire of manufacturers to maintain their profit margins.[23] It's interesting to ponder, however, whether one day a textbook might use cigarettes—just as this text uses lead—as a mechanism for introducing the precautionary principle.

NOTE[23]

It's interesting, disturbing, and well worth considering that companies engaged in manufacturing and selling products that negatively impact health (junk food, cigarettes, etc.) appeal to the American desire for "freedom of choice"; when a company trumpets that you have the right to decide for yourself what you will and won't do with your body—and that the government has no business legislating such decisions—odds are they're selling something that science tells us is harmful. Odds are, too, that if we paid more attention to the science behind the product (and stopped worrying so much about whether the government was going to limit our access to it), we'd decide for ourselves it was probably a product best avoided. Generally speaking, the more loudly a company declares that all it wants is to help defend your freedom to make your own choices, the more questionable the product it's selling.

SUMMARY OF MAIN POINTS

- CO is a product of incomplete carbon combustion and reduces oxygen delivery to tissues.
- Catalytic converters help reduce CO emissions, but produce CO_2, a potent greenhouse gas.
- NOx are the products of high-temperature reactions of N_2 with O_2.
- NO_2 is a respiratory irritant.
- VOCs are largely indoor pollutants and are toxic.
- PAHs are products of incomplete fuel combustion and are carcinogenic.
- Ozone (O_3) is a pollutant in the troposphere and a secondary pollutant formed via the reaction of NO_2 with sunlight.
- O_3 is a respiratory irritant.
- $PM_{2.5}$ and PM_{10} are aerosolized particulate matter from various sources and are respiratory irritants.
- SO_2 is produced naturally (in small quantities) by volcanoes and produced anthropogenically (in much larger quantities) by coal combustion.
- SO_2 is a respiratory irritant.
- Mercury pollution is due in large part to coal combustion, and is an air, land, and water contaminant.
- Mercury concentrates in the food chain, particularly through consumption of aquatic organisms. The EPA recommends limiting consumption of predator fish due to high mercury levels.
- Lead is mostly found in and around old buildings (due to lead paint) in the United States. It is a potent neurotoxin.
- Rain is naturally slightly acidic due to dissolved CO_2.
- Rain with a pH of less than 5 is considered to be acid rain.
- Acid rain is due to dissolved SOx and NOx in raindrops, producing H_2SO_4 and HNO_3 (both strong acids).

- Acid rain damages terrestrial and aquatic environments, enhances rust formation, and damages man-made structures.
- The Industrial Revolution was the point in history during which anthropogenic pollution levels became significant.
- Regulation of pollution in the United States began in earnest with the Clean Air Act, creation of the EPA, and amendments to the Act.
- Coal scrubbing results in the reduction of SOx emissions from coal burning, but produces contaminated sludge.
- The EPA determines acceptable levels of airborne pollutants by considering two important factors for each: toxicity and exposure.
- To simplify air-quality reporting, the EPA created the Air Quality Index (AQI).
- Perceptions of risk change over time: Lead was a common additive to gasoline through the 1970s but is now highly regulated by the EPA.
- Corporations producing products may be highly motivated to prevent the publicizing of studies demonstrating toxicity of their product.
- The precautionary principle can be summarized to state that in the absence of evidence, it's better to be safe than sorry. This principle is appropriate to take into consideration in making major decisions as a society.

QUESTIONS AND TOPICS FOR DISCUSSION

1. Why do you suppose most tents have a label printed inside that warns occupants not to use a stove in the tent?

2. Home carbon monoxide detectors are more common in cold parts of the country than they are in warm parts. Why do you suppose that might be?

3. What are the benefits of a catalytic converter? What are the drawbacks?

4. Why do you think that CO's ability to interact with hemoglobin depends in part upon the similarity of its shape to that of O_2?

5. If a sample of hydrocarbon (say, octane) were burned slowly, do you think the proportion of CO to CO_2 in the combustion products would be higher, lower, or similar to that produced if the octane were burned rapidly?

6. Natural gas produces low levels of CO compared to larger hydrocarbons. Compare the combustion reactions of CH_4 and C_8H_{18}, and propose a reason for this.

7. Carbon monoxide is described in the text as causing "chemical suffocation." How can a person suffocate if their ability to breathe is unhindered (assuming normal air pressure and 21% oxygen)?

8. What does it mean to say that nitrogen is inert?

9. Why are VOCs generally considered an indoor rather than an outdoor pollutant? Which other pollutants might reach concentrations of concern indoors, and why?

10. Why is stratospheric ozone not considered a pollutant?

11. What is a free radical?

12. The lungs consist of airways that branch and grow smaller as they get deeper in the lung tissue. Based upon this information, suggest why $PM_{2.5}$ might be more harmful than PM_{10}.

13. Asthmatics are said to have "reactive airway disease," whereby the small airways become constricted. Why might an asthmatic be more susceptible to PM exposure than a nonasthmatic?

14. Why is the EPA concerned with visibility in national parks?

15. Natural gas is considered the cleanest burning of the common fuels. Why?

16. Despite claims (by the natural gas industry) that natural gas is the cleanest burning of the common fuels, its combustion still produces many pollutants. Which are most likely? Which are least likely?

17. Why might natural gas combustion result in less NOx emission than coal combustion?

18. In discussing sources of NOx and SO_2, why is it important to differentiate natural from anthropogenic sources?

19. A coal lobbyist argues that NO_2 and SO_2 shouldn't be regulated as pollutants because they are naturally occurring: NO_2 is produced in lightning storms, and SO_2 is emitted by volcanoes. Is this argument valid? Why or why not?

20. Why is it accurate to say that mercury concentrates up a food chain? Why is this effect more significant in aquatic food chains as opposed to terrestrial food chains? (Hint: think about the number of "links" in each chain.)

21. Why is a shark likely to have higher mercury levels than, for instance, a sardine?

22. Why are the EPA's recommendations for seafood consumption more conservative for pregnant women than for women who aren't pregnant?

23. Would it be safe to paint over the old paint in a house built before 1978? Why or why not?

24. In *Alice's Adventures in Wonderland*, Lewis Carroll describes a character called the Mad Hatter; in fact, the expression "mad as a hatter" is quite old. A hatter, or person who makes hats, traditionally worked with $Mg(NO_3)_2$ (mercuric nitrate) to separate animal fur from the skin. From where might the historic expression—and Carroll's character—have come?

25. What are the effects of acid rain on terrestrial, aquatic, and man-made environments?

26. Destruction of sculptures and buildings doesn't really constitute an "environmental" or "toxicity" effect of acid rain, yet these effects are generally of great concern. Why is destruction of man-made structures of any concern?

27. What are SOx and NOx, and why are they of environmental concern?

28. Where is acid rain most prevalent in the United States? Why do you think this is so?

29. Why do you think $CaCO_3$ dissolves in acid, even though it doesn't dissolve in water?

30. What are the major ways in which the Industrial Revolution changed society? Think about this both in terms of positive and negative effects.

31. Why do you think initial regulation of pollution was so difficult to achieve?

32. In what ways were the Clean Air Act and its amendments successful? In what ways were they unsuccessful?

33. Do you feel the practice of emissions trading should have been allowed for by the Clean Air Act Amendment of 1990? Why or why not?

34. Coal is a highly controversial fuel in the United States—despite the fact that it is plentiful, it is a very dirty fuel. How is coal dirtier than other common fuels?

35. How do you think coal's importance as a U.S. mine product has impacted the success of legislation creating pollution caps and limiting emissions? Do you think pollution regulations might be stricter if the US had major natural gas—but not coal—reserves?

36. How does coal scrubbing help reduce airborne pollution? What problem does it create? Which do you think is worse?

37. Look up the AQI for your city on www.airnow.gov. What are your air quality values today? What sorts of pollution does your area struggle with?

38. Why was the AQI an important communication tool created by the EPA?

39. Why was lead a major airborne pollutant by the EPA in the 1970s, and why is it not commonly encountered in the United States now?

40. How did the fuel industry affect public opinion of lead exposure risk in the 1920s?

41. Provide an example of an industry attempting to impact public opinion regarding the risk of a product or behavior today. How is the industry engaging the public? Are they appealing to desire for freedom? Withholding information?

42. Where do you draw the line between allowing individuals freedom of choice and ensuring that they remain safe? Consider that regulation reduces individual freedom, but also consider that in the absence of regulation, industries may withhold information, conduct poorly designed studies, provide misinformation, and so forth.

43. What is the precautionary principle? What are the benefits and drawbacks of abiding by it? (Hint: think about both industrial and environmental perspectives.)

44. Why are independently funded or government-funded (as opposed to industry-funded) scientific studies sometimes ignored by the public (as in the case of the lead studies in the early and mid-1900s)?

45. Why is it important to consider the funding source for a study in evaluating its efficacy? Where did funding contribute to public opinion of lead safety?

PROBLEMS

1. Write a balanced combustion reaction (idealized—CO_2 and H_2O products) for propane (C_3H_8), a common gas grill fuel.

2. Write an alternate combustion reaction (CO and H_2O products) for propane (C_3H_8), a common gas grill fuel. Compare to the reaction from question 1; explain why this reaction requires less oxygen.

3. Compare the idealized combustion reactions of butane (C_4H_{10}) and methane (CH_4); which is more likely to be a significant source of CO, and why?

4. What is the anthropogenic source of CO?

5. What are the effects of exposure to CO?

6. What is a catalytic converter?

7. Why is N_2 so chemically inert?

8. What is the anthropogenic source of NO_2?

9. What are the effects of exposure to NO_2?

10. What are VOCs and where are they commonly encountered?

11. What are the effects of exposure to formaldehyde, a common VOC?

12. What are PAHs and where do they come from?

13. What are the effects of exposure to PAHs?

14. What are mutagens and carcinogens?

15. Write a Lewis dot structure for ozone.

16. What is an allotrope? Give an example.

17. In what portion of the atmosphere does ozone occur naturally? In what portion is it a pollutant?

18. What is the anthropogenic source of ozone?

19. What are the effects of exposure to ozone?

20. Why is ozone referred to as a "secondary" pollutant? Which other pollutants are involved in ozone production?

21. What is the difference between PM_{10} and $PM_{2.5}$? Which is more damaging?

22. Are the PMs a single substance, or a mixture? If a mixture, of what are they comprised?

23. What are the effects of PM exposure?

24. Name a natural source of SO_2 and an anthropogenic source of SO_2. Which is more significant, with regard to quantity produced?

25. Which fuel is most likely to result in significant SO_2 production?

26. What are the effects of exposure to SO_2?

27. What is the current source of most lead exposure in the United States?

28. What are the effects of exposure to lead?

29. What are the effects of exposure to mercury?

30. Why is rain naturally acidic?

31. Write out the chemical reaction explaining the natural acidity of rain.

32. What is the approximate pH of normal rain?

33. What pH constitutes acid rain?

34. What is the chemical source of acid rain? How does it form?

35. Why are aquatic invertebrate shells and fish skeletons susceptible to acidification of water?

36. Why are marble and limestone damaged by acid rain?

37. What is the AQI and how is it used?

38. Which two factors are used by the EPA to determine acceptable pollutant concentrations in air?

39. Define toxicity and exposure.

40. What is tetraethyl lead and what was it used for? Is it still used?

1. $C_3H_8 + 5\,O_2 \rightarrow 3\,CO_2 + 4\,H_2O$

3. $2\,C_4H_{10} + 13\,O_2 \rightarrow 8\,CO_2 + 10\,H_2O;\ CH_4 + 2\,O_2 \rightarrow CO_2 + 2\,H_2O.$ Butane will be a more significant source of CO, because the idealized combustion requires a butane:oxygen ratio of 2:13, while the idealized combustion of methane requires a methane:oxygen ratio of 1:2. This means there's more likely to be enough oxygen for "ideal" combustion of methane than for ideal combustion of butane.

5. CO exposure reduces oxygen delivery to tissues.

7. N_2 contains a very strong triple bond between atoms.

9. NO_2 is a respiratory irritant.

11. Formaldehyde irritates mucous membranes.

13. PAHs are mutagens and carcinogens.

15.

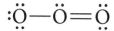

17. Ozone occurs naturally in the stratosphere and is a pollutant in the troposphere.

19. Ozone is a respiratory irritant.

21. PM_{10} is larger and somewhat less damaging than $PM_{2.5}$.

23. Lung irritation and possible cardiovascular symptoms result from PM exposure.

25. Coal is the fuel most likely to result in significant sulfur dioxide production.

27. Most lead exposure in the United States is through old (pre-1978) paint in homes or in the dust around older homes.

29. Mercury is a neurotoxin.

31. $CO_2 + H_2O \rightarrow H_2CO_3 \rightarrow H+ + HCO_3^-$

33. Acid rain has a pH less than 5.

35. Aquatic invertebrate shells and fish skeletons are largely composed of $CaCO_3$, which reacts with acid to form H_2CO_3, which dissociates to form H_2O and CO_2. The shells and skeletons dissolve.

37. The AQI normalizes pollutant concentrations onto a scale from 0–500. AQI values are divided into several categories based upon health hazard (good to hazardous) so that citizens can easily assess the quality of air without having to read and interpret individual pollutant concentrations.

39. Toxicity is the intrinsic health threat of a pollutant. Exposure takes into consideration both the concentration of a pollutant and the amount of time it is in the atmosphere.

Industrial Revolution

Jacobson, M. (2002). *Atmospheric pollution: History, science, and regulation.* New York, NY: Cambridge University Press.

Pollution and Health

Bell, M. L., Peng, R. D., Dominici, F., *et al.* (2009). Emergency hospital admissions for cardiovascular diseases and ambient levels of carbon monoxide results for 126 United States urban counties, 1999–2005. *Circulation, 120*(11), 949–69.

Chabot, F., Gomez, E., Guillaumot, A., *et al.* (2009). Acute exacerbations of chronic obstructive pulmonary disease. *La Presse Médicale, 38*(3), 485–95.

Chiu, H. F., Yang, C. Y., (2009). Air pollution and emergency room visits for arrhythmias: are there potentially sensitive groups? *Journal of Toxicology and Environmental Health, 72*(13), 817–23.

Dales, R. E., Cakmak, S., and Vidal, C. B. (2009). Air pollution and hospitalization for headache in Chile. *American Journal of Epidemiology, 170*(8), 1057–66.

Grant, W. B. (2009). Air pollution in relation to U.S. cancer mortality rates: an ecological study; likely role of carbonaceous aerosols and polycyclic aromatic hydrocarbons. *Anticancer Research, 29*(9), 3537–45.

Peng, R. D., Bell, M. L., Geyh, A. S., *et al.* (2009). Emergency admissions for cardiovascular and respiratory diseases and the chemical composition of fine particle air pollution. *Environmental Health Perspectives, 117*(6), 957–63.

Srebot, V., Gianicolo, E. A. L., Rainaldi, G., *et al.* (2009). Ozone and cardiovascular injury. *Cardiovascular Ultrasound, 7*, Article 30.

Wingspread Conference on the Precautionary Principle. (1998). *The Wingspread consensus statement on the precautionary principle.* Retrieved from http://www.sehn.org/wing.html

Common Air Pollutants

www.epa.gov

VOCs and PAHs

Finlayson-Pitts, B. J., and Pitts, J. N. (1997). Tropospheric air pollution: Ozone, airborne toxics, polycyclic aromatic hydrocarbons, and particles. *Science, 276*(5315), 1045–51.

Gulyurtlu, I., Abelha, P., Gregorio, A., *et al.* (2004). The emissions of VOCs during co-combustion of coal with different waste materials in a fluidized bed. *Energy and Fuels, 18*(3), 605–10.

Hannigan, M. P., Cass, G. R., Penman, B. W., *et al.* (1997). Human cell mutagens in Los Angeles air. *Environmental Science and Technology, 31*(2), 438–47.

Leiter, J., Shimkin, M. B., and Shear, M. J. (1942). Production of subcutaneous carcinomas in mice with tars extracted from atmospheric dusts. *Journal of the National Cancer Institute, 155*(3), 167–74.

NOx, VOCs, and Ozone

Ryerson, T. B., Trainer, M., and Holloway, J. S. (2001). Observations of ozone formation in power plant plumes and implications for ozone control strategies. *Science, 292*(5517), 719–23.

Mercury

Dufault, R., LeBlanc, B., Schnoll, R., *et al.* (2009). Mercury from chlor-alkali plants: Measured concentrations in food product sugar. *Environmental Health, 8*(1).

http://www.montereybayaquarium.org/cr/seafoodwatch.aspx

Acid Rain

Jacob, D. J., (1999). *Introduction to atmospheric chemistry.* Princeton, NJ: Princeton University Press.

Coal Scrubbing

Dunham, J. T., Rampacek, C., and Henrie, T. A. (1974). High-sulfur coal for generating electricity. *Science, 184*(4134), 346–51.

www.ucsusa.org/clean_energy

Air Quality Index

www.airnow.gov

Lead

Kaye, S., and Reznikoff, P. A comparative study of the lead content of street dirt in New York City in 1924 and 1934. *Journal of Industrial Hygiene and Technology, 29*(3), 178–79.

Kovarik, Bill. (1994, rev. 1999). *Charles F. Kettering and the 1921 discovery of tetraethyl lead in the context of technological alternatives.* Presented at the Society of Automotive Engineers Fuel and Lubricants Conference, Baltimore, Maryland.

Moore, Colleen. (2003). *Silent scourge: Children, pollution, and why scientists disagree.* New York, NY: Oxford University Press.

www.cdc.gov/nchs/nhanes

Smoking

Rabinoff, M., Caskey, N., Rissling, A., et. al. (2007). Pharmacological and chemical effects of cigarette additives. *American Journal of Public Health, 97*(11), 1981–91.

http://www.pmintl-technical-product-information.com

CREDITS

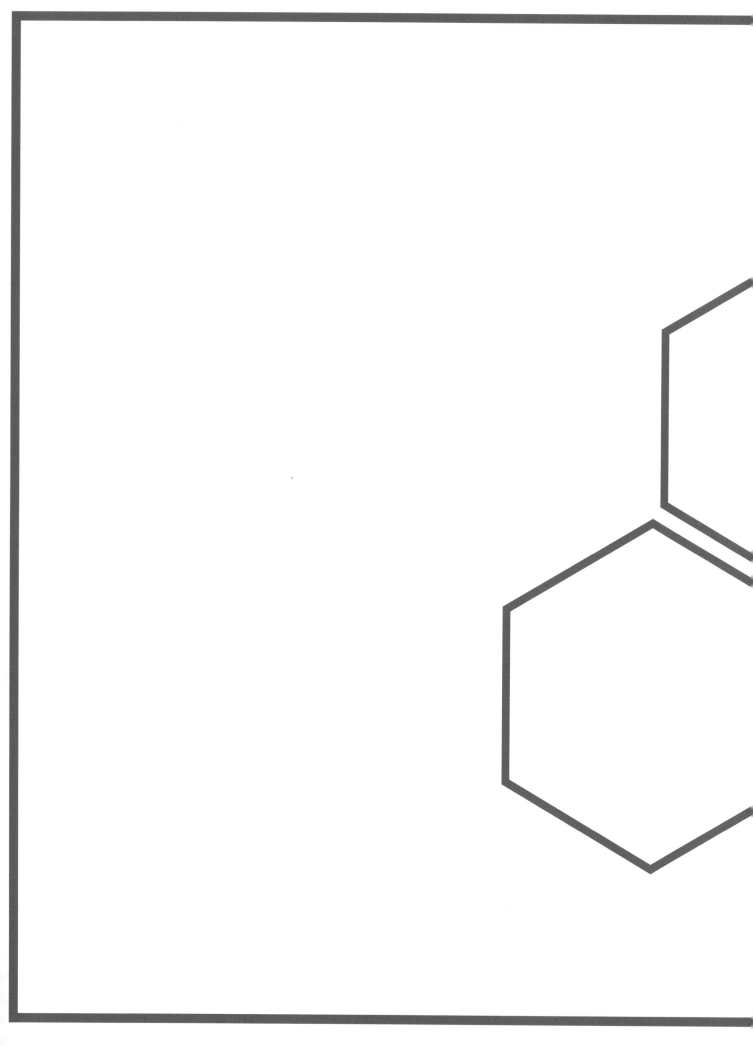

Unit 3

PERSO N AL
C He MISTRY

Pharmaceuticals and Poisons

7

IMAGE 7.1. The bark from the white willow tree (*Salix alba*) can be used to make analgesic tea.

Almost everyone has benefited at times from the healing power of pharmaceutical compounds. We don't often stop to ponder the chemical nature of the medications we're taking, how they were discovered, or what exactly they do; we simply take the pills we're prescribed, and we generally get better. There's a fine chemical line, however, between a miracle cure and a deadly poison—a line we generally try not to cross, occasionally flirt with, and sometimes completely disregard. Considering the nature of this line leads us to consider some important questions. Chemically, what are pharmaceuticals? How are they discovered? What differentiates pharmaceuticals from natural compounds, or from toxins? In this chapter, we will explore the chemistry of compounds that heal and those that kill, and the very fine line that separates the two.

7.1   From Tea to Tablets

Throughout history, humans have sought out ways to reduce our pain and treat our illnesses. Until the middle of the nineteenth century, many treatments came in the form of herbs known by reputation to have healing effects. For instance, the ancient Greeks used a tea made from white willow (*Salix alba*) bark to reduce headache, inflammation, pain, and fever (Image 7.1).

Because making tea from willow diluted whatever ingredient in the bark was responsible for its effect, chemists began to wonder whether there might be a way to concentrate the active ingredient in order to yield a more powerful and effective compound. In the mid-1800s, a chemist named Johannes Buchner discovered that it was possible to isolate yellow crystals[1] of a substance that he called salicin, since it was extracted from *Salix alba*. Further investigation of the compound, both by Buchner and others, revealed that salicin was actually a glucose (sugar) unit bonded to a hydrocarbon

NOTE[1]

ring to make one large molecule. Chemical removal of the sugar yielded a second pure substance, named salicylic acid for its parent compound and its acidity:

Salicylic acid is a very large molecule, and its Lewis structure is somewhat cluttered and unwieldy. This is a common problem we face in portraying large molecules; in particular, organic compounds tend to be quite large. **Organic chemistry is the study of organic compounds, which are carbon-based molecules made up of the major and minor elements of life.** The fact that these molecules are always carbon based allows us to simplify the manner in which they are drawn. In organic molecules, carbon always makes four bonds, so as to fill its octet. Remembering this, we can actually leave the hydrogen atoms that are bonded to carbon out of our organic structures; we simply mentally fill in the requisite number of hydrogens when looking at a molecule. Because our discussion of acids in Chapter 5 showed us that the number of bonds formed by oxygen, nitrogen, and other elements may vary somewhat depending upon the circumstances, we can't make the same simplification with regard to hydrogen atoms bonded to elements other than carbon; those hydrogens must be explicit. Using our hydrogen atom simplification, salicylic acid would look like this:

Note that the hydrogen atom attached to each oxygen atom is still shown. The hydrogens attached to carbon are not shown, but are still part of the molecule; we're merely rendering the molecule quicker and easier to draw by making the hydrogen atoms implicit rather than explicit. **In order to determine how many hydrogen atoms are attached to a particular carbon, we count the explicit bonds and then mentally fill in enough hydrogen atoms to complete carbon's set of four bonds:**

This carbon has four explicit bonds; it does not have any implicit hydrogens.

This carbon has three explicit bonds; it must have one bond to hydrogen to complete its octet.

There is one further simplification we can make in drawing organic molecules: knowing that organic compounds are carbon based, we can eliminate the atomic symbol for carbon from our structure. **Atoms of elements other than carbon are** denoted by their atomic symbols, and anywhere in the molecule that an atomic symbol is omitted, we assume the element in question is carbon. Portrayed thus, our salicylic acid molecule looks much less cluttered:

 NOTE[2]

It's impossible and/or impractical to draw very small molecules in line-angle. Methane (CH_4), for instance, would disappear entirely!

Note that the bonds make it quite clear where the carbon atoms are located, and that despite our simplifications, the molecular structure above (called a **line-angle formula**) allows a significant amount of information to be conveyed by a relatively uncluttered visual aid. Any compound with at least one bond from carbon to a nonhydrogen atom[2] can be drawn in line-angle. In order to help make sense of the line-angle formulas, it's important to use correct bond angles. In a small molecule like ethane (CH_3CH_3), the line-angle diagram requires no portrayal of bond angles; ethane simplifies to a mere line:

In line-angle, nothing of ethane remains but the C-C bond.

Notice that in this example, the full Lewis structure of ethane is drawn without the wedge-dash notation we've been using to indicate tetrahedral geometry. Instead, plain bonds are drawn where the wedge and dash lines would have gone, and are taken to imply wedge and dash. Simplifications of structure, such as implying rather than explicitly drawing the wedge and dash lines, are common as we progress through an understanding of chemistry. Even without explicit wedge-dash notation, we are still meant to understand that a tetrahedral structure, with one hydrogen from each carbon coming out of the page toward us, and one hydrogen from each carbon headed away from us, is being represented. This is another common way to portray organic structures. It's not nearly as concise as line-angle, but it provides some simplification and is useful in drawing very small molecules such as ethane that don't lend themselves particularly well to line-angle.

A larger molecule, like pentane ($CH_3CH_2CH_2CH_2CH_3$), would be impossible to make sense of if we tried to simplify it without attending to bond angles. Imagine the (incorrect) line-angle diagram that would result if we simplified a flat Lewis structure of pentane:

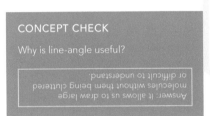

Without attending to bond angles, our pentane would be nothing more than a line. The would incorrectly suggest that, rather than five carbons, the molecule in question consisted of only two (like ethane).

Instead, if we are careful to portray pentane with appropriate 109.5° bond angles, we get a sensible, readable line-angle simplification:

In an appropriately drawn line-angle diagram, pentane's five carbons are clearly evident.

Armed with an easier way to draw and interpret organic molecules, let's return to our discussion of willow bark extract. It was determined through experimentation that salicylic acid, like salicin, was an anti-inflammatory and analgesic. Unfortunately, it was also terribly irritating to the stomach. Felix Hoffmann, a chemist working for the Bayer Company, chemically modified the molecule.³ His modified molecule was called acetylsalicylic acid (Image 7.2).

Hoffmann's molecule, when tested on human patients, proved to be both effective and far less irritating to the stomach than salicylic acid. Bayer named and marketed the drug as Aspirin.⁴ Aspirin's initial popularity and subsequent ubiquity have worked together to make it the best-selling pharmaceutical of all time.

In the years since the synthesis of aspirin, other pharmaceuticals have been developed in this same way: a natural product is observed to have a desirable effect, but side effects are often noticed as well. In an attempt to amplify the positive while reducing or eliminating the negative, pharmaceutical companies purify out the active ingredient and often modify it. Examples of drugs obtained or developed from natural sources are shown in Image 7.3, and include quinine from cinchona tree

IMAGE 7.2. Acetylsalicylic acid and an early advertisement for aspirin.

IMAGE 7.3. Cinchona (a), foxglove (b), and Pacific yew (c)—some of the many plants from which pharmaceuticals have been extracted or derived.

bark (used to treat malaria), digitalis from foxglove (used to treat heart conditions), and paclitaxel from the Pacific yew (a potent chemotherapeutic)

7.2 Pharmaceutical Design

When Felix Hoffmann modified the willow derivative salicylic acid to the less irritating compound known as aspirin, how did he know what to do? Was he simply playing "mad scientist," mixing chemicals in his lab and hoping for the best? Or was he completely sure of what needed to be done in order to produce an effective pharmaceutical? The reality lies somewhere between these extremes, and is typical of the development of pharmaceuticals. In Hoffmann's case, the structure of salicylic acid was known. Further, enough was known about the functional groups on the molecule to reasonably conjecture which of them was responsible for the side effects. **A functional group is a characteristic arrangement of atoms in a molecule that lends a predictable set of properties to that molecule.** We've already come across one functional group: the carboxylic acid. Note that salicylic acid is a carboxylic acid, with the characteristic –COOH atomic arrangement. Not only does the presence of the carboxylic acid provide us with information about the molecule (it's acidic), it also allows us to condense these atoms further in the line-angle diagram of salicylic acid, as shown below:[5]

In addition to being acidic, carboxylic acids typically smell unpleasant or bracing. Vinegar, an aqueous solution of acetic acid (CH_3COOH) is one example of the characteristic, biting odor of these molecules. Another carboxylic acid, butyric acid ($CH_3CH_2CH_2COOH$), contributes to the fragrances of rotting butter and human body odor. There are many functional groups besides that of the carboxylic acid. A benzene ring is one such example. **This functional group consists of six carbons arranged into a hexagon, with alternating single and double bonds between them.** Benzene rings (also called aromatic rings because the molecules containing them are often highly fragranced—sometimes for better, sometimes for worse) are very common in pharmaceuticals. The salicylic acid molecule contains a benzene ring. In addition, salicylic acid contains a functional group that was of particular interest to Hoffmann. **An –OH functional group is generally referred to as an alcohol.** Pictured in Image 7.4, common household alcohols include methanol (wood alcohol), ethanol (grain alcohol—the kind in beer, wine, and liquor), and isopropanol (rubbing alcohol).

In the particular case of an alcohol that is attached directly to a benzene ring, a new compound functional group is formed. This group is called a phenol, and

NOTE[6]

Phenol-containing compounds were among the first antiseptics, used to wash and cover wounds in order to prevent infection. In addition to killing pathogens, unfortunately, they also had a tendency to burn the skin chemically.

IMAGE 7.4. (From left to right) Methanol, ethanol, and isopropanol structures and line-angle diagrams.

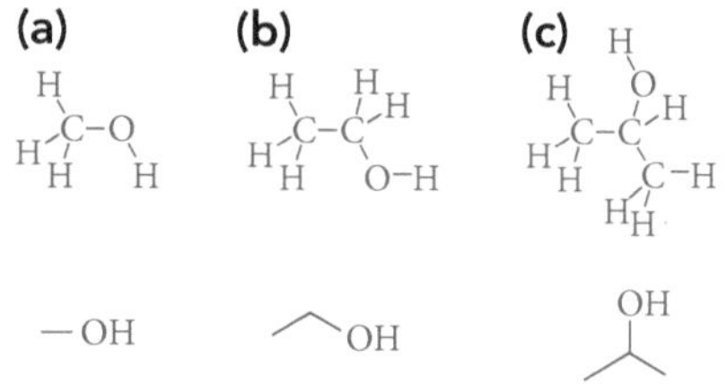

(a) **(b)** **(c)**

phenol groups were known by Hoffmann and his contemporaries to be very irritating to the skin and mucous membranes.[6] Hoffmann hypothesized that it might be the phenol functional group that was responsible for irritation of the stomach lining. By modifying rather than simply removing it, he was able to maintain much of the molecule's original shape (an important consideration, as we'll see later in this chapter). As you can see, knowledge of functional groups and organic chemistry was the driving force behind Hoffmann's ability to manipulate willow extract to produce the far more effective pharmaceutical, aspirin.

Organic molecules sometimes contain a single functional group, but far more commonly they contain many. Methanol, as we saw above, has only the alcohol functional group (carbons singly bonded to other carbons and carbons bonded to hydrogens are not considered functional groups—rather, these are backbone elements). For this reason, we can say that methanol is an alcohol. Salicylic acid, though, is a much larger and more complicated molecule. It's in possession of two separate functional groups: a phenol and a carboxylic acid. As such, we wouldn't say *salicylic acid* is *a phenol*; rather, it is correct to say *salicylic acid* contains *a phenol*.

In addition to the functional groups relevant to salicylic acid, there are several others worth discussing. **Amines are nitrogen-containing groups.** They are common in pharmaceuticals—we'll see them later in this chapter as one of the functional groups in morphine, for instance. Amines often have dead, unpleasant, or fishy odors. Ammonia is one of the least objectionable-smelling of the amines, and still, few people would care to smell it for any length of time. Putrescine and cadaverine, also amines, lend rotting cadavers their putrid odor (Image 7.5).

You may be sensing that organic nomenclature is quite a bit more complex than the binary compound nomenclature discussed in Chapter 1. A complete discussion of the organic naming system is well beyond the scope of this text. In general, though, organic compounds often have two names—one systematic (as in *methanol*), and one common (as in *wood alcohol*). Sometimes, the common names sound like systematic names. Putrescine, for instance, sounds like a chemical name, but is in fact the common name for a compound systematically called 1,4-diaminobutane. The name putrescine shares the twin benefits of being less of a mouthful to pronounce than the systematic name and being representative of a notable source of the compound (putrid flesh). Common names are encountered frequently in organic chemistry and are often informative (β-*carotene*, for instance, is found in and named for carrots) and sometimes amusing (a large organic molecule with a shape similar to that of the geodesic dome designed by architect Buckminster Fuller is called *buckminsterfullerene*).

Esters, like phenols, are compound functional groups; they combine a carboxylic acid with an alcohol. Unlike phenols, in which the benzene ring and alcohol

(a) **(b)** **(c)**
NH_3 H_2N~~~~NH_2 H_2N~~~~~NH_2

IMAGE 7.5. (From left to right) Ammonia, putrescine (1,4-diaminobutane), and cadaverine (1,5-diaminopentane) exhibit the characteristic unpleasant odor of amines to varying degrees.

can't easily be separated chemically, esters are literally synthesized from (and can be decomposed into) their constituent parts:

A generic alcohol. As in the generic carboxylic acid structure, "R-prime" (to differentiate it from the "R" used in the acid structure) means this can be any alcohol.

This bond joins the elements of the parent carboxylic acid to the parent alcohol.

A generic carboxylic acid. The "R" is simply taken to mean "any hydrocarbon backbone" so as to indicate that this is not a particular carboxylic acid, but is instead any molecule containing the COOH functional group.

An ester contains elements of the parent carboxylic acid and the parent alcohol.

The resulting ester is neither a carboxylic acid (it is not acidic) nor an alcohol, but has its own unique set of properties. Esters are highly fragrant molecules, but unlike the unpleasant carboxylic acids and amines, these compounds generally have very pleasing odors. In fact, esters are responsible for many fruit and floral fragrances (Image 7.6).

It's worth remembering that the ester is, at its chemical heart, the combination of a carboxylic acid and an alcohol, and under the right circumstances the ester can decompose into its constituent parts. In fact, the decomposition of an ester takes place readily in the presence of water and a very small amount of acid or base. The amount of acid or base required for esters to decompose is so small that when exposed to surfaces or air (which are always contaminated with small quantities of mildly acidic or basic compounds), esters will generally react fairly quickly. This explains why expensive perfume fragrances tend to wear true throughout the day,[7] while inexpensive perfume fragrances change (often becoming unpleasant) over time. Perfumes are made using scented molecules, many of which are esters. More expensive perfumes use oil as the solvent in which fragrance molecules are dissolved. The oil protects the delicate esters from water and keeps them from reacting chemically. Less expensive than oil, however, is the alcohol and water solvent used in cheaper scents. In the bottle, protected from acids and bases, the esters do not react. Sprayed on the skin, however, esters are exposed to the environment, and react in short order. The resulting carboxylic acids tend to smell far less pleasant than the original esters.

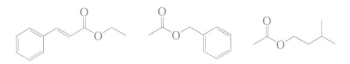

IMAGE 7.6. (From left to right) Ethyl cinnamate (cinnamon), benzyl acetate (a component of the fragrance of pear and strawberry), and isopentyl acetate (a component of banana) are esters.

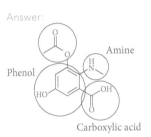

CONCEPT CHECK

What functional group does aspirin
have that salicylic acid lacked?

Answer: Aspirin has an ester group.

7.3 Wonder Drugs

An **ether** is yet another functional group. Unlike the highly reactive esters, ethers are fairly inert. **As a group, they consist of two hydrocarbon backbones—these can be two of the same hydrocarbon or they can be different from one another—joined by a common oxygen.** As in the ester example above, R and R' are taken to mean *hydrocarbon of some sort*, and are used in portraying a generic ether:

Most of us have heard the word ether before: it's commonly used to refer to a particular molecule more appropriately called by its chemical name, *diethyl ether*:[8]

Diethyl ether gained popularity as an early anesthetic. Before the middle of the nineteenth century, no anesthetics were known. Surgical procedures were completed on fully (or at least mostly) conscious patients. Prior to surgery, patients were often encouraged to consume large quantities of alcohol in the hope that this would reduce their sensation of pain somewhat, or at least render them less aware of their milieu. Regardless, surgical assistants were sometimes required to forcibly restrain patients.[9] It goes without saying that the horror of surgery in the days before anesthetic made it a last resort medically; surgeons had to work quickly on their suffering patients, making procedures sloppier than those practiced as a near art form today, and patients often succumbed to shock from the pain (not to mention the prevalence of postsurgical infection). Diethyl ether, first used by the dentist William Morton in 1846, was found to reduce consciousness so that surgery could be performed on a motionless (and mercifully pain-free) patient. Not only did this increase the popularity of surgery as a therapy, it also allowed for the honing of surgical skills and development of techniques that increased the effectiveness of procedures. Despite the fact that ether was abandoned by the medical world rather shortly after its introduction—it had the potential to be explosive and left patients feeling very ill upon waking—it was replaced with a series of increasingly advanced anesthetics. A modern physician has at his or her disposal a variety of anesthetics—so many, in fact, that anesthesiology has become a medical specialty in its own right—that can be selected based upon the procedure, patient allergies, or a mere matter of preference.

NOTE[8]

It's not unusual for the most common (or first discovered) member of a group of molecules to be referred to in popular parlance by the functional group name. *Ethanol*—grain alcohol—is usually simply called *alcohol*. The simplest of the phenols, an otherwise unsubstituted benzene ring with an OH group attached, is just *phenol*.

NOTE[9]

Old surgical accounts suggest that the surgeons detested inflicting these surgeries as much (or nearly so) as their patients detested experiencing them; it was not unknown for surgeons to brace for an operation with a bit of alcohol themselves!

7.4 Drug Solubility

As noted in Chapter 4, the polarity of a compound produces its solubility. Highly polar compounds (particularly ionic and hydrogen bonding molecules) dissolve well in water, while nonpolar compounds are hydrophobic and dissolve well in other nonpolar molecules, including fats. As we turn our discussion to organic molecules, most of which are quite large, the issue of polarity becomes somewhat more complex. Take, for instance, the aspirin molecule. This molecule has some areas that are highly polar and other areas in which the hydrocarbon backbone predominates:

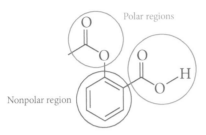

In a case such as this, how do we ultimately define the polarity of the molecule? It turns out that solubility is not as rigorously dichotomous as implied earlier. Molecules need not be either entirely **hydrophilic** (which means water loving) or entirely hydrophobic. Rather, solubility is a continuum, with polar functional regions lending hydrophilicity, and nonpolar regions lending hydrophobicity. **To determine solubility, we compare the predominance of polar and nonpolar regions and come to a judgment. Because the continuum makes it difficult to judge in absolute terms the polarity of a compound, in most cases we'll simply use relative terms to compare two or more compounds' solubilities.** For instance, a compound with more OH groups relative to its number of carbons is likely to be more soluble in water than a compound with fewer OH groups relative to its number of carbons (Image 7.7).

In designing pharmaceuticals, water solubility is of great importance—a drug must dissolve in water in order to be absorbed from the digestive tract or dissolve in the aqueous media of the body. While some drugs are appropriately hydrophilic on their own, others (including many of the large amines) are only slightly soluble in water. Morphine, a potent analgesic shown below, has very limited water solubility, making it difficult to administer in its pure form:

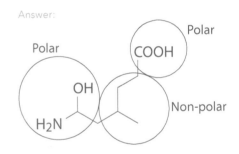

(a)   **(b)**   **(c)**

IMAGE 7.7. Ethylene glycol (left) is likely to have greater water solubility than ethanol (center), which is likely to be more water-soluble than hexanol (right).

TRY THIS

Determine which areas of the molecule below are polar, and which are nonpolar.

Answer:

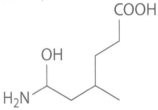

TRY THIS

Which is likely to be more water-soluble and why?

Answer: The molecule on the right will be more water-soluble—it has more polar functional areas (-COOH is polar) compared to the number of carbons.

In order to make morphine water-soluble, it's possible to take advantage of the chemical properties of one of its functional groups. Ammonia, as we saw in Chapter 5, is basic. Many of the related amines are basic as well. **It really doesn't matter whether a nitrogen atom has three hydrogen atoms, three hydrocarbon groups, or a combination of the two attached to it—the chemical property of basicity is consistent across the amines. It's possible to take advantage of this by neutralizing the amine group on morphine with an acid (often HCl). The resulting ionic salt (written *morphine HCl*) has greatly enhanced water solubility:**

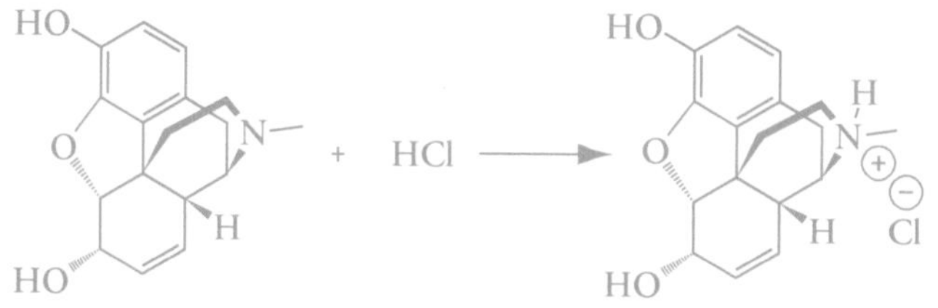

It's quite common to see acidic salts of pharmaceuticals: a few common examples from the medicine cabinet include diphenhydramine HCl (an antihistimine used to reduce allergy symptoms), pseudoephedrine HCl (a decongestant), and dextromethorphan HBr (a cough suppressant). In each case, the parent pharmaceutical is rather hydrophobic, while the amine salt is significantly more water-soluble.

The primary goal of salt formation from pharmaceuticals is to enhance water solubility, but there are additional benefits. Shelf lives of salts are longer than those of the parent compounds, and odor is reduced—a major consideration for many amine-containing pharmaceuticals!

7.5 Receptors and Pharmacophores

In addition to being an interesting example of a pharmaceutical often administered as a salt, morphine has a long and intriguing history. Produced naturally by the opium poppy (*Papaver somniferum*), morphine's analgesic, sleep-inducing, and hallucinogenic properties have been appreciated for hundreds, perhaps thousands, of years. Unfortunately, morphine is highly addictive, which limits its pharmacological utility. As a result, pharmaceutical chemists have long sought to produce alternative compounds that might replicate the activity of morphine without the addictive side effects. Two compounds that produce similar analgesia (with greatly reduced hallucination and potential for addiction) are codeine and meperidine (marketed under the trade name Demerol). While the chemical similarity between codeine and morphine is clear from the structures as they're most commonly drawn, at first glance, meperidine looks quite different:

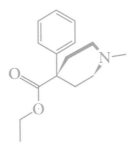

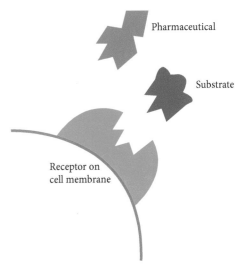

Drawn from a slightly different three-dimensional perspective, however, the similarities between meperidine and the other two compounds become more clear:

IMAGE 7.8. A natural substrate fits into the receptor site. Any pharmaceutical binding to the same receptor would be expected to have a functional area (pharmacophore) similar to that of the natural substrate. The rest of the molecule, however, could look quite different.

The similarity in shape among these molecules is no coincidence: **pharmaceuticals operate in the body by mimicking, enhancing, or repressing the action of the body's own products. In essence, pharmaceuticals act as artificial "keys" for the body's own "locks," where the "locks" are more properly called receptors, and are large protein-based molecules on body cells.** A substrate is a compound that fits into a receptor like a key, producing some effect when it binds. Morphine fits into a receptor whose native substrate is an *endorphin*, a pain-reducing compound produced by the body's cells. Receptors may have one or more natural substrates, but any molecule with a shape similar to that of the receptor's natural substrate will be able to bind. **With regard to morphine and compounds of similar activity, meperidine shows us that it isn't necessary for a molecule to be nearly identical to morphine (as codeine is) to produce the morphine effect; in fact, only a few key functional areas appear to be important.** This is quite typical of pharmaceuticals. Generally, a core functional area called a pharmacophore is responsible for binding to the receptor. The pharmacophore will be identical or nearly identical in all substrates that can bind to a given receptor. The rest of the molecule, however, can have any of a number of shapes (Image 7.8). Areas apart from the pharmacophore are responsible for various pharmaceutical properties, but do not bind to the receptor. It is often through manipulation of the nonpharmacophore molecular structure that chemists achieve a reduction in side effects and otherwise manipulate the properties of drugs.[10]

While the morphine-like molecules codeine and meperidine are frequently used today—as is morphine, though with great caution—a very early attempt to synthesize a less addictive morphine analog produced unexpected results. Shortly after Felix Hoffmann acetylated salicylic acid to produce aspirin, he was asked

NOTE[10]

That said, though, manipulation of the pharmacophore itself—provided the shape isn't changed so drastically as to prevent binding to the receptor—can change the substrate's *affinity* for the receptor. That is, changes to the pharmacophore can change how tightly a drug clings to the receptor and impact variables such as the strength and length of its effect in the body.

by the Bayer Company to work on producing a less addictive form of morphine. Probably suspecting he was going to end up with codeine (which was known at that time to be less addictive but which had not yet been synthesized), Hoffmann acetylated morphine. The result was diacetylmorphine:

Bayer marketed the new drug extensively as a nonaddictive morphine substitute and a cough suppressant. Unfortunately, it was soon discovered that diacetylmorphine was quickly and readily converted to morphine in the body and was, in fact, even more active than its parent compound. It was also more addictive. While most of us probably aren't familiar with the chemical name *diacetylmorphine*, almost everyone has heard of Bayer's trademark name for the drug: Heroin (Image 7.9). The heroin episode was quite embarrassing for Bayer, for obvious reasons.

IMAGE 7.9. Bayer Heroin didn't enjoy the success that Felix Hoffman, developer of Bayer Aspirin, had hoped it would as a nonaddictive cough syrup.

NOTE[11]

The cost of the legal battle was significant, however. Coca-Cola ended up settling out of court and did eventually reduce the caffeine in their product. Cocaine, however, remained an ingredient for some time.

7.6 Early Pharmaceutical Regulation

With the rapid advancements in pharmacological chemistry taking place in the late nineteenth and early twentieth centuries, clearly some regulation was in order. **In 1906 President Theodore Roosevelt signed the Food and Drug Act, the primary stipulations of which prevented the transport of goods across state lines unless they were correctly labeled and free from contaminants.** There were no provisions for safety or efficacy; the proverbial "snake oil" salesman was free to sell any substance he liked—provided that it was pure and correctly labeled—whether it was toxic, ineffective, or both. The weaknesses of the 1906 Food and Drug Act are readily apparent: Coca-Cola, for instance, won a lawsuit (*United States vs. Forty Barrels and Twenty Kegs of Coca-Cola*, 1909) in which they were charged with putting excessive caffeine (as well as cocaine) into their product. Thanks to accurate labeling and lack of contamination, however, Coca-Cola was spared legal sanctions.[11]

A major improvement in food and drug safety legislation came in 1938, but only after a tragic incident demonstrated beyond argument the need for further control. In 1937 a pharmaceutical company found what appeared to be an excellent solvent for the antibiotic sulfanilamide. While quite effective, sulfanilamide has exceedingly low solubility in water. It's also very bitter, making it difficult to administer to children. The company determined that the compound dissolved readily in diethylene glycol. The solvent had the added advantage of being sweet, increasing palatability to children. Unfortunately, diethylene glycol is also quite toxic (a fact that was known to chemists at the time, but was somehow overlooked by the pharmaceutical company), and no animal testing preceded the marketing of the new elixir.

Of a group of 353 children to whom the drug was administered, 100 died of kidney failure. This created a firestorm of public, scientific, and political response, the result of which was the 1938 Food, Drug, and Cosmetic Act.[12] **This act required truthful labeling and proven safety of all drugs.** Manufacturers of pharmaceuticals had to request a new drug application for each proposed product, to be reviewed by the FDA (which had been created about a decade prior). The FDA required that the new drug application include submission of toxicity studies.

It is worthy of mention that while safety standards were made more stringent via the 1938 FD&C Act, proof of efficacy was still not required. A major concern associated with a legislative act that allows for pharmaceuticals to be produced and marketed with some proof of safety but none whatsoever of efficacy is that the consumer is unable to formulate a risk-to-benefit ratio for purposes of decision-making. If a drug is slightly risky but highly effective, such that the potential for benefit far outweighs the potential for harm, a consumer might choose to use the medication. If a drug is risky and minimally or rarely effective, a consumer might decide against using the drug. Because no drug is completely free from safety concerns, it is only when armed with information about the potential benefit of a substance that a consumer is able to use risk–benefit analysis in determining the best course of action. **Put more simply, it is impossible to do a risk–benefit analysis if one does not have information regarding both the risks *and* the benefits.** The lack of an efficacy requirement was a major weakness of the 1938 legislation. And then came thalidomide, and everything changed.

IMAGE 7.10. Sulfanilamide elixir, with its toxic diethylene glycol solvent, was the driving force behind more stringent safety regulations.

 NOTE[12]

We see evidence of this regulatory measure every time we read the ingredient label on a food or drug containing artificial coloring. FD&C Red 40, for instance, is a red color approved under the Food, Drug, and Cosmetic Act.

7.7 Through the Looking Glass—Healer or Killer?

Beginning in the late 1950s, the pharmaceutical thalidomide was marketed as a mild sedative and antinauseant. It was particularly appreciated by women in the early months of pregnancy because it helped relieve the symptoms of morning sickness. The drug was prescribed extensively in Germany and England, but was refused an application for release in the United States by FDA inspector Frances Oldham Kelsey, who felt that more safety testing was necessary before the drug should be brought to market.[13] Of course, pregnant women in the United States were outraged (temporarily) at the FDA's refusal to allow them access to the "wonder drug" available in Europe. The later realization that Kelsey's decision had spared U.S. babies the fates of their European counterparts converted her, in popular opinion, from villain to heroine. Shortly after thalidomide made its debut in Europe, physicians became aware of an increased incidence of what had been an historically rare birth defect called *phocomelia*, meaning "flipper limb" (Image 7.11). During certain key periods of

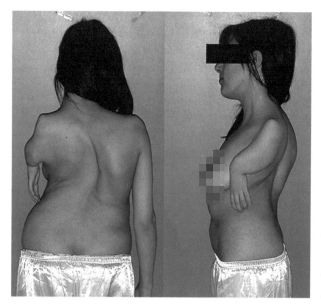

IMAGE 7.11. An adult with phocomelia.

NOTE¹³

(a)   (b)

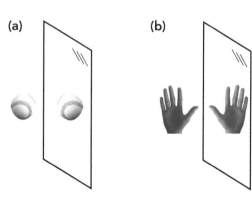

gestation, thalidomide inhibits normal embryonic development, resulting in severe physical deformities. Hundreds of thousands of children were born with these defects during the 1950s and 1960s, mostly in Europe. Kelsey's refusal to allow marketing of the drug spared the United States from widespread effects.

There are two interesting issues here. The first is that despite Kelsey's feeling that the drug's safety was not adequately established, it had, in fact, undergone testing. Adult studies of thalidomide revealed it to be safe and low in side effects. Unfortunately, developing embryos (and fetuses) are often much more sensitive to drugs than adults. The *teratogenic*, or development-related, effects of thalidomide were unknown because drug studies are not traditionally done on pregnant women; the risk is felt to be too great. **A second interesting bit of information about thalidomide is that it exists in two forms that are mirror images of each other. One form is a safe sedative; the other is a teratogen and causes birth defects.**

How is it possible for a molecule to have a mirror image? To answer this question requires some three-dimensional visualization. Imagine, first, a symmetric object like a baseball. If a baseball is held up to a mirror, the mirror image of the baseball is identical to the original. An asymmetric object, however, such as a hand, is not identical to its mirror image (Image 7.12). The mirror image of a right hand is a left hand. They have many similarities, and can do many of the same things, but they are not **superimposable**, meaning that where the left hand has a thumb, the right has a pinkie finger and vice versa. If we flip one hand over so that thumbs and pinkies are aligned, now the left hand has a back where the right has a palm, and so forth. A hand is not superimposable upon its mirror image.

Chemicals, depending on their symmetry, can be superimposable upon their mirror images (like baseballs), or nonsuperimposable upon their mirror images (like hands). Chloroform, an early anesthetic, is an example of a symmetric molecule. If a chloroform molecule is reflected in a mirror and the mirror image is rotated, we can see that the mirror image is identical to the original (Image 7.13). Remember that molecules, like all physical objects, can be turned over and rotated; this does not change the molecule chemically. We can turn a mirror image of a molecule any way we need to in order to superimpose it upon the original. We mustn't break bonds or rearrange the molecule, however; to do that would be to make a new, different molecule.

Symmetric molecules like chloroform, which are superimposable upon their mirror images, are called **achiral** molecules. Alternately, a molecule may be asymmetric. The compound in Image 7.14 is an example of such a molecule:

There is no way, short of breaking bonds and rearranging the molecule, to superimpose the molecule in Image 7.14 on its mirror image. Molecules like this,

IMAGE 7.12. Mirror images of symmetric objects are identical to (superimposable upon) the original object. This is not so with asymmetric objects.

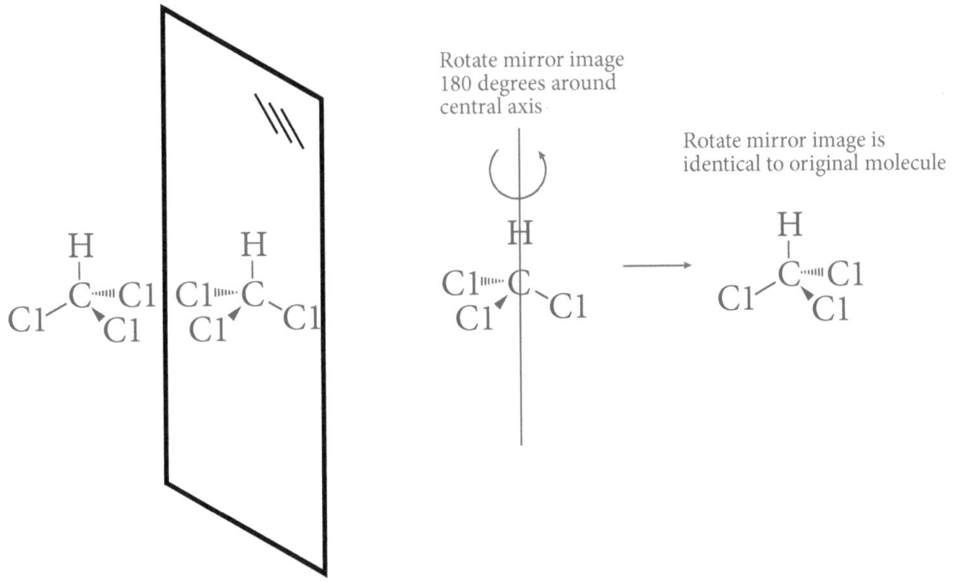
Rotate mirror image
180 degrees around
central axis
Rotate mirror image is
identical to original molecule

IMAGE 7.13. Chloroform is identical to its mirror image.

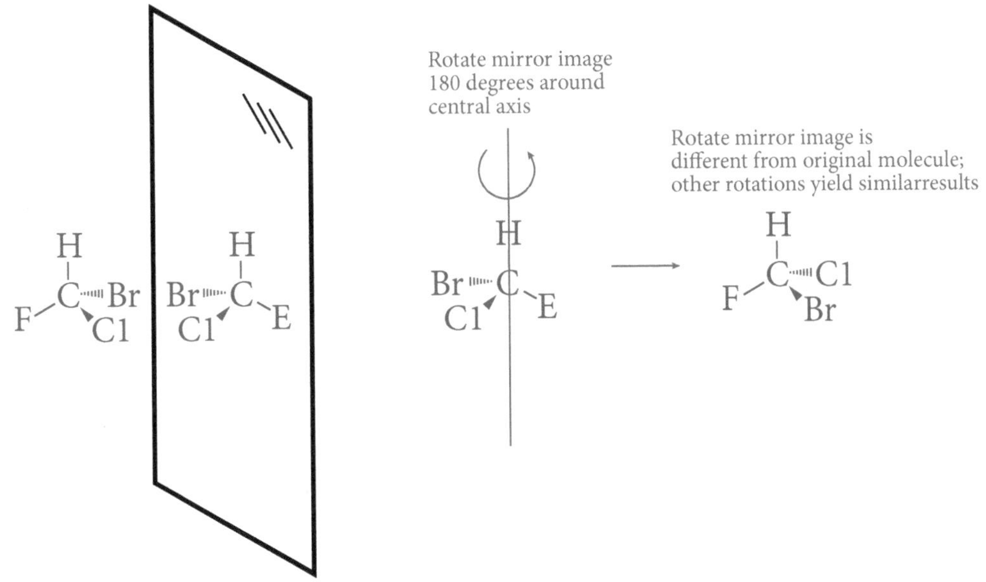
Rotate mirror image
180 degrees around
central axis
Rotate mirror image is
different from original molecule;
other rotations yield similarresults

IMAGE 7.14. The molecule on the far left is not identical to its mirror image.

which are nonsuperimposable on their mirror images, are called chiral molecules. The original molecule and its mirror image are referred to as a pair of enantiomers. **Enantiomers have identical physical properties—they have the same melting point, boiling point, color, and so forth—but they react very differently in the body.** Let's return to thinking about the substrate-receptor interaction that takes place in the body, in which the substrate fits into the receptor like a key into a lock. If you look closely at your house key, you'll notice it's asymmetric. That is, its mirror image would not be identical to—or superimposable upon—the original key. If you had the key's enantiomer made, you wouldn't be able to fit it into the locks on your front door. The same is true of substrates: if a substrate is chiral (and most are), only the correct enantiomer will be able to fit into the receptor and produce an effect.

TRY THIS

Indicate the carbon(s) with four
different substituents on the following
molecule:

Answer:

Where this becomes a problem for pharmaceutical chemists is that, owing to the incredible similarity of a pair of enantiomers, it's very time consuming and difficult to produce only one of the two. One of the amazing things about living cells is that they can do quite easily what is very difficult for chemists: cells generally only make *one* enantiomer of a pair as a natural product. As a result, the receptors for that product are specialized to fit only the proper enantiomer. A way to think about this is in terms of hands (substrate) and gloves (receptor). If the body makes "right-hand" molecules as a natural product, then the body will only produce "right-glove" receptors for that substrate. If a pharmaceutical chemist wants to make an artificial substrate, he or she generally synthesizes a drug containing *both* the "right-hand" and the "left-hand" molecules. A mixture like this, composed of 50% of one enantiomer and 50% of the other, is called a racemic mixture. **The "right-hand" molecules from the racemic mixture will bind to receptors in the body. The "left-hand" molecules will not. They might be harmlessly excreted, they might cause a side effect, or they might be highly toxic: it depends upon the molecule, and it's important for a chemist to be aware of this when designing a drug.**

It can be difficult to ascertain at a glance whether or not a molecule is chiral; that is, whether it will have a nonidentical mirror image. An easy shortcut is to search the molecule for a carbon atom with four different substituents, where a substituent is any group bound to the carbon. **Regardless of the size of the molecule, even *one* carbon atom with four different groups bonded to it produces asymmetry, meaning the molecule is chiral, nonsuperimposable upon its mirror image, and has an enantiomer.** As possessing even a single carbon with four different substituents renders a molecule chiral, large molecules—and many biomolecules are large—are more likely to be asymmetric than small ones. Each of the molecules below is chiral; the carbon with four different substituents is indicated in each case:

This carbon has a CH₃, an OH,
an H, and a benzene ring attached

This carbon has a CH₃, an NH₂,
an H, and a CH₂CH₃ attached

This carbon has a NH₂, an H, a COOH,
and CH₂CH₂CH₂CH₃ attached

This carbon has a CH₃, an OH,
an H, and a COOH attached

As mentioned earlier, where chirality becomes a major consideration for the pharmaceutical industry is in the disposition of the nonsubstrate enantiomer; it's critical to know whether this enantiomer will have any physiological activity, and if so, what that activity is. For example, take a chemical called *methorphan*. This compound is chiral, and the two enantiomers are distinguished from one another in nomenclature by the addition of prefixes to their names: one form is called *levomethorphan*, while the other is *dextromethorphan*.[14]

NOTE[14]

Dextromethorphan   Levomethorphan

There are many ways to indicate through nomenclature that a compound is chiral. One is the use of the prefixes *levo-* and *dextro-*; another is through using the prefixes *R-* and *S-*. Yet another mechanism puts the symbol (+) in front of one enantiomer and (–) in front of the other. Why one mechanism is used over another depends upon many factors, not the least of which is simply the preference of the pharmaceutical company. A further frustration is that many compounds (like morphine) are chiral but carry no telltale prefix in front of their names. (In actuality, there is a levomorphine and a dextromorphine, but these names are rarely used.) As a result, a discussion of enantiomer nomenclature is well beyond the scope of this text. That said, you'll see the common prefixes and symbols used to indicate chirality frequently if you look at the literature that accompanies prescription pharmaceuticals, and now you know what those prefixes and symbols mean!

Dextromethorphan is a nonaddictive, effective cough suppressant. Levomethorphan, however, is an addictive opiate; if you compare the structure to that of morphine from earlier in the chapter, you'll notice many similarities between the shapes. Dextromethorphan, while appearing to have many key chemical similarities to morphine, contains a mirror image of the morphine pharmacophore and can't bind to the morphine receptor. When a pharmaceutical company synthesizes methorphan, both the dextro- and levo- forms are produced. Because it's known that the levo- form is not safe, the enantiomers must be separated from one another and the levomethorphan removed. This process is time consuming and expensive, but in this particular case, it's quite necessary.

A much more favorable situation is presented by ibuprofen. Like methorphan, ibuprofen is synthesized as a combination of two enantiomers. Dexibuprofen is an analgesic and anti-inflammatory. Levibuprofen has no activity in the body. Luckily for the pharmaceutical company, however, a liver enzyme converts levibuprofen to the active form. This means the drug can be produced inexpensively as a racemic mixture.

Levibuprofen   Dexibuprofen

Thalidomide presents a third scenario: like methorphan and ibuprofen, thalidomide is chiral. As mentioned previously, one enantiomer is a safe, effective sedative. The other is teratogenic. Unlike methorphan, however, pharmaceutical companies can't simply purify the drug and give only the safe form; the safe enantiomer is actually converted into the teratogenic enantiomer by the liver. As such, there is no safe way to administer thalidomide—at least, not to relieve symptoms of morning sickness.

In 1962, following the thalidomide tragedy, an amendment to the Food, Drug, and Cosmetic Act made the safety restrictions on pharmaceuticals even tighter than they already were. **Provisions included that before human trials of a drug could begin, an Investigational New Drug application (IND) had to be submitted, replete with results of animal testing. A further stipulation provided that, for all drugs introduced after 1962, both proof of efficacy and documentation of safety were required.** While new, stricter safety standards were not retroactively applied to drugs submitted under the Food, Drug, and Cosmetic Act between 1938 and 1962, the amendment nevertheless required that manufacturers of drugs introduced during this period of time supply proof of efficacy in order to maintain their pharmaceutical on the market. Of course, a major result of the new legislation was to increase the cost of pharmaceutical development into the hundreds of millions of dollars per prescription drug (including the cost of mandatory post-marketing surveillance required by the FDA). This has had a direct effect upon the cost of prescription drugs to consumers, as pharmaceutical companies have to recoup their development costs. A second result of the more stringent regulation is that the time necessary to take a drug from the lab bench to the market is, on average, eight to ten years. Patients and families of patients with illnesses for which there are currently no appropriate pharmaceuticals find the wait intensely frustrating, and understandably so. However, **the FDA must balance public safety with pressure from both the pharmaceutical companies and the public to bring new drugs to market quickly.**

In the end, thalidomide has found an application as a useful drug for treating symptoms of Hansen's disease (leprosy), some forms of cancer, and a small variety of other ailments. It is a highly controlled substance, and is never administered if there is any possibility that the patient might be or become pregnant. The thalidomide case, however, highlights the utility of having not only safety information for a pharmaceutical, but also efficacy information. In two separate applications, the risk-to-benefit ratio of using thalidomide turns out very differently. For a pregnant woman (high risk), taking thalidomide for morning sickness (beneficial, but symptoms are not life threatening) represents an unacceptable risk-to-benefit ratio. For a nonpregnant patient (minimal risk) with a painful and disfiguring disease (high benefit), the risk-to-benefit ratio clearly points in the direction of using the drug. **Since the 1962 amendment, we have both safety and efficacy information at our disposal, allowing for risk–benefit analysis in decision-making.**

To many individuals concerned with health, welfare, and wellbeing, it's important to avoid "chemicals" or the "artificial," and to seek out the "natural." The avoidance of chemicals, as we saw in Chapter 1, is simply not possible. All matter is composed of elements, compounds, or mixtures—in short, chemicals. Our bodies are made of nothing but chemicals, water is a chemical, the air we breathe is chemical, and the food we eat is entirely chemical. Dispensing with the notion that we can in any way avoid chemicals, let's instead examine the difference between "natural" and "artificial," or man-made, chemicals.

A good place to start is with flavoring agents: some of us, in choosing which foods to eat and which to avoid, tend to steer clear of products containing artificial flavors. Perhaps as we read those words we picture scientists mixing chemicals in a lab, producing completely synthetic and falsified (and possibly dangerous!) versions of the naturally scented compounds found in nature. By contrast, when we read the words *natural flavor*, perhaps images of fruits, vegetables, and spice plants come to mind. Unfortunately, the reality is not nearly so dichotomous. Natural and artificial flavors are often identical chemicals; the natural flavors are extracted from foods, and the artificial are made in the lab. While that definition seems to support the typical notions of flavors and their origins, let's look a bit deeper. Natural flavors don't have to—and in fact, often don't—come from the food they taste like. While some of the chemicals used in natural banana flavor, for instance, may be extracted from banana, natural almond flavor comes from the pits of stone fruits (like peaches). Furthermore, the artificial almond flavor made in the lab is the *same* chemical as the natural almond flavor extracted from peach pits (an organic molecule called benzaldehyde). That is to say, whether one is consuming natural or artificial almond flavor, one is consuming identical molecules of benzaldehyde, and in neither case were almonds the original source of those molecules. **As it turns out, and surprisingly, artificial flavors are often safer and more pure than natural flavors, since the chemicals have been controlled for purity in the lab and are much more likely to be free from contaminants.** Natural almond flavor, for instance, generally contains trace amounts of cyanide, a toxic chemical that occurs naturally in peach pits; artificial almond flavor never does. Lest you become fearful of natural almond flavor, a general principle of toxicology that will be addressed further shortly is that the dose makes the poison. There is not enough cyanide in natural almond flavor to hurt you. The point is academic more than it is practical. That said, one cannot truthfully make the argument that natural almond flavor is in any way better—and certainly not safer—than artificial. Neither should we say that since natural flavor contains the same chemicals as artificial flavor, which is made in a lab, they should both be avoided—remember, chemicals *can't* be avoided, by definition. The point here is simply that upon even mild scrutiny, the idea of equating *natural* with *good* and *artificial* with *bad* (or even trying to discriminate between the two) falls apart.

IMAGE 7.15. The death cap mushroom (a), poison dart frog (b), and castor beans (c) are just a few examples of organisms producing completely natural chemicals that are deadly poisons.

Going further, there are many completely natural chemicals synthesized by plants and animals—often as a form of self-defense—that are deadly poisons for humans. *Amanita phalloides*, commonly called the "death cap," is probably the most poisonous mushroom in the world. Ingestion leads to nausea, vomiting, diarrhea, and often to liver and kidney failure.

Other examples of deadly natural substances include a variety of toxic alkaloids from the poison dart frog, ricin from castor beans, and atropine and scopolamine from deadly nightshade, to name just a few.

Occasionally, our desire for the "natural" leads us to eschew man-made pharmaceuticals in favor of herbal or traditional medicines. We should remember that many pharmaceuticals—including aspirin, digitalis, and antimalaria drugs—had their roots (so to speak!) in herbal extracts, so to draw a line between pharmaceuticals and herbals in the first place is exceedingly difficult. Additionally, in much the same way that artificial flavor is frequently of higher purity than natural flavor, so too man-made pharmaceuticals are subjected to much greater scrutiny and regulation than are herbals.

While it's difficult to classify a substance as purely toxic (just as it's difficult to classify a substance as purely safe—the same digitalis that is a life-saving cure in one situation can be deadly if misused) some herbals are so inherently toxic as to have no safe dose. For example, the traditional Chinese herb *aristolochia*, used for weight loss, causes kidney failure. With regard to pharmaceuticals, we count on the FDA to regulate drug safety, and expect that any preparation prescribed to us by a physician or available on drugstore shelves will not do us harm. Unfortunately, we don't have the FDA's assurance—nor that of any other regulatory agency—with regard to the safety or efficacy of herbals.[15] The lax government regulation of herbs means that important safety information is not necessarily disclosed to consumers, and potentially toxic herbs are sold legally. In the particular case of aristolochia, many unwitting U.S. customers who purchased the herb ended up requiring dialysis or kidney transplant. The difference in the way pharmaceuticals and herbals are regulated has its roots in the 1962 amendment to the Food, Drug,

NOTE[15]

As FDA blaming (like the blaming of other federal agencies) is very nearly a recreational sport in this country, it should be stated that herbals are unregulated by the FDA not because the agency *won't* keep an eye on them, but because they *can't*. The FDA's powers are strictly limited by Congress, and they are not given regulatory control over dietary supplements—herbals included—with regard to checking safety and efficacy.

and Cosmetic Act. The amendment classified herbals as dietary supplements and provided that together, supplements and herbals were to be held to lower safety (and no efficacy) standards. The **Dietary Supplement Health and Education Act (1994)** further defined (that is to say, limited) the role of the FDA in regulating herbals and dietary supplements. Provisions of this act included exempting dietary supplements and herbals from premarketing evaluation by the FDA (requiring only that new supplements/herbals be "reasonably" expected to be safe) and grandfathering supplements/herbals marketed prior to 1994, meaning they were not required to have a reasonable expectation of safety. **In essence, the FDA was disallowed from regulating dietary supplements and herbals.** A further legislative order passed in 2007 required dietary supplement and herbal manufacturers to ensure—but not prove—lack of product contamination and correct labeling. **A cursory glance back at the history of FDA regulation of pharmaceuticals reveals that current supplement and herbal regulation is no more stringent than the pharmaceutical regulation mandated by the original Food and Drug Act of 1906, before the diethylene glycol and thalidomide tragedies forced further legislation.** The FDA requests that supplement manufacturers track and report adverse reactions and evidence of toxicity, but reporting is still on a voluntary basis at this time.

It appears that in the end, any attempt to equate *natural* with *safe* (or to equate *artificial* with *dangerous*) falls apart—even when it comes to government regulation. **The FDA regulations of pharmaceutical drugs help provide us with the information we need (with the help of our physicians) to assess the risk-to-benefit ratio of using a drug.** There is by no means assurance that simply because a pharmaceutical is available for purchase or prescription, it is 100% safe—such a notion runs completely counter to the principle that the dose makes the poison. Many pharmaceuticals are quite toxic even in small doses, and all are toxic in sufficient dose. Still, safety and efficacy information allow for the determination of an appropriate dose, such that the potential for harm from the drug is far outweighed by the potential that it will do us good, when used as intended. With regard to "natural" substances (whether a tea made from tree bark, a flavoring agent extracted from a seed, or a field herb), we rarely have safety and efficacy information available to us.[16] **Our ability to weigh the potential risk against the potential benefit of, say, an herb, is nonexistent in the absence of scientific data.** Some man-made chemicals are, as we've already seen, highly toxic. It's important to remember, though, that the deadliest toxins aren't the ones we produce, but the ones carefully crafted by nature—which has proven itself to be the most accomplished of synthetic chemists.

CONCEPT CHECK

Does the current herbal supplement legislation provide the information necessary for making a risk-benefit analysis?

Answer: No. Efficacy information is not provided, and safety information is not required.

NOTE[16]

The National Center for Complementary and Integrative Health (NCCIH), an arm of the National Institutes of Health (NIH), funds and maintains a database of safety and efficacy information for alternative medicine, herbals, and dietary supplements. It's possible to look up the safety and efficacy data (if any exists) for supplements and the like on their website at www.nccih.nih.gov.

7.9   The Last Word—The Paracelsus Principle

Paracelsus, a Renaissance botanist, is credited with first articulating a major principle of toxicology. While today we might simply say that the dose makes the poison, Paracelsus eloquently stated, "All substances are poisons; there is none that

is not a poison. The right dose differentiates a poison from a remedy." Throughout this chapter, we've seen examples of pharmaceuticals that are safe in one application, and harmful in another, like thalidomide. Further, almost all pharmaceuticals are safe in the correct circumstances and dosage, but highly toxic if misused. This doesn't come as any particular revelation to most people; in general, we approach pharmaceuticals and other "man-made" chemicals with caution (though hopefully this chapter has at least partially unseated the myth that "natural" means "safe" and "man-made" means "harmful"). Perhaps the most striking exhibition of the Paracelsus Principle's veracity, though, comes through examining two very different substances, one widely regarded as the safest, most life-giving chemical there is, and the other infamous as the deadliest poison known to man.

Clostridium botulinum is a bacterial species that produces the botulinum toxin, considered the deadliest poison on Earth. Exceedingly small doses of the substance are lethal.[17] While *C. botulinum* can infect wounds (leading to a condition known as gas gangrene, which was far more common in the days before the advent of antiseptics), it's also possible to contract botulism through the ingestion of contaminated foods, particularly improperly canned goods. Symptoms include muscle weakness and paralysis, which eventually spread to the respiratory muscles and preclude the ability to breathe. Left untreated, most botulism is fatal. Despite the incredibly deadly nature of botulinum toxin, however, it has found medical utility in a number of applications. Medical botulinum toxin (under the trade name Botox) is used in cosmetic procedures. Injected into facial muscles, the toxin acts as a paralytic, preventing contraction and reducing the appearance of existing—and development of new—wrinkles. Further, Botox is used to reduce excessive sweating, ease the muscle spasms and associated pain of temporomandibular joint syndrome (TMJ) and other spasmodic muscular disorders, and ease symptoms of diabetic neuropathy (damage to the peripheral nervous system, particularly common in the feet of diabetic individuals), among a host of other applications. In the right dose and for the right reason, deadly botulinum toxin becomes a valuable tool in the physician's arsenal.

On the flip side of the coin, water (as we saw in Chapter 4) is indispensable, unavoidable, and necessary for survival. We consume it daily, surround ourselves in it regularly, and consider it so safe a substance that the general population doesn't think of it as a "chemical" at all. Of course, like all matter, water is a chemical—and like all chemicals, water is a poison in the right (wrong?) dose. While water intoxication and water toxicity occur rarely (since the dose required to render water toxic is quite large), they do occur. Overconsumption of water can lead to imbalances in electrolytes that cause a host of symptoms, culminating in brain function abnormalities that can be harmful or fatal. Water intoxication is observed most commonly among athletes,[18] occasionally among the mentally ill (who may consume large amounts of water without regard to thirst), and in at least one bizarre incident, in a highly competitive individual out to win a contest. In 2007, a radio talk show offered a video game prize to the contestant who could drink the most water in the shortest period of time. Jennifer Strange, a California woman, died of

water toxicity after drinking approximately two gallons in about an hour. While tragic, this incident and others like it serve to remind us that just as toxic substances can find safe medicinal applications, "safe" substances can be toxic. As Paracelsus said, "The right dose differentiates a poison from a remedy."

SUMMARY OF MAIN POINTS

- Salicin, isolated from white willow, was chemically modified to amplify its effectiveness while reducing its side effects.
- Organic chemistry is the study of organic molecules, which are (often large) compounds with carbon skeletons.
- Organic molecules are simplified into line-angle diagrams, in which hydrogen atoms bonded to carbon are implicit, as are the atomic symbols for carbon.
- Functional groups are characteristic arrangements of atoms with predictable chemical properties, and are of importance to synthetic chemists
- Important pharmaceutical functional groups include carboxylic acids, benzene rings, alcohols, phenols, amines, esters, and ethers.
- The solubility of large compounds is complicated—compounds often have areas of hydrophobicity and hydrophilicity, where solubility is the additive combination of different functional areas.
- Amines are common in physiologically active compounds, but often have poor water solubility, so are reacted with acid to make a freebase, or ionic salt.
- Pharmaceuticals interact with receptors for natural substrates in the body.
- Pharmaceuticals (and natural substrates) fit into a receptor like a key into a lock. As a result, compounds (natural or man-made) that fit into the same receptor must have areas of similar shape.
- The major functional area of a molecule that contributes to its activity in the body is called the pharmacophore.
- Early drug regulation included the Food and Drug Act (1906), which required correct labeling and lack of contamination of substances transported across state lines.
- The 1938 Food, Drug, and Cosmetic Act added that proof of safety was required for new drugs.
- Thalidomide, used in Europe in the 1950s and 1960s for morning sickness, was highly teratogenic, but was not approved for use in the United States.
- Symmetric molecules are superimposable on their mirror images and are called "achiral."
- Asymmetric molecules (identified as molecules containing one or more carbons with four different substituents) are nonsuperimposable on their mirror images and are called "chiral."
- A molecule and its mirror image are called "enantiomers"; a 50:50 mix of a pair of enantiomers is a racemic mixture.
- Thalidomide is chiral; one form is safe, the enantiomer is teratogenic.

- Drug manufacturers must be aware of enantiomers when synthesizing drugs; generally only one of the enantiomers is active as a pharmaceutical, but the other may have any one of a number of effects in the body.
- It's cheaper to synthesize a drug as a racemic mixture, but it cannot be marketed as such if one enantiomer is toxic.
- The 1962 amendment to the Food, Drug, and Cosmetic Act added proof of efficacy to the FDA's requirements of a drug. This amendment made it possible to use risk-benefit analysis in decision making.
- Natural does not always mean "good" or "safe," and "man-made" does not always mean "bad" or "toxic"; many natural compounds are quite toxic even in low doses.
- Herbal compounds are nowhere near as stringently regulated as pharmaceuticals; legislation requires only that they be "reasonably" expected to be safe, correctly labeled, and free from contamination.

QUESTIONS AND TOPICS FOR DISCUSSION

1. Why was Johannes Buchner interested in trying to isolate the active compound from willow bark?

2. Explain the two simplifications made in converting a Lewis structure into a line-angle diagram. Why do these simplifications make sense—that is, why is it still easy to read line-angle diagrams despite all the "missing" pieces?

3. What kinds of molecules can and can't be drawn in line-angle?

4. Why are bond angles so important when drawing in line-angle?

5. How do we know where the carbons are in a molecule drawn in line-angle?

6. How do we know how many hydrogens are attached to a given carbon in line-angle?

7. Why can't we make hydrogens attached to atoms other than carbon implicit in line-angle?

8. What was the purpose of modifying salicylic acid?

9. Why do so many pharmaceuticals have their roots in herbals or natural products? Why would it be hard for a pharmaceutical chemist to invent a drug *de novo*, meaning without anything to use as a template?

10. What is the purpose of keeping track of functional groups in organic chemistry?

11. List the key characteristics of each of the following functional groups: carboxylic acids, benzene rings, alcohols, phenols, amines, esters, and ethers.

12. Why do perfumes start to stink if exposed to water and acid or base? What about them, chemically, causes this?

13. Why can we say *ethanol is an alcohol*, but we can't say *salicylic acid is a phenol*?

14. Why was diethyl ether such an important molecule from a social perspective?

15. Why is it more difficult to determine the solubility or polarity of a large organic compound than of a small compound? What does it mean to refer to areas of hydrophilicity or areas of hydrophobicity on an organic molecule?

16. Which functional groups are likely to contribute to water solubility? Which are likely to contribute to fat solubility?

17. Why are some amine-containing pharmaceuticals reacted with HCl or HBr?

18. What is a pharmacophore and why is it important to pharmacological chemistry?

19. In order to have some function in the body, what characteristic(s) must a pharmaceutical have?

20. Can you determine the morphine pharmacophore simply from looking at the structure of morphine? What additional information do you get if you also look at the structure of codeine? Of morphine, codeine, and meperidine together?

21. Why are receptors and substrates sometimes explained in terms of locks and keys?

22. Where did the street drug heroin come from? Why was it produced?

23. Describe the 1906 Food and Drug Act. What did it require of pharmaceuticals?

24. What led to the passing of the 1938 Food, Drug, and Cosmetic Act? What did this act require of pharmaceuticals?

25. What led to the passing of the 1962 amendment to the Food, Drug, and Cosmetic Act? What did this amendment require of pharmaceuticals?

26. What is a risk-to-benefit ratio? Why were consumers unable to use risk–benefit analysis prior to 1962?

27. Why do some molecules have nonidentical mirror images, while others don't? Think of a nonmolecular (macroscopic) example of a "chiral" object and an "achiral" object (don't use the examples that were used in the text).

28. Why do pharmaceutical chemists prefer to market a drug as a racemic mixture? Under what circumstances can they do so? Under what circumstances must a drug be marketed as a single enantiomer?

29. Why is a drug's enantiomer not expected to behave in the same manner as the drug in the body?

30. Why are stereoselective receptors (receptors for only one enantiomer of a pair) so common in the body, even though it's hard for chemists to produce a single enantiomer of a pair?

31. Sketch dextromethorphan and levomethorphan. Sketch morphine. What do you notice about your sketches? Would you be able to predict (even without being told) which methorphan is the addictive opiate? How can you tell?

32. Describe the risk-benefit analysis that might be used to decide whether to take a drug in each of the following cases: a drug has a 50% chance of treating a fatal illness, but a 10% chance of causing a fatal toxic reaction in the patient; a different drug has a 50% chance of treating acne, but a 10% chance of causing a fatal toxic reaction in the patient.

33. Why can't we make the assumption that natural things are safe?

34. Explain why the general assumption that natural flavors are safer or "better for you" than artificial flavors is erroneous.

35. An herbalist tells you that an herb has the ability to treat headaches, but has no side effects or toxicity. Do you believe the herbalist? Why or why not?

36. Compare and contrast herbal and supplement regulation with pharmaceutical regulation. What are the similarities and differences in the law?

37. Is it possible to use risk–benefit decision making in determining whether to use an herbal in the United States? Why or why not?

1. Convert the following to line-angle:

2. Convert the following to line-angle:

3. Convert the following to full Lewis structures:

4. Convert the following to full Lewis structures:

5. How many hydrogens and carbons do each of the following molecules contain?

6. How many hydrogens and carbons do each of the following molecules contain?

7. Circle and label all the functional groups in the molecule below:

8. Circle and label all the functional groups in the molecule below:

9. Describe each of the following functional groups in terms of their characteristic arrangement of atoms: carboxylic acids, benzene rings, alcohols, phenols, amines, esters, and ethers.

10. Circle and label regions of polarity and nonpolarity on the molecule below:

11. Circle and label regions of polarity and nonpolarity on the molecule below:

12. Arrange the following molecules in order of increasing solubility:

13. Arrange the following molecules in order of increasing solubility:

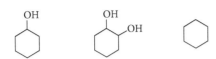

14. Define: pharmacophore and substrate.

15. Define: chiral; achiral; enantiomer; and racemic mixture.

16. If the following items were molecules, determine whether they would be chiral or achiral: foot; basketball; die (singular for dice); human; and an ear.

17. Indicate the carbon(s) with four different substituents on the following molecule:

18. Indicate the carbon(s) with four different substituents on the following molecule:

19. Draw a simple molecule consisting of only a carbon with the following groups attached: X, X, Y, Z. Draw the mirror image of the molecule. Are the two molecules superimposable? Is your molecule chiral?

20. Draw a simple molecule consisting of only a carbon with the following groups attached: W, X, Y, Z. Draw the mirror image of the molecule. Are the two molecules superimposable? Is your molecule chiral?

21. Determine whether the following molecules are chiral or achiral:

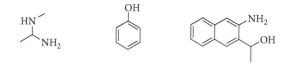

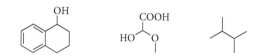

22. Determine whether the following molecules are chiral or achiral:

23. Why must methorphan be sold as a single enantiomer, while ibuprofen can be sold as a racemic mixture?

24. Given that one enantiomer of thalidomide is safe and nonteratogenic, why can't pregnant women take the safe enantiomer for morning sickness?

25. Why can thalidomide be used as a pharmaceutical in some cases, even though it's highly teratogenic?

26. Did the 1906 Food and Drug Act require pharmaceutical companies to provide consumers with information regarding drug safety? Drug efficacy?

27. Did the 1938 Food, Drug, and Cosmetic Act require pharmaceutical companies to provide consumers with information regarding drug safety? Drug efficacy?

28. Did the 1962 amendment to the Food, Drug and Cosmetic Act require pharmaceutical companies to provide consumers with information regarding drug safety? Drug efficacy?

29. Does the 1994 Dietary Supplement Health and Education Act require manufacturers to provide consumers with information regarding dietary supplement/herbal safety? Efficacy?

30. Per 2007 legislation, what is required of dietary supplements and herbals sold in the United States? Which pharmaceutical legislation does this resemble?

1.

3.

5. 8C, 8H. 7C, 12H.

7.

9. Carboxylic acids contain a COOH group; benzene rings are six-carbon rings with alternating single and double bonds; alcohols contain an OH group; phenols are alcohols attached directly to a benzene ring; amines are nitrogen containing; esters are a combination of a carboxylic acid and an alcohol, or a C double bonded to an O and single bonded to another O that has an R group attached; ethers are two R groups connected by a common O.

11.

13.

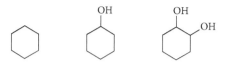

15. A chiral molecule is asymmetric; an achiral molecule is symmetric; enantiomers are a pair of nonsuperimposable mirror images; a racemic mixture is a 50:50 mixture of two enantiomers.

17.

19. Drawings will vary; molecules are superimposable, achiral.
21. Chiral; achiral; chiral.
23. Methorphan has one safe enantiomer and one enantiomer that is an addictive opiate, so only the safe one can be taken as a pharmaceutical. Ibuprofen has one active (safe) enantiomer, and one inactive enantiomer, but the inactive enantiomer is converted into the active form by the body.
25. Thalidomide has medical applications in certain diseases, but can only be used if there is no chance the patient is or will become pregnant.
27. The 1938 Food, Drug, and Cosmetic Act required pharmaceutical companies to provide (some) safety, but not efficacy, information to consumers.
29. The 1994 Dietary Supplement Health and Education Act does not require pharmaceutical companies to provide information regarding either safety or efficacy.

SOURCES AND FURTHER READING

Le Couteur, Penny, and Burreson, Jay. (2003). *Napoleon's buttons*. New York, NY: Penguin.

Schlosser, Eric. (2001). *Fast food nation*. New York, NY: HarperCollins.

Timbrell, John. (2005). *The poison paradox*. New York, NY: Oxford University Press.

www.fda.gov

Paracelsus Principle

www.cdc.gov

Associated Press. (2007, January 13). *Woman dies after water-drinking contest*. Retrieved from MSNBC website: http://www.msnbc.msn.com/id/16614865/

Dietary Supplements

www.nccih.nih.gov

CREDITS

The Chemistry of What We Eat

8

Every living thng needs to get energy from its environment. Some organisms get energy from the sun, while others consume food. Humans, of course, fall into the second category: we need to eat. Unlike pandas, which are able to extract all the nutrition they need from bamboo, or marmots, which eat nothing but grass and still manage to store enough fat to support eight months of hibernation, we require a complex and varied diet. What is it that we need from our food in order to stay alive and healthy? Why must we concern ourselves not simply with the energy content of what we consume, but the vitamin and mineral content as well? Most importantly, what do we actually do with our food, from a chemical standpoint, once it's chewed and swallowed?

8.1 Energy and Chemical Bonds

We've learned that atoms form molecules because in so doing, they are stabilized in a number of ways. Bond-formation allows atoms to fill their valence shells, and additionally, electrons in bonds are stabilized by the presence of two attractive nuclei. In this way, we have a qualitative sense of the stabilization that occurs when molecules form. We can also look at the process of bond formation—and the opposite process of bond dissociation—in a quantitative sense; the bonds that form when atoms assemble into a molecule result in specific amounts of energetic stabilization. Examining the energetic processes that molecules undergo as they form, break apart, and rearrange is called the study of thermodynamics. While there are several principles and laws of thermodynamics, only a few are critically important to our discussion. First, we must understand that stabilization is simply a reference to energy. **High-energy molecules (or atoms) are less stable than lower-energy molecules or atoms.** Atoms with incomplete valence shells are less stable (that is to say, have more energy) than atoms with

IMAGE 8.1. A yellow-bellied marmot (*Marmota flaviventris*) in Great Basin National Park builds up its energy stores for hibernation by eating nothing but plenty of grass (photo © Scott Lefler, 2009). Humans need— and crave—a variety of foods.

Bond Type	Energy Released (kcal/mol)
H-H	104.2
C-H	99
O-H	111
N-H	93
C-C	83
C-O	85.5
C-N	73
C=C	146
C=O	192
O=O	119
N≡N	226
C≡C	200

TABLE 8.1. Bond energies for common bonds in organic molecules.

TRY THIS

How much energy is released if three moles of H atoms and one mole of N atoms formed one mole of NH_3?

Answer: 279 kcal/mol would be released.

Solution:
3 N-H bonds = 3 (93 kcal/mol) = 279 kcal/mol

full octets. As atoms form bonds, they become more stable and lower in energy. The extra energy—the energy difference between the higher- and lower-energy states—is released, and can be quantified. There's a characteristic amount of stabilization (i.e., energy release) for each bond that forms, where the energy released depends both upon the identities of the atoms and the type of bond. Further, the total amount of stabilization (energy released) when a molecule forms is the sum of the energies released in the formation of each individual bond. Table 8.1 shows the amount of energy (in kcal/mol[1]) released for each type of bond that forms (these are in molar quantities).

We can use this information to tabulate how much energy, for instance, would be released if two moles of H atoms and one mole of O atoms formed one mole of H_2O. When water forms from atoms, each hydrogen atom forms a bond to oxygen, such that two new O-H bonds form:

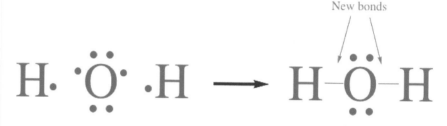

New bonds

Per table 8.1, there is a release of energy of 111 kcal/mol for each O-H bond, so the total energy released is calculated:

$$2 \text{ O-H bonds} = 2 \ (111 \text{ kcal/mol}) = 222 \text{ kcal/mol}$$

In other words, each mole of water that forms from elements will release 222 kcal of energy.

What happens to the released energy? Well, that depends upon the circumstances. Let's take a nonchemical example that we can visualize easily. If I lift a box into the air, I use energy. Very formally, energy is defined as the potential to do work, where work moves an object against a force. So I have moved an object (the box) against a force (gravity). Therefore, I've done work, and doing that work must have taken energy. What's happened to the energy I expended? I've given it to the box, which now has more energy (more potential to do work) than it had before. In particular, the box has gravitational potential energy, meaning it has the potential to fall to the ground. How can we tell that the box has energy? If I drop the box it can do work (like perhaps break my toe if I don't get out of the way quickly enough). This process involves several energetic transformations. The gravitational potential energy the box has while in the air is transformed into kinetic energy—the energy of motion—as the box falls. That kinetic energy then exerts a force on the bones of my toe, breaking them. Lifting a box and winding up with a broken toe is an example of a series of energy transfers (from me, to the box, to my toe). Another possibility is that after lifting (and releasing) the box, I manage to move my foot so that the box falls harmlessly to the floor. In this case, I'd hear a "thud" sound as the box landed, and nothing further would happen. **Here, it appears at first as though the energy has disappeared, but in fact, that is not possible. The first law of thermodynamics says that the energy of the universe is constant: that is, energy can never be created or destroyed.** When I lifted the box, the energy it gained came from me; after lifting the box, it had more energy, and I had less. When the box landed on the floor, the energy it lost[2] must have gone somewhere. In the second scenario, the box made a noise ("thud"), which means molecules in the air were made to vibrate (they were given energy), and their vibrating caused my eardrums to vibrate (my eardrums were given energy), which is how I heard the sound. The floor probably shook a little (also a vibration, so the floor got some energy), and all the vibrating molecules in the air and the floor created a little bit of friction, meaning that the molecules heated up. Heat is a form of energy—warmer molecules have more energy than cooler ones—so we can see that with all the vibrating and warming, our box's energy hasn't been lost at all; it's merely been transferred and transformed, as energy always is.

NOTE[2]

How do we know the box has lost energy? Because a box sitting on the ground has no potential to fall toward the ground (no gravitational potential energy). Additionally, if it's motionless, it has no kinetic energy either.

NOTE[3]

Heat is not always "useless" energy (entropy). In the case of a coal-fired power plant, for instance, heat released by combustion is used to produce steam, which is used to generate electricity. In the human body, however, no heat-utilization system exists, so heat released through combustion reactions is no longer usable energy. Regardless of the form it takes, energy that can no longer be harnessed for use is entropy.

8.2   Entropy

It's important to note that while *total energy* is always conserved, *usable energy* isn't. In the second box-falling scenario, we saw that gravitational potential energy was transformed into kinetic energy, which was then transformed into the energy of vibration and heat. However, since we can't harness the vibration or heat to do any work in this case, the energy from the box has been transformed into an unusable form. This unusable, disordered energy is called **entropy**, and it's an important concept in thermodynamics.[3] **While the first law holds true—the energy of the universe is constant—the second law of thermodynamics states that for any process, the entropy of the universe increases, meaning that the usable energy in the universe decreases (Image 8.2).**

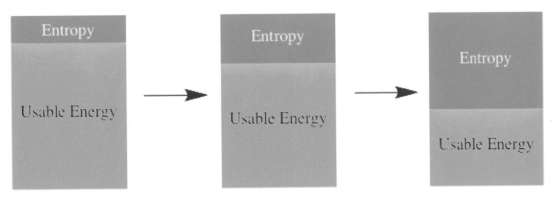

IMAGE 8.2. While the energy of the universe stays constant, entropy increases over time (at the expense of usable energy).

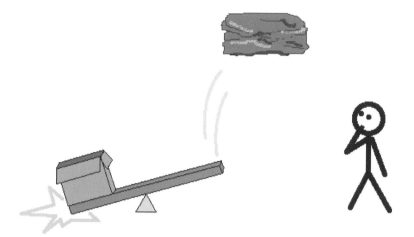

IMAGE 8.3. A lever system allows the transfer of energy from a falling box to a rock (clearly, there was something really heavy in that box!).

Let's imagine one more box-dropping scenario. This time, instead of letting the box hit the ground (or break my toe), I drop it onto a lever system (Image 8.3) with a rock sitting on the other end. Now, as the box falls, it transfers its energy (via the lever system) to the rock, which is launched into the air (gaining gravitational potential energy), and which can then transfer its energy to something else.

Here, it's tempting to imagine that we've avoided losing any energy to entropy, but unfortunately, this is not so. The rock launched by the falling box won't have as much energy as the box did. When the box landed on the lever, disordered energy was produced in the form of molecular vibration, and not all the box's energy was transferred to the rock. **In the end, we're left with a fundamental truth—every transfer of energy results in a "loss" of some usable energy as entropy, so no energy transfer can ever be 100% efficient.** As our box examples demonstrate, however, different systems have different efficiencies of energy transfer. Our cleverly designed lever system, for instance, was more efficient at capturing the box's energy than the case in which the box simply landed on the floor.

To return to the original question from several pages back (we wanted to know what happens to the energy released when a bond forms), there are several possibilities, depending upon the circumstances. **The energy could be released as heat energy, or it could be stored as bond energy in another molecular system. We'll see both of these take place in our discussion of nutritional chemistry.**

8.3 Energy of Chemical Reactions

The last principle of thermodynamics important to this discussion is that processes (and chemical reactions) are reversible: I can pick up a box, or I can let a box fall. In one case, energy is required. In the reverse process, energy is released. This always holds true. **For any process or chemical reaction, if the reaction releases energy in one direction, it requires energy to go in the other direction. Furthermore, the** amount of energy required in one direction is always equal to the amount of energy released in the other direction. To return to our water example, what happens, thermodynamically speaking, when water is broken into its constituent atoms? First of all, since energy was released when two moles of hydrogen and a mole of oxygen formed O-H bonds to make a mole of water, breaking water into atoms must require energy. Second, the energy required to break bonds is equal to the energy released when bonds are made. For a mole of water, the energy required would be:

$$2 \text{ moles of O-H bonds (2 bonds per } H_2O, 1 \text{ mole of } H_2O)$$
$$= 2 \ (111 \text{ kcal/mol}) = 222 \text{ kcal/mol}$$

In other words, each mole of water that is broken into atoms requires 222 kcal of energy.

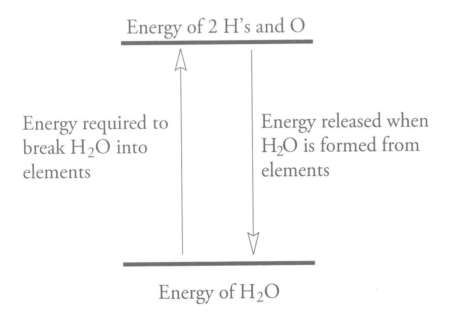

Energy of 2 H's and O

Energy required to break H_2O into elements

Energy released when H_2O is formed from elements

Energy of H_2O

Since forming a bond is always stabilizing, it always releases energy. Since breaking (dissociating) a bond is the reverse of forming a bond, it is always destabilizing, and always requires energy.

We very rarely need to think about forming molecules from atoms (or breaking molecules into atoms) in real life. Generally, we'll be looking at reactions in which molecules are rearranged into new ones. Reactions like these require that some bonds be broken and some new bonds be formed. By looking at which bonds break (and how much energy is required) and which bonds form (and how much energy is released), we can figure out whether a reaction releases or requires energy overall. For example, let's determine whether the combustion of methane in oxygen requires or releases energy. It's easiest to make such a determination by following a series of steps:

TRY THIS

How much energy is involved (and is it released or required) when 1 mole of CH_4 is broken into atoms?

Answer: 396 kcal/mol required.

Solution:
4 C-H bonds = 4 (99 kcal/mol) = 396 kcal/mol; breaking bonds always requires energy.

1. Balance the chemical reaction to accurately determine how many bonds of each type are involved. Since our bond energies are in kcal/mol, we'll think of our coefficients as molar quantities rather than molecular/atomic quantities.

$$CH_4 + 2\,O_2 \rightarrow CO_2 + 2\,H_2O$$

2. Drawing Lewis structures if necessary, determine what kinds of bonds (and how many of each) are broken in the reaction. Molecular shapes aren't really relevant to this discussion, so to keep things simple, we'll use plain Lewis structures:

Bonds that need to break:
4 C–H bonds, 2 O=O bonds

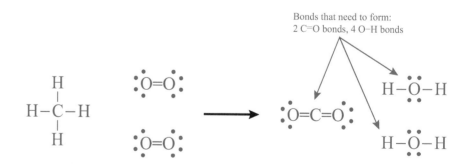

4 C-H bonds and 2 O=O bonds are broken

3. Using Table 8.1, determine how much total energy is required to break the bonds. By convention, these values are written as positive to indicate that the energy is entering the molecule(s).

4 (99 kcal/mol) + 2 (119 kcal/mol) = 634 kcal/mol required

4. Drawing Lewis structures if necessary, determine what kinds of bonds (and how many of each) are formed in the reaction.

Bonds that need to form:
2 C=O bonds, 4 O–H bonds

2 C=O bonds and 4 O-H bonds are formed

5. Using Table 8.1, determine how much total energy is released when bonds form. By convention, these values are written as negative to indicate that the energy is leaving the molecule(s).

$$4 (-111 \text{ kcal/mol}) + 2 (-192 \text{ kcal/mol}) = -828 \text{ kcal/mol released}$$

6. Add the energy required (always a positive value) to the energy released (always a negative value). If the sum is positive, that's how much energy is *required* by the reaction. If the sum is negative, that's how much energy is *released* by the reaction.

$$634 \text{ kcal/mol} + -828 \text{ kcal/mol} = -194 \text{ kcal/mol}$$

Since more energy was released by forming bonds than required for breaking bonds, combustion of methane releases 194 kcal of energy for every mole of methane burned.

Remember, if energy is released by a reaction, it can be released as heat, it can do work, or some combination thereof; the disposition will depend upon the circumstances. **However, by convention, we refer to a reaction that releases energy as exothermic (from the Greek for *heat out*), and a reaction that requires energy as endothermic (from the Greek for *heat in*).**

TRY THIS

Determine whether the reaction of H_2 with N_2 to form NH_3 is exothermic or endothermic, and how much energy (kcal/mol) is released or required.

Answer: Exothermic; 19.4 kcal/mol released (-19.4 kcal/mol)

Solution:
$3 H_2 + N_2 \rightarrow 2 NH_3$
Energy to break bonds = 3 (104.2 kcal/mol) + 226 kcal/mol = 538.6 kcal/mol
Energy released when bonds form = 6 (-93 kcal/mol) = -558 kcal/mol
Total = 538.6 kcal/mol + -558 kcal/mol = -19.4 kcal/mol

8.4 How Molecules Provide Energy

All of this helps us understand how molecules can provide us with energy. We eat molecules and transform them, through chemical reactions, into different molecules. As long as these reactions are exothermic—which, for food molecules, they always are—the reactions release energy. Some of that energy is released as heat (and helps us maintain our body temperature), but we also use some of the energy to do things like build structural molecules that make up our cells, run chemical reactions that require energy, and so forth.

The nutrients we take in as food fall into three major macronutrient categories: carbohydrate, protein, and fat. The word macronutrient is used to indicate that these three nutrients are used in large amounts by the body—that is, they are the energy-providing molecules. Depending upon the type of nutrient, we get a characteristic amount of energy (kcal) per quantity of food. Carbohydrates, for instance, provide us with about 4 kcal per gram. Protein, similarly, provides 4 kcal/g. Fat is much more nutritionally dense, at 9 kcal/g. This is one of the reasons we crave the taste of fatty foods; evolutionarily, they represented a great source of energy. Even though it's completely appropriate to refer to the energy supplied by a food in terms of kcal, it's a bit more common in the United States to see the unit **Calorie** (always capitalized) instead. A Calorie (Cal) is equal to a kcal, which is

1,000 cal. The nutritional information on food packages always provides energy values in Cal.[4] It's important to know that not all of a food's weight comes from energy-yielding nutrients. For instance, lean ham is almost entirely protein with regard to its macronutrient composition; it contains very little fat and almost no carbohydrate. However, a 100-g slice of ham does not contain 400 Cal, because in addition to protein, there's quite a bit of water and indigestible matter in the ham. Determining the caloric (or nutritional) content of a food can be quite difficult, since most foods contain a combination of nutrients, and weight alone doesn't allow an accurate measurement of the amount of nutritive material in the food. Thankfully, packaged foods are required to be labeled with nutritional information so that interested individuals can determine exactly what it is they're eating.

8.5 Protein

The word **protein** comes from a Greek word meaning "of primary importance," because protein is more than just a source of energy. **Large protein molecules form the structural components of cells and extracellular material. Proteins produced by the body help us digest our food, interact with our environments, and allow our cells to communicate with one another.**

Proteins are made up of **amino acids,** which are small organic molecules. Amino acids consist of a short carbon skeleton, attached to which are two functional groups—one amine and one carboxylic acid. Additionally, each amino acid has a third characteristic group, called a side chain, attached to one of the carbons. The composition of the side chain determines the identity of the amino acid. There are 20 naturally occurring amino acids, each with a distinctive and unique side chain. The generic structure for an amino acid (with the side chain simply denoted as "R") is shown below:

IMAGE 8.4. Structural representations of two common proteins. In each case, the protein chain is represented as a "ribbon" (individual atoms are not visible at this resolution). Hemoglobin (a) in red blood cells carries oxygen from the lungs to the tissue, while collagen (b) adds elasticity to skin and epithelial tissue.

(a)

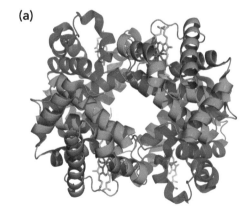

(b)

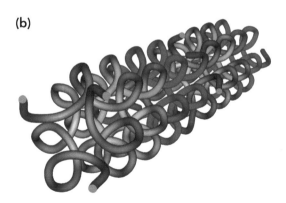

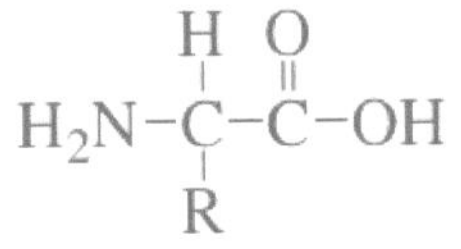

$$H_2N-\overset{\overset{\displaystyle H}{|}}{\underset{\underset{\displaystyle R}{|}}{C}}-\overset{\overset{\displaystyle O}{||}}{C}-OH$$

Amino acid structure

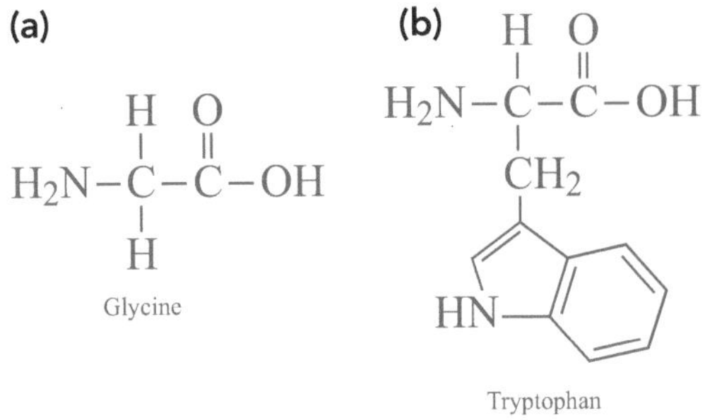

(a)

$$H_2N-\overset{\overset{\displaystyle H}{|}}{\underset{\underset{\displaystyle H}{|}}{C}}-\overset{\overset{\displaystyle O}{||}}{C}-OH$$

Glycine

(b)

$$H_2N-\overset{\overset{\displaystyle H}{|}}{\underset{\underset{\displaystyle CH_2}{|}}{C}}-\overset{\overset{\displaystyle O}{||}}{C}-OH$$

Tryptophan

IMAGE 8.5. Two very different amino acids. There are 18 other amino acids, each with a unique side chain.

Side chains can be made up of a wide variety of atoms and functional groups. The smallest amino acid, glycine, has a very simple side chain consisting of nothing but an atom of hydrogen. Tryptophan, the largest amino acid, has a much more complex side chain, formed of two carbon rings (Image 8.5). Some side chains are purely hydrocarbon, while other amino acids have side chains that contain N, O, and S.

Amino acids are protein building blocks—they are chemically linked together to produce long molecular chains called polypeptides (*poly* meaning "many" and *peptide* meaning "protein unit"). A polypeptide is a subcategory of the large class of organic molecules called polymers (*mer* means "unit," so a polymer is any large organic molecule made up of repeating units). Amino acids link together through the chemical removal of water, which is removed in pieces—H from the amine end of one amino acid, OH from the carboxylic acid end of the other amino acid—and the formation of a new bond. A chemical reaction like this, whether it takes place in the lab or in nature, is called a dehydration synthesis, since the atoms comprising a molecule of water are removed:

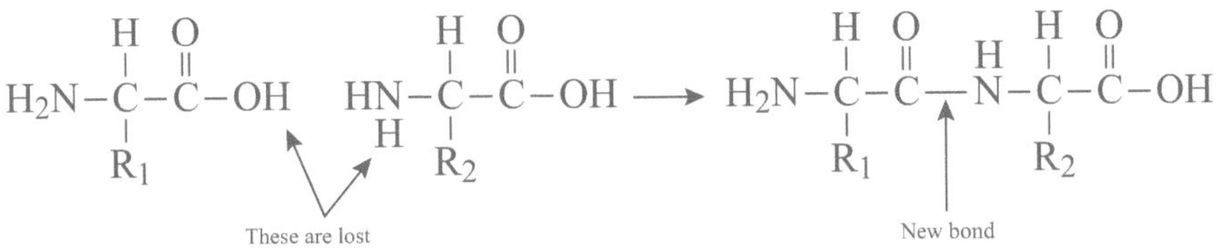

These are lost

New bond

The new bond is called a peptide bond because it joins the units of a polypeptide. For a polypeptide to become a fully functional protein, it must be processed further. This includes some chemical modification and folding, a full discussion of which is beyond the scope of this text. Regardless, at its heart, a protein is a chain of amino acids linked by peptide bonds. **Proteins can vary in length from just a few amino acids to several thousand. The amino acids used to produce the protein, the order in which they are placed, and the length of the protein chain collectively determine the structure and function of the resulting protein.** The short peptide shown in Image 8.6, called *met-enkephalin*, is an *endogenous opioid*—one of the body's own natural painkillers. This particular peptide consists of five total amino acids: tyrosine, two units of glycine, phenylalanine, and the sulfur-containing amino acid, methionine.

Due to the myriad possible combinations of amino acids, proteins have the most structural and functional variability of any class of biomolecules. Receptors, such as those discussed in Chapter 7, are made of protein. Some proteins are

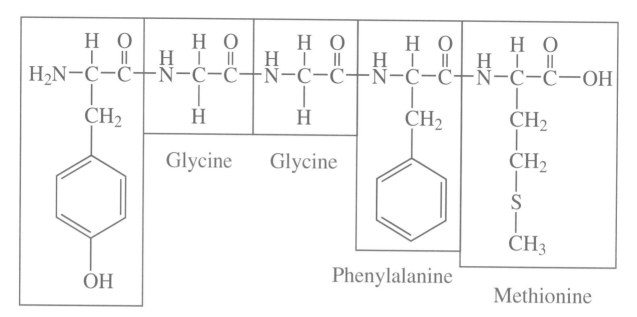

Glycine   Glycine

Phenylalanine

Methionine

Tyrosine

enzymes, which speed up otherwise slow chemical reactions. Proteins can also be structural—our hair and nails are made of the tough, fibrous protein *keratin*, while the elasticity of our skin is maintained by the springy protein *collagen*. Muscle is made up of protein specialized for movement. Even certain **hormones** are made of protein. A hormone is a chemical (some, but not all of which are peptides) used for cell-to-cell communication in the body. Examples of peptide hormones include *insulin*, which helps us to regulate blood sugar, and *oxytocin*, which is important to childbirth and the emotional experience of love and attachment. Clearly, the variety of function that can be achieved by linking amino acids together is immense!

8.6   Protein in Food

Proteins are digested through a reversal of dehydration synthesis: water is added in pieces across the peptide bond, restoring the missing H and OH. This process, called **hydrolysis** (which literally means *to split with water*), breaks the peptide bond

(a)

(b)

(c)

IMAGE 8.7. Fish and other lean meats, beans, and eggs are good sources of dietary protein.

Tryptophan   Serotonin

IMAGE 8.8. In this side-by-side comparison, the structural similarities between serotonin and its parent compound, tryptophan, are apparent.

NOTE[5]

It has been hypothesized that the sleepy feeling many people experience after eating large quantities of turkey (which contains tryptophan) on Thanksgiving is due to production of serotonin. However, in all likelihood, the real explanation is a bit simpler. Overeating—as many people tend to do on holidays—leads to diversion of blood away from the brain and toward the gut. The reduced blood flow to the brain produces a sensation of sleepiness. Further, turkey (like all animal protein) contains tryptophan, but not in particularly impressive amounts.

and releases the amino acids. **When the protein we consume is digested, meaning broken into its constituent amino acids, the amino acids are taken up by cells and can be used as building blocks for new body proteins or broken down further to provide us with energy.**

The body can also use certain amino acids for a third purpose: they can be chemically modified into related molecules with various functions. One example is that of the neurotransmitter *serotonin* (Image 8.8). Our brains use neurotransmitters to gather information about the outside world and relay that information to the brain, to produce movement, and to experience emotion. Serotonin is a neurotransmitter involved in emotion; when released in the appropriate amount, it produces feelings of calm and wellbeing. It's synthesized in the body from the amino acid tryptophan, and some neurochemists speculate that consuming foods high in tryptophan may lead to relaxation or sleepiness.[5]

8.7 Carbohydrates

Carbohydrates include sugars, starches, and indigestible dietary fiber. Even though sugars taste sweet to us while starch does not, they are very similar to one another chemically, and in fact are processed similarly by the body. Carbohydrates contain the elements C, H, and O, generally arranged into a ring, with many alcohol functional groups as substituents to the ring. A ring unit is called a **monosaccharide**, *mono* meaning "one" and *saccharide* meaning "sugar." Monosaccharides taste sweet, and are the building blocks of larger carbohydrates. A ubiquitous monosaccharide, **glucose**, appears in Image 8.10.

While it's certainly OK to draw glucose in line-angle, sugars are often shown from a slightly different perspective, which makes it

IMAGE 8.9. Carbohydrates are found in honey and other sweets, in fruits and many vegetables, and in breads and other grain products.

IMAGE 8.10. A line-angle diagram of glucose (left), and a Haworth projection (right). The dark line at the bottom of the Haworth projection is taken to mean that the ring is lying perpendicular to the page (parallel to the ground, if the page is held up in front of you), with the darkened edge of the ring closer to the reader.

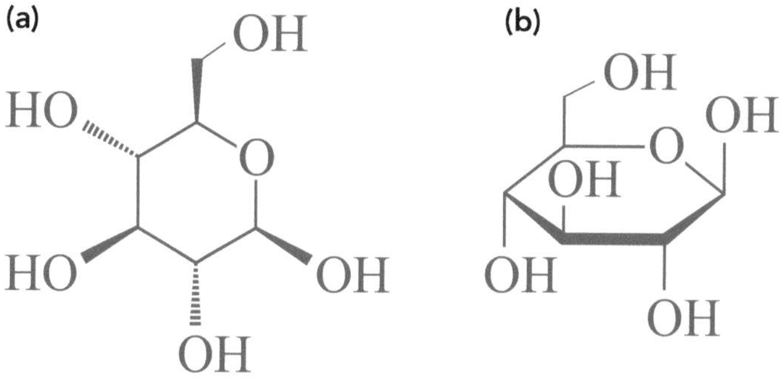

(a)

(b)

IMAGE 8.11. The Haworth projections of two monosaccharides. Both taste sweet and are chemically very similar to glucose.

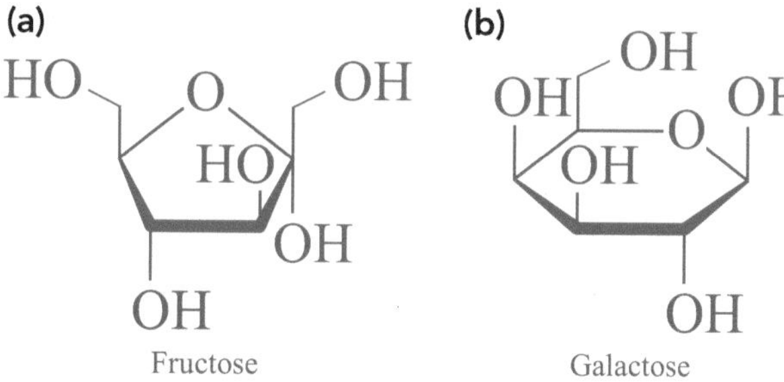

(a)

(b)

Fructose

Galactose

NOTE[6]

Notice that like monosaccharides, disaccharides have names ending in -ose. This pattern will hold for larger carbohydrates as well.

much easier to visualize the joining of monosaccharide units to form more complex carbohydrates. Tipping the sugar (imagine shoving the line-angle diagram so that it falls over backward) results in what is called a *Haworth projection* and is a common way to portray carbohydrates. Glucose is not the only monosaccharide—others include **fructose** (fruit sugar) and galactose (a constituent of milk sugar) (Image 8.11).

To form slightly larger carbohydrates, two monosaccharides are chemically linked together. As with the peptide bond that links amino acids, sugar bonds result from the removal of the atomic constituents of water from two monosaccharides— H from one, OH from the other—and the formation of a new bond in their place. This produces a **disaccharide** ("two sugars") via dehydration synthesis:

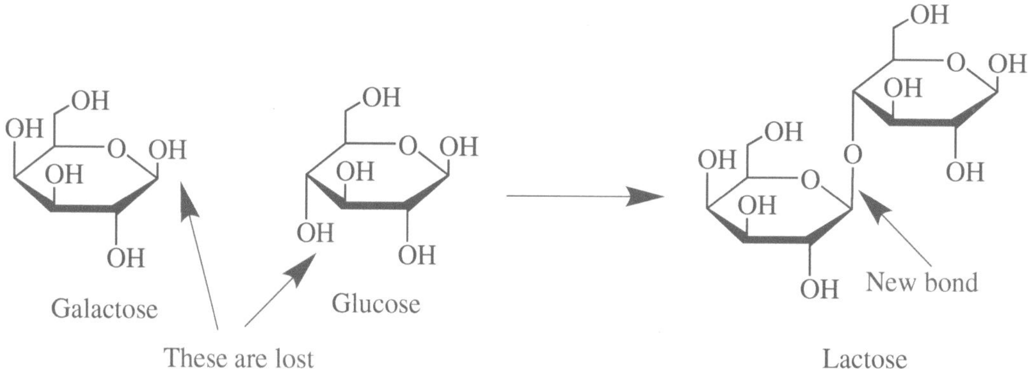

Galactose

Glucose

These are lost

New bond

Lactose

Common disaccharides include **sucrose** (table sugar), **lactose** (milk sugar), and maltose (malt sugar) (Image 8.12).[6] These molecules, like monosaccharides, have a sweet taste. In common parlance, both mono- and disaccharides are called *sugars*.

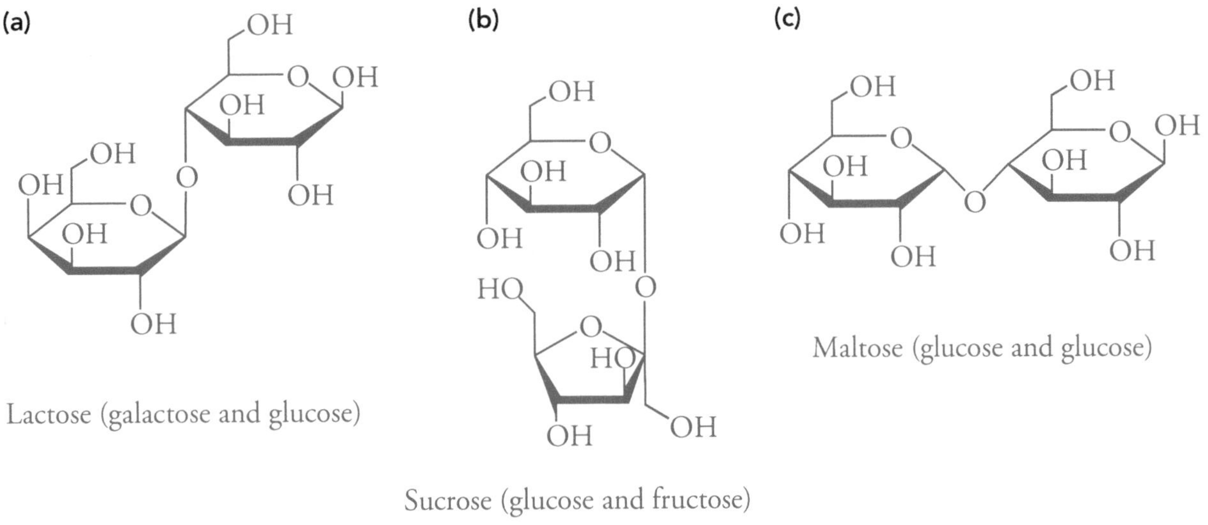

(a)

Lactose (galactose and glucose)

(b)

Sucrose (glucose and fructose)

(c)

Maltose (glucose and glucose)

IMAGE 8.12. Some disaccharides. Like monosaccharides, they taste sweet.

When we eat starchy foods like bread, corn, or potatoes, we don't taste the sweet flavor of sugar. Regardless, we're consuming carbohydrate, this time in the form of long chains of monosaccharides. Like disaccharides, these chains (called **polysaccharides**, for "many sugars") are formed by dehydration synthesis. They may consist of hundreds to thousands of monosaccharide units (generally glucose). Depending upon the orientation of the bond linking the glucose units, the resulting polysaccharides have slightly different shapes, as we observe with **cellulose** (plant fiber) and **amylose** (plant starch):

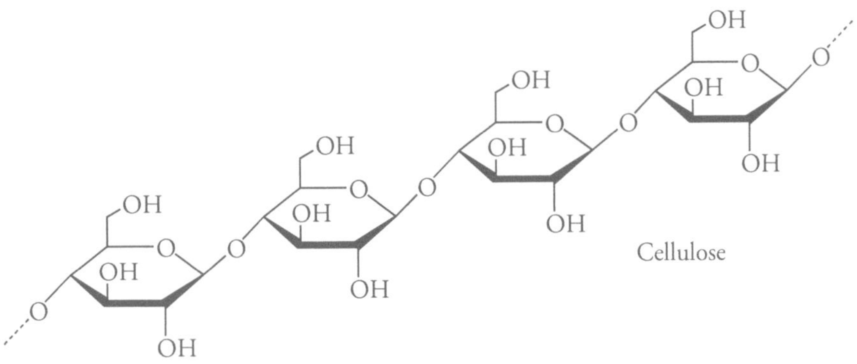

CONCEPT CHECK

How is the formation of a sugar bond similar to that of a peptide bond?

Answer: Both are formed by dehydration synthesis.

Cellulose

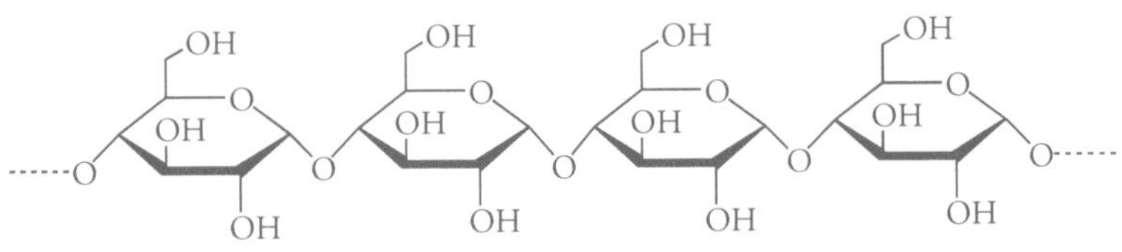

Amylose

NOTE[7]

A variety of digestive enzymes are used in the breakdown of protein and fat as well as carbohydrate.

NOTE[8]

Animals that depend solely on fibrous plants for energy, like cattle, use symbiotic bacteria to help them digest their food. The bacteria produce cellulase, so are capable of breaking the sugar linkages in fiber.

The sugar bond, like the peptide bond, can be broken via hydrolysis. The elements of water are added across the sugar bond, breaking it and releasing the constituent monosaccharides. This is important, because **the digestive tract is capable of absorbing only monosaccharides; larger carbohydrates must be broken down through digestive processes before absorption can occur.** While this process will occur slowly on its own if carbohydrates are exposed to water, digestive enzymes are used to speed the reaction up significantly.[7] Enzymes interact with their substrates in the same "lock and key" manner that we saw exhibited by receptors in Chapter 7. As such, the enzyme responsible for hydrolysis of lactose (called lactase) is not the same as the enzyme for sucrose (called sucrase) or amylose (called amylase). This helps us understand phenomena like *lactose intolerance*, a condition in which consumption of dairy leads to uncomfortable digestive symptoms. Lactose-intolerant individuals do not make lactase, and are unable to break lactose into its monosaccharides. As such, the lactose passes through the small intestine without being digested or absorbed. Lactose is easily digested by native bacteria in the large intestine, however, and as these bacteria process the lactose, they produce large quantities of gas. The gas buildup in the intestine leads to the bloating and cramping symptoms of lactose intolerance. People with lactose intolerance still produce all the other normal carbohydrate-digesting enzymes, so they can eat other sugars and starches without difficulty.

The only difference between cellulose and amylose is the orientation of the linkages, but this is significant enough that different enzymes are required to digest the two polysaccharides. Humans and other plant-eating animals produce amylase, but lack the cellulose-digesting enzyme (cellulase). As a result, **animals are unable to digest plant fiber, so it passes through the digestive tract without being absorbed and provides no energy.**[8] Despite the fact that we can't digest fiber, however, it serves several important dietary purposes. Fiber provides our mechanical digestive tract with bulk—it's easier for the intestines to move food (and eventually food waste) if there's a fair amount of it. Low-fiber diets tend to produce less waste, but the waste spends more time in the lower digestive tract as a result of the difficulty associated with moving small amounts of material. In this case, toxins can build up as undigested food begins to decompose before being eliminated. Furthermore, fiber can physically bind to toxins (fiber binds to cholesterol too, as we'll soon see!), helping to prevent them from being absorbed by the body. Finally, as we'll shortly see, fiber serves the important role of helping to regulate the rate of nutrient absorption.

NOTE[9]

As already discussed, glucose is not the only monosaccharide. However, it's the most commonly occurring and, for the sake of simplicity, it is the one to which this text will refer in discussions of monosaccharide absorption and processing. Other monosaccharides are processed in an analogous manner.

As we've seen, monosaccharides[9] require no chemical digestion prior to absorption from the gut, while disaccharides and amylose must be hydrolyzed. This process, however, does not significantly slow the rate at which sugar from starch can be absorbed relative to sugar from a monosaccharide source. The hydrolysis of carbohydrate is quite rapid, such that even the amount of time required to hydrolyze amylose into pure glucose is trivial. This, it turns out, is an important consideration when it comes to looking at food from a chemical perspective. For instance, the amylose molecule is visually more similar to cellulose than it is to sucrose. Both amylose and cellulose are long chains of glucose, and neither has a sweet flavor. Sucrose, on the other hand, is a small, sweet-tasting disaccharide. From a sensory perspective, sucrose and amylose seem very different indeed. Chemically, however, amylose acts nearly identical to sucrose once swallowed. It, like sucrose, is quickly and efficiently hydrolyzed into its constituent monosaccharide units, which are then absorbed. **From a cellular and chemical perspective, amylose is nothing more than sugar that doesn't happen to taste sweet.** On the contrary, for all the apparent similarity between the cellulose and amylose molecules, to our digestive tracts, they could not be more different. We can digest amylose into glucose, but can't digest cellulose. Amylose provides us with energy, while cellulose does not.

The similarity between amylose and table sugar has major nutritional ramifications. To make sense of this, we must first understand that the human body works hard to maintain homeostasis, or a steady set of physiological conditions. We have mechanisms to keep our temperature, blood pH, salt levels, hydration level, and many other parameters constant over time. This is because our cells can only survive in a very specific set of conditions. If we get too hot or cold, too acidic or basic, too low on water or too high, the cells quickly begin to lose function and can die. **One important parameter maintained by the body is the concentration of blood glucose, often just called blood sugar.** A relatively constant blood sugar level provides plenty of available energy for the cells—too little blood sugar, called hypoglycemia, can lead to feelings of weakness and even unconsciousness. On the other hand, too much blood sugar, called hyperglycemia, is not safe for the cells; severe acute hyperglycemia can cause result in dehydration, leading to symptoms

IMAGE 8.13. Chemically, white rice (b, composed largely of amylose) has much more in common with table sugar (a) than it does with brown rice (c), which contains lots of cellulose in its fibrous bran.

Digestive enzymes have to physically contact starch and disaccharides to hydrolyze them, but with no way to "see" one another, the enzymes and carbohydrates bump around blindly and contact each other only by chance. With lots of large fiber molecules in the gut, also bumping around blindly, it takes significantly longer for enzymes to "find" and digest all the sugar and starch.

of shock and possibly a coma. Chronic hyperglycemia can cause kidney and heart disease, damage to the eyes, and nerve cell death. For this reason, a healthy body maintains a relatively constant level of blood sugar. This requires the use of several hormones, including **insulin**, which lowers blood sugar by causing the cells to take up glucose from the bloodstream, and **glucagon**, which raises blood sugar by causing the release of some glucose from stores in the liver. The activity of the hormone insulin, in particular, is relevant to our discussion of carbohydrate absorption. It turns out that cells are very sensitive to the amount and timing of insulin released in response to food—we could call this the "strength" of an insulin signal. Sugar from foods that contain lots of fiber, like brown rice, whole grain, and most fruit, is absorbed slowly. This is because the cellulose, which is indigestible, literally gets in the way of digestive enzymes in the gut as they attempt to hydrolyze starches and disaccharides.[10] Further, the cellulose gets in the way of the proteins that are responsible for the absorption of monosaccharides into the bloodstream. As a result, the monosaccharides from high-fiber foods are absorbed very slowly, meaning that blood sugar levels increase very slowly. **In response to a slow increase in blood sugar, an organ called the pancreas slowly begins to release insulin. This "normal" insulin signal tells cells to take up blood sugar in accordance with their energy needs.** If body cells have all the sugar they need to fulfill their requirements, extra glucose is stored. Glucose storage can occur through either of two different mechanisms: it can be converted to fat, or it can be polymerized (made into a polysaccharide) in the liver and muscles. The storage polysaccharide, called **glycogen**, bears a strong resemblance to amylose: it's made of glucose units connected in the same way that the amylose units are connected, meaning that it can later be hydrolyzed to release glucose if blood sugar starts to fall. The only significant difference between glycogen and amylose is that the former is highly branched, which allows for rapid mobilization of glucose in cases of need:

Glycogen

While muscle glycogen is an important source of energy for movement (as we'll see in Chapter 9), it's liver glycogen that is really critical to the organism as a whole. In the instance of hunger or fasting, cells begin to run out of energy quickly. The hormone glucagon (which, like insulin, comes from the pancreas) signals the liver to hydrolyze glycogen and release the glucose units into the bloodstream, raising blood sugar. **This means that liver glycogen is a surrogate source of food for the body**

in the absence of a meal. The liver only stores about 2,000 Cal of glucose in the form of glycogen, so it can't provide too many meals' worth of sugar to the body cells, but glycogen is nevertheless an important and rapid source of energy in a pinch—particularly in comparison with fat, which may be used for energy, but which is slower to access, making it less than ideal for energy emergencies. Further, some cells (including those in the brain) consume glucose preferentially, making it very important for the body to keep stored glucose on hand at all times. **A physiological benefit of the "normal" insulin signal associated with slow-to-digest sources of carbohydrate that are rich in fiber is that it encourages the storage of extra glucose as glycogen rather than fat. This helps to ensure that the liver is full of glycogen and that valuable sources of emergency energy are not converted to fat—essentially placed in long-term storage—unnecessarily.**

Protein and fat also slow digestion. This phenomenon will be discussed in further detail in Chapter 9.

The chemical process described above is in sharp contrast with that which prevails if a fast-digesting source of carbohydrate is consumed. Sugars and amylose are hydrolyzed and absorbed very quickly in the absence of cellulose. As a result, blood sugar rises sharply and dramatically. This rapid increase in blood sugar can lead to feelings of overexcitation—the proverbial "sugar high." The pancreas responds with a rapid and abundant insulin release—a "strong" insulin signal. This particular type of insulin release appears to signal the body to store sugar as fat rather than allowing cells to burn it to satisfy their immediate energy requirements. Further, the strong and excessive cellular response causes blood sugar to be depleted far more rapidly (and to a greater extent) than normal, causing hypoglycemia. This leads to feelings of weakness, fatigue, depression, and hunger—a "sugar crash." Usually, individuals experiencing a sugar crash are inclined to eat, which raises blood sugar again and relieves the symptoms. We should note, though, that if the crash is treated with more fast-digesting carbohydrate, the entire cycle starts again. In addition, because sugar is being removed from the bloodstream and stored as fat (rather than being used for immediate energy or stored as glycogen), it takes many meals of fast-digesting carbohydrate to sustain the cells for the same amount of time that they would be sustained by one meal of slow-digesting carbohydrate. This means that more total calories (most of which are stored as fat) are consumed in a given period of time. **Meals consisting of significant amounts of amylose and sugar in the absence of cellulose or other digestion-slowing molecules[11] generally lead to unstable blood sugar and overstorage of body fat.**

Which carbohydrates tend to be slow digesting and which tend to be fast digesting? In general, it's rare to find fast-digesting foods in nature; natural sources of pure sugar are few and far between. Table sugar, for instance, can be obtained from plants such as sugar cane or beet, but must be purified out of the rest of the plant material. Fructose is found in fruit, but most fruit is also high in cellulose, which leads to slow absorption of the sugar. Honey is essentially the only naturally occurring source of pure sugar that would have been available to humans before the advent of agriculture and food refinement. Amylose apart from cellulose, similarly, is quite rare in nature. Most sources of starch, including grains, beans, and vegetables, also contain significant amounts of cellulose. Refined grains such as white rice and white flour have had the cellulose (found in the **bran** casing around the grain) physically removed. The reason for this has nothing to do with nutrition and everything to do with sensation and gustation—people discovered that removing the bran from grain produced stickier rice and finer flour, which they found pleasant to the taste and (in the case of flour) easier to bake with. Additionally, because the bran-removing process was time consuming and expensive, refined flours were more expensive and became associated with wealth, increasing their desirability. Unfortunately, however, removing the bran from wheat or other grain leaves behind a core of nearly pure amylose, which, as we've learned, is chemically nothing more than so much sugar. It's interesting to think about the effects our gustatory desires have upon our body chemistry: the human system isn't really designed to handle significant quantities of plain sugar or refined grain, as apparent from the paradoxical insulin response to the consumption of such foods. While our individual and cultural preferences lead us to manipulate foods in order to produce particular flavors or sensations, altering foods from their natural forms can affect not only the taste and texture, but the chemistry within the body as well!

8.10 Fats

Fat, unfortunately, has a very bad reputation as a nutrient class: many people consider fat as something to be avoided. It's important to understand, however, as we begin our discussion of the third macronutrient class, that fat in food and on the body is critical to survival. With regard to the food we eat, fat is a dense source of energy: remember that it provides 9 Cal/g consumed, as opposed to 4 Cal/g from protein and carbohydrate. Prehistorically, when food was scarce and had to be gathered or hunted, fat was valuable and difficult to come by; most plant foods are quite low in fat. A strong desire to consume fatty foods and the pleasure we

IMAGE 8.14. Olive oil and other liquid vegetable oils, nuts, and seeds provide healthy sources of fat. Animal fat (with the exception of the fats of cold-water fish such as salmon and tuna) is much less healthy.

experience when eating them are both holdovers from a powerful evolutionary drive to seek out these rich sources of energy. In modern society, with food (and fat) plentiful and readily available, our drive to consume fat can lead us to take in too much of a good thing, leading to weight and health concerns, which will be addressed in Chapter 9. Regardless, fat remains an important part of the modern diet when consumed in appropriate amounts. **Fat allows for maintenance of normal weight and body composition, is necessary to maintain immune and other body system functions, and allows for the uptake of fat-soluble vitamins from the digestive tract.** It's important to note that just because a nutrient is consumed doesn't necessarily mean it is absorbed. While a vegetable salad is chock-full of vitamins, some of these are soluble in water, and others are soluble in fat (this will be discussed in detail later in the chapter). The vitamins that dissolve in fat can't be absorbed efficiently unless fat is being absorbed from the digestive tract at the same time.

Dietary fats fall into the class of molecules called lipids, which are large molecules, but which (unlike protein and carbohydrate) are not polymers. Furthermore, while the structures of proteins and carbohydrates are quite predictable—proteins are chains of amino acids, while carbohydrates are one or more sugar units linked together—lipids include a large variety of molecules. Some lipids consist of hydrocarbon rings, while others are open chains of atoms. Some are simple, and others are much more complex. Regardless, lipids share one definitive trait—they are all primarily nonpolar, or hydrophobic. The simplest lipid is called a fatty acid and is nothing more than a very large carboxylic acid:

Steric acid, an example of a fatty acid

Note that while the carboxylic acid functional group—called the fatty acid's *polar head*—lends some polarity to the molecule, the majority of the structure is hydrocarbon in nature. The hydrocarbon chain is referred to as the molecule's *nonpolar tail*; due to its length in comparison to the polar head, the molecule is largely hydrophobic. Free fatty acids are not commonly found in nature. They are, however, often combined with an alcohol molecule called glycerol to make a complex lipid. Stoichiometrically, three fatty acids combine with each glycerol molecule, and at each junction, H from the glycerol and OH from the fatty acid's polar head are removed—H_2O in each case—making this reaction a dehydration synthesis:

NOTE [12]

We saw earlier in the chapter that the liver stores glycogen, a carbohydrate, for consumption by body cells, but compared to the average human's fat stores, glycogen represents a small percentage of total stored energy. A healthy 150-pound male human stores approximately 2,000 Calories (enough for about one day of average activity) of glycogen in the liver, and about 90,000 Calories worth of fat on the body as a whole. Given, though, that some fat is essential to normal function (which we'll discuss in Chapter 9), not all of the fat a body stores is actually available to burn under normal circumstances. Nevertheless, the tremendous difference in stored carbohydrate vs. fat energy is apparent.

Lost atoms shown in burgundy

Glycerol

The resulting molecule, called a **triglyceride**, is even less polar than the original fatty acids:

New bonds shown in burgundy

A triglyceride

Triglycerides are the molecules of which animal and plant fats and oils are composed. All plants and animals store fat as a source of energy (though this fat is not evenly distributed throughout the organism), and when we consume an organism, we can digest its stored energy. In addition to digesting fat for energy, we can store it ourselves: our personal fat stores are also composed of triglycerides. There are two major reasons that fat (as opposed to protein or carbohydrate) is the primary energy storage molecule.[12] First, fat contains more than twice as much energy for the weight as do either of the other macronutrients; storing fat allows us to pack the same amount of energy into a much smaller and lighter package, which is convenient for an active animal. Second, like all lipids, triglycerides are nonpolar. This means they neither dissolve in water nor do they attract it. The glycogen stored by the liver and muscles is highly polar (because of all the OH groups), and attracts tremendous amounts of water, which is co-stored with it. This increases both the weight and the bulk of the stored carbohydrate. **Fat is the lightest, most compact way to store a lot of energy for times of need.** Incidentally, while carbohydrate can be converted into fat, fat can't be converted into carbohydrate.

In the previous section of the chapter, we saw the stearic acid molecule, which is a common component of animal triglycerides. Stearic acid is an example of a saturated fatty acid, meaning that it has no double bonds in its hydrocarbon tail. This is an important consideration, because it affects the molecule's melting point and physical state. Since fats are nonpolar, the primary attractions between them are van der Waals forces. While these attractive forces are weak in small molecules, they can be quite strong in larger molecules like fatty acids. **Generally, saturated fatty acids and triglycerides composed of saturated fatty acids (called saturated fats) are soft solids at human (and other warm-blooded animal) body temperature.** The reason for this is that the long fatty acid tails can stack quite effectively on top of one another, almost like sticks in a woodpile (Image 8.15), making molecular packing very efficient. Recall from Chapter 2 that the more closely packed molecules are, the more solid their physical state.

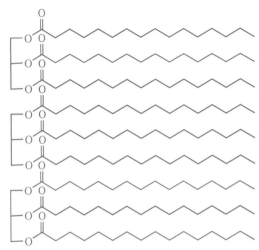

IMAGE 8.15. Saturated fatty acids form triglycerides that can pack together tightly, producing soft solids at room temperature.

Some fatty acids have one or more double bonds in the tail. These are called unsaturated fatty acids, which combine with glycerol to form unsaturated fats; they have melting points significantly lower than those of saturated fats. Table 8.2 compares the structures and melting points of four fatty acids. Note that increasing the number of double bonds in a fatty acid decreases the melting point. The same is true of unsaturated fatty acids in triglycerides.

The melting point of unsaturated fats is lower than that of saturated fats simply because the triglycerides can't pack as closely together in space (Image 8.16). **This produces a physical state less dense than that observed in saturated fats—most unsaturated fats are liquid at body (and even at room) temperature.**

The double bonds in unsaturated fatty acids all put kinks in the tails—this is quite important. There are two possible types of double bonds in organic molecules—one called a *cis* double bond and one called a *trans* double bond:

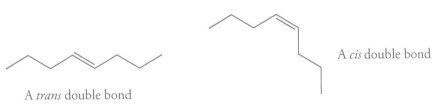

A *cis* double bond

A *trans* double bond

While the *cis* double bond creates a kink in a hydrocarbon chain, a *trans* double bond does not. **Almost without exception, naturally occurring unsaturated fatty acids have *cis* rather than *trans* double bonds.** Remember that the reason plants and animals store fat is for energy. In order to be an effective energy store, fat needs to be light and compact, which we've already learned it is. It must also be solid without being rigid—it would be quite difficult for an animal to move around efficiently with either liquid or completely rigid energy stores all over its body. Further, in addition to being incorporated into triglycerides as energy storage, fatty acids are also

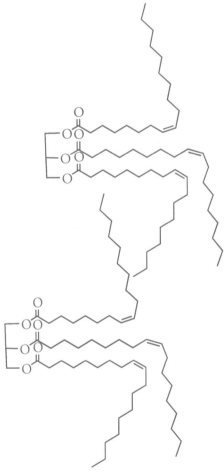

IMAGE 8.16. Unsaturated fats pack much less densely and are therefore liquid at room temperature and human body temperature.

Name Melting Point	Structure
Stearic acid **69.9°C**	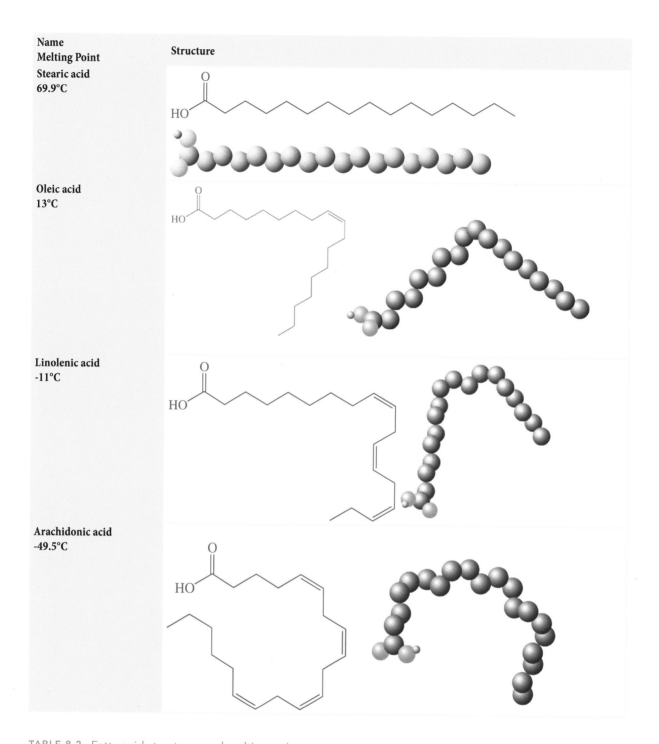
Oleic acid **13°C**	
Linolenic acid **-11°C**	
Arachidonic acid **-49.5°C**	

TABLE 8.2. Fatty acid structures and melting points.

part of each cell's surrounding membrane. To function properly, cell membranes must be sturdy but malleable. For this reason, an animal that maintains a body temperature in the range of 37°C (like humans and other mammals) needs a fat that will be a soft solid at that temperature. Stearic acid, with its melting point of 69.9°C, and other saturated fatty acids work well. From this, we can surmise that most of our triglycerides are composed of saturated fatty acids. Further, the fat we consume from warm-blooded animal sources is mostly saturated fat. Tropical plants, such as the coconut, are also likely to contain saturated fat, owing to the very warm environment in which they live.

Consider, however, a cold-water fish such as the salmon. These animals are not able to regulate their body temperature like mammals. In the near-freezing water in which a salmon lives, saturated fat stores would solidify to the point that swimming would become impossible! As a result, salmon evolved the ability to produce and store unsaturated fat, which, because of the kinks produced by the *cis* double bonds, is a malleable solid at much lower temperatures. This evolutionary adaptation allows salmon and other cold-bodied animals to survive in environments unsuited to saturated fatty acids comprising triglycerides and cell membranes. Besides cold-water fish, dietary sources of unsaturated fats are temperate-weather plants such as corn and soybean. **The easiest way to visually determine whether a fat is saturated or unsaturated is to note its physical state at room temperature: saturated fats are solids, while unsaturated fats are oils.**

Note that no such adaptive advantage would be conferred to an organism by the evolution of fatty acids with *trans* double bonds. *Trans* double bonds do not bend the fatty acid tails (Image 8.17), and therefore triglycerides incorporating them would pack identically to (and have melting points identical to) saturated fats. An organism with *trans* double bonds in its fats would be able to survive only in the same environments as those available to organisms with saturated fats. **Because the *trans* fatty acid does not confer an adaptive advantage, there was no evolutionary pressure for it to evolve, and so it did not.**

Despite the fact that *trans* fatty acids didn't evolve in nature, they do sometimes appear in food. **Particularly common in highly processed foods, triglycerides containing *trans* fatty acids (called trans fats) come from unsaturated plant oils that have been chemically reacted so as to remove some of the double bonds via a reaction called *hydrogenation*.** A side effect of this reaction is that some of the double bonds are not removed, but instead are *isomerized*, or changed from *cis* to *trans*. The resulting triglycerides, called partially hydrogenated oils, pack much more tightly—and therefore have much higher melting points—than the parent oils used to produce them. This fat-modifying process was developed in order to solidify vegetable fats for cooking—shortening and margarine are examples.[13] While the health effects of trans fats will be discussed later in the chapter, it is telling to note that they are not utilized efficiently by bacteria, which is yet another reason for their ubiquity in processed foods: trans fats have long shelf lives.

Two final important classes of fatty acids are omega-3 and omega-6. In fatty acid nomenclature, *omega* is used to refer to the carbon furthest from the carboxylic acid end. An omega-3 fatty acid has one or more double bonds—like other naturally occurring fatty acids, the double bonds will be *cis*—the furthest of which is three carbons from the omega end. An omega-6 fatty acid has its most distant double bond six carbons from the omega end.

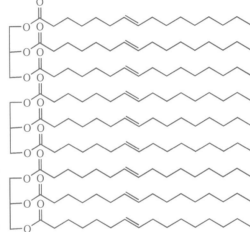

IMAGE 8.17. The packing of triglycerides with *trans* double bonds in them is identical to that of saturated fats.

NOTE[13]

The rationale behind converting vegetable oils into solids is that animal fat, like lard, works well for baking, but it is expensive. Vegetable oils are too runny for many baking purposes; cookie batter, for instance, becomes very liquid and cookies lose their shape in the oven as they're heated. By converting vegetable oil into a soft solid, a cheap and abundant material is modified such that it has the necessary properties for cooking purposes.

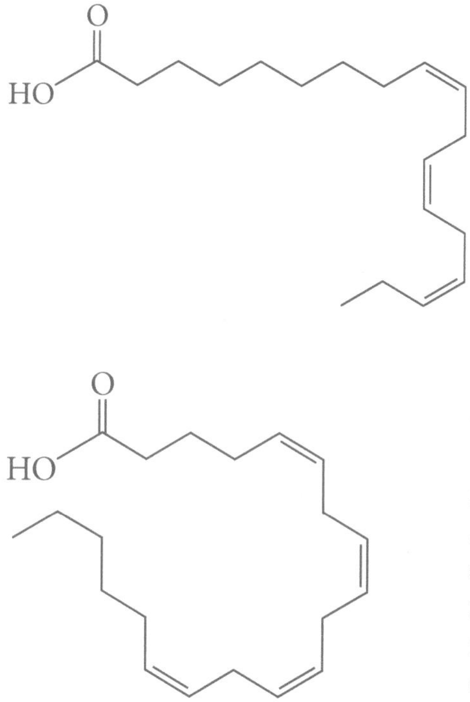

IMAGE 8.18. Linolenic acid (up) is an omega-3 fatty acid, while arachidonic acid (down) is an omega-6.

NOTE[14]

Food marketers, however, take advantage of the fact that, while many people know that omega-3 fats are a healthy component of diet, most do not know that we can't actually utilize the omega-3 fats in flax and nuts. Many cereals, breads, and so forth will proudly advertise that they are excellent sources of omega-3 fats, a fact that, while technically true, is functionally misleading.

Both omega-3 and omega-6 fatty acids, in addition to being healthy unsaturated fats, are necessary components of diet. They appear to play important roles in normal development, brain and immune system function, and metabolism (the sum of all chemical reactions in the body), among other things. While both are essential to health, research suggests that an optimal diet should include omega-6 and omega-3 fatty acids in a 2:1 to 4:1 ratio. Typical Western diets, however, have up to 25 times more omega-6—common in cereals, grains, and most vegetable oils—than omega-3, which is found mostly in cold-water fish. There are vegetarian sources of omega-3 fats, including flax seeds and nuts, but these contain alpha-linolenic acid (ALA), which the human body does not utilize. Humans require the omega-3 fatty acids docosahexaenoic acid (DHA) and eicosapentaenoic acid (EPA). While we can theoretically convert ALA into DHA and EPA, the conversion is so inefficient (if it happens at all) as to render the omega-3 fats from vegetable sources essentially unusable.[14] Research suggests that a balanced intake of omega-6 and omega-3 fatty acids (somewhere closer to the 2:1 to 4:1 range) may help to reduce the incidence and severity of chronic inflammatory conditions, as omega-6 fatty acids are (among other things) chemical precursors to *prostaglandins*, which are inflammatory molecules. Other therapeutic benefits of omega-3 fatty acids well supported by research include reduction in symptoms or prevention of cardiovascular disease and high blood pressure. Further benefits, including relief from symptoms of asthma and cancer prevention, are under investigation.

8.12 Cholesterol and Steroid Hormones

Cholesterol, a lipid on the basis of its polarity, is not composed of fatty acids. Instead, the molecule consists of four connected rings (Image 8.19). **An important biomolecule, cholesterol is the starting material for a number of synthesis reactions that take place in the body, with products including vitamin D and the steroid hormones.** The latter, like peptide hormones, are important in cell-to-cell communication. Some of the steroids synthesized from cholesterol are pictured in Image 8.19 and include cortisol (regulates metabolism), testosterone (male sex hormone), estradiol (female sex hormone), and progesterone (a pregnancy hormone). Note that cholesterol and the steroid hormones, while containing different functional groups, all share the same ring structure.

In addition to its role as a biosynthetic starting material, cholesterol serves the crucial function of helping to maintain appropriate fluidity in cell membranes. As discussed earlier, cell membranes must be malleable without being either too rigid or too fluid. In part, appropriate rigidity is maintained by incorporation of fatty acids with the appropriate degree of saturation. However, cholesterol is also incorporated into membranes and acts as a stiffening agent. Cells take up cholesterol from the bloodstream in accordance with their needs, but are able to make cholesterol if none is available to them. In other words, while we need cholesterol in order

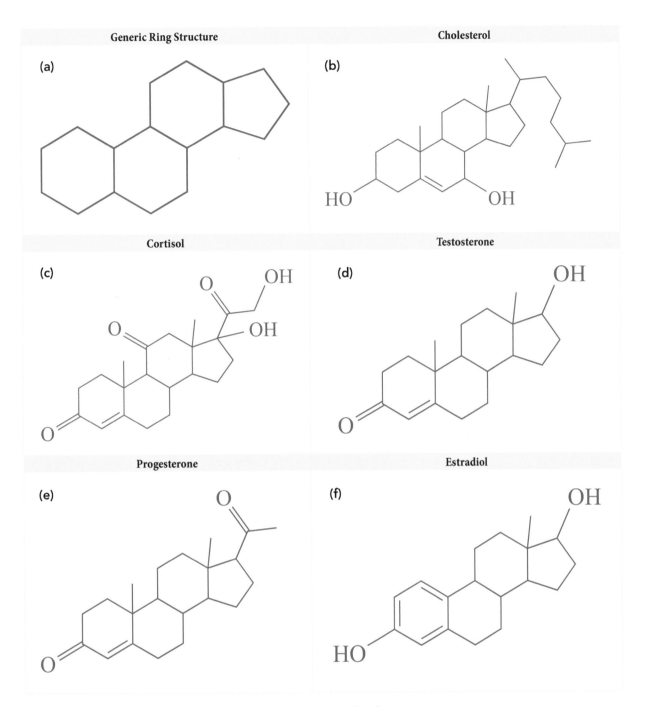

IMAGE 8.19. Cholesterol and some important derivative molecules.

to survive, we don't have to consume it; we can synthesize it. A normal, healthy individual really can't be cholesterol deficient, but it's quite easy to overconsume or overproduce cholesterol. **High cholesterol has severe ramifications for cardio-vascular health and leads to atherosclerosis, or hardening of the arteries.** In order to understand why cholesterol can cause heart disease, it's necessary to understand how it travels around the body.

NOTE¹⁵

Alcohol consumption also increases
HDL, which researchers are
considering as one of the reasons
for the "French Paradox"—the
observation that some cultures that
eat fatty foods but consume wine
regularly have lower instances of heart
disease than Americans. It's important
to note, though, that anything above
moderate alcohol consumption
quickly leads to significant negative
health effects—including liver disease
and increased risk of cancer—that
outweigh any benefit.

CONCEPT CHECK

Which has the most negative effect
upon cardiovascular health: saturated,
unsaturated, or trans fat?

Answer: Trans fat.

It is a common misconception that there are two kinds of cholesterol, one of which is "good" and one of which is "bad." In fact, cholesterol (like water and glucose) is the name of a single molecule—there is only one kind of cholesterol. **There are, however, two types of cholesterol transporters: high-density lipoprotein (HDL) moves excess cholesterol from the cells to the liver for excretion, and low-density lipoprotein (LDL) transports cholesterol to the cells.** The key to this system is that HDL and LDL, both containing identical cholesterol and both traveling through the bloodstream, move cholesterol in different directions because of the location of their receptors. LDL is made in the liver and is packed with cholesterol absorbed from food. Its receptors are located on body cells. Cells carefully control what they take in and what they excrete. Therefore, a cell with plenty of cholesterol doesn't need to take in any more. In order to regulate cholesterol intake, only cells with a need for cholesterol produce LDL receptors. Cells with sufficient cholesterol stores do not produce the receptors, and therefore can't accept cholesterol from the bloodstream. This protects the cells, but can lead to the cardiovascular effects mentioned earlier. If an individual consumes a diet high in cholesterol or cholesterol precursors, LDL production is high. The abundant cholesterol quickly fulfills cellular needs, and in response, the cells decrease their production of LDL receptors. **As a result, LDL is not removed from the bloodstream, and the cholesterol from the LDL can collect in the arteries, forming plaques and stiffening the arterial walls.** The plaques narrow arteries, reducing blood flow. Additionally, plaques can break off, resulting in the formation of clots. These clots can then travel through the bloodstream, potentially becoming lodged in small vessels elsewhere in the body. If a clot occludes a vessel that carries blood to the heart, a heart attack results. If a vessel in the brain is occluded, the result is a stroke. Clots occluding vessels feeding muscles can lead to difficult moving, pain, muscle death, and gangrene. HDL, on the other hand, lowers the risk of cardiovascular disease. It is produced in body cells, and is packed full of cholesterol that the cell doesn't need. The HDL receptors are on the liver, which takes up the cholesterol, modifies it, and excretes it via the digestive tract. By packaging and removing excess cholesterol from cells, HDL encourages them to produce LDL receptors so that LDL-contained cholesterol is removed more efficiently from the bloodstream.

The American Heart Association sets forth recommendations regarding cholesterol levels: **LDL (the so-called "bad" cholesterol) levels should be low and HDL (the "good" cholesterol) levels should be high.** Not only does cholesterol intake influence the amount of LDL in the bloodstream, overall fat intake and the type of fat consumed also have an effect. Saturated fats tend to increase LDL levels. Unsaturated fats have long been thought to have a neutral effect on both LDL and HDL, though some recent evidence suggests that they may improve the LDL/HDL ratio by decreasing LDL and increasing HDL. Trans fats appear not only to increase LDL, but also to decrease HDL. For this reason, most nutritionists and the American Heart Association suggest avoiding them entirely. The most efficient route to healthy levels of HDL appears to be regular, vigorous exercise.¹⁵

In addition to the nutritional relevance of fatty acids and other lipids, they are also the molecules used in the production of soap. A possibly apocryphal tale tells of the discovery of soap in ancient times, when women would take dirty clothes to the river to rinse them in the water and rub dirt off on the rocks. Above the river was a hill, and on the hill there was a temple where animals were sacrificed and burned. Women of the lower classes had to use the undesirable portion of the riverbank where the remains of the sacrifices, including blood, fat, and ash from the fires, ran into the water. However, it was soon discovered that these women got their clothes cleaner than those of more privileged classes. The reason for this phenomenon is that animal fat, if reacted with ash or lye (both of which contain NaOH), produces a foamy mixture that removes dirt and grease from clothing and skin far more effectively than water alone. By itself, water is far too polar to dissolve grease and oil (both of which are hydrophobic) and the dirt that clings to them.

Recall that animal fat is composed of triglycerides, which consist of glycerol and three fatty acid molecules. Since triglycerides are formed through dehydration synthesis, they can be taken apart again through hydrolysis. However, this reaction is quite slow. NaOH makes the hydrolysis much faster and more efficient, easily separating the three fatty acid molecules from glycerol. Because the fatty acids are acidic, however, and NaOH is basic, a further reaction occurs—the neutralization of the fatty acids to produce their sodium salts:

The molecule above is even more dichotomous with regard to its polarity than an ordinary fatty acid: its ionic head is highly water-soluble, while the tail remains quite hydrophobic. **Because the fatty acid salt has distinct regions of polarity, it neither dissolves entirely in water nor in fat; instead, fatty acid salt dissolves best when a combination of water and fat are present in a mixture.** Molecules displaying behavior like this are called amphiphilic, meaning that they are both water- and fat-loving. They are also perfectly suited to removing grease from dirty surfaces. When a bit of grease, some soap, and some water are mixed together, the fatty acid salts arrange themselves in a sphere called a micelle around the grease. The tails cluster together, embedding themselves in the grease, while the polar heads remain on the outside of the micelle, exposed to water (Image 8.20). The grease-containing micelle, because of the polar exterior, is very water-soluble and dissolves with ease, carrying the grease away.

Today's soaps are a bit more chemically complex than the ancient variety; animal fat soaps made with lye tend to be quite drying to the skin and generally don't smell very pleasant. Current soap technology incorporates pleasant fragrance,

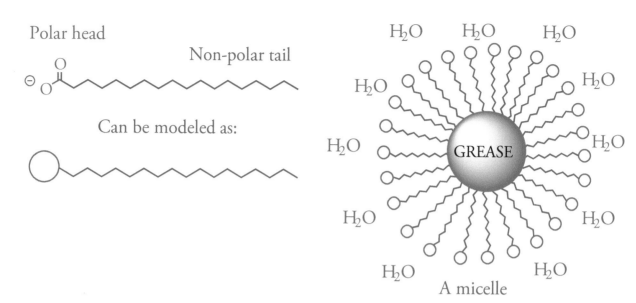

Polar head

Non-polar tail

Can be modeled as:

H₂O H₂O H₂O

H₂O H₂O

H₂O GREASE H₂O

H₂O H₂O

H₂O H₂O

A micelle

IMAGE 8.20. A micelle forms naturally when fatty acid salts are dissolved in water (shown here in cutaway; a micelle is spherical in reality).

moisturizers for the skin, and other advancements. At heart, though, today's soaps share their chemistry with that of their ancient counterparts: they are amphiphilic molecules whose dichotomous polarity allows them to make fat soluble in water.

8.14   Vitamins

In addition to the macronutrients, we require some **micronutrients**. These are chemicals necessary in very small amounts that perform vital functions but do not yield energy. **Vitamins** are one of the classes of micronutrients. They are organic molecules, some of which are water-soluble and some of which are fat-soluble. The solubility of a vitamin has significant ramifications upon the frequency with which it must be consumed and its potential for toxicity. **The majority of the water-soluble vitamins aren't stored to any significant extent in the body, so they must be eaten on a regular basis. Since water-soluble vitamins consumed in excess of the body's needs are excreted, it is much more difficult to consume a toxic dose of these molecules (within reason—the Paracelsus Principle holds true here, as always)[16] than it would be to overdose on a fat-soluble vitamin.**

Vitamin C

NOTE[16]

While our cells require oxygen, it's nevertheless a very reactive molecule. O_2 is capable of engaging in reactions in the body that damage structural proteins and our genetic material, DNA. Vitamin C and other antioxidants react preemptively with a variety of reactive molecules, environmental radiation, and so forth; this destroys the antioxidants (which is why we need to consume a regular supply), but protects our biomolecules.

Vitamin C, or ascorbic acid, is a water-soluble antioxidant. Antioxidants help to protect the cells from damage resulting from exposure to environmental toxins, radiation (including the sun's damaging ultraviolet light), and even oxygen.[17] Vitamin C is also critical to healthy immune system function (though the popular notion that very large doses of it will prevent or cure a cold has been shown to be untrue in multiple scientific studies[18]) and is critical to the body's ability to produce collagen. Vitamin C deficiency is rare in modern cultures, because it's so easily obtained from fresh fruits and vegetables—especially citrus fruit, berries, and leafy greens. However, deficiency results in a disease called scurvy, which at one time was epidemic among sailors. Scurvy causes a variety of symptoms, including susceptibility to infections, bleeding gums, and generalized weakness and malaise. Sailors' diets consisted almost entirely of dried meat and bread; these items resisted spoilage and were high in calories relative to their weight. Fruits and vegetables on ships were almost unheard of. Eventually, however, it was discovered that scurvy could be cured (and more importantly, prevented!) by the regular consumption of fruit or fruit juice, which from that point on was included in the sailors' rations. In fact, the term *limey*—slang for a British person—was derived from the British practice, introduced by Captain James Cook (an 18th-century explorer who, among other things, mapped the Hawaiian Islands), of consuming lime juice while at sea to hold scurvy at bay.

Vitamin B

There are actually many B-vitamins; they are referred to individually either by name or number (B1 is called thiamine, B2 is called riboflavin, and so forth). **As a class, the B-vitamins play supportive but critical roles in metabolism.** Many are **coenzymes**, meaning that they help metabolic enzymes function appropriately. **The B-vitamins are all water-soluble, so as a general rule, they are not stored in the body.** The only exception to this is B12, called cyanocobalamin, which can be stored in the liver for several years. While B-vitamins are found in many foods, including grains, fruit, and vegetables, B12 is found only in animal products (and one very esoteric species of yeast). This presents a challenge to vegans—without meat, eggs, or milk, the liver's stores of the vitamin become depleted. Since B12 is critical to synthesis of red blood cells, depletion leads to a condition called *pernicious anemia*.

NOTE[17]

While our cells require oxygen, it's nevertheless a very reactive molecule. O_2 is capable of engaging in reactions in the body that damage structural proteins and our genetic material, DNA. Vitamin C and other antioxidants react preemptively with a variety of reactive molecules, environmental radiation, and so forth; this destroys the antioxidants (which is why we need to consume a regular supply) but protects our biomolecules.

NOTE[18]

It's a relatively common misconception that if we need Substance X to maintain normal function, then more of Substance X will allow for "super function." In most cases, this is untrue. For instance, we require a certain amount of vitamin C for normal immune function, but there's no truth whatsoever to the claim that taking more than that amount will supercharge (boost) the immune system. In fact, since vitamin C is water-soluble, amounts in excess of our daily requirement are simply urinated out. Given the American love affair with vitamin supplementation and the fact that so many people take vitamin doses hugely in excess of their body's need for and ability to utilize the vitamins, other cultures joke on occasion that Americans have the most expensive urine the world.

For this reason, vegans must supplement their diet with B12, either in the form of vitamin pills or vitamin-enriched food products.[19]

The fact that a vegan diet must be supplemented with vitamin B12 is evidence that refutes the not uncommon proveganism argument that it is a more "natural" diet for humans than one incorporating animal products. As the sole nonanimal source of B12 (a species of yeast) is so rare and esoteric as to have been unavailable to the majority of early humans, we would not have been able to survive on a vegan diet prior to our ability to manufacture vitamin B12 in the lab. As such, while there are belief-related reasons that individuals might choose to consume a vegan diet, from a scientific perspective, one cannot make the argument that such a diet is natural for our species.

Vitamin A

Also called retinol, vitamin A can be obtained directly from food or biosynthesized from its precursor compound, beta-carotene. **Among its other functions in the body, retinol is crucial to vision: deficiencies can lead to reduced visual acuity, particularly in low-light conditions. Because vitamin A is fat-soluble, it does not need to be consumed daily—it can be stored in the body.** In particular, it tends to concentrate in the liver. While this is beneficial in times of vitamin shortage in the diet, the ramification is that toxicity can result from an acute or chronic overdose. Retinol can be obtained from food in two different ways. Fruits and vegetables containing beta-carotene (particularly common in yellow-orange produce) allow for biosynthesis of the vitamin. Alternately, animal sources contain retinoids, which are partially synthesized vitamin A made by the animal. Common animal sources include eggs and dairy (as retinoids are required for the developing chick or growing calf), as well as organ meats like kidney and liver—the very same organs in which we store vitamin A!

Vitamin D

This vitamin is fat-soluble and crucial to the absorption of calcium (a mineral necessary for bone production and maintenance, in addition to normal cell, nerve, and muscle function) from the digestive tract. It can be obtained in two ways: either from food sources or via biosynthetic modification of cholesterol in the skin upon exposure to sunlight. In most cases, about fifteen minutes of sunshine a few times a week is sufficient, though sunlight at extreme northern or southern latitudes is not strong enough to cause the reaction. Compared to the other vitamins, there are few food sources of vitamin D. Some fish, including salmon, tuna, and cod, as well as the liver oils of these fish, are good sources. Beef liver, eggs, and some cheeses also contain the vitamin. Furthermore, many dairy products and cereals are supplemented with vitamin D in order to prevent deficiencies, which can be quite serious, especially in children. Childhood vitamin D deficiency can produce severe skeletal abnormalities and deformities due to incompletely hardened bones; this condition is called rickets, and was relatively common in the United States before vitamin D fortification of dairy and cereal products became standard in the 1930s. Adult deficiency of the vitamin is associated with osteoporosis (weakening of the bones) and autoimmune disease, and research is beginning to hint that the vitamin has roles in many other areas of normal body function as well.

Vitamin E

Like vitamin C, **vitamin E is an antioxidant responsible for protecting cells from reactive oxygen species, free radicals, and the like. Vitamin E, however, is fat-soluble, meaning that between the two antioxidants, all cellular environments are protected.** The best sources of the vitamin include nuts, seeds, and vegetable oil. Leafy greens, like spinach and broccoli, are also good sources. Mild deficiency in the vitamin rarely causes symptoms, and severe deficiency is quite rare, since vitamin E–rich foods are so common—generally, only those with malabsorption syndromes (inability to absorb nutrients from the digestive tract) are likely to be vitamin-E deficient.

Vitamin K

Primarily known as the vitamin necessary to produce some of the proteins required for blood clotting, vitamin K (which is fat-soluble) also has a role in building bone. Deficiencies result in inability to clot blood, which can lead to bleeding disorders. Vitamin K can be obtained from vegetable sources—particularly leafy greens. In addition, native bacteria in the digestive tract, including *E. coli*, produce the vitamin, which we can then absorb from the intestine.

8.15   Minerals

Like vitamins, minerals are micronutrients. They do not supply us with energy, and we need only consume small amounts of them. As with vitamins, though, mineral deficiencies can be very serious. **The minerals are trace elements, many of them metal or halide ions, and serve various purposes in the body.** Iron, for instance, must be incorporated into hemoglobin for the protein to be capable of delivering oxygen to tissues. Calcium is critical to production and maintenance of the skeletal system and also plays key roles in muscle function, maintenance of the heartbeat, nervous system function, and cellular communication. Sodium and potassium ions have important metabolic and neurological function: they are critical to fluid balance, neural transmission, muscle movement, and production of energy. As we saw in Chapter 7, severe sodium deficiencies can cause weakness and even death. The chloride ion is necessary to production of stomach acid (HCl), and is also involved in maintaining fluid and electrolyte balance. Copper is essential to the body's ability to build hemoglobin and many other proteins. Iodine maintains normal thyroid gland function, which helps to regulate metabolism. Minerals are generally obtained from food. A balanced diet composed of all the food groups, such as we'll discuss in Chapter 9, satisfies mineral requirements.

Throughout this chapter, we've examined the chemistry of both macro- and micronutrients in the human body: why they're needed, how they're utilized, and how they affect health. One theme that begins to develop through a discussion of nutritional chemistry—and which will be expanded upon in Chapter 9—is that while the chemistry that takes place in the body is quite complex, it is also quite predictable. When reactants are supplied and conditions are favorable, reactions proceed. When reactants are not supplied, reactions can't proceed. As complex as human nutritional requirements seem—the marmots and koalas of the world, with their very simple diets, have it much easier—everything we take in is toward the single purpose of providing reactants and energy for cellular reactions. In the end, we're really just big thermodynamic machines, running chemical reactions and harvesting the energy they release.

8.16   The Last Word—Vitamin Enriched!

A cursory glance at drugstore shelves stocked with personal care products reveals scores of bottles touting special ingredients: amino acid–enriched conditioners, shampoos containing vitamins, and even shampoos containing lotion or conditioner. While it's tempting to believe television and magazine ads that encourage us to buy the newest and most "scientifically advanced" products, reevaluating these items in light of the principles of chemistry provides a different perspective.

With regard to hair, while shampoo and conditioner manufacturers tirelessly attempt to persuade us that tresses need to be provided with nutrition, the simple fact is that hair is composed primarily of the protein keratin—the same protein that forms the fingernails. In a manner highly analogous to the formation of skin, living cells from deep within the hair follicle rapidly divide and move upward, losing internal structures and functions as they fill with keratin until they are no longer living cells, but simply membrane sacs full of protein. Because the cells composing hair are neither alive nor metabolically active, they require no nutrition. We've seen in this chapter that vitamins allow living cells to run chemical reactions, an activity in which nonliving structures (such as membranous sacs of protein) do not engage. Live and actively dividing follicular cells, located within the scalp, do require both macro- and micronutrients in order to produce desirable quantities of strong, well-formed hair. However, these nutrients must be supplied in the same manner as they are to any cell: from the digestive tract via the bloodstream. In the end, there are only two things that can be accomplished by rubbing products on hair: keeping it clean (a function of soap), and keeping it from getting dry and fragile (a function of some sort of moisturizer or conditioner).

On a similar topic, many products claim to provide vitamins to the skin if applied topically. While it is not a stretch to imagine that vitamin E, for instance, could penetrate the skin's outer layers, a chemist struggles to envision how the highly hydrophilic vitamin C might do any good at all. The membranes of all cells are composed of lipids; fatty acid complexes, cholesterol, and other nonpolar molecules surround the cell, separating its aqueous interior from the aqueous extracellular environment. As a result, while nonpolar molecules (particularly small ones) cross the membrane with ease, polar molecules are excluded. In the case of the skin, this function is highly protective—it keeps us from absorbing chemicals (particularly toxins) from our environments. Very few compounds have the ability to penetrate the skin. Some research has suggested that if chemically modified (by attaching a series of very long hydrocarbon tails through ester linkages), vitamin C can be made to pass through the skin. Other studies indicate that at low pH (around 3.5), vitamin C can be absorbed transdermally. Neither of these conditions, however, are met in the average hair- or skin-care product, meaning that vitamin C in shampoo ends up where it is most soluble—dissolved in the shower water, running down the drain!

SUMMARY OF MAIN POINTS

- Thermodynamics is the study of the energetics of chemical reactions.
- The formation of bonds between atoms always releases energy, while breaking bonds always requires energy.
- Reactions are rearrangements of bonds—the formation of bonds that are lower in energy (more stable) results in a release of energy (and vice versa).
- The first law of thermodynamics states that energy must be conserved in all processes.
- Entropy is disordered (unusable) energy.
- The second law of thermodynamics says that entropy must increase for all processes.
- The energy released (or required) by a chemical reaction is the sum of energy required to break bonds minus the sum of energy released to make new bonds.
- Exothermic reactions release energy, while endothermic reactions require energy. The macronutrients we eat are broken down in exothermic reactions, and we harvest that energy to do work.
- Macronutrients are nutrients we need in large quantities, and include protein, carbohydrate, and fat.
- Food energy is measured in Calories (Cal), where 1 Cal = 1 kcal.
- Proteins are polymers of amino acids, which can be metabolized for energy, used to make new protein, or converted into other molecules.
- Protein is a major component of every cell: enzymes, receptors, and fibrous structural materials are all protein.

- Carbohydrates include sugars (mono- and disaccharides), starch polysaccharides, and indigestible fiber.
- Different enzymes break disaccharides and starches into their monosaccharide units, but we lack the enzyme to digest cellulose.
- Amylose starch breaks down quickly into pure glucose, so from a chemical perspective, it's sugar.
- Sugar consumption elevates blood sugar, which is regulated by the hormones insulin (causes cells to take up glucose) and glucagon (causes release of glucose).
- Sugar and amylose cause rapid elevation of blood sugar and rapid, strong insulin responses; this leads to low blood sugar and enhanced fat storage.
- We store excess sugar as glycogen or convert it into fat.
- Fat is an important dietary component. It's nutritionally dense, maintains immune and other system functions, and allows for absorption of fat-soluble vitamins.
- Fats are consumed (and stored) as triglycerides, which are made of a glycerol molecule and three fatty acids.
- We store most of our energy as fat because it's energetically dense and doesn't co-store with water, since it's nonpolar.
- Fatty acids with no double bonds are called saturated fats and are malleable solids at human body temperature.
- Naturally occurring fatty acids with double bonds are called unsaturated fats, and the double bonds are all *cis*.
- Fatty acids with *trans* double bonds don't occur naturally—they are synthesized by partial hydrogenation and are common in processed foods.
- Omega-3 and omega-6 fatty acids have many double bonds (all *cis*), the last of which is three (or six) carbons from the tail of the fatty acid.
- An appropriate dietary ratio of omega-3:omega-6 (about 1:2) is associated with good health.
- Cholesterol is a lipid made of four connected rings. It is important for biosynthesis of steroid hormones and vitamin D, and it also appears in cell membranes.
- Too much systemic cholesterol can result in atherosclerosis and cardiovascular disease.
- LDL is a cholesterol transporter that takes cholesterol from the digestive tract to the cells; high levels are associated with cardiovascular disease.
- HDL is a cholesterol transporter that takes cholesterol from the cells to the liver for excretion; high levels are associated with good cardiovascular health.
- Saturated and trans fat consumption increase LDL, trans fat lowers HDL, and unsaturated fats have no effect on HDL or LDL levels. Most consumed fat should be unsaturated for these reasons.
- In the presence of base, triglycerides break down into fatty acid anions. These are soaps.

- Soaps are amphiphilic, since they have a polar region and a nonpolar region. They spontaneously form micelles in water, and can make grease water-soluble.
- Vitamins and minerals are micronutrients—we don't need to consume large quantities of them.
- Water-soluble vitamins are C (antioxidant) and the B-vitamins (mostly coenzymes).
- Fat-soluble vitamins are A (involved in vision), D (involved in calcium absorption and regulation), E (antioxidant), and K (blood clotting).
- There are many important minerals, including iron (for hemoglobin), calcium (for bones), sodium and potassium (for fluid balance and nerve/muscle function), and iodine (for thyroid).

QUESTIONS AND TOPICS FOR DISCUSSION

1. What does it mean to say that energy is always transferred or converted (rather than created or destroyed)?

2. Think of an example of a process (chemical or otherwise) that involves at least three energy conversions/transfers. Describe and explain.

3. Explain why forming bonds releases energy, while breaking bonds requires energy. Does this seem reasonable from an atomic perspective? (Think in terms of subatomic particles and stability.)

4. Why do double (and triple) bonds have higher bond energies than single bonds?

5. Explain why breaking a given bond requires exactly the same amount of energy that would be released if that same bond formed.

6. In terms of total energy and entropy, how is burning sugar in a beaker and letting the heat escape different from burning sugar in a beaker and using the flame to heat water?

7. If forming bonds required energy, how would the universe be different?

8. Why is food energy counted in Cal rather than cal? What is the relationship between Cal and cal?

9. Does it make sense that polymers made by dehydration synthesis are digested by hydrolysis? What do each of those mean chemically?

10. Why might a nutritionist say that starch is really just sugar? Is that reasonable?

11. From a nutritional standpoint, does a bowl of white rice have more in common with a bowl of brown rice or a bowl of sugar? Explain.

12. What is the difference (with regard to digestion and blood sugar) between whole grain and refined grain?

13. Why is a rapid increase in blood sugar detrimental from a chemical perspective? What does it lead to?

14. From an evolutionary perspective, why are our bodies "confused" by sources of pure sugar or pure starch?

15. Why is fiber important, even if we can't digest it?

16. Why can't humans digest fiber?

17. Why can someone with lactose intolerance still eat other carbohydrate-containing foods?

18. What generally happens to glucose when it's absorbed into the bloodstream slowly? Quickly?

19. Why didn't trans fat evolve as a storage fat in nature?

20. Why are most of our stored calories in the form of fat? What advantages does it have? Why is our storage fat mostly saturated?

21. A disease called familial hypercholesterolemia results from a genetic inability to produce LDL receptors. What sorts of symptoms do you think disease sufferers experience?

22. In popular culture, HDL is sometimes called "good cholesterol" and LDL is called "bad cholesterol." Are these accurate descriptions? Why or why not?

23. After a doctor's appointment including blood work, a friend tells you that he has high cholesterol. You find out that his LDL is in the normal range, but his HDL is very high. What might you say to him? Are you concerned for his health?

24. Early soaps were made with animal fat and lye. In what ways are modern soaps superior? In what ways must they be the same?

25. Bile salts are amphiphilic molecules released into the intestine by the gall bladder. They are used to aid in the digestion of fats. How do you think they achieve

this, given that they do not chemically react with fats at all? (Hint: digestive enzymes are all water-soluble!)

26. Why do you think organ meat (like liver) is very high in vitamins A and D?

27. A friend tells you that he is taking megadoses (10 times the daily requirement) of vitamin C to boost his immune system. Are you concerned for his safety? What is happening to all the vitamin C in his body? Is it likely boosting his immune system?

28. A friend tells you that he is taking megadoses (10 times the daily requirement) of vitamin A to boost his night vision. Are you concerned for his safety? What is happening to all the vitamin A in his body? Is it likely boosting his night vision?

PROBLEMS

1. What is the first law of thermodynamics, and what does it mean?

2. What is the second law of thermodynamics, and what does it mean?

3. Calculate the energy released or required to form 1 mole of O_2 from atoms.

4. Calculate the energy released or required to form 1 mole of N_2 from atoms.

5. Calculate the energy released or required to form 1 mole of C_2H_4 from atoms (hint: draw a Lewis structure!).

6. Calculate the energy released or required to form 1 mole of COH_2 from atoms (hint: C is in the middle—draw a Lewis structure!).

7. Calculate the energy released or required to form 2 moles of CO_2 from atoms (hint: remember that Table 8.1 lists energies in kcal/mol).

8. Calculate the energy released or required to form 2 moles of CH_3OH from atoms (hints: draw a Lewis structure, and remember that Table 8.1 lists energies in kcal/mol).

9. Calculate the energy released or required to break 1 mole of H_2 into atoms.

10. Calculate the energy released or required to break 1 mole of O_2 into atoms.

11. Calculate the energy released or required to break 1 mole of C_2H_6 into atoms (hint: draw a Lewis structure!).

12. Calculate the energy released or required to break 1 mole of CH_3NH_2 into atoms (hint: draw a Lewis structure!).

13. Calculate the energy released or required to break 2 moles of H_2O into atoms (hint: remember that Table 8.1 lists energies in kcal/mol).

14. Calculate the energy released or required to break 2 moles of C_2H_2 into atoms (hints: draw a Lewis structure, and remember that Table 8.1 lists energies in kcal/mol).

15. Determine how much energy is released or required for the reaction shown. Is this reaction exothermic or endothermic?

$$2\ H_2 + O_2 \rightarrow 2\ H_2O$$

16. Determine how much energy is released or required for the reaction shown. Is this reaction exothermic or endothermic?

$$CO_2 + 2\ H_2O \rightarrow CH_4 + 2\ O_2$$

17. Determine how much energy is released or required for the reaction shown. Is this reaction exothermic or endothermic?

$$C_2H_4 + 3\ O_2 \rightarrow 2\ CO_2 + 2\ H_2O$$

18. Determine how much energy is released or required for the reaction shown. Is this reaction exothermic or endothermic?

$$C + O_2 \rightarrow CO_2$$

19. Fill in the blanks in the following table with equivalent values across rows:

40 kcal	_____Cal	_____cal
_____kcal	550 Cal	_____cal
_____kcal	_____Cal	10,700 cal

20. Fill in the blanks in the following table with equivalent values across rows:

_____kcal	487	_____cal
_____kcal	_____Cal	13,600
320 kcal	_____Cal	_____cal

21. Is protein structural, functional, or both? Explain.

22. What's a polymer? Which of the macronutrients are polymers and which are not?

23. Define: protein; amino acid; peptide bond; enzyme.

24. How can the digestion products of proteins be used in the body?

25. What are the general categories of carbohydrates, and how do they differ from one another?

26. What are the roles of the hormones insulin and glucagon? How does insulin affect body chemistry if it's released slowly? Quickly?

27. How is excess glucose stored in the body?

28. Why are fats important to the diet?

29. What is the difference between saturated and unsaturated fat?

30. Why is unsaturated fat considered much healthier than saturated or trans fat?

31. Does the term "unsaturated fat," used in common parlance, refer to fat with *cis* or *trans* double bonds?

32. Are omega-3 and omega-6 fatty acids saturated, unsaturated, or trans fats? Are they considered healthy?

33. Why is cholesterol an important biomolecule? What is it used for?

34. What are HDL and LDL, and what do they do in the body?

35. What factors affect HDL and LDL levels, and how?

36. For optimal heart health, should LDL be high or low? Should HDL be high or low?

37. What is a micelle? What does it do?

38. Which vitamins are water-soluble? Fat-soluble? Which can be stored in the body?

39. Fill in the following table for vitamins:

Vitamin	Good Source	Function in the Body
A		
B's		
C		
D		
E		
K		

40. Fill in the following table for minerals:

Mineral	Function in the Body
Iron	
Calcium	
Sodium and potassium	
Chloride	
Copper	
Iodine	

41. Why are protein, carbohydrate, and fat called macronutrients, but vitamins and minerals are called micronutrients?

42. Do vitamins and minerals provide energy? If not, why do we need them?

ANSWERS TO ODD-NUMBERED PROBLEMS

1. Energy can neither be created nor destroyed; energy of the universe is constant.
3. 119 kcal released.
5. 542 kcal released.
7. 768 kcal released.
9. 104.2 kcal required.
11. 677 kcal required.
13. 444 kcal required.
15. 116.6 kcal/mol released; exothermic.
17. 313 kcal/mol released; exothermic.
19. Fill in the blanks in the following table with equivalent values across rows:

40 kcal	40 Cal	40,000 cal
550 kcal	550 Cal	550,000 cal
10.7 kcal	10.7 Cal	10,700 cal

21. Protein is both structural (forms structural elements of cells, structures like hair, etc) and functional (enzymes, receptors, etc.).

23. A protein is a polymer made up of amino acids; an amino acid is a small molecule consisting of a carboxylic acid group, an amine group, and a side chain; a peptide bond is the bond between amino acids in a protein; an enzyme is a protein that speeds up a chemical reaction.

25. Sugars (mono- and disaccharides) are rapidly absorbed; starches are polysaccharides that can be broken down into sugars quickly; fiber is a polysaccharide that we can't digest.

27. Excess glucose can be polymerized into glycogen in the liver and muscles or converted to and stored as fat.

29. Saturated fat has no double bonds and has a much higher melting point. It is a malleable solid at human body temperature. Unsaturated fat has *cis* double bonds (one or more) and is liquid at human body temperature.

31. Unsaturated fat has *cis* double bonds.

33. Cholesterol regulates the stiffness of cell membranes and is the biosynthetic precursor for vitamin D and steroid hormones.

35. Saturated fats increase LDL, trans fats increase LDL and lower HDL. Vigorous exercise and moderate alcohol increase HDL.

37. A micelle is the spherical arrangement of fatty acid salts in aqueous solution—it can sequester grease to the inside of the micelle, essentially making grease balls water-soluble.

39. Fill in the following table for vitamins:

Vitamin	Good Source	Function in the body
A	Fruits and vegetables, some animal sources	Vision, especially low-light
B's	Grains, fruits, vegetables. Animal only for B12	Various—mostly coienzymes
C	Many fruits	Antioxidant
D	Fish and fish oils, sunlight	Calcium absorption
E	Nuts, seeds, vegetable oils	Antioxidant
K	Leafy greens and intestinal bacteria	Blood clotting

41. We need protein, carbohydrate and fat in large quantities (relatively speaking), but vitamins and minerals are only needed in small quantities.

Fats

Benson, M. K., Kshama, D., and Fattepur, S.R. (2008). Studies on SFA/TFA (saturated/trans fatty acid) rich dietary fats on lipid profile and antioxidant enzymes in normal and stressed rats. *Pharmacognosy Magazine, 4*(16), 320–28.

Berbert, A. A., Kondo, C. R., Almendra, C. L., *et al.* (2005). Supplementation of fish oil and olive oil in patients with rheumatoid arthritis. *Nutrition, 21*(2), 131–36.

Crupkin, M., Zambelli, A. (2008). Detrimental impact of trans fats on human health: Stearic acid-rich fats as possible substitutes. *Comprehensive Reviews in Food Science and Food Safety, 7*(3), 271–79.

Dry, J., and Vincent, D. (1991). Effect of a fish oil diet on asthma: Results of a 1-year double blind study. *International Archives of Allergy and Applied Immunology, 95*(2–3), 156–57.

Erkkila, A. T., Lichtenstein, A. H., Mozaffarian, D., *et al.* (2004). Fish oil is associated with a reduced progression of coronary artery atherosclerosis in postmenopausal women with coronary artery disease. *American Journal of Clinical Nutrition, 80*(3), 626–32.

Lecerf, J. M. (2009). Fatty acids and cardiovascular disease. *Nutrition Reviews, 67*(5), 273–83.

Zevenbergen, H., deBree, A., Zeelenberg, M., *et al.* (2009). Foods with a high fat quality are essential for healthy diets. *Annals of Nutrition and Metabolism, 54,* 15–24.

http://www.nlm.nih.gov/medlineplus/druginfo/natural/patient-fishoil.html

Cholesterol

www.americanheart.org

http://www.nlm.nih.gov/medlineplus/cholesterol.html

Vitamins

http://www.nlm.nih.gov/medlineplus/vitamins.html

http://www.hsph.harvard.edu/nutritionsource/what-should-you-eat/vitamins/

Vitamin Enrichment

Farris, P. (2005). Topical vitamin C: A useful agent for treating photoaging and other dermatologic conditions. *Dermatological Surgery, 31,* 814–18.

Lin, J. Y., Selim, M. A., Shea, C. R., *et al.* (2003). UV photoprotection by combination topical antioxidants vitamin C and vitamin E. *Journal of the American Academy of Dermatology, 48*(6), 866–74.

CREDITS

The Body as a Beaker—Nutrition, Health, and Wellness

9

IMAGE 9.1. Incorporation of exercise into daily life, as these fun-run participants are doing, has a positive impact on health.

NOTE[1]

Since the first edition of this text, the phenomenon of increasing obesity and media exposure to impossible standards of perfection is starting to show up worldwide in industrialized nations. Within the next decade, I suspect, we'll need to rename the paradox being described here to reflect the global nature of the problem.

While it's difficult to generalize across the board, the trend in many industrialized Western countries in recent decades has been toward increasing obesity. Simultaneously, citizens of industrialized nations are exposed more than ever before to airbrushed and computer-generated images portraying impossible (or nearly impossible) standards of perfection with regard to musculature, athleticism, slenderness, and shape. At the juxtaposition of these two observations lies what we will call the **Western Paradox**[1]—the fact that Western societies are simultaneously consuming a generally atrocious diet and putting a premium on physical perfection. This has spurred the rapid growth of the weight-reduction industry, made room for questionable diet gimmicks touting rapid results, and all the while failed to slow the pace at which obesity has been overtaking smoking as the number one cause of preventable death. People all over the world share the need to provide their bodies with appropriate reactants (in the form of food) so as to synthesize necessary cellular products, generate energy, and maintain life. Along the way, however, we are faced with a variety of challenges. Some individuals are unable to secure enough food to meet their energy needs, while others are tempted with an incredible array of available food at every turn. The lifestyles of some expose them, by choice or by necessity, to regular vigorous exercise; others lead more sedentary existences. As a result, the world is full of all manner of different body habitus—some lean and others plump, some bulkily muscular and others sinewy. This chapter focuses on the ways in which the molecules of nutrition affect body composition, health, and composition-dependent illness. How do our diets affect how we look and feel? What is the rationale behind the U.S. government–recommended nutritional requirements? For individuals striving to lose weight, what is the chemistry of various weight-loss strategies? Most importantly, how does a health-conscious individual construct a balanced, healthy lifestyle amid desk jobs, socially prescribed notions of appropriate body composition, and a plethora of fast food restaurants?

9.1 Body Fat—Too Much, Too Little, and Just Right

Before we begin a discussion of the disease processes associated with being overly fat, let's first examine the reasons the human body needs fat at all. From Chapter 8, we know fat is necessary in the diet: It provides a valuable source of energy, allows for absorption of fat-soluble vitamins, and maintains immune system function, among other things. In a culture in which periods of plenty of food are followed by times of scarcity, body fat serves as a crucial store of many weeks' worth of energy. In a society in which food is never farther away than a walk to the refrigerator or a trip to the neighborhood grocer, however, is body fat even necessary? Could a very strict diet and exercise regimen strip all the fat off a human body? And if it did, what would be the ramifications? **It turns out that fat on the body is as critical to health as fat in the diet: we simply can't live without it.**

First, while it might seem that fat energy reserves are of little utility in a society with ready access to food, we nevertheless draw upon these reserves frequently. Periods of sleep, a long day at work without a lunch break, and extended moderate-intensity exercise are all examples of times during which we rely on mobilization of the energy from our fat stores—without them, we would awaken frequently to eat, become frazzled and fatigued at work, and be unable to complete a workout. In addition, fat is more than simply the vehicle by which fat-soluble vitamins are absorbed from the digestive tract. It's also a repository for these vitamins in the body. Fat-soluble vitamins build up in fatty tissue, meaning that they need not be consumed daily, since reserves are available. Water-soluble vitamins, on the other hand, must be consumed with greater frequency in order to prevent deficiency. Steroid hormones, as we saw in Chapter 8, are cholesterol derived. Like cholesterol, they're nonpolar, and like fat-soluble vitamins, they concentrate in the fatty tissues. This is important to normal body function, as concentrations of one hormone often influence the production of others. Finally (and eminently practically), fat provides physical padding. The soles of the feet are cushioned by fat pads, as are the palms of the hands. Fat in the lower back helps to protect the kidneys and other posterior organs that lie, in all or in part, below the protective enclosure of the rib cage but above the pelvic girdle. As you read this book, you're likely seated—a prospect which would be significantly less comfortable without a bit of fat padding the sitting bones! Too little body fat is not compatible with life.

On the other hand, excess body fat (referred to as either overweight or obesity, depending upon the degree to which the condition exists) is quite dangerous and associated with a number of disease processes. High blood pressure and high cholesterol, both of which are risk factors for cardiovascular disease and stroke, occur more frequently among the overweight and obese. Sleep apnea, a temporary inability to breathe during sleep, can result from fat in the neck collapsing the back of the throat as muscles relax. In order to take a breath, an individual with apnea must partially awaken (often hundreds of times a night) to reopen the airway. Sleep apnea is associated with fatigue and general poor health due to chronic exhaustion. Arthritis, a degeneration of the joints, occurs among the overweight and obese due to the physical stress of carrying weight in excess of the body's innate biomechanical capacity. Some forms of cancer are more common in this demographic as well. **One of the most prevalent disease processes in the overweight and obese population, however, is type 2 diabetes.** Unlike type 1 diabetes, which results from an inability to produce the blood sugar–regulating hormone insulin (and which is treated with injected insulin), type 2 diabetes is a cellular insensitivity to the hormone, which is often (but not always) a result of excess body fat. As a result, blood sugar builds up after meals, and can often become dangerously high. This leads to symptoms of chronic hyperglycemia. Further, since cells are unable to take up glucose in the absence of insulin, symptoms of cellular starvation are also evident. Type 2 diabetes is severe, and left unchecked, can be fatal. Since insulin production occurs in type 2 diabetics to the same extent that it occurs in the nondiabetic population, injected insulin is not a helpful treatment regimen. Rather, disease sufferers are counseled to control their diet, reduce intake of sugars and refined carbohydrates, exercise regularly, and, most importantly, lose weight.

The overweight and obesity epidemic in the United States is of great proportion, so to speak. **From 2001–2004 National Institutes of Health (NIH) data, two thirds of U.S. adults are overweight, with half of these (one third of total U.S. adults) falling into the obese category.** Less than one third of U.S. adults are of a healthy weight. Minorities, in particular, are afflicted by the obesity epidemic, as are those of lower socioeconomic standing, due in part to the availability and low cost of high-fat, highly processed foods, the consumption of which is associated with high body fat. Obesity levels have increased dramatically in recent decades; in 1989, no U.S. state reported higher than a 15% obesity rate, while in 2008, only one state reported *less* than 20–25% obesity (Image 9.2).

IMAGE 9.2. Between 1989 (top) and 2008 (bottom), the incidence of obesity in the population has increased astronomically, according to the Centers for Disease Control and Prevention records.

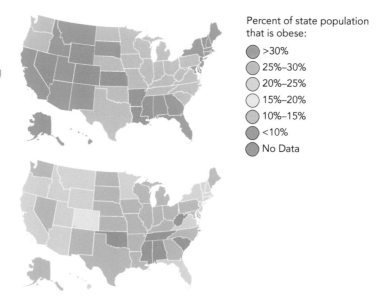

Percent of state population that is obese:

- >30%
- 25%–30%
- 20%–25%
- 15%–20%
- 10%–15%
- <10%
- No Data

From a social perspective, obesity is problematic beyond the individual level. Of course, individuals are directly affected by their own health. With regard to the larger community, however, the cost of obesity is worth considering. Direct health care costs include preventive and diagnostic medicine, as well as medical treatment. These are footed by the government and recouped through taxes in the case of subsidized care, and are footed by insurance companies and recouped via increased premiums across the board in the case of privately funded care. Indirectly, lost wages and work time represent significant individual and group costs. Clearly, obesity is not simply a problem faced by the obese: as members of society, we all pay the price.

9.2 Body Composition Metrics

Various health-monitoring organizations, including the World Health Organization (WHO) and NIH, set forth guidelines regarding the percentage of body fat (by mass) appropriate to a healthy human. **For men, essential body fat, meaning the minimum percentage body fat by mass required to sustain normal function, is 2–4% of body weight. For women, this value is higher, at 10–12%.** The reason for a higher essential body fat in women has to do with the female hormone cycle; in order to produce normal menses and maintain reproductive function, estrogen and progesterone must build up in fatty tissues. Women with very low body fat often experience *amenhorrea*, or lack of a menstrual cycle. As a result, these individuals are unable to become pregnant. Normal body-fat values are considered to be in the range of 6–26% of body weight for men and 14–31% for women, with athletes and fit individuals falling near the lower end of this range. Body fat percentages higher than these indicate overweight or obesity, depending upon the value. A major reason to monitor body composition (which falls primarily under the auspices of health care professionals) is to provide appropriate medical care and counseling to individuals who might be at risk for diseases associated with being under- or overweight. For this reason, it's quite common for a physician to check patient weight and height during annual visits. However, from a practical perspective (and despite the prevalence of electronic scales claiming to do the job), it's quite difficult to measure body-fat percentage accurately, even in a physician's office. Some metrics provide for a very rough approximation of percent body fat from **wrist, neck, and waist circumference**. However, particularly extensive (or absent) musculature significantly decreases the degree to which these measurements accurately reflect body composition. Alternately, three- or seven-site **skin fold tests** may be used to estimate body fat, the theory being that fat deposits *subcutaneously* (just under the skin) in proportion to total body fat. Calipers, which isolate a pinch of skin and the fat lying just under it, are used to take thickness measurements at various points on the body. Unfortunately, slight variations in caliper placement or technique, hydration level, and other factors can significantly affect the accuracy of the measurement.

The "gold standard" of body fat percentage determination is the **immersion test**, which requires that an individual be weighed while completely submerged in water, having blown all the expellable air out of their lungs.[2] In addition to the discomfort and cost associated with the procedure, it's rather impractical. For this reason, **physicians and other health care providers generally rely on a simple metric based upon weight and height to approximate body composition; the number obtained is called body mass index (BMI).** This metric is calculated as follows and correlated to body composition as per Table 9.1:

$$BMI = \frac{weight\ (kg)}{[height\ (m)]^2}$$

As techniques for approximating weight status go, BMI is both easy to calculate and relatively accurate for most individuals. Where it fails is with the exceedingly muscular (muscle, far denser than fat, can make an individual with normal body fat weigh more than average for their height, resulting in a higher BMI). Further, very inactive, ostensibly skinny individuals with little muscle tend to have BMI values that do not accurately reflect body composition. The reason for this is that, as fat is less dense than muscle, a very slight person (these individuals are often female) may actually be carrying a high percentage of body fat while appearing slender and weighing very little, provided that there is minimal muscle density to add to the weight. Despite their appearance and healthy-appearing BMI, these "skinny fats" are nevertheless susceptible to diseases of obesity. As such, BMI values for individuals falling into these two groups must be regarded with a certain amount of skepticism, or at least verified through a secondary measurement technique.

BMI	Weight Status
<18.5	Underweight
18.5–24.9	Normal
25.0–29.9	Overweight
≥30.0	Obese

TABLE 9.1. BMI scores as they correlate to body-weight status.

TRY THIS

Calculate the BMI for an individual who is 1.83 m tall, and who weighs 75 kg. Into which weight status category does this individual fall?

Answer: 22.39; normal.

Solution:

$$BMI = \frac{weight}{height^2} = \frac{75\ kg}{(1.83\ m)^2} = 22.39$$

9.3   Nutritional Recommendations

NOTE[3]

They also sometimes reflect the
efforts of special-interest lobbyists—
for instance, the current standards
are based upon a 2,000 Cal/day diet,
but some information regarding a
2,500 Cal/day diet is also listed on
nutritional labels. This is the result of a
compromise made between the FDA
(who wanted the 2,000 Cal/day label)
and the USDA. The latter organization,
looking out for the interests of meat
and dairy farmers—groups that
didn't want consumers thinking their
products were unreasonably high in
fat or calories relative to nutritional
recommendations—sought the higher
value.

NOTE[4]

The exceptions include spices, water,
and alcoholic beverages; the latter
was the result of intense lobbying
by the industry, presumably because
they feared falling sales if the caloric
content and nutritional vacuity of
their products were made clear to
consumers.

Armed with the knowledge that we need protein, carbohydrate, fat, vitamins and minerals to stay healthy, coupled with the information that we should eat enough to stay fit without overeating and gaining excess body fat, how do we go about making informed decisions about our food? **The U.S. Department of Agriculture (USDA) and Food and Drug Administration (FDA) work jointly to provide the public with nutritional guidelines from which we can extract information for decision making. These guidelines and standards reflect current research and scientific thinking regarding human nutritional needs.**[3] From the "Basic Four," unveiled in the 1950s and recommending that people consume foods from each of four food groups (meat, breads and grains, dairy, and fruit and vegetables) each day, through a few iterations of the USDA pyramid, and on to the new USDA visual, called "My Plate," some guidelines have changed, while others have remained the same.

While nutritional recommendations have historically been uniform across gender, age, and activity level, recent USDA recommendations reflect differences between individuals. This is a nod to research revealing that caloric needs generally decrease with age (among adults), are higher for men than for women, are higher among more active individuals, and are higher for those looking to maintain or gain weight than for those looking to shed some body fat. For instance, a 20-something male wishing to maintain his weight would need significantly more calories each day than a 60-something woman interested in losing a few pounds. The USDA "My Plate" site (www.choosemyplate.gov) has also been praised for avoiding nebulous references to "servings"—an earlier pyramid advised, for instance, the consumption of six to 11 servings per day from the bread, cereal, rice, and pasta group, but no indication was made as to what constituted a "serving." New guidelines provide recommended amounts in weight and volume measurements, reducing ambiguity. The latest guidelines also distinguish between healthy (unsaturated, plant-based) and unhealthy (saturated or trans) fat and encourage the consumption of a "rainbow" of fruits and vegetables each day as a means of maximizing the quantity and variety of micronutrients ingested. By entering personal data, including gender, age, and activity level, "My Plate" users can get personalized nutritional recommendations (as well as access to a tracking website, which serves as a food journal of sorts) that allow each individual to put together a daily meal plan that will provide for their unique needs.

In addition to providing daily consumption guidelines, the USDA and FDA also require that most commercially prepared and processed foods be printed with ingredient lists (in order of prevalence by weight) and nutritional facts (Image 9.4).[4] These nutrition labels are based upon a 2,000 Calorie/day diet, and provide both the mass of each nutrient and the percentage that mass represents of the recommended daily quantity, called the Percent Daily Value. Of course, for individuals who generally consume significantly more or fewer than 2,000 Calories each day,

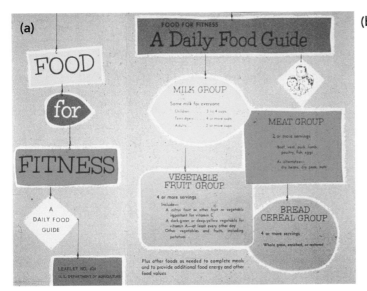

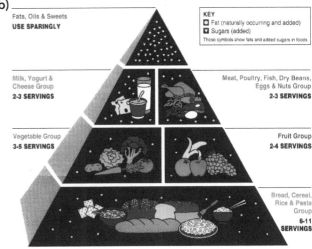

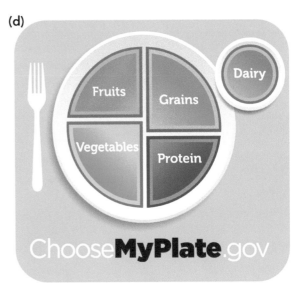

IMAGE 9.3. The "Basic Four" focused equally on milk, meat, grains, and produce (a). An older USDA food pyramid (b) put bread and grains at the base, fats and sweets up top. A later pyramid (c) tried to move away from the idea that one particular food group was more important (the "base" of the pyramid, so to speak). It also incorporated a representation of exercise. The latest visual, a plate, emphasizes produce while sectioning a dinner plate into easy-to-interpret fractions (d).

these percentages will be inaccurate. Regardless, a consumer can obtain at least a general idea of what exactly is in their food by using the information provided.

A Nutrition Facts Label may contain more information than that shown in Image 9.4, but by law, it may not contain less. It's important to understand that the label does not necessarily tabulate all the nutrients in a given package: If the package is meant by the manufacturer to be subdivided into servings, the serving size and number of servings per container must be specified. In addition, some macronutrient subdivision is required—not only total fat, for instance, but also

Nutrition Facts

Serving Size 1 cup (228g)
Servings Per Container 2

Amount Per Serving

Calories 250   Calories from Fat 110

	% Daily Value*
Total Fat 12g	**18%**
Saturated Fat 3g	**15%**
Trans Fat 3g	
Cholestrol 30mg	**10%**
Sodium 470mg	**20%**
Total Carbohydrate 31g	**10%**
Dietary Fiber 0g	**0%**
Sugars 5g	
Protein 5g	

Vitamin A	4%
Vitamin C	2%
Calcium	20%
Iron	4%

* Percent Daily Values are based on a 2,00 calorie diet. Your Daily Values may be higher or lower depending on your calorie needs.

	Calories	2,000	2,500
Total Fat	Less than	65g	80g
Sat Fat	Less than	20g	25g
Cholesterol	Less than	300mg	300mg
Sodium	Less than	2,400mg	2,400mg
Total Carbohydrate		300g	375g
Dietary Fiber		25g	30g

IMAGE 9.4. A sample Nutrition Facts Label, color coded for emphasis. Items highlighted in yellow are generally nutrients to limit, while those highlighted in blue are generally nutrients of which sufficient quantities are necessary.

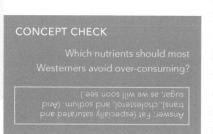

CONCEPT CHECK

Which nutrients should most Westerners avoid over-consuming?

Answer: Fat (especially saturated and trans), cholesterol, and sodium. (And sugar, as we will soon see.)

quantities of saturated and trans fat must appear so that consumers seeking to limit these are able to avoid them. Note that cholesterol and sodium, both of which are prevalent and commonly over-consumed in the Western diet, are listed directly under fat. **This conveniently puts all the nutrients that should be limited in the same place for ease of reading.** Carbohydrate is always subdivided into fiber (indigestible but important) and sugar. Note that sugar is not subdivided into naturally occurring and added. In other words, the Nutrition Facts Label for whole fruit—if fruit came with such a label—would reveal the presence of sugar. Vitamins and minerals appear at the bottom of the label. Information regarding vitamin A, vitamin C, iron, and calcium is required by law. Other vitamins and minerals may be listed or omitted at the discretion of the manufacturer. Finally, the label contains a brief synopsis of the USDA and FDA recommendations for maximum quantities of fat, cholesterol, and sodium to be consumed each day, as well as minimum recommended carbohydrate and fiber.

The inclusion of trans fat quantities on nutrition labels is relatively new; the legislation requiring this information took effect in 2008. As discussed in Chapter 8, research indicates that even more so than saturated fat, trans fat is detrimental to cardiovascular health. For this reason, new guidelines recommend limiting consumption to fewer than 2.5 grams daily. Unfortunately, the government allows the food industry a few loopholes in trans fat reporting. For instance, a value of zero may be reported for all quantities of trans fat less than 0.45 g per serving. A bit of simple math reveals that if one were to consume six servings a day of food containing 0.45 g of trans fat per serving, one would quickly approach the daily recommended maximum—all without realizing any trans fat had been consumed at all! **Astute label readers watch for the words "partially hydrogenated" in a food's list of ingredients (for instance, "partially hydrogenated soybean oil"); if those words appear, trans fat is present, regardless of what the nutrition label says.**

A second technique for producing trans-like fat has recently become more common in the food industry. This technique avoids hydrogenation (and therefore the use of the words "partially hydrogenated" in food labels), but fats produced in this way can be identified by the word *interesterified* on ingredient lists (as in "interesterified soybean oil"). **Preliminary research suggests that interesterified fats not only act like trans fat in food, but they also act like trans fat in the body: they appear to increase LDL, decrease HDL, and negatively affect cardiovascular health.** One further trans fat pitfall lies in the chemistry of the double bonds themselves. The *cis* double bonds present in naturally occurring unsaturated (and therefore healthy) oils are less stable than *trans* double bonds. As a result, exposure to intense heat over long periods of time can cause the *cis* bonds to break and reform as *trans* bonds. As a result, **oils that have been kept at high temperature for long periods (such as in commercial deep-fat fryers) often contain heat-produced trans fats.** Because the oil that went into the fryer was unsaturated (and trans fat–free), the government allows the food industry to claim that the product is trans fat–free—despite the fact that it may contain many grams of trans fat per serving. While home fryers are unlikely to reach sufficient temperatures (over sufficient time periods) to result in the production of trans fats in any significant amount, restaurant and commercially fried food is often a source of trans fat in the diet.

9.4   What We're Eating—The Typical American Diet

The average American gets approximately half of his or her calories from carbohydrate, about 10% from protein, and the remaining 40% from fat (Image 9.5). They eat out about five times per week, three meals of which generally consist of hamburgers and French fries. While some sources cite the typical American intake nearing 4,000 Cal/day, this fails to account for spoilage, discarded food, and loss through other mechanisms. In reality, estimates closer to 2,750 Cal/day are more likely (averaged across men and women)—a value still significantly higher than the recommended 1,800 Cal/day or 2,400 Cal/day for relatively sedentary females and males respectively.

Purely on the basis of its macronutrient breakdown, the American diet doesn't look too bad—it's higher in fat than most nutritional experts recommend, and lower in protein, but not particularly deviant with regard to the percent of carbohydrate consumed versus the percent that is recommended. **Why, then, is our national consumption pattern so detrimental to health? The answer lies more in the source of the carbohydrate, protein, and fat (and in the amounts!) than in the percentages.**

Fast food and processed food have become staples of the American way of life. Unfortunately, these foods are high in saturated and trans fat, refined carbohydrate, and salt, while being low in fiber, protein, and micronutrients. Some processed foods (particularly those that advertise themselves as "healthy" options, like energy bars) boost their vitamin and mineral content through additives, but the efficacy of these added micronutrients is questionable (see Section 9.14, The Last

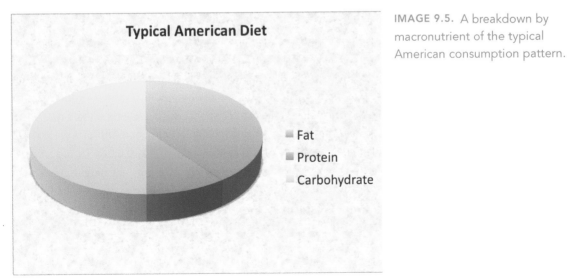

IMAGE 9.5. A breakdown by macronutrient of the typical American consumption pattern.

NOTE[5]

You can get an idea of how much of each ingredient is in a food by noting the order in which they appear on the label—by law, ingredients must be listed from greatest quantity (by mass) to least.

CONCEPT CHECK

Why are calorie-containing drinks problematic in terms of diet?

Answer: They generally contain calories without other nutrients, and they don't provide lasting fullness or satisfaction.

Word—Vitamin Supplements). Further, even those fast-food options that purport to be healthy often present nutritional pitfalls lurking beneath a wholesome-appearing veneer. The FDA regulations allow for considerable leeway with regard to advertising statements made by food companies. For instance, a label identifying "whole wheat bread" means only that the bread contains whole wheat flour among the ingredients—even if it's just a small percentage of the total flour.[5] A fast-food company's "healthy" sandwich, consisting of grilled chicken on a whole wheat bun, could easily be composed of inexpensive, fattier chicken cuts on a bun made mostly of refined flour. The claim that a fried food is free of trans fats holds no assurance whatsoever of legitimacy, as we've already explored. Eating as many meals away from home as we do, we have little knowledge of (or control over) what goes into our food. Most of us get less than half the fiber necessary to maintain health, and three times the recommended amount of salt. Where guidelines call for four to five cups of fruits and vegetables a day, most of us get less than one and a half. Even those individuals who would like to control consumption and attempt to keep track of how much they're ingesting have difficulty with meals prepared outside the home (for which nutritional information is not available): Studies show that people routinely underestimate the caloric content of a meal by 35–50%. This effect is more significant in larger meals, as those served in restaurants tend to be.

Caloric beverages, too, present a problem. **According to USDA data, the average American gets more than half their daily calories from soda and other sweetened beverages.** These represent a huge source of empty calories—meaning that there is little nutritional benefit apart from energy—and also present the problem of being nonsatiating. It seems that our ability to sense the caloric content of food, and the intrinsic mechanisms that promote feelings of hunger in response to a caloric deficit and feelings of satiety in response to caloric balance, are derailed by calorie-containing beverages. Sugared beverages also raise blood sugar quickly, resulting in a strong insulin signal and everything that follows. This combines to make them a nefarious source of calories: they provide no nutritional benefit, not even lasting energy, since the sugar is quickly taken up and stored as fat. The quantity of sugar in a soda is staggering—a 20-oz. bottle typically contains 17 teaspoons of added sugar, or 69 grams. Lately, fancy coffee drinks have become commonplace. While coffee itself is calorie free, the calories in fancy drinks containing added sugars, full-fat dairy, and whipped cream add up quickly. A 16-oz. blended coffee and cream drink topped with whipped cream can contain a whopping 500 calories—and that's for the smallest size available at many coffee shops! In simple terms, that's one beverage providing more than a quarter of a woman's (or about a fifth of a man's) daily recommended calories—without any lasting sense of satisfaction or reduction in hunger. **All of this adds up to an approximately 2-lb.-per-year weight gain among average American adults, making it no small wonder that an estimated 80 million people in the United States go on a weight-loss diet each year.**

In an interesting twist of language, the word *diet* has taken on an additional (and in many cases quite disparate) meaning than that listed as the primary definition in most dictionaries. Diet, from its etymological origins, refers to the food and drink that is taken regularly or habitually—in other words, one's diet is the sum of what one routinely eats and drinks to support life and health. In common parlance, however, the expression *on a diet* connotes a reduction in or departure from one's regular consumption pattern, often for the purpose of losing weight. Further, the "diet" is often designed to produce weight loss quickly—to starve the body of nutrients in order to force it to consume itself. **Because self-consumption is not sustainable over time (one can't survive habitual nutritional deprivation), being "on a diet" is incompatible with the concepts of regularity and habit.**

Semantics aside, however, diets (and in deference to common parlance, this section of the text will use the word in the colloquial manner) and diet books, diet advice and diet methods, abound. While the concept behind weight loss is mathematically simple—when energy expended surpasses energy consumed, the body is forced to make up the difference by burning stored fat—it's chemically quite complex. **The reason for the complexity (and the reason weight loss is so difficult for many individuals to achieve) is that the body is designed to maintain homeostasis, or balance.** Just as some mechanisms help maintain body temperature and pH, others regulate habitus. To a certain extent, when we consume more than we expend energetically, cells become "sloppy" with their energy use, and calories are wasted. For this reason, occasional overeating doesn't lead to significant weight gain. Likewise, when we expend more than we consume, cells respond by increasing their fuel economy—energetic needs decrease somewhat—to avoid having to burn fat. It is only in the case of regular energy overabundance or deficiency that significant weight change occurs. Further complicating this equation is the fact that the body, evolutionarily engineered to survive periods of famine by storing fat during periods of plenty, much more quickly and easily gains fat than loses it. These factors combine to produce a human body that gains weight relatively easily in a society in which food is plentiful, and loses it with difficulty. Because the most realistic and effective weight-loss technique—which involves a minor reduction in calories consumed coupled with an increase in calories expended, and which must take place slowly and gradually—does not appeal to many individuals, diet fads promising quick and dramatic weight loss abound. Some of these rely on "miracle pills" promising results with no exercise or dietary restriction, while others encourage altered consumption patterns. The chemistry of these diets is quite interesting and worth examining.

9.6 The Low-Fat Diet

Based upon the theory that fat is the most calorically dense of the nutrients, low-fat diets often work toward overall caloric reduction through limitation of fat consumption to less than 15% of total calories (Image 9.6).

Unfortunately, there are a number of short-term and long-term flaws in any attempt to significantly reduce dietary fat. In the short term, fat is an important **satiety signal**, meaning that its presence in the digestive tract helps to communicate a sense of fullness and satisfaction to the brain. This is accomplished through the action of a hormone called *cholecystokinin (CCK)*, which is only released in response to ingested fat. While there are other satiety signals, including the sensation of fullness that comes from the physical stretching of the stomach, these are generally delayed[6] and weaker than the CCK signal. Fat also slows digestion and absorption, meaning it helps to regulate blood sugar—in this regard, it has the same effect fiber does. Furthermore, our evolutionary drive to seek out foods with high energy content leads us to derive sensory pleasure from fat, in much the same way that we derive pleasure from sugar. In fact, studies designed to quantify gustatory pleasure in subjects eating ice cream or whipped cream show that as the fat content of a food decreases, its sugar content must increase in order for its consumption to result in a similarly pleasurable experience. Manufacturers must therefore add far more sugar to low-fat versions of foods for the foods to taste good to consumers. This, of course, increases the caloric content of these foods. **The importance of fat as both a satiety signal and as a provider of gustatory satisfaction means that dieters employing a low-fat strategy often take in more calories—and far more sugar—than they would if they consumed a normal nutritional distribution while reducing total caloric intake.** Because sugar can be converted to fat in the body, a low-fat/high-sugar diet still leads to fat (and weight) gain if overconsumption of calories occurs. An amusing and poignant maxim on that note is *low-fat in the box doesn't mean low-fat in the body!*

CONCEPT CHECK

Which would be more satisfying, a 500-calorie meal containing some fat or a 500-calorie meal containing no fat? Why?

Answer: The meal containing fat, because it is an important satiety signal.

IMAGE 9.6. Low-fat diets are composed largely of carbohydrate. Most nutritionists recommend more protein, more fat, and much less carbohydrate than the typical low-fat diet provides.

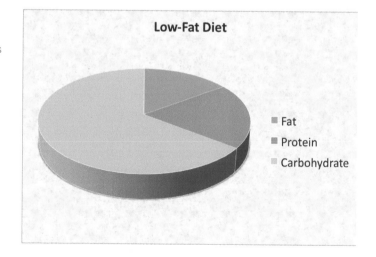

Low-Fat Diet

- Fat
- Protein
- Carbohydrate

While FDA-approved pharmaceutical weight-loss aids certainly exist (and are becoming more common), here we will focus on over-the-counter diet pills that are advertised to produce rapid results without lifestyle modification. Exceedingly popular for their promise to produce weight loss (and increased energy) without exercise or dietary restrictions, diet pills can be found everywhere from specialty nutrition stores to the local drugstore. They're sometimes recommended by alternative healthcare practitioners and are often advertised on the Internet and late-night television. A stunning variety of ingredients are featured in these pills, and a complete discussion of the array of chemical strategies they employ (and chemical results they produce) is beyond the scope of this text. However, it's worth spending some time on one compound that was nearly ubiquitous in commercial weight-loss agents until 2004, when it was banned from dietary supplement use in the United States: ephedrine, a chemical compound isolated from herbs of the *Ephedra* family, is a sympathomimetic alkaloid.

Sympathomimetic agents are ones that mimic adrenaline and stimulate the *sympathetic nervous system*, or the branch of the nervous system designed to produce the "fight or flight" response to life-threatening situations. Adrenaline is a major hormone of the sympathetic nervous system, and the symptoms of its release in the body, including pounding heart, heightened response time, and generalized agitation, are also produced by sympathomimetic agents. **While ephedrine-containing herbs have been used in Chinese medicine for thousands of years to treat symptoms of asthma and bronchitis,[7] their adrenaline-like action can also be used to produce weight loss, since activation of the sympathetic nervous system requires the expenditure of large amounts of energy.** As a result, overall energy expenditure increases dramatically—even without increased activity level or exercise—and if caloric consumption remains unchanged while caloric expenditure increases, weight loss occurs.

There is no question that, in sufficient doses, ephedrine produces weight loss. The reason for ephedrine's mimicry of adrenaline is revealed through a cursory glance at their structures; the shared pharmacophore is quite evident (Image 9.8). Disturbingly, however, the increasingly popular street drug methamphetamine also shares this pharmacophore—in fact, like ephedrine, methamphetamine produces

IMAGE 9.7. An Ephedra plant, the source of ephedrine.

NOTE[7]

Ephedrine, like adrenaline, stimulates responses that ready the body for action, including dilating the airways. Asthma and bronchitis produce wheezing through irritation and constriction of airways, so the dilating effect of ephedrine helps to alleviate symptoms.

IMAGE 9.8. Adrenaline, also called epinephrine (left), shares much in common structurally with ephedrine (center) and methamphetamine (right).

CONCEPT CHECK

Why do the similar shapes of adrenaline, epinephrine, and methamphetamine suggest that they'll have similar effects on the body?

Answer: The shared pharmacophores mean they'll bind to the same receptors. (See Chapter 7 to review this concept.)

rapid and dramatic weight loss. The similarity among these compounds is also evident in the undesirable side effects of ephedrine, which include rapid and irregular heartbeat, nausea, confusion and hallucinations, pulmonary edema, and metabolic disturbances, all of which are side effects of methamphetamine use as well.

The side effects of ephedrine can be life threatening at worst, and even at best, adversely affect quality of life. Pill users commonly report feeling nervous, faint, and unable to concentrate. Additionally, like many street drugs, ephedrine use results in *habituation* over time, meaning that it is necessary to take higher and higher doses of the drug in order to produce the same response. Both because of the habituation and because sympathomimetics stimulate release of dopamine, a neurotransmitter that produces feelings of pleasure, addiction is common. Withdrawal from ephedrine can lead to lethargy, depression, and (most disturbing to many dieters) rebound weight gain.

Due to the FDA's determination in the last decade that ephedrine is not a safe dietary supplement, its inclusion in over-the-counter pills has been banned. Still, supplement manufacturers utilize the principles of pharmacology to produce new sympathomimetic agents containing the adrenaline pharmacophore. These substances are often added to the banned-compound list shortly after they become available for purchase, but supplement manufacturers just as rapidly respond by synthesizing new ones, meaning diet pills containing ephedrine-like compounds are always available. Simply put, supplements promising rapid weight loss and increased energy without the need for caloric reduction or exercise generally contain ephedrine-like compounds—and produce ephedrine-like adverse effects.

9.8 Low-Carbohydrate Diet

While lower-carbohydrate diets show a great deal of promise in effecting weight loss (particularly if the consumed carbohydrates fall into specific categories—see

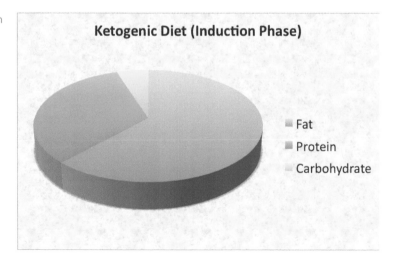

IMAGE 9.9. The caloric distribution in the induction phase of a ketogenic diet is vastly different than that recommended by most nutritionists.

Ketogenic Diet (Induction Phase)

- Fat
- Protein
- Carbohydrate

Section 9.10, Eating Pattern Sustainability, below), extremely low-carbohydrate diets (<20 g carbohydrate per day during early phases, Image 9.9) can have some odd biochemical effects upon the body. Occasionally called ketogenic diets in research and literature, these diets call for a drastic reduction in carbohydrate consumption for the duration of the weight loss attempt.

The theory behind ketogenic diets is that strong insulin signals, which result from consumption of sugars and refined carbohydrates, increase body fat storage, as discussed in Chapter 8. Avoiding carbohydrate, then, eliminates the possibility that a strong insulin signal will be produced by a meal. **Proponents of these diets suggest that the body can be "trained" to burn fat, and that reducing carbohydrate intake (which reduces glycogen stores) forces cells to rely more heavily upon fat metabolism to meet energy needs.** This, in theory, trains the body to utilize fat to a greater extent even upon cessation of the diet. Unfortunately, while both these theories are correct to a point, they fail to fully address the biochemistry of extreme carbohydrate reduction.

Upon beginning a ketogenic diet, individuals are generally pleased with the rapid and dramatic weight loss they experience. **A closer look at the chemistry of carbohydrate deprivation, however, reveals that very little of the lost weight is actually fat. Most, in fact, is water.** The reason is that carbohydrate is an important energy source for almost all cells in the body. In particular, glucose is important to proper brain function; while most other cells can utilize fatty acids in the absence of carbohydrate, the brain cannot. As carbohydrate deprivation ensues, blood sugar begins to drop. A functioning brain requires upwards of 100 g of glucose daily; with carbohydrate consumption in the region of 20 g per day, glycogen must make up the difference. The hormone glucagon causes the breakdown of liver glycogen into glucose, which is released into the bloodstream and travels to the brain. Without a constant source of glucose from food, muscles are also forced to break down glycogen stores (they have their own, though under normal conditions, glucose from muscle glycogen is used locally and not released into the bloodstream). Both liver and muscle glycogen become depleted rapidly, and because glycogen is highly polar and co-stores with water in both liver and muscle, the stored water (many pounds of it) is released and excreted as glycogen breaks down.

Early symptoms of glycogen depletion are common among extremely low-carbohydrate dieters and include fatigue and an inability to focus their attention. Muscles tend to burn and feel weak with even slight exertion. Exercise is simply not possible in most cases, or is, at best, extremely unpleasant. As glycogen stores disappear entirely, the body shifts its chemistry in order to provide an alternate source of nutrition to the starving brain.

In the absence of glucose, there is one other energy source that the brain can utilize. Ketone bodies, a group of chemicals including acetone (the compound that gives fingernail polish its familiar smell), are produced through metabolism of fatty acids. Generally, they are mere intermediates in fatty acid breakdown, and do not enter systemic circulation. In the case of extreme glucose deprivation, however, they can enter circulation and be taken up by brain cells.[8] While diet-induced

NOTE[8]

The other circumstances in which circulation of ketone bodies would take place include starvation and uncontrolled diabetes. It's important to note that, in general, ketones only circulate when something is terribly wrong.

NOTE[9]

Lab safety protocol in those days obviously wasn't what it is now, as evidenced by the fact that he traced down the source of the sweetness by returning to the lab and tasting each of the chemicals he'd been working with!

ketosis has not been shown to be significant enough to be harmful (people generally find this experience so uncomfortable that they increase their carbohydrate intake before any real damage is done), it nevertheless takes a toll on the body. From a purely aesthetic perspective, ketone bodies in the bloodstream dissolve in exhaled air, giving dieters an odd, chemical smell. Additionally, research suggests that ketosis is interpreted as a starvation signal, causing cells to down-regulate their energy usage and making weight gain upon cessation of the diet much more likely. There is also evidence that depression can occur with a higher prevalence among these dieters—low blood glucose results in decreased serotonin production, which results in depressed mood and decreased feelings of well-being. Regardless, most ketogenic diets begin allowing increased amounts of carbohydrate shortly after ketosis is achieved, meaning that this extremely odd chemistry is short lived.

Research studies on ketogenic diets show mixed results—some suggest they are effective weight-loss strategies (at least in the short term), while others find that they are no more effective than simple caloric restriction. Most studies reveal that the strict limits on consumption of carbohydrate-containing foods result in severe nutritional deficiencies. In particular, the avoidance of fruits and whole grains leads to low fiber intake, which can result in severe constipation. Further, while some studies suggest that initial weight loss is greater with ketogenic diets than low-fat or simple calorie-reduction diets, all three diet types show similar effectiveness with regard to maintained weight loss over time (six months or longer). Generally speaking, the diet most likely to be effective is the one an individual is most likely to stick to, and of the fad diets discussed here, none lend themselves well to the long term.

9.9 Sugar Substitutes

Nearly all weight-focused diets that involve a modification of consumption agree on one thing—sugar provides empty calories, and should be limited. Unfortunately, the avoidance of sugar is at odds with human instinct; it's contrary to our tendency to seek out pleasurable tastes and eating experiences. For this reason, the **nonnutritive** (meaning they do not provide us with calories) **sweeteners** are hot chemical commodities. The first of these to be developed was *saccharin*, which was discovered serendipitously in 1879 when a chemistry student who'd been in the lab all morning licked his fingers during lunch and noticed they tasted sweet.[9] While the saccharin molecule is shaped differently from sucrose, it nevertheless has the key components of the pharmacophore in the appropriate positions, giving it the ability to bind to sweetness receptors on the tongue (Image 9.10).

In fact, saccharin binds very tightly to sweetness receptors, making it intensely sweet—about 500 times as sweet as sucrose—such that only a tiny amount need be consumed to provide a similar flavor sensation to that of table sugar. It's often mixed with glucose (which is a bit less sweet than sucrose) to dilute its sweetness so that consumers can measure an appropriate amount into their food or drink. Because

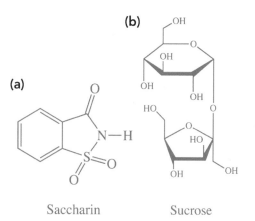

(a) Saccharin   **(b)** Sucrose

IMAGE 9.10. The saccharin molecule compared to sucrose.

IMAGE 9.11. A can of saccharin-containing soda bearing a warning label.

the molecule is not one we evolved to consume, we have no metabolic pathways in place that allow us to extract energy from it. As such, it has no caloric value. Some researchers have expressed concern that saccharin may be carcinogenic (commercially marketed packages carry a warning to this effect, as in Image 9.11), as the molecule has been shown to cause cancer in lab rats. The rats in question, however, were fed tremendous amounts of saccharin in a short period of time, and evidence shows that analogous carcinogenic activity is not observed in people. Despite this, saccharin has lost popularity in recent years as other nonnutritive sweeteners have become available. The newer sweeteners are more stable, don't carry cancer-warning labels (which, despite the lack of evidence to support them, understandably concern consumers), and don't share saccharin's unpleasant, metallic aftertaste.

One such sweetener, *aspartame*, was discovered (again serendipitously) by Jim Schlatter in 1965. Schlatter was working toward the development of a tetrapeptide (four-amino acid) ulcer drug, which he was synthesizing piece by piece. The story goes that he discovered that a modified two-amino acid piece of the drug tasted sweet when he licked his fingers to turn the pages of his lab notebook (again, a case in which poor lab protocol led to a lucrative discovery). Like saccharin, aspartame is ostensibly a very different molecule from sucrose (Image 9.12). However, again, a few key atoms in a few key locations give aspartame the ability to bind to the sweetness receptor quite tightly—it's 200 times sweeter than sucrose. Researchers have investigated a possible link between aspartame consumption and cancer, and while these studies have indicated a possible connection in rats, there is no evidence for concern in humans. Cancer aside, concerns over potential aspartame toxicity come from the fact that aspartame is not stable over long periods of time, and is particularly unstable if exposed to heat, which is why aspartame-containing diet soft drinks lose their sweetness over time. For this reason, not only the parent molecule, but its breakdown products as well, have been (and continue to be) scrutinized for potential toxicity. Of all the aspartame breakdown products, the one that seems to receive the most negative attention is formaldehyde, which is indeed produced as aspartame is metabolized. The human body, however, easily converts formaldehyde to CO_2, meaning that small amounts of it, when ingested, do no harm. As always, the Paracelsus Principle (Chapter 7) holds true: the dose makes the poison. With regard to the formaldehyde we're exposed to as a result of aspartame consumption, cup for cup, tomato juice contains six times as much formaldehyde as diet soda.

While aspartame is not required to carry a cancer warning, it does have its own warning label, instructing individuals with the genetic disease *phenylketonuria (PKU)* to avoid it. Those with PKU are unable to metabolize phenylalanine, as they lack the liver enzyme responsible for its breakdown. Unmetabolized phenylalanine

IMAGE 9.12. Aspartame consists of two amino acids (aspartic acid and phenylalanine), with a methoxy (-OCH^3) group forming an ester on phenylalanine instead of the usual OH, for a carboxylic acid. For shape comparison, sucrose is shown on the right.

(a)

(b)

Aspartyl-phenylalanine methyl ester   Sucrose

accumulates in the blood and disrupts brain development, leading to developmental delay and intellectual disability. To avoid effects of the disease, individuals with PKU must not consume foods rich in phenylalanine, particularly during childhood.[10] Meat, fish, cheese, legumes, and nuts are among the foods that must be avoided—as, of course, are aspartame-containing foods and sweets.

Among the newest of the synthesized sweeteners is *sucralose*, a chlorinated disaccharide. This molecule is much more structurally similar to sucrose than are the other artificial sweeteners (Image 9.13). In fact, the only differences (although they are crucial, as they render the molecule indigestible[11] and therefore nonnutritive) are that the glucose unit is replaced with galactose, and three of the OH groups are replaced with chlorine atoms.

The differences between sucralose and sucrose also render the former 600 times sweeter than table sugar, by virtue of causing it to bind more tightly to sweetness receptors. Unlike aspartame, sucralose has the benefit of being heat- and pH-stable, meaning that it can be used in baked goods and has a longer shelf life. Sucralose has been evaluated for toxicity and carcinogenic potential by several studies, none of which has found evidence of a risk to human health in reasonable doses. As new as this sweetener is, however, it may take some time before long-term effects of routine consumption (should there be any) begin to reveal themselves.

As for any major known health effects of artificial sweeteners, research reveals two items of concern. There is research evidence that artificial sweeteners (of any type) may cause a paradoxical weight gain by increasing the desire for sweet foods, thus increasing the consumption of sugar-sweetened items. Evidence is also beginning to accumulate to suggest that the role of gut bacterial populations in human health is vastly larger than previously understood and may be altered by artificial-sweetener consumption. For instance, obese individuals have gut bacterial population profiles very different than those found in the nonobese. Disturbingly, it appears that aspartame consumption alters the gut bacterial profile: mice fed aspartame ended up with gut bacterial populations very different than those they had prior to aspartame consumption; further, aspartame consumption altered the

NOTE[10]

The effects of PKU can be devastating, but simply by controlling diet, affected individuals are able to develop completely normally. As such, all babies born in the United States are routinely tested for the disease at birth by pricking a toe and testing a drop of blood. This public-health measure has saved countless lives and vastly improved prognosis and quality of life for PKU-positive infants.

NOTE[11]

In fact, most of the sucralose consumed isn't absorbed into the bloodstream and passes unaltered through the digestive tract; the small amount that is absorbed is not metabolized and is excreted as sucralose in the urine.

bacterial populations so that they resembled those of obese mice who had not consumed the artificial sweetener. While this research is relatively new and has not yet been conducted on humans, it is worth monitoring as it develops. In sum, nutritional experts are split in their opinions on the topic of artificial sweeteners. Some believe that a desire to consume sweet food is inevitable, and that replacing caloric sweeteners with nonnutritive alternatives is beneficial. Others believe that reduced sugar intake reduces the desire to consume sweet food, and suggest that sweetened foods should be largely avoided, regardless of the source of the sweetness. From a chemical perspective, it is reasonable to posit that man-made molecules to which we are not evolutionarily adapted might be expected to have strange and unanticipated effects on the body; trans fats provide evidence of this. A cost-to-benefit ratio, therefore, might be of use to an individual in determining whether or not to use these chemicals in food.

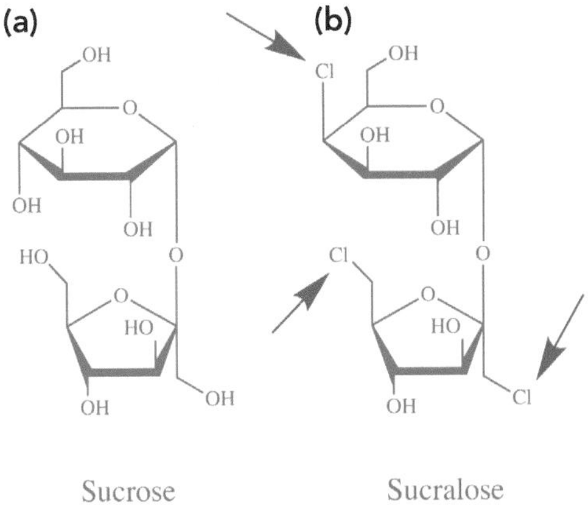

IMAGE 9.13. Sucrose and sucralose, with arrows to indicate differences between the molecules.

9.10 Eating-Pattern Sustainability

While only a few weight-loss diets and strategies are addressed in this chapter, there are hundreds of them being marketed. Most, however, share a fatal flaw: they focus upon weight rather than health. Another way of saying this is that they attempt (by design) to produce rapid weight loss through unsustainable eating and/or lifestyle practices. **Scores of research studies show that, diet strategy notwithstanding, the single greatest predictor of weight-loss success is the duration of an eating pattern: the longer it can be maintained, the more successful it will be.** Since most diet strategies are too drastic and uncomfortable to be maintained for long, they tend to be abandoned as soon as the desired weight loss (or some of it, anyway) is achieved. Most plans do not provide any instruction whatsoever regarding appropriate eating practices for weight maintenance, and for this reason, weight rebound after cessation of a diet plan is a common phenomenon. Rebound also results from a variety of other factors, including a desire to consume foods that were restricted during the diet and an increased storing of extra calories as fat (and increased efficiency of metabolism) as a result of the starvation signals received by the brain during many diets. Of course, there are times when excess weight becomes an emergency, and a physician might recommend a drastic (temporary) eating plan or interventional procedure to achieve a rapid reduction. For most individuals, however, the best road to a healthy weight (and to maintenance of that weight) is an eating pattern that can be happily and healthily sustained over time—a balanced *diet*, in the true sense of the word.

> **CONCEPT CHECK**
>
> Why do some of the nonnutritive sweeteners taste sweeter than regular sugar?
>
> Answer: They bind more tightly to sweetness receptors on the tongue.

Most research suggests (and many nutritionists agree) that a well-balanced diet consists of about 45–65% carbohydrate, 10–35% protein, and 20–35% fat, with plenty of fruits and vegetables. Macronutrient percentages in these ranges provide enough carbohydrate for glycogen production and brain function, sufficient protein to provide for metabolic needs while avoiding unnecessary hardship to the kidneys

Metabolic waste (urea) from protein degradation is filtered into the urine by the kidneys. Effective elimination of urea is extremely water dependent, meaning that high protein consumption can easily result in dehydration. Since kidneys function suboptimally in low-water conditions, overconsumption of protein can stress these delicate organs.

(the organs responsible for protein waste management),¹² and enough fat to meet needs and serve as a satiety signal without overindulgence. There is, of course, quite a bit of variability in those guidelines; different people have slightly different energy needs and personal biochemistry, and what works well for one may not be identical to what works well for another. Regardless, the profile of a balanced diet broken down by macronutrient (Image 9.14) looks quite different from that of any of the weight-loss diets discussed earlier.

Another common thread among nutritionists is the agreement that not all fats and not all carbohydrates are equally healthy. While 20–35% of calories consumed should be from fat, the vast majority of these should be from unsaturated fat. Some even recommend emphasizing certain sources of unsaturated fat (cold-water fish, for instance) on the grounds that they are high in omega-3 fatty acids. The theory here is that some oils work toward shifting the omega-3/omega-6 ratio in a healthier direction—toward a bit more omega-3 and a little less omega-6 than most Western diets provide. Other nutritionists, however, feel that simply avoiding saturated and trans fats improves the diet so significantly over the traditional American fare that one need not worry about the source of unsaturated fat, as long as it comprises the majority of fat in the diet. In many ways, it comes down to the individual—some people have a deep dedication to crafting the healthiest diet they can and are happy to keep track of sources and types of oils in order to achieve this goal. Others find it challenging (and time consuming) enough just to avoid fast food, and simply aren't willing or able to microanalyze their nutrients.

With regard to carbohydrates, again, most nutritional experts agree that some are superior to others. In Chapter 8, we discussed the important role of fiber in regulating the digestive system, slowing digestion, and decreasing the rate at which blood sugar increases in response to a meal. The latter, critical to optimizing fuel utilization and preventing unnecessary fat storage, is actually quantified through what is called a food's **glycemic index (GI)**. GI measures the rate at which a carbohydrate food enters the bloodstream as sugar and compares this to the rate for glucose, which has a defined GI of 100. To determine a food's GI, human subjects

IMAGE 9.14. Most nutritionists agree that a healthy, balanced diet breaks down to around 45–65% carbohydrate, 10–35% protein, and 20–35% fat, depending somewhat upon the individual.

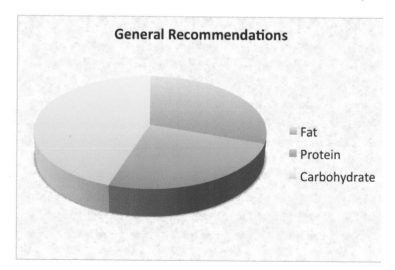

TABLE 9.2. GI values for
some common foods.

Low GI (up to 55)		Moderate GI (56–69)		High GI (70 and higher)	
Plain yogurt	14	Orange juice	57	White bagel	72
Cherries	22	White flour pasta	58	Watermelon	72
Kidney beans	27	Cheese pizza	60	Jelly beans	80
Bran cereal	30	Oatmeal	61	White rice	80
Multigrain bread	31	Canned apricots	64	Corn flake cereal	92
Orange	33	Raisins	64	French baguette	95
Ice cream (full fat)	36	Whole-grain cracker	67	Baked potato	111
Brown rice	55	Croissant	67	Tofu frozen dessert	115

consume a standard-size sample of food (all foods being tested must contain the same net mass of carbohydrate). Over a period of two hours, their blood sugar is tested repeatedly; at the end of that time, the change in blood sugar is compared to the change that would have resulted from an equal-mass sample of glucose. If the food increases blood sugar at half the rate observed for glucose, the food is assigned a GI of 50. A GI of 25 would correspond to an increase in blood sugar at one quarter the rate observed for glucose, and so forth. Foods with GI values greater than 100 increase blood sugar at a rate greater than that of pure glucose. Some common foods and their GI values are listed in Table 9.2. **Many nutritionists suggest that lower-GI foods are higher in quality than higher-GI foods, although as we'll soon see, there are caveats to this.**

Note that while the list of foods in the table is not exhaustive, some trends are nevertheless apparent. Refined versions of foods (white rice, white bagel) have significantly higher GI values than their unrefined counterparts (brown rice, multigrain bread). This is due to the role of fiber in slowing the rate at which sugar enters the bloodstream. In addition, fat or protein in food slows absorption of sugar—cheese pizza, for instance, has a lower GI than white bread, despite the fact that the pizza's crust is essentially identical in composition to the latter. Of course, one must weigh the positive effect of the lower GI value against the negative effect of the saturated fat in cheese when determining whether the pizza is a healthy food choice. Some so-called "health foods," like tofu-based ice cream substitute, have surprisingly high GI scores. While traditional ice cream contains saturated fat, it also has a much lower glycemic index than the tofu version.

All of this information (which sometimes seems contradictory) can get confusing: How does a health-conscious consumer make appropriate food choices? In the end, it helps to remember that as complex as human bodies (and their nutritional requirements) seem, they're really nothing more than large beakers in which many different reactions are running at any one time. Consumption of food is a mechanism for supplying the reagents for these reactions. Given that, we can look at nutrition as a chemical equation that needs to be balanced. The reagents are high-quality carbohydrate, protein, and high-quality fat. While a low GI may indicate that a food contains high-quality carbohydrate, if the GI is low because the food contains low-quality carbohydrate coupled with lots of saturated fat, it's probably not a good reagent! Of course, psychology and the desire to enjoy the food we eat also play roles in all of this—the healthiest diet in the world becomes problematic if it consists of foods that an individual doesn't find satisfying; in such a case, the diet will either be abandoned or will strip some of the enjoyment from life. As omnivores, we have many more options at our disposal than an animal without the ability to choose from a variety of foods, cultivate foods, and seek out those items it prefers most. With a little ingenuity, label reading, and consideration of the chemistry of nutrition, it becomes quite feasible to put together a healthy and tasty diet. It's noteworthy, too, that nutritional experts agree on one more thing: even the best and healthiest diet still has room for the occasional treat, caloric splurge, or nutritionally devoid (but very tasty) dessert!

9.11 The Other Half of the Equation—Caloric Expenditure

While a complete discussion of the biochemical pathways that require caloric input is beyond the scope of this text, it's worthwhile to spend a little time on the other half of the caloric balance equation: Where do the calories go? **The vast majority of our caloric expenditure goes toward maintaining cellular function.**[13] The number of calories an individual burns each day on maintenance alone is called **resting metabolic rate (RMR)**, and varies with age, gender, body composition, and many other factors. While there are several equations that can be used to approximate RMR, most do not take body composition into account, making them less accurate for individuals with higher- or lower-than-normal body-fat percentages. Because even at rest, muscle burns as many as five times more calories than fat, pound for pound, two individuals of the same weight and gender will have different RMRs if one is leaner than the other. Assuming that body habitus is relatively normal, however, the Harris-Benedict equation (which has been used for some time and has demonstrable accuracy) for calculation of RMR is as follows:

$$Women: \ RMR = 655 + (4.35 \times weight \ in \ pounds)$$
$$+ (4.7 \times height \ in \ inches) - (4.7 \times age \ in \ years)$$
$$Men: \ RMR = 66 + (6.23 \times weight \ in \ pounds)$$
$$+ (12.7 \times height \ in \ inches) - (6.8 \times age \ in \ years)$$

Note that RMR does not take into account any sort of physical activity; exercise and even normal daily activity result in the burning of additional calories. **In individuals of normal activity level, RMR accounts for about 70% of calories burned each day.**

A second (albeit very minor) caloric expenditure that is independent of activity results from the digestion of food. Called the **thermic effect of food**, this phenomenon arises from the fact that it takes a certain amount of energy to digest and process the nutrients we consume. On average, this accounts for a loss of about 5–10% of total consumed calories, with fat having a very low thermic effect and protein having the highest effect of the three macronutrients.

Physical activity energy expenditure (PAEE) is the term for calories burned as a result of activity. This includes both routine daily activity (cooking, working, walking to and from the car) and exercise. In normally active individuals, PAEE will account for about 20% of total calories burned. Of course, highly active individuals or habitual exercisers can burn significantly more than this. Even movements that don't serve a particular purpose, such as pacing while thinking, tapping a foot while sitting, or twiddling a pencil while in class, result in an increased PAEE. The pacers, tappers, and twiddlers of the world burn demonstrably more calories each day than their somewhat less restless counterparts.

9.12   Exercise

Despite minor squabbles in the arenas of optimal diet, vitamin supplementation, and so forth, government organizations, physicians, and health experts unite on the subject of exercise: Americans (and citizens of many industrialized Western societies) simply don't get enough! Surveys from the late 1990s and early 2000s show that about 60% of U.S. adults do no physical activity at all, apart from whatever is required by their jobs. About a quarter of U.S. adults engage in a total of 30 minutes of vigorous non-work-related physical activity each week. By contrast, the NIH recommends at least 150 minutes per week of exercise or vigorous physical activity. **These numbers are unequivocal: most adults aren't getting even close to the amount they need.**

Exercise is important to maintaining health for several reasons, not the least of which is that it burns calories, helping to maintain energy balance (or in the case of those needing to lose weight, to shift the balance toward a caloric deficit). Many of us are familiar with the concept that calories are burned during exercise—the more vigorous the activity, the greater the caloric burn. **However, there are actually many mechanisms through which exercise produces caloric expenditure; the so-called direct effect—the calories burned during the activity itself—is only one of these.** That actually comes as good news to most people, since the number of calories burned during activity can appear shockingly paltry. For instance, a 170-lb. man running 11.5 minute miles (5.2 mph) burns approximately 600 Cal/hr, meaning that to burn off the calories in a single meal consisting of a fast-food quarter-pound burger, large fries, and large soda, he'd have to run for two solid hours![14] For this

NOTE[14]

Actually, most fitness machines claiming to tally caloric expenditure are quite inaccurate. Some allow the user to input weight, but for those that don't, caloric estimates are inaccurate purely on the basis that caloric expenditure is directly proportional to weight. Furthermore, most machines don't allow the user to indicate gender, and a man burns more calories during exercise than a woman of the same weight. This can result in a 30% (or more!) systematic error in the fitness machine's direct effect calculations; generally, the error is on the side of overestimation, rather than underestimation.

Activity	Cal/hr (M)	Cal/hr (F)	Activity	Cal/hr (M)	Cal/hr (F)
Aerobics	630	480	Skiing (Nordic)	636	480
Cycling (10 mph)	444	336	Swimming (fast)	720	522
Racquetball	642	456	Swimming (slow)	288	210
Running (7.5 mph)	900	678	Weights (moderate)	360	264
Running (5.2 mph)	600	456	Walking (3.5 mph)	354	270

TABLE 9.3. Approximate Cal/hr burned due to direct effect of various activities for an average (170-lb.) male and average (123-lb.) female. Values will be higher than indicated for individuals of greater weight and lower for individuals who weigh less or are in excellent physical condition.

NOTE[15]

TRY THIS

same man to burn the 3,500 Calories in a pound of body fat, he'd have to run for five hours. Values for the direct effects of other exercise activities are listed in Table 9.3. Thankfully, the direct effect is only a small part of the overall contribution of exercise toward health and caloric balance.

A second, very significant effect of exercise is that it temporarily increases resting metabolic rate. This elevated post-exercise metabolic rate can last for quite some time—research suggests that RMR can be up to 10% higher than usual the morning following exercise. If, for instance, a man with an 1,800 Cal/day RMR lifts weights for half an hour, he burns 180 Cal. If his RMR increases by 10% for a period of 24 hours, he burns another 180 Cal—without any further exercise. It's worth noting that while this effect is approximately equivalent among different exercise activities several hours post-exercise, it's particularly emphasized in the first few hours following weight lifting (presumably because of the significant stress placed upon muscles by strength training). For this reason, many fitness experts encourage individuals attempting to lose weight to incorporate not only aerobic exercise (owing to its cardiovascular benefits—see below), but also weight training into their regimen. **A final (and the most significant) effect of exercise with regard to caloric balance is that it increases muscle mass.** Since muscle is highly metabolically active even at rest, increased muscle increases RMR—not just post-exercise, but all the time. This is yet another reason that many fitness experts encourage weight training, since it tends to increase muscle mass very efficiently.[15]

Apart from its contribution to negative caloric balance, exercise positively impacts cardiovascular and general health. Like any muscle, the heart needs to be exercised. Aerobic exercise has been found to be the most effective means of maintaining good cardiovascular health, for a host of reasons. Aerobic exercise is defined from a metabolic perspective. **The lower end of the aerobic zone is the heart rate at which the body begins working hard enough that metabolic waste products start to build up in the blood.** It's quite difficult to determine this heart rate without metabolic testing, but for most people, it occurs at about 65% of maximum heart rate (HR_{max}), which is the highest heart rate an individual can sustain and is generally related to age:

$$HR_{max} = 220 - age$$

The upper end of the aerobic zone is again defined metabolically—it's the point at which the body starts working so hard that waste products in the blood increase dramatically—but again, it's difficult to determine precisely without testing. For most people, this occurs at 85% of HR_{max}, meaning that 75% of HR_{max} is right in the middle of the aerobic zone for most individuals.

The benefits of aerobic exercise are many. First, forcing the muscles to work harder increases their demand for energy and oxygen. This requires faster circulation, which in turn causes the heart muscle to beat faster and harder, exercising it. Cardiovascular exercise helps reduce risk of stroke, heart attack, atherosclerosis, and other cardiovascular diseases. Also of benefit to the heart, exercise increases

HDL production, which favorably affects blood cholesterol, as discussed in Chapter 8. Further, to saturate the blood with oxygen, breathing becomes deeper and faster. *Capillaries* (tiny blood vessels) in the lungs open up to allow blood to flow in more efficiently. Vessels in the muscles also dilate, allowing more efficient blood flow. Enhanced blood flow to the muscles means waste products are flushed away to a greater extent than usual.[16] Natural painkillers are released into the bloodstream. These reduce any sensation of pain and, in conditioned individuals, can result in a feeling of well-being (the so-called "runner's high").[17] The immune system is activated, accounting for the fact that regular exercisers get sick less frequently than sedentary individuals. There's even evidence that exercise is good for maintaining a sharp mind, a finding of importance to aging adults. A final, and significant, benefit of aerobic exercise lies in the fuel source utilized. It turns out that either of the two major stored fuels—fat and glycogen (which is broken into glucose)—can be burned for energy in the presence of oxygen as per the equations below (the fat equation is not balanced since different fatty acids require different quantities of oxygen and produce products in different ratios):

$$fat + x\ O_2 \rightarrow y\ CO_2 + z\ H_2O + energy$$
$$C_6H_{12}O_6 + 6\ O_2 \rightarrow 6\ CO_2 + 6\ H_2O + energy$$

Because the body stores so much more energy in the form of fat than it does in the form of carbohydrate, it preferentially relies on fat metabolism to fuel energy expenditure when possible. The word *aerobic* means "with air," and a second metabolic definition of the aerobic zone is that it is of an intensity high enough to require the body to work hard but low enough that we're able to supply the muscles with enough oxygen for efficient metabolism. Of course, when we engage in aerobic exercise, we breathe harder than normal (because the muscles are working harder and need more oxygen to burn more fuel), but we're not truly short of breath—it's possible to get enough air. Because the muscles are able to get plenty of oxygen, they can burn their preferred fuel: fat. Muscles preferentially burn fat when possible simply because there's so much more of it available than there is glycogen. The muscles also prefer to save the glycogen for emergencies, as we'll see shortly. **During aerobic exercise, then, the major source of fuel is fat: exactly the source of fuel most people would like to have a little less of.** While many people who are new to exercise feel out of breath and uncomfortable even during moderate efforts—mostly because the feeling of exertion is new to them—a relatively fit individual can tell when they're in their aerobic zone even without checking their heart rate, simply because they're used to the feeling: their breathing will be faster and deeper than usual but not labored. In fact, a great "quick and dirty" exertion test is that when you're in your aerobic zone, you can talk but not sing.[18]

As we have all experienced from time to time, there is a level of exertion above the aerobic zone. This is referred to as the anaerobic zone, and during anaerobic efforts, muscles are working so hard that the body can't supply sufficient oxygen for efficient metabolism (*anaerobic* means "without air"). Despite our best efforts to

NOTE[16]

You've noticed this effect if you've ever observed that a light workout the day after a hard one can help ease muscle soreness from exercise. The soreness comes in part from built-up metabolic waste products, and these are flushed from the muscles during exercise.

NOTE[17]

It has traditionally been thought that the natural painkillers released during exercise are endorphins, but recent research is starting to suggest that instead, the responsible compounds are endocannabinoids. These chemicals, when they bind to their receptors, result in decreased sensation of pain and increased relaxation and well-being. In fact, cannabinoids from sources outside the body—such as the active compounds in marijuana—bind to these same receptors and result in a similar experience.

NOTE[18]

Singing requires prolonged exhalations, which aren't possible at aerobic intensities—if you can sing, you're below your aerobic zone and aren't reaping the benefits. Being unable to talk, or able to talk only in short bursts, means that the body is struggling to get enough oxygen. By definition, this means you're outside your aerobic zone.

TRY THIS

If a 20-year-old woman is exercising with
a heart rate of 175 BPM, is she likely in an
aerobic or anaerobic zone? Can she likely
talk? Sing?

Answer: She is anaerobic. She likely can't talk
(except perhaps in short bursts) or sing.

Solution:
$$HR_{max} = 220 - age = 200 \text{ BPM}$$
$$\frac{175}{200} \times 100 = 87.5\%; > 85\% = anaerobic$$

NOTE[19]

Further, this notion that the body
must be "trained" to burn fat is
fallacious; the human body, like that
of all animals, is the result of billions
of years of evolution. The metabolic
mechanisms in place do not require
"activation" or "training" in order
to function. When conditions are
right—when muscles are working in
the presence of ample oxygen—fat is
burned. When conditions are not right,
such as in the absence of appropriate
amounts of oxygen, fat is not burned.

breathe fast and the heart's efforts to pump blood fast and hard, it simply isn't possible to get enough oxygen to the muscles to allow for aerobic fat and carbohydrate burn during very intense efforts. As a result, the muscle cells must switch over to an anaerobic metabolic strategy—they must burn fuel without oxygen. Only glucose can be burned in this way, meaning fat is not utilized as an energy source during very intense efforts. Furthermore, anaerobic carbohydrate burn is not nearly as efficient as aerobic metabolism—the energy yield of a given quantity of glucose is approximately 15 times less under anaerobic conditions. This means lots of glucose is burned very quickly to provide energy, and anaerobic efforts result in rapid depletion of fuel:

$$C_6H_{12}O_6 \rightarrow 2 \; lactic \; acid + small \; amount \; of \; energy$$

Note also the production of lactic acid during anaerobic metabolism. This is a metabolic waste product that can build up in muscles and in the blood. It is partially responsible for the burning sensation in muscles during very hard efforts, as well as some of the soreness that follows. Generally, the anaerobic zone is identified as any effort about 85 percent of HR_{max}, but an individual accustomed to exercise can tell by feel when they enter this zone, as breathing becomes much more labored and talking becomes difficult (or impossible).

While anaerobic exercise is incredibly effective at burning large numbers of calories in small periods of time, it's not often emphasized by fitness experts as much as aerobic exercise. There are several reasons for this. First, it is incredibly uncomfortable. While it provides many of the same fitness benefits as aerobic exercise, the discomfort of anaerobic exercise can make it unenjoyable. Second, it is much harder on the heart and body than aerobic exercise, making it inappropriate for novices or out-of-shape individuals. While some "experts" claim that aerobic exercise does a better job of training the body to burn fat[19] (since anaerobic metabolism relies purely on glycogen), research shows that the tremendous caloric deficit created by anaerobic work leads to high levels of fat metabolism post-exercise, so this is not a legitimate deterrent. **Regardless, the best exercise program—like the best diet—is the one a person is able to stick with.** For that reason alone, aerobic exercise is far more popular; after a brief acclimation phase, most individuals find it to be not only tolerable but enjoyable.

9.13 Muscle Types

As part of our discussion of exercise and muscles, let's take a moment to ponder the chemistry of why conditioned athletes in some sports tend to look very sinewy, while in other sports they appear much bulkier. Even within the sport of running, differences can be striking: marathon runners tend to have very slender, long muscles, while sprinters almost look like bodybuilders (Image 9.15). The reason for this lies in the different muscle types that these athletes have cultivated.

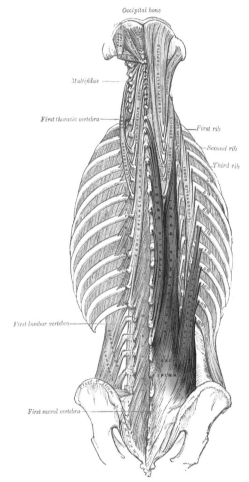

IMAGE 9.15. Despite their similarly low body-fat percentages, a marathon runner (a) looks very thinly built compared to the bulk of a sprinter.

Muscle can be divided into different categories based on the type of metabolism upon which it primarily relies. Type I muscle fibers, also called slow twitch fibers, are largely aerobic in their metabolic strategy. As a result, they are found in muscles that engage in endurance efforts, either as a result of genetics (and possibly training) or their position in the body. All of us, for instance, have lots of type I fibers in the long muscles that run down either side of the spine and maintain our upright posture (Image 9.16). Because we sit and stand for long periods of time every day, these muscles are constantly working against gravity. They don't often engage in intense effort, but they do have to maintain the effort for long periods of time. It's a bit difficult to tell (especially since trying to do so risks injury), but they're also not as powerful as other similarly sized muscles—their capacity for strength is much less than their capacity for endurance.

On the other hand, type II muscle fibers, also called fast twitch fibers, rely primarily upon anaerobic metabolism. They are capable of extremely fast (and extremely powerful) movements, but are unable to sustain those efforts for long. We have type II muscle fibers in our hands and eyelids: they allow us to make very quick and precise motions. We also find them elsewhere in the body, where they function in explosive movements (like jumping) and power efforts (like lifting a very heavy box). At this point, there is no conclusive research to indicate whether muscle fiber prevalences within an individual are purely genetic or can be trained. That is, accomplished marathon runners have a very high proportion of type I muscle fibers, though whether this is because their muscle fiber profile gave them a predisposition to endurance sport greatness, or because they trained in endurance efforts resulting in a proliferation of type I fibers, is not known.

Regardless of whether genetics, training, or a combination of the two influences an individual's muscle fiber profile, the fact remains that muscle analysis of accomplished endurance athletes shows enrichment in type I fibers compared to the average individual, and analysis of accomplished power athletes shows enrichment

IMAGE 9.16. The erector spinae muscles (highlighted) are among the postural muscles that hold the human body upright, maintaining low-intensity efforts over long periods of time with type 1 (slow twitch) muscle fibers.

It's worthwhile to note that just because a muscle specializes in anaerobic respiration doesn't mean it doesn't need any oxygen—all cells use oxygen to maintain normal function. Instead, the working muscle doesn't require supplemental oxygen for fuel combustion above and beyond its ordinary requirements (which are provided for by oxygen in the bloodstream).

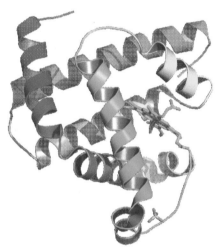

IMAGE 9.17. Like hemoglobin, myoglobin is a protein (shown as a curled ribbon). Also like hemoglobin, it contains a heme group (the molecule in green and red in the center of the protein). The heme is responsible for binding to oxygen.

CONCEPT CHECK

Why don't type II fibers contain significant myoglobin?

Answer: Type II fibers specialize in anaerobic efforts, which burn fuel in the absence of oxygen.

in type II fibers. **It is the difference in the predominant fiber type that leads to the observable difference in *morphology* (shape) between these athletes.** Because type II fibers are primarily involved in anaerobic efforts,[20] they store large amounts of glycogen. Since glycogen is highly polar and co-stores with water, the stored fuel takes up quite a bit of space. This plumps up muscle groups containing large amounts of type II fibers and gives them the bulky look typical of a bodybuilder or sprinter. Type I fibers, however, are primarily aerobic and are structured to allow for maximum availability of their preferred fuel (fat) and the oxygen needed to burn that fat. All muscle cells store small amounts of fat for use during aerobic efforts, but type I fibers store a particularly large amount since they specialize in fat metabolism. Since fat is hydrophobic and doesn't co-store with water, the stored fuel takes up very little space in the muscle. This results in muscle groups composed of predominantly type I fibers having a long, lean look to them, as observed in endurance athletes. Additionally, type I fibers contain large quantities of the chemical **myoglobin** (Image 9.17).

Like its close chemical relative hemoglobin, myoglobin has a reddish color and binds oxygen. Unlike hemoglobin, however, it doesn't circulate through the body in red blood cells. Instead, it is localized in the muscles—primarily type I fibers. The job of myoglobin is to take oxygen from circulating hemoglobin and hold that oxygen until the muscle needs it. Essentially, myoglobin provides a muscle with its own personal oxygen supply, allowing aerobic metabolism to proceed even during efforts hard enough that breathing alone can't supply all the oxygen a muscle requires.

In addition to producing the distinct physical appearance of athletes who are enriched in one type of muscle fiber, the differences between fiber types also affects the appearance and nutritional value of muscle that we consume for food. For example, many of us know that chicken breast is lower in fat and calories than a similar quantity of chicken drumstick or thigh. The difference lies in the fat content of the muscle—a given mass of chicken breast is higher in protein and lower in fat than the same mass of chicken thigh (Table 9.4).

Based upon an understanding of muscle fibers and the fuel they store, we can posit that chicken breast contains a higher proportion of type II fibers, while chicken thigh contains more type I fibers. In fact, that is exactly the case. Further, the dark color of chicken leg meat ("dark meat") comes from the reddish-brown myoglobin in the type I muscle fibers. Without the need for rich stores of myoglobin, the type II fibers that predominate in chicken breast meat appear pale ("white meat"). All of this makes even more sense when we stop to think about the daily routine of a chicken (Image 9.18). As ground birds, chickens don't do much flying. Instead, they spend most of their time walking around, pecking for food. Their legs don't need to produce extraordinary bursts of speed (chickens rarely run), but they do need to support the chicken as it goes about its business. The thighs and drumsticks, therefore, are specialized for low-to-moderate effort endurance efforts, and are composed primarily of type I fibers. When a chicken is startled, it flies a short distance. These flights are intense efforts of limited duration, and so the

(a)   (b)

IMAGE 9.18. Chickens and ducks have different habits, thus the composition of their muscles (particularly their flight muscles) is very different. This affects not only their abilities, but also the nutritional profile of their meat.

flight muscles—the pectorals, or what we call chicken "breast"—are specialized for power, and contain mostly type II fibers.

Ducks, on the other hand, are migratory birds capable of flying thousands of miles. They would not do well with a chicken's type II fiber-heavy pectorals. Instead, ducks have primarily type I fibers in their muscles of flight. From a nutritional perspective, one might conjecture that duck breast is dark meat—and probably fatty—which, in fact, it is.

	Chicken Breast	Chicken Thigh
Protein (g)	43	36
Fat (g)	5	15
Total Calories	231	293

TABLE 9.4. Nutritional information for 140 g serving of chicken meat.

9.14 The Last Word—Vitamin and Mineral Supplements

Nutrition is a heavily researched field. Not only is it of interest to researchers across many areas and specializations of science; its study is also funded by countless special interest groups and food and supplement manufacturers for product promotion and marketing purposes. While mainstream experts may express minor differences in recommendations regarding quantity and quality of the macronutrients, nowhere in nutrition are opinions as varied and disparate as they are with regard to supplementation. On one hand, it's no secret (nor is there any debate) that the typical Western diet is low in vitamins and many minerals; for this reason, some experts suggest that a daily multivitamin is simply good insurance for health maintenance. On the other hand, however, some experts feel that supplementation encourages a "Band-Aid" treatment of a poor diet, providing the veneer of good health without encouraging individuals to change their eating habits in order to provide for all their macro- and micronutrient needs. Complicating all of this, there is little conclusive evidence to support long-term supplementation. For each study that shows a positive effect, there's another study that shows no effect whatsoever. A cursory scan of the supplement literature turns up the following batch of contradictory results:

- Higher folic acid consumption decreases risk of breast cancer (Rohan *et al.*, 2000).
- A supplement including folic acid, vitamin B_6, and vitamin B_{12} does not affect breast cancer risk (Zhang *et al.*, 2008).

CONCEPT CHECK

Why is "dark meat" higher in fat than "white meat"?

Answer: It is made up of primarily type II fibers, which store fat for energy.

- Vitamin E supplementation shows a 30–40% reduction in risk of heart disease in women (Stampfer *et al.*, 1993).
- Vitamin E supplementation shows no reduction in risk of heart disease in women (Lee *et al.*, 2005).
- Vitamin C supplementation shows a reduction in heart disease risk (Knekt *et al.*, 1994).
- Vitamin C supplementation shows no reduction in risk of heart disease (Sesso *et al.*, 2008).
- Selenium supplementation helps prevent prostate cancer in men (Duffield-Lillico *et al.*, 2003).
- Selenium supplementation has no effect upon prostate cancer incidence in men (Lippman *et al.*, 2009).

Further muddying the waters are recent studies demonstrating that supplementation with synthetic vitamins may actually increase mortality. The mechanism for this is as yet unclear, but some researchers suspect that the proportion of vitamins relative to one another may be more important than the quantity, such that artificial supplementation (which often delivers vitamins in proportions not normally encountered in nature) could produce unintended chemical signals, and therefore unanticipated chemical consequences. While this theory has yet to be proven in the case of supplemental micronutrients, it certainly holds true in other areas of physiological chemistry—trans fats, for instance, while similar to saturated fats in their physical properties, have odd and unanticipated chemical properties in the body that probably stem from the fact that we didn't evolve the ability to process them normally. Where does this leave us with regard to supplementation? From a chemical perspective, empirical evidence suggests that we don't fully understand all the reactions taking place in the body: some have layers of complexity that we haven't yet uncovered. As a result, trying to adjust the extent to which reactions happen by supplementing (or withholding) a reagent can have unanticipated, and potentially severe, consequences. Ideally, of course, we'd all get the nutrients we need from the sources we evolved with. Ideally, too, we'd still live the physically active lives our evolutionary ancestors did—though hopefully with the improvements of medicine and technology afforded by our modern existence. Obviously, such an idyllic marriage of evolutionary roots and the modern world is often simply not possible. Our jobs force us into sedentary lifestyles, we have little time to cook wholesome meals from scratch, and a combination of psychology and food science has provided food manufacturers with all the information they need to tempt us into purchasing their products. The vitamin quandary—to supplement or not to supplement—is just one in a long line of questions for which there are no real answers, at least not yet. Perhaps the best compromise in the meantime can be achieved through an understanding of personal chemistry: we're really just big beakers. Where we lack reagents, we can't produce products. If there is a product we need, or a reaction we need to run, we must supply the reagents (and if there is a product we want to avoid, we must avoid supplying them). We bear in mind that the best sources of reagents are those that occur in nature, and from there, we do the best we can.

SUMMARY OF MAIN POINTS

- The Western Paradox is the idea that Western societies are increasingly consuming an atrocious diet while putting a premium on physical perfection.
- An appropriate amount of body fat (neither too much nor too little) is necessary for normal function.
- Excess body fat leads to disease—overweight and obese are two subcategories of excess body fat and are associated with apnea, arthritis, cardiovascular disease, and type 2 diabetes.
- Essential body fat is the minimum required to sustain normal function.
- Body fat percentage is monitored by health care professionals to ensure an appropriate amount of body fat—circumference techniques, skin-fold techniques, and BMI are commonly utilized metrics.
- The USDA and FDA are responsible for providing nutritional guidelines for healthy eating; current recommendations are outlined in the "My Plate" visual.
- These organizations require most foods to carry Nutrition Facts labels to help consumers make informed choices.
- Trans fat in quantities less than 0.45 g/serving may be listed as 0 g on a nutrition facts label. Trans fat in a food is identified through the words *partially hydrogenated* or *interesterified* in the ingredients list.
- Unsaturated fats can chemically rearrange to become trans fats under intense heat conditions, as in a commercial deep-fat fryer.
- Americans (and many Westerners) are eating too much fat, refined carbohydrate, sugar, and salt, as well as too little fiber, lean protein, and micronutrients.
- Eating meals away from home makes it difficult to keep track of what (and how much) we're eating.
- Sugared beverages provide lots of empty calories and don't create lasting feelings of fullness; they are associated with obesity.
- Many weight-loss diets attempt to produce rapid loss through unsustainable practices that can be detrimental to body chemistry.
- Low-fat diets are generally too high in sugar and—because fat is an important satiety signal—in calories.
- Ephedrine-containing stimulant diet pills can have adverse, methamphetamine-like effects and are addictive.
- Low-carbohydrate diets produce ketosis, which affects mood, energy level, and ability to focus.
- No conclusive evidence links nonnutritive sugar substitutes to cancer in humans, but concern over their long-term effects abounds.
- The best health-promoting diet is one that includes appropriate portions of all macronutrients, as well as plenty of fruits and vegetables.
- Many nutritional experts suggest that balance diets should consist of 45–65% carbohydrate, 10–35% protein, and 20–35% fat. The fat should be mostly unsaturated, and the carbohydrate mostly low to moderate GI (glycemic index).

- A food's GI is a measure of the rate at which it increases blood sugar—higher-quality carbohydrates have lower GI values.
- RMR (resting metabolic rate) is the daily caloric expenditure required to maintain cellular and organ function. It's higher for taller, heavier, younger people, and higher in males than in females.
- The thermic effect of food is the daily caloric expenditure required to digest food.
- PAEE (physical activity energy expenditure) is the daily caloric expenditure from routine activity and exercise.
- Exercise burns calories through a direct effect as well as elevated post-exercise metabolic rate and increased muscle mass.
- Aerobic exercise occurs at moderate intensity and burns fat primarily. It is beneficial to health in many regards.
- Anaerobic exercise occurs at high intensity and burns sugar primarily.
- Type I muscle fibers specialize in endurance efforts, contain high amounts of fat, and are rich in myoglobin.
- Type II muscle fibers specialize in power efforts and contain high amounts of glycogen.

QUESTIONS AND TOPICS FOR DISCUSSION

1. What observations can you make to support (or challenge) the idea of the Western Paradox?

2. What do you think are the primary reasons people in the United States (and other Western industrialized cultures) are becoming fatter? Why do you think this isn't happening in other parts of the world?

3. Given that food is plentiful for most Westerners, could we survive without body fat? Support your answer.

4. Why is being overweight/obese a society-wide problem rather than simply an individual problem?

5. Why is essential body fat lower in men than in women?

6. Young female athletes like Olympic gymnasts have a tendency to look much less physically mature than other females their age. Using an understanding of body fat, comment on why this might be.

7. Compared to young female Olympic athletes, who tend to look much less physically mature than expected for their age, young male Olympic athletes don't look noticeably different (with regard to physical maturity) from their

less athletic counterparts. Using an understanding of body fat, comment on why this might be.

8. Why does BMI fail to accurately estimate weight status in certain portions of the population? For what sorts of individuals does BMI fail?

9. Why do healthcare practitioners tend to use BMI (instead of other mechanisms) to keep track of body composition in their patients? What is the purpose of tracking a patient's body composition?

10. Many Western industrialized societies currently hold a "thinner is better" attitude toward body composition, particularly in women. What are some of the health concerns associated with being underweight?

11. A very slender, inactive young woman is confused when her doctor tells her that she is at risk for type 2 diabetes and other conditions associated with excess body fat, despite having a BMI of 19. Might she truly be at risk? If so, why?

12. When Arnold Schwarzenegger was a body builder, he weighed around 240 lbs. during the competition season (109 kg). He is 6 feet tall (1.83 meters), putting his then-BMI at 32.5. Comment on the accuracy of this metric in Schwarzenegger's case; explain why his BMI accurately (or inaccurately) reflected his body composition.

13. Go to www.choosemyplate.gov. Click "Daily Food Plan" under Popular Topics (or search for "Daily Food Plan") and enter information requested. Do you think the suggested diet plan would be easy to conform to? Why or why not?

14. Go to www.choosemyplate.gov. Click "Daily Food Plan" under Popular Topics (or search for "Daily Food Plan") and enter information requested. How many calories does the plan suggest you consume each day? Based upon this, is a standard Nutrition Facts label going to over- or underestimate the percentages of each nutrient in what you eat, compared to recommendations?

15. Write a paragraph in which you explain—on a level understandable to a fourth grader, how to put together a healthy plate of food. What considerations are important? How much of each type of food should the plate hold? How should carbohydrates, proteins, fats, and produce be selected?

16. What are some of the benefits of the new USDA food guidelines (www.choosemyplate.gov) over the "Basic Four" and the pyramids?

17. How does the evolution of the USDA guidelines reflect changing needs and a changing understanding of nutrition?

18. Many nutrition experts took issue with the last of the USDA food pyramids—the one that showed a person climbing stairs—because they felt it was the result of food-industry lobbying and did consumers a disservice. Why might they have felt this way? Is a set of stairs on the side of the pyramid a good idea or a bad idea? Explain your position.

19. How do you feel about the influence lobbying and special interest groups can have upon government organizations and their nutritional suggestions? Do you feel you can trust the recommendations of the USDA?

20. How do you feel about the fact that alcoholic beverage companies don't have to put Nutrition Facts labels on their products? Would you want access to that information?

21. Is it possible to consume the maximum recommended quantity of trans fat in a given day even if all foods consumed say 0 g trans fat on the Nutrition Facts label? How?

22. How is the typical American (Western) diet different from what is commonly recommended?

23. Why are sweetened beverages considered such strong contributors to obesity?

24. What are some ways in which food manufacturers can trick a consumer into thinking a food is healthier than it really is? What are ways to avoid these pitfalls?

25. Discuss low-fat, ephedrine, and ketogenic diets. What are the unique problems associated with each? What is a common problem shared by all three?

26. How would you counsel a friend who came to you asking your opinion on a good weight-loss diet?

27. What might your advice be to a friend who asked your opinion on diet soda?

28. A friend of yours puts lots of butter on his white bread toast. When you ask him about it, he says he is trying to reduce the GI of the toast. What is your opinion on his choice? What might you recommend?

29. While studying late one night, you decide you really need a sweet treat. Your local convenience store has three options—full-fat ice cream, fat-free fruit sorbet, and a frozen tofu dessert. Assuming your objective is to make the best nutritional choice, which do you pick and why?

30. What are the health benefits of aerobic exercise?

31. A friend goes to the gym and spends an hour on the treadmill. She tells you she just burned 643 calories according to the machine. Does this number likely underestimate or overestimate her true expenditure (direct effect only)?

32. A friend doesn't want to weight lift because it doesn't burn as many calories per hour as running. Comment on this logic.

33. A friend who doesn't exercise asks you for advice on exercise frequency and intensity. What do you tell her?

34. A friend who exercises regularly and would like to improve his caloric balance asks you for advice on exercise frequency and intensity. What do you tell him?

35. Why is weight training considered such a positive addition to many exercise programs?

36. Why do marathon runners and sprinters look so different?

37. In your opinion (based upon your understanding of chemistry), what is the best way to construct a healthy, balanced, sustainable lifestyle in terms of diet and exercise?

PROBLEMS

1. Why is body fat important? What role(s) does it serve?

2. What are the major health risks of being overweight/obese?

3. What is the difference between type 1 and type 2 diabetes?

4. What is the biggest risk factor for type 2 diabetes?

5. What are the major treatment/management strategies for type 2 diabetes?

6. What percentage of U.S. adults are overweight? Obese?

7. What are the three mechanisms addressed in the text to measure body fat percentage? Which is the most (and least) accurate?

8. What is BMI? How is it calculated?

9. A woman weighs 41 kg and is 1.65 m tall. What is her BMI? What is her weight status?

10. If you found out the woman in Problem 9 led a very sedentary lifestyle, how might you reevaluate the information obtained from her BMI?

11. A man weighs 114 kg and is 1.72 m tall. What is his BMI? What is his weight status? Do you trust this weight status on the basis of the available information?

12. For what daily caloric consumption is a standard Nutrition Facts label written?

13. For what daily caloric consumption does a Nutrition Facts label *also* include limited information?

14. What information is required by law to be on a Nutrition Facts label? Of these, which are nutrients most people get too much of? Which are nutrients most people need more of?

15. Why can deep-fat fryers produce trans fats?

16. What words does a savvy consumer looking to avoid trans fat look for on food labels?

17. What makes fat an important satiety signal?

18. Why is it no surprise to a chemist that ephedrine, adrenaline, and methamphetamine all have similar effects in the body?

19. Define: satiety signal; sympathomimetic; ketogenic; ketone bodies.

20. Imagine that a friend starts a low-carbohydrate diet and loses 10 pounds in two weeks. Are you impressed? What might you tell her about the weight loss?

21. The friend from problem 20 notices that she is feeling lethargic and her muscles burn when she walks up stairs. She asks you to explain what is happening. What would you say?

22. The friend from problem 20 manages to stay on her diet for three weeks, and you start to notice an odd smell, a little like nail polish, when she talks. What is happening?

23. Even though saccharin and aspartame are shaped very differently than sucrose, they fit into the sweetness receptor. Explain this phenomenon.

24. What specific behavior, with regard to the sweetness receptor, causes a molecule to taste sweeter than table sugar?

25. Why must individuals with PKU avoid aspartame?

26. Sucralose is very similar, structurally, to sucrose. Why is it nonnutritive (hint: think enzymes)? Why is it sweeter than sucrose?

27. What two potential negative effects of artificial sweetener consumption are most scientifically plausible?

28. What is the single greatest predictor of weight loss or weight-maintenance success?

29. What do most nutritional experts agree comprises a balanced diet?

30. What is meant by high-quality fat? What kinds of fat are high quality?

31. What is meant by high-quality carbohydrate? What kinds of carbohydrate are high quality?

32. What is a glycemic index? How is it measured?

33. A food increases blood sugar at twice the rate of glucose—what is its GI?

34. A food increases blood sugar at one-third the rate of glucose—what it its GI?

35. In general, what sorts of foods have low GI values? High GI values?

36. Why does brown rice have a lower GI than white rice?

37. A high-GI food can be "turned" into a low-GI food through the addition of fat or protein. Why is this so? What are the nutritional concerns associated with this?

38. Define: RMR; thermic effect of food; PAEE.

39. Calculate the expected RMR for a 19-year-old, 5' 2", 110-lb. female.

40. Calculate the expected RMR for a 45-year-old, 5' 8", 300-lb. male.

41. Under what circumstances would you expect RMR calculations to be inaccurate?

42. What percentage of total caloric expenditure does RMR normally account for? Under what circumstances would this be different?

43. What percentage of total caloric expenditure does the thermic effect of food normally account for?

44. What percentage of total caloric expenditure does PAEE normally account for? Under what circumstances would this be different?

45. Define: Direct effect and elevated post-exercise metabolic rate.

46. Why is increased muscle mass an important effect of exercise?

47. What is aerobic exercise? What is the mathematical way to determine the upper and lower limits of the aerobic zone? What's a "quick and dirty" way to determine if exercise is aerobic?

48. What is anaerobic exercise? What is the mathematical way to determine the lower limit of the anaerobic zone? What's a "quick and dirty" way to determine if exercise is anaerobic?

49. What fuels are used during aerobic exercise? During anaerobic exercise?

50. Calculate HR_{max} and the limits of the aerobic zone for an 18-year-old female.

51. Calculate HR_{max} and the limits of the aerobic zone for a 75-year-old male.

52. A 25-year-old female is exercising with a heart rate of 120 BPM. In what type of exercise is she engaging?

53. An 85-year-old female is exercising with a heart rate of 120 BPM. In what type of exercise is she engaging?

54. What are the differences between type I and type II muscle fibers? Where in the body might the average individual have each?

55. What kind of muscle fiber would you expect a power lifter to be enriched in? A long-distance cyclist?

56. What is myoglobin? In which type of muscle fiber is it found?

57. What does "dark meat" mean in terms of nutritional content of the meat? In terms of type of muscle fiber? What about "white meat"?

1. Body fat cushions the organs and stores fat-soluble vitamins and hormones (and, of course, acts as a source of stored energy).

3. Type 1 diabetes (juvenile onset) occurs because the body can't produce insulin. It is treated with insulin injections. Type 2 diabetes occurs because the cells become insensitive to insulin. It is treated through diet and weight management.

5. Those with type 2 diabetes need to control their diet, reduce intake of refined sugars and carbohydrates, exercise regularly, and lose weight.

7. Wrist, neck, and waist circumference is least accurate. Skin fold is more accurate, and immersion is the most accurate.

9. BMI = 15.1; she is underweight.

11. BMI = 38.5; he is obese. It's hard to know whether he's truly obese without more information. He could have low body fat but be incredibly muscular.

13. Limited information is provided for 2,500 Cal/day.

15. Trans fats are more stable than unsaturated fats; the *cis* double bonds can rearrange to *trans* double bonds given heat and time.

17. Fat causes release of cholecystokinin (CCK), which communicates a feeling of fullness.

19. A satiety signal is a chemical signal indicating fullness; sympathomimetic means that a molecule mimics adrenaline and stimulates the sympathetic nervous system; ketogenic means producing ketosis; ketone bodies are produced through the metabolism of fats, and can be used to provide for the brain's energy needs during glycogen deprivation.

21. She is lethargic and her muscles burn because she is depleted of glycogen. She's also probably feeling lethargic because her brain is being deprived of glucose.

23. Saccharin and aspartame still share the sucrose pharmacophore, but it's not easy to see. They're flexible molecules and have the right atoms in the right places to fit into the sweetness receptor.

25. Aspartame contains phenylalanine, which individuals with PKU can't metabolize. It builds up in their body and can cause brain damage.

27. It appears that artificial sweeteners can increase a desire for sweet foods (thereby increasing calorie-containing sweets), and may also change the population of bacteria in the gut, which has unknown consequences at this time, but which is concerning.

29. Most nutritional experts recommend about 45–65% carbohydrate, 10–35% protein, and 20–35% fat, with plenty of fruits and vegetables.

31. High-quality carbohydrate generally has a low GI—it's high in fiber and low in refined carbohydrate and sugar.

33. GI = 200.

35. In general, fruits and whole grains have low GI. Refined carbohydrates and sugary foods have high GI.

37. Protein and fat turn high-GI foods into low-GI foods by interfering with (and slowing the rate of) carbohydrate hydrolysis and absorption. As long as the

added fat is healthy (unsaturated) and the protein is lean, this is a reasonable strategy. However, it would be questionable to add saturated fat, for instance, to white bread. Not only does this introduce extra saturated fat into the diet, it also encourages eating a less healthy choice (white bread), where a better option might be whole-grain bread.

39. 1,335.6 Cal/day.

41. RMR calculations are likely inaccurate for individuals with extremely low or high amounts of lean tissue as compared to the average individual.

43. The thermic effect of food accounts for about 10% of calories expended.

45. Direct effect is the calories burned during an exercise; elevated post-exercise metabolic rate is the higher-than-normal RMR that occurs for some time after exercise.

47. Aerobic exercise is enough exertion to cause an increase in metabolic waste products but is at moderate enough intensity that adaptations (breathing, heart rate) can supply sufficient oxygen to muscles. It occurs at 65–85% of HR_{max}. Exercise is aerobic if the individual can talk but not sing.

49. Aerobic exercise uses fat and glycogen—fat is preferred. Anaerobic uses glycogen only.

51. HR_{max} = 145 BPM, aerobic zone = 94–123 BPM.

53. She's at 89 percent of HR_{max}, so she's anaerobic.

55. A power lifter will have more type II than average, while a long-distance cyclist will have more type I than average.

57. Dark meat is primarily type I; it is higher in fat and calories than white meat (and lower in protein for a given mass). White meat is primarily type II; it is lower in fat and calories and higher in protein than the same mass of dark meat.

SOURCES AND FURTHER READING

Body Fat

http://www.nhlbisupport.com/bmi/

http://www.nlm.nih.gov/medlineplus/obesity.html

http://win.niddk.nih.gov/statistics/index.htm

National Institutes of Health. (1998). Clinical guidelines on the identification, evaluation, and treatment of overweight and obesity in adults. *The Evidence Report*. No. 98–4083.

Nutritional Recommendations

Grandgirard, A., Sebedio, J. L., and Fleury, J. (1984). Geometrical isomerization of linoleic acid during heat treatment of vegetable oils. *Journal of the American Oil Chemists' Society*, *61*(10), 1563–68.

Nestle, M. *Food politics: How the food industry influences nutrition and health*. (2002). Berkeley, CA: University of California Press.

Sims, L.S. (1998). *The politics of fat: Food and nutrition policy in America*. Armonk, NY: M.E. Sharp, Inc.

www.choosemyplate.gov

Typical American Diet

Lansky, D., and Brownell, K. D. (1982). Estimates of food quantity and calories: Errors in self-report among obese patients. *American Journal of Clinical Nutrition, 35*(4), 727–32.

Pollan, M. (2006). *The omnivore's dilemma.* New York, NY: Penguin Group.

Schlosser, E. (2001). *Fast food nation.* New York, NY: HarperCollins.

Schulze, M. B., Manson, J. E., Ludwig, D. S., *et al.* Sugar-sweetened beverages, weight gain, and incidence of type 2 diabetes in young and middle-aged women. *JAMA, 292*(8), 927–34.

Wansink, B., and Chandon, P. (2006). Meal size, not body size, explains errors in estimating the calorie content of meals. *Annals of Internal Medicine, 145*(5), 326–32.

Diets

Bravata, D. M., Sanders, L., Huang, J., *et al.* (2003). Efficacy and safety of low-carbohydrate diets: A systematic review. *JAMA, 289*(14),1837–50.

Foster, G. D., Wyatt, H. R., Hill, J. O., *et al.* (2003). A randomized trial of a low-carbohydrate diet for obesity. *New England Journal of Medicine, 348*(21),2082–90.

Freedman, M. R., King, J., and Kennedy, E. (2001). Popular diets: A scientific review. *Obesity Research, 9,* 1S–40S.

Kirby, R. C. (2005). Atkins diet—discussion paper. *Discussion paper developed for the AAFP Commission on Public Health.*

Staszkiewicz, J., Horswell, R., and Argyroupolos, G. (2007). Chronic consumption of low-fat diet leads to increased hypothalamic agouti-related protein and reduced leptin. *Nutrition, 23*(9), 665–71.

Wansink, B., and Chandon, P. (2006). Can "low-fat" nutrition labels lead to obesity? *Journal of Marketing Research, 43*(4), 605–17.

Westman, E. C., Mavropoulos, J., Yancy, W. S., *et al.* (2003). A review of low-carbohydrate ketogenic diets. *Current Atherosclerosis Reports, 5,* 476–83.

Sugar Substitutes

Blundell, J. E., and Hill, A. J. (1986). Paradoxical effects of an intense sweetener (aspartame) on appetite. *Lancet, 1*(8489), 1092–93.

Cohen, S. M., Aral, M., Jacobs, J. B., *et al.* (1979). Promoting effect of saccharin and DL-tryptophan in urinary bladder carcinogenesis. *Cancer Research, 39*(4), 1207–17.

Gurney, J. G., Pogoda, J. M., Holly, E. A., *et al.* (1997). Aspartame consumption in relation to childhood brain tumor risk: Results from a case-controlled study. *Journal of the National Cancer Institute, 89*(14), 1072–74.

Olney, J. W., Farber, N. B., Spitznagel, E., *et al.* (1996). Increasing brain cancer rates: Is there a link to aspartame? *Journal of Neuropathology and Experimental Neurology, 55*(11), 1115–23.

Prodolliet, J., and Bruelhart, M. (1993). Determination of aspartame and its major decomposition products in foods. *Journal of AOAC International, 76*(2), 275–82.

Rodero, A. B., Rodero, L. D., and Azoubel, R. (2009). Toxicity of sucralose in humans: A review. *International Journal of Morphology, 27*(1), 239–44.

Soffritti, M., Belpoggi, F., Tibaldi, E., *et al.* (2007). Life-span exposure to low doses of aspartame beginning during prenatal life increases cancer effects in rats. *Environmental Health Perspectives, 115*(9), 1293–97.

Suez, J., *et al.* (2014). Artificial sweeteners induce glucose intolerance by altering the gut microbiota. *Nature, 514*(7521), 181–86.

Tsuda, H., Fukushima, S., Imaida, K., *et al.* (1983). Organ-specific promoting effect of phenobarbitol and saccharin in induction of thyroid, liver, and urinary bladder tumors in rats after initiation with N-nitrosomethylurea. *Cancer Research, 43*(7), 3292–96.

Eating Pattern Sustainability

http://www.glycemicindex.com/

Fosterpowell, K., and Miller, J. B. (1995). International tables of glycemic index. *American Journal of Clinical Nutrition, 62*(4), S871–90.

Caloric Expenditure

D'Alessio, D. A., Kavle, E. C., Mozzoli, M. A., *et al.* (1988). Thermic effect of food in lean and obese men. *Journal of Clinical Investigation, 81*(6), 1781–89.

Rosa, A. M., and Shizgal, H. M. (1984). The Harris Benedict equation reevaluated: Resting energy requirements and the body cell mass. *American Journal of Clinical Nutrition, 40*(1), 168–82.

Exercise

Bielinski, R., Schutz, Y., and Jequier, E. (1985). Energy-metabolism during the post-exercise recovery in man. *American Journal of Clinical Nutrition, 42*(1), 69–82.

Burleson, M. A., O'Bryant, H. S., Stone, M. H., *et al.* (1998). Effect of weight training exercise and treadmill exercise on post-exercise oxygen consumption. *Medicine and Science in Sports and Exercise, 30*(4), 518–22.

Fuss, J., *et al.* (2015). A runner's high depends on cannabinoid receptors in mice. *Proceedings of the National Academy of Sciences,* epub ahead of print Oct 5.

Lethbridge-Çejku M., and Vickerie, J. (2005). Summary health statistics for U.S. adults: National Health Interview Survey, 2003. National Center for Health Statistics. *Vital Health Statistics, 10*(225).

Melby, C., Scholl, C., Edwards, G., *et al.* (1993). Effect of acute resistance exercise on post-exercise energy expenditure and resting metabolic rate. *Journal of Applied Physiology, 75*(4), 1847–53.

Reebok Instructor News. (1991). 4(2).

Vitamin Supplements

Bjelakovic, G., Nikolova, D., Gluud, L. L., *et al.* (2007). Mortality in randomized trials of antioxidant supplements for primary and secondary prevention; systematic review and meta-analysis. *JAMA, 297*(8), 842–57.

Duffield-Lillico, A. J., Dalkin, B. L., Reid, M. E., *et al.* (2003). Selenium supplementation, baseline plasma selenium status and incidence of prostate cancer:

An analysis of the complete treatment period of the Nutritional Prevention of Cancer Trial. *BJU International, 91*(7), 608–12.

Knekt, P., Reunanen, A., Jarvinen, R., *et al.* (1994). Antioxidant vitamin intake and coronary mortality in a longitudinal population study. *American Journal of Epidemiology, 139*(12), 1180–89.

Lee, I. M., Cook, N. R., Gaziano, J. M., *et al.* (2005). Vitamin E in the primary prevention of cardiovascular disease and cancer: The Women's Health Study: A randomized controlled trial. *JAMA, 294*(1), 56–65.

Lippman, S. M., Klein, E. A., Goodman, P. J., *et al.* (2009). Effect of selenium and vitamin E on risk of prostate cancer and other cancers: The selenium and vitamin E cancer prevention trial. *JAMA, 301*(1), 39–51.

Rohan, T. E., Jain, M. G., Howe, G. R., *et al.* (2000). Dietary folate consumption and breast cancer risk. *Journal of the National Cancer Institute, 92*(15), 266–69.

Sesso, H. D., Buring, J. E., Christen, W. G., *et al.* (2008). Vitamins E and C in the prevention of cardiovascular disease in men: The Physicians' Health Study II randomized controlled trial. *JAMA, 300*(18), 2123–33.

Stampfer, M. J., Hennekens, C. H., Manson, J. E., *et al.* (1993). Vitamin E consumption and the risk of coronary disease in women. *New England Journal of Medicine, 328*(20), 1444–49.

Zhang, S. M., Cook, N. R., Albert, C. M., *et al.* (2008). Effect of combined folic acid, vitamin B6, and vitamin B12 in women: A randomized trial. *JAMA, 300*(17), 2012–21.

CREDITS

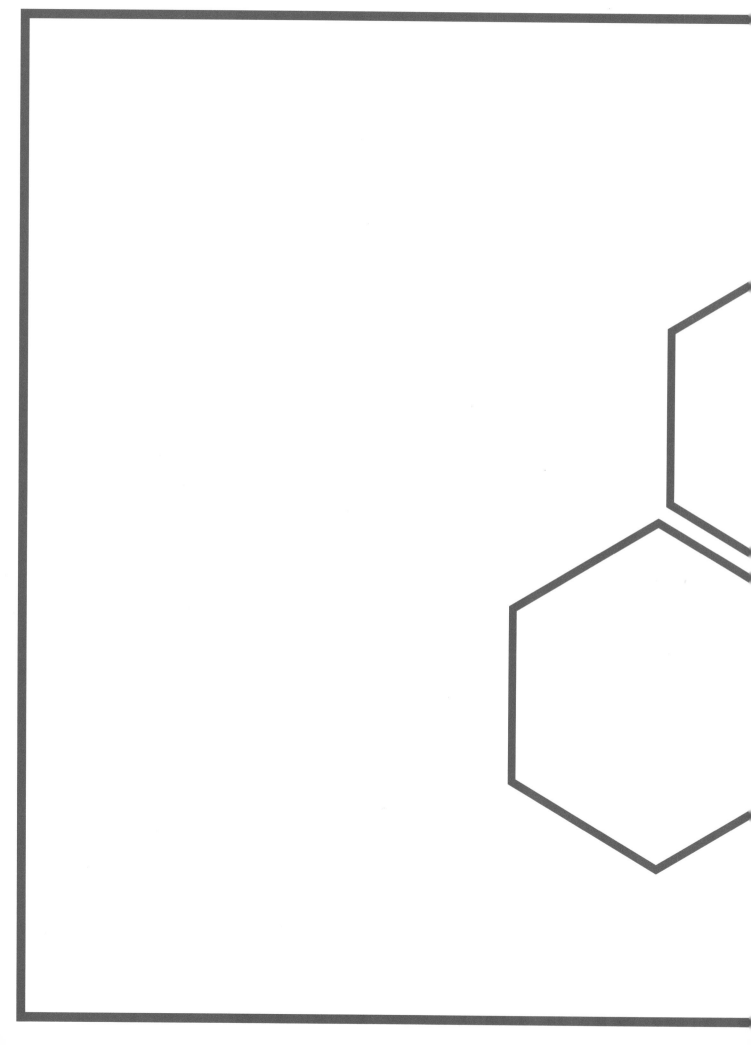

GLOBaL CHEMISTRY

Changing the World with Chemistry

10

We've spent the last three units of this text discussing the myriad ways in which chemistry affects our communities and our bodies. There's no doubt, at this point, that the local water and air quality and the type of food we eat affect our lives and those of the people around us. But can chemistry impact the entirety of the global community? Are there discoveries and processes so momentous that their effects are felt around the planet? In this chapter, we'll explore some of the major ways in which chemistry has changed the world—for better and for worse—and the discoveries and issues that affect us all, no matter where we live. We'll also begin to discuss the challenges associated with these issues, particularly where it comes to making global decisions and changes. How, for instance, do we portion out global responsibility for effecting change to developed and developing nations, given their very disparate economic welfares and technological capabilities? How do we know whether lab-derived evidence applies in the real world, and how do we strike a balance between too much caution and not enough when it comes to implementing (or withdrawing) a technology worldwide? Most importantly, where do we draw the line between concern for the individual and utilitarianism, or the responsibility to do the most good for the most people?

10.1   Refrigeration

Hopefully we're all comfortable at this point with the notion that science isn't simply what takes place in the lab: the relationship between the lab and the real world is reciprocal and inextricable. Scientific exploration in the lab models the real world, allowing researchers to make predictions. Animal studies, for instance, are used to determine the safety of a new pharmaceutical before it's ever tested on humans. Unfortunately, the lab has its limitations. It's not the "real world," but is instead a microcosm thereof—an incomplete facsimile that allows us only a glimpse of the full potential and ramifications of a technology. **When a chemical, an invention, a discovery, or a technology graduates from the lab and debuts in the world, we are often greeted with unanticipated effects and outcomes.**

The unanticipated global ramifications of refrigeration technology illustrate this concept. Until the twentieth century, refrigeration as a household commodity was virtually unknown. The theory behind it was firmly established, however, and was based upon some well-understood laws regarding the behavior of liquids and gases. The common experience of sweating gives us insight into some of these laws. When we exert ourselves or are exposed to high ambient temperatures, we secrete liquid sweat onto the surface of the skin. The water component of this liquid (which also contains dissolved salts) evaporates into a gas. As it does so, it absorbs a great deal of heat, lowering the temperature of the body. The heat is carried away with the evaporating water, leaving us feeling much cooler. The reason an evaporating liquid absorbs heat is that the process of vaporization is endothermic—it requires energy. We discussed in Chapter 8 that for every process, a reverse process is possible. Furthermore, if a process requires energy, the reverse will release it. We can conjecture, then, that if water vapor were to condense into liquid (on the skin or elsewhere), it would release heat. In fact, we are subjected to this phenomenon during the unpleasant experience of a steam burn. Boiling water, if it splashes onto the skin, burns. Steam from the boiling water, however, burns much more severely. The steam released from boiling water is the same temperature as the water; once it escapes the pot, it doesn't continue to heat up. However, in addition to transferring the heat energy of the water vapor to the skin, energy from the condensation process is also transferred as vapor turns to liquid. **For any substance, the process of vaporization absorbs energy (heat), while the process of condensation releases it.**

This theory is put to use in the design of a refrigeration device. A liquid is confined to a closed tubing system, part of which is inside (sandwiched between inner and outer walls) and part of which is outside a sealed refrigeration box. The liquid circulates through the tubing, and as it enters the refrigeration box, it passes through an expansion valve (Image 10.1). **The expansion causes the liquid to vaporize, a process that absorbs heat and decreases the temperature inside the refrigeration box.** Why does expansion cause vaporization? While a full discussion of gas behavior is beyond the scope of this text, in short, matter expands to fill available space. This phenomenon is echoed in the maxim *Nature abhors a vacuum.* If the volume of a tubing system is suddenly greater than the volume of liquid in the tubing, the molecules comprising the liquid will move away from one another to fill the space. As the molecules move away from one another, liquid turns to gas. The chemical (now a vapor) continues circulating through the tubing, eventually exiting the refrigeration box. Upon exiting, it is condensed back into a liquid by a compressor, which forces the molecules closer together. During this process, heat is released. However, since the condensation takes place outside the refrigeration box, the heat is dissipated to the air—you can actually feel the heat dissipating from the coils on the back of the refrigerator. The liquid then circulates back to the expansion valve, and the cycle starts again.

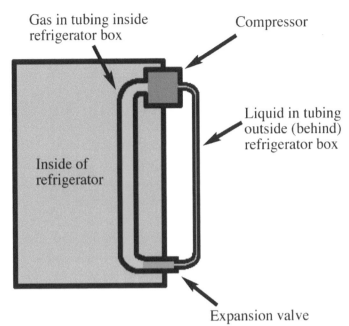

Gas in tubing inside
refrigerator box

Compressor

Liquid in tubing
outside (behind)
refrigerator box

Inside of
refrigerator

Expansion valve

IMAGE 10.1. A simple refrigerator schematic.
The tubing is coiled rather than straight in
an actual fridge to allow more surface area.
This increases the extent to which the cold
gas absorbs heat from the inside of the box
and the extent to which the warm liquid
dissipates its heat outside of the box.

The exact temperature that can be achieved through a vapor-compression cycle depends upon the boiling point of the chemical in the tubing. **For a refrigeration system to be able to freeze water—an essential quality of a freezer—the boiling point of the refrigerant must be lower than the melting (freezing) point of water (0°C).**

Several early refrigerants with appropriate boiling points had been identified in the late 1800s and early 1900s—in particular, SO_2 (-10°C), NH_3 (−33°C), and CH_3Cl (methyl chloride, -24°C) were used. Unfortunately, these were toxic, flammable, or both, resulting in many accidents. In 1928 Frigidaire (a division of General Motors) began supporting development of a new refrigerant. Thomas Midgley[1] was given the task of identifying or developing a refrigerant that would be stable, nontoxic, nonflammable, noncorrosive, cheap, and—of course—have a boiling point lower than 0°C. Midgley prepared substituted methane and ethane molecules, in which the usual hydrogen atoms were replaced by fluorine and chlorine. He named his compounds *Freon*, differentiating them from one another by use of a cryptic numbering system. In fact, all the Freons belong to a larger class of molecules called chlorofluorocarbons (CFCs).

The new refrigerants seemed to fit the bill nicely: they had appropriate boiling points and were evidently quite safe. Midgley even went so far as to demonstrate the safety of his compounds at a meeting of the American Chemical Society with a stunt reminiscent of his tetraethyl lead hand-washing display: he inhaled a lungful of Freon (to demonstrate that it was nontoxic), and then exhaled it upon a candle

NOTE[1]

Remember him? He brought us
tetraethyl lead in 1921.

(a)

H—C—H
with H top and H bottom

Methane

(b)

F—C—Cl
with F top and Cl bottom

Freon 12

(c)

Cl—C—Cl
with F top and Cl bottom

Freon 11

(d)

H—C—C—H
with H H top and H H bottom

Ethane

(e)

F—C—C—F
with F F top and Cl Cl bottom

Freon 114

(f)

Cl—C—C—F
with Cl F top and Cl F bottom

Freon 113

IMAGE 10.2. Methane, ethane, and some of Midgley's Freon molecules.

flame, which promptly went out, demonstrating that the Freon was not flammable. **Inexpensive to produce, CFCs were responsible for making home refrigerators and air conditioning (both of which relied upon the same technology) standard amenities.** The Sun Belt, a strip of the southern United States characterized by mild temperatures in the winter and uncomfortably hot summers, experienced a population explosion as air conditioning rendered the region at least somewhat tolerable. Owing to their inert chemical nature, CFCs found applications outside of the refrigeration industry. They were used in fire extinguishers to displace oxygen from a flame, they made excellent solvents for cleaning purposes, and they were condensed and used as propellants in spray paint, hair spray, and even asthma inhalers.

The heyday of CFCs continued until the early 1970s, when several researchers made some disturbing discoveries about the compounds. **Evidence began to amass that while CFCs are, in fact, nontoxic and nonreactive in the lower atmosphere, they are quite active in the stratosphere, because it is there that they encounter one of the few things with which they do react—UV light.**

10.2   UV Light and EMR

It turns out that our sun emits many different kinds of radiation: some types are responsible for lighting the planet and allowing us to see color, others provide us with warmth, and still others can burn us. Understanding the different types of radiation and why some are damaging while others are not necessitates examining the nature of the sun's rays.

What we call "light" is actually a form of electromagnetic radiation (EMR). Formally, it's a traveling electrical and magnetic disturbance, but **EMR is more simply defined as radiant energy.** The energy has some wavelike properties and some particle-like properties; this dual nature is one of the fundamental tenets—and enigmas—of physics. Suffice it to say, EMR behaves in some regards like an ocean wave, with measurable wavelike properties. For instance, just as it would be possible to measure the size of an ocean wave, it's possible to measure that of an EMR wave. In either case, the distance from one wave crest to the next is called the

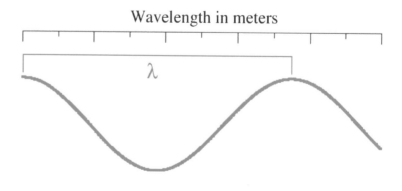

IMAGE 10.3. The wavelength of an EMR wave, like that of an ocean wave, is the distance from one crest to the next.

wavelength (λ), which is measured in meters (Image 10.3). EMR wavelength can vary tremendously depending upon the kind of radiation; it can be as short as a few femtometers (fm, 10^{-15}m), or billions of meters long.

A second measurable property of EMR is its **frequency (ν)**. This is the number of crests that pass a given point in a given period of time. If we continue the comparison to ocean waves, the frequency would be the number of waves per second that crash on the shore. Frequency is measured in units of Hertz (Hz, 1/s, s^{-1}). For instance, if three crests passed a given point in one second, the frequency of the wave would be 3 Hz. As with wavelength, the frequency of EMR depends upon the kind of radiation, and varies greatly—anywhere from less than 1 Hz to more than 10^{24} Hz.

An interesting property of EMR waves that makes them very distinct from ocean waves is that regardless of their wavelength or frequency, they all travel at exactly the same speed in a vacuum. In fact, all EMR travels at the speed of light, which is given the symbol c in equations (c = 3×10^8 m/s).[2] A ramification of this is that for all EMR, wavelength and frequency are inversely related—EMR of longer wavelength has lower frequency, and vice versa. To get an idea of why this is so, take a look at the three waves in Image 10.4, and imagine them all moving at equal speed toward the right. If you were to count the number of crests of the red wave (which has the longest wavelength) that passed a given point in one second, you'd find it to be less than the number of crests of the green wave and much less than the number of crests of the blue wave (with the shortest wavelength) passing the same point in one second. If the three waves travel at the same speed, the one with the longest wavelength has the lowest frequency, and the one with the shortest wavelength has the highest frequency.

This relationship is captured in the wave equation:

$$c = \lambda \nu$$

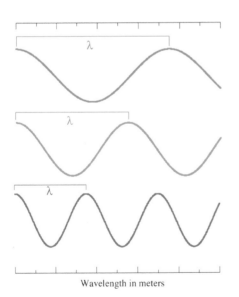

IMAGE 10.4. Wavelength and frequency are inversely related for waves traveling at the same speed.

Wavelength in meters

As we see in the equation above, wavelength (λ) and frequency (ν) are inversely related, and must always multiply to give the same number—c, the speed of light. For any EMR wave, it is possible to calculate either frequency or wavelength, as long as the other value is known.

Example: What is the wavelength of an EMR wave with a frequency of 10^6 Hz?

$$c = \lambda \nu$$

$$3 \times 10^8 \text{ m/s} = \lambda \ (10^6 \text{ Hz})$$

$$\lambda = 300 \text{ m}$$

NOTE[2]

Technically, this is the speed of light in a vacuum; when travelling through media (including air), the speed of light is somewhat different than the value given here. For simplicity's sake, however, we will use c = 3×10^8 m/s as the speed of light for all calculations.

TRY THIS

What is the frequency of an EMR wave with a wavelength of 5×10^{-10} m?

Answers:
ν = 6×10^{17} Hz.

Solution:
c = λν
3×10^8 m/s = (5×10^{-10} m) ν
ν = 6×10^{17} Hz

TRY THIS

Which has higher frequency, an EMR wave with a wavelength of 3×10^8 m or one with a wavelength of 7×10^4 m?

Answers:
Because 7×10^4 m represents a shorter wavelength, this wave will have a higher frequency.

NOTE[3]

When we calculate the energy of a "wave" of EMR, we're actually calculating the energy of an individual particle of radiation with a given frequency and wavelength. How can a particle have a wavelength (or frequency, for that matter)? Remember that EMR has a dual nature—it is both wavelike and particle-like. A particle of light is called a *photon*.

TRY THIS

What is the frequency of a wave with energy equal to 5.4 × 10−12 J?

Answers:
n = 8.1 × 10²¹ Hz.

Solution:
E = hν
5.4 × 10⁻¹² J = (6.626 × 10⁻³⁴ J·s)(ν)
ν = 8.1 × 10²¹ Hz

NOTE[4]

In fact, X-rays penetrate soft tissue but are reflected by dense tissue like bone. This has led to their utility as a diagnostic tool in medicine.

A mathematical discussion of EMR leaves us with the (somewhat deceptive) sense that all types of radiation are very similar, differing only in wavelength and frequency. While it is true that the fundamental nature of a high-frequency EMR wave is identical to that of a low-frequency EMR wave, these waves affect their surroundings in very different ways. This is because frequency is also related to energy,[3] a relationship described by another wave equation:

$$E = h\nu, \text{ which can also be written as } E = h(c/\lambda), \text{ since } c = \lambda\nu.$$

In this equation, E is the wave's energy (measured in Joules), and h is called **Planck's constant** (h = 6.626 × 10⁻³⁴ J·s). This constant is used in several different equations in physics and chemistry, but we'll see it here only in the context of relating frequency and energy to one another. **The relationship between frequency and energy is directly proportional—higher frequency waves have higher energy and vice versa, and we can calculate either value from the other.**

Example: What is the energy of a wave with a frequency of 3.00 × 10¹⁶ Hz?

$$E = h\nu$$

$$E = (6.626 \times 10^{-34} \text{ J·s})(3.00 \times 10^{16} \text{ Hz})$$

$$E = 1.99 \times 10^{-17} \text{ J}$$

Since we know that wavelengths and frequencies of EMR span many orders of magnitude, we can conjecture that the energies of EMR waves must also encompass a large range. This is quite true, and furthermore, the physical effect of EMR upon the world (and the way in which we experience the radiation) depends entirely upon its energy. Image 10.5 shows the spectrum of EMR, from very high energy radiation (high frequency, short wavelength) like gamma rays, to very low energy radiation (low frequency, long wavelength) like radio waves.

Gamma rays, which we'll see again in Chapter 12, are a form of extremely high- energy radiation that can be released during nuclear reactions. They are very damaging to living tissue, and can cause radiation sickness, mutations, and death. Slightly lower in energy are **X-rays**. They are not quite as penetrating as gamma rays,[4] but sufficient exposure can still lead to tissue and biomolecular damage, which can result in mutation. **Ultraviolet radiation (UV)**, lies between X-rays and visible light on the EMR spectrum. The penetrating power of UV light is less than that of X-ray radiation; generally, UV can penetrate the skin but can't go deeper into the body. Still, it can damage living tissue and cause mutation. Why do these three types of radiation possess the ability to damage tissue, and why do mutations result? The answer lies in the mechanism by which the radiation interacts with molecules in cells. **UV, X-rays, and gamma rays are collectively called** ionizing radiation.

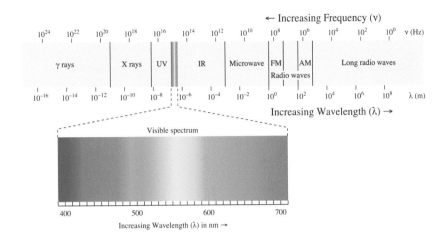

TRY THIS

What is the energy of a wave with λ = 2.3 × 10⁻¹⁰ m?

Answers:
E = 8.6 × 10⁻¹⁶ J.

Solution:
c = λν
3 × 10⁸ m/s = (2.3 × 10⁻¹⁰ m) ν
ν = 1.3 × 1018 Hz

E = hν
E = (6.626 × 10⁻³⁴ J·s)(1.3 × 10¹⁸ Hz)
E = 8.6 × 10⁻¹⁶ J

IMAGE 10.5. Types of EMR. Higher-energy radiation is on the left end of the spectrum.

When such radiation strikes a molecule, it has the ability to weaken and break bonds (Image 10.6).[5]

It's useful to model bonds as springs when talking about the effect of radiation. In a very real way, bonds are actually much like springs—they have a natural frequency at which they stretch and compress, bringing atoms closer together and allowing them to drift further apart while retaining their attachment to each other. Bonds can bend and flex a bit without breaking—which to some extent they do naturally—but outside forces can exacerbate this. It's not far from the truth to imagine that UV and other ionizing radiation stresses and stretches a bond with such force that it causes it to break, much as a spring breaks under sufficient strain. In fact, as discussed in Chapter 8, breaking bonds always requires energy, and ionizing radiation has sufficient energy to effect this process.[6] Of course, we must keep in mind that breaking a bond can result in the freeing of atoms or parts of molecules, which are now free to form new bonds and rearrange into different molecules.

The biological ramifications of exposure to ionizing radiation stem directly from its ability to break bonds in molecules: if the broken bonds occur in DNA (our genetic material), mutations and cancer can result. The type of cancer that occurs is related

NOTE⁵

The name *ionizing radiation* comes from the fact that these wavelengths can also cause an electron to be knocked out of a molecule, leaving the molecule positively charged (ionic). Since electrons are normally found in pairs, the loss of one electron leaves a lone electron—a free radical—behind. As we've seen, free radicals are highly reactive species.

NOTE⁶

Visible light and other EMR with frequency lower than that of UV does not have sufficient energy to break bonds. These types of radiation can affect molecules, but not in a way that destroys molecules or generates free radicals. We'll see the effects of these different types of radiation later in this chapter and throughout the rest of the text.

UV Radiation

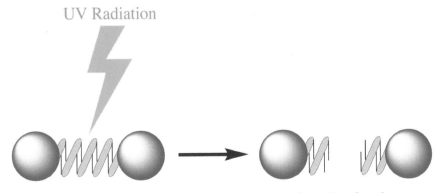

IMAGE 10.6. The effect of UV (and other ionizing radiation) on bonds.

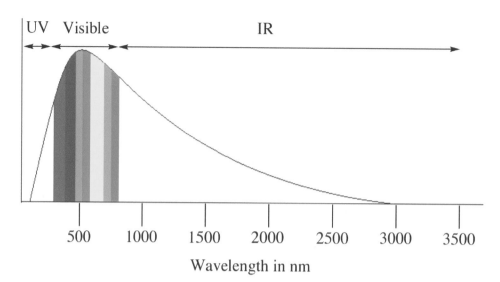

IMAGE 10.7. The Solar spectrum.

to the type of ionizing radiation. Very penetrating radiation, like gamma rays, can cause deep tissue mutations, including leukemia and other blood cancers. The far-less-penetrating UV radiation can't cause mutations in deep tissues, but can certainly lead to skin cancer. DNA isn't the only molecule susceptible to ionization—proteins and many other biomolecules can be affected as well. For instance, the protein collagen is responsible for maintaining the skin's elasticity. Over time and with age, this protein begins to degrade, leaving the skin with less resilience and elasticity (and more wrinkles). This natural process is dramatically accelerated by repeated exposure to UV radiation: sunbathing can lead to prematurely wrinkled skin.

As anyone who's ever had a bad sunburn can attest, cellular damage from UV exposure can be acute: severe sunburn results in the death of several layers of skin cells. This provides us some insight into the importance of protection from UV: if we were exposed to the full strength and spectrum of the sun's rays (almost 10% of which are UV—Image 10.7), all living organisms would be forced to retreat underground or underwater to avoid catastrophic radiation exposure. **Thankfully, the ozone layer protects us from most of the UV light that is incident upon the earth.**

10.3   The Ozone Layer

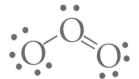

IMAGE 10.8. The ozone molecule; ozone is an allotrope of oxygen.

We first encountered the ozone molecule in Chapter 6 as a secondary product of combustion. **Tropospheric ozone is not naturally occurring; it's a pollutant. In the stratosphere, however, ozone is a natural component of the atmosphere, and serves the vital function of protecting us from the sun's ultraviolet radiation.** As we learned in Chapter 6, ozone is an allotrope of oxygen, but its odd and asymmetric arrangement of atoms (Image 10.8) makes it much less stable and much more reactive than O_2.

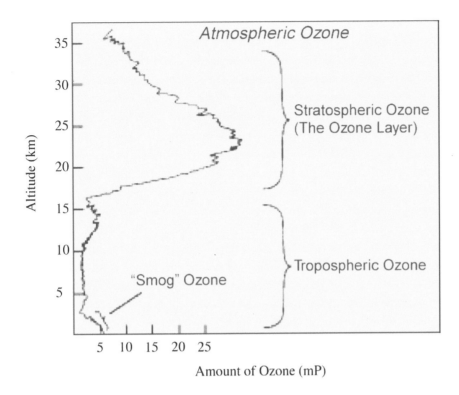

Atmospheric Ozone

Stratospheric Ozone
(The Ozone Layer)

Tropospheric Ozone

"Smog" Ozone

Altitude (km)

Amount of Ozone (mP)

IMAGE 10.9. The ozone layer is a band of higher ozone concentration occurring in the stratosphere, between 15 and 35 km above the earth's surface. Tropospheric ozone is a pollutant; it does not occur naturally.

A layer of ozone 20 km thick spans the stratosphere (Image 10.9). Since atmospheric pressure is much lower at those elevations, however, gases are far less dense and molecules much farther apart than they are near Earth's surface. In fact, if the entire ozone layer were brought down to sea level, it would be a mere 3 millimeters in thickness! Regardless, the ozone layer is an important shield—one without which life would be very different.

Sydney Chapman, a British chemist, determined the mechanism by which the ozone layer shields us from UV light, which is now known as the Chapman Cycle (Image 10.10). In fact, the chemistry of the cycle is somewhat complex, involving two separate processes, each of which takes place in two steps:

Destruction of ozone:[7]

$$O_3 + h\nu \rightarrow O + O_2$$

$$O + O_3 \rightarrow 2\,O_2$$

Production of ozone:

$$O_2 + h\nu \rightarrow 2\,O$$

$$O + O_2 \rightarrow O_3$$

NOTE[7]

Recall that in chemical equations, $h\nu$ is taken to mean "light of the appropriate energy." Now we know why: the wave equation $E = h\nu$ reminds us that $h\nu$ is, literally, energy from light.

In one process, ozone absorbs UV radiation and a bond is broken, producing an oxygen atom and O_2. The oxygen atom is highly reactive, and can combine with more ozone to form two additional molecules of O_2. A chemical species that is formed in one step of a process and consumed in the next step is called an intermediate. While we wouldn't expect to find large quantities of atomic oxygen in the stratosphere (since it's used as a reactant in the second step of the process), it is present in small concentrations due to the fact that the Chapman Cycle is continually occurring, temporarily producing atomic oxygen for each turn of the cycle. The other Chapman Cycle process requires the absorption of UV by O_2, which splits the molecule into two atoms of oxygen; here again, atomic oxygen is an intermediate. Each of these atoms can react with more O_2, resulting in the production of ozone. In each case, UV light is providing the energy necessary to break bonds and allow for molecular rearrangements.

The Chapman Cycle is a dynamic equilibrium, meaning that while individual molecules of ozone are constantly being destroyed, new molecules of ozone are constantly being formed. These two processes occur at the exact same rate in a healthy atmosphere, such that ozone concentration is constant over time.

While both processes of the Chapman Cycle require the absorption of UV light, destruction of ozone depends upon a different wavelength (and energy) of radiation than does its production (Image 10.10). The reason for this is that ozone is much less stable than O_2, meaning that it takes more energy (shorter wavelength radiation) to break the bond in O_2 than to break the bond in O_3. Lower-energy UV—radiation with wavelengths from about 310–242 nm—is absorbed by ozone, resulting in the production of O_2. Higher-energy UV, with wavelengths less than 242 nm, is absorbed by O_2, initiating the reactions that culminate in the production of ozone. **Since both O_2 and ozone are capable of absorbing UV, and since both are present in the stratosphere, much of the UV radiation incident upon the earth never reaches its surface; most is absorbed in the upper atmosphere through the reactions of the Chapman Cycle.**

UV light can actually be divided into three categories: **UV-A** has the longest wavelength (400–315 nm) and is the least damaging, **UV-B** has an intermediate

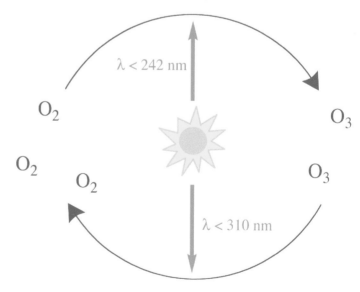

IMAGE 10.10. The Chapman Cycle, for which a simplified chemical reaction can be written: $3 O_2 \rightarrow 2 O_3$, $2 O_3 \rightarrow 3 O_2$.

wavelength (310–280 nm), and UV-C has the shortest wavelength (280–100 nm) and is most damaging to living organisms. Thankfully, UV-C is absorbed in both the destruction and production of ozone. As a result, no UV-C penetrates the upper atmosphere under normal circumstances. UV-B is partially absorbed in the stratosphere—it has a wavelength appropriate for effecting the destruction of ozone. While the ozone layer shields us from most UV-B, a small amount still reaches the earth. Because this form of radiation is so penetrating, even small amounts are capable of causing tremendous damage: UV-B is responsible for most sunburns and cancers. UV-A is not significantly absorbed in the upper atmosphere—it's too low in energy to break bonds in either ozone or O_2—and the majority of it passes through the atmosphere and reaches the earth's surface. While UV-A is not as penetrating as the other forms of UV radiation and doesn't cause sunburn, it nevertheless produces changes in DNA, and can cause cancer.

10.4   CFCs and UV Light

As we've seen, UV light has the ability to break bonds in lots of different molecules, including O_2, O_3, DNA, proteins, and a wide variety of other species. Generally, it's absorbed by and reacts with the first appropriate molecule it encounters (with the proviso that high-energy UV is capable of breaking stronger bonds than low-energy UV). Thanks to the ozone layer, not much UV reaches the troposphere, but it is intense in the upper stratosphere. **To return to our discussion of CFCs, their nonreactive nature in the troposphere means that they have a long lifetime—they enter the global atmospheric cycle, and eventually drift upward into the stratosphere. Here they encounter intense UV, with which they quickly react, resulting in the release of free atoms of chlorine** (Image 10.11).

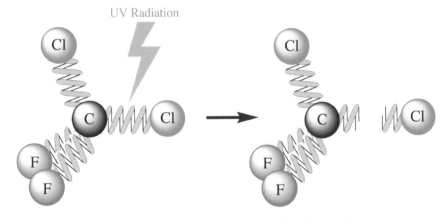

IMAGE 10.11. A CFC molecule reacts with UV light. The result is the liberation of an atom of Cl, a highly reactive species.

The chlorine atom split from its parent CFC is very high in energy, and goes on to react with the first unstable molecule it encounters, generally O_3:

$$Cl + O_3 \rightarrow ClO + O_2$$

The products of this reaction are O_2 and chlorine monoxide, ClO, itself a reactive species. The ClO is capable of reacting with an atom of oxygen (recall that oxygen atoms are intermediates in both parts of the Chapman Cycle, so they are present in the upper atmosphere), producing more O_2 and releasing the original chlorine atom.

$$ClO + O \rightarrow O_2 + Cl$$

The ramifications of these reactions are not immediately disturbing—it appears on the surface that chlorine simply reacts with ozone, producing two molecules of O_2 as a result. The problem, however, is that the chlorine atom responsible for initiating the ozone destruction was reformed in the process, meaning that it can go on to react with another molecule of O_3 (producing more O_2), and then another, and so forth. A reactant that is regenerated in this fashion is called a catalyst, which means a chemical species that accelerates a reaction without being consumed itself. **Small amounts of the chlorine catalyst are capable of reacting with large numbers of ozone molecules very quickly. In fact, a single molecule of CFC in the upper atmosphere can destroy tens and even hundreds of thousands of ozone molecules.** The result is a massive perturbation, or disturbance, of the natural ozone–oxygen cycle that is normally held in dynamic equilibrium by the sun's radiation. CFCs accelerate ozone destruction without affecting the rate of ozone formation.

The theory that CFCs possessed the ability to react with ozone arose in the early 1970s. Paul Crutzen, an atmospheric chemist, was studying the reaction of NOx with O_3. He also reported that oxides of other elements, including chlorine, could theoretically react with ozone and destroy it. However, the relevance of this report went unrecognized, as there was no known source of ClO or other chlorine oxides in the upper atmosphere. A few years later, the National Aeronautics and Space Administration (NASA) asked researchers to evaluate the potential for and ramifications of chlorine released into the upper atmosphere by space shuttle exhaust. It was determined that there was, indeed, a mechanism by which ClO could react with ozone, and that the reaction was catalytic, dramatic, and rapid. **An alarming breakthrough followed shortly thereafter when Mario Molina and Sherwood Rowland, a pair of researchers working in California, proposed that CFCs were a plentiful source of stratospheric chlorine.** Further, they found that there were no natural processes by which CFCs were removed from the environment other than that by which they reacted with UV light, resulting in ozone destruction.[8] In response to these findings, the *New York Times* ran a front-page story about the connection between CFCs—ubiquitous in refrigeration devices and aerosols—and ozone depletion. In response, Americans voluntarily cut their purchase and use of CFC-propelled aerosols by 67%.

NOTE[8]

For this accomplishment, they, together with Crutzen, received the 1995 Nobel Prize in chemistry.

For several years thereafter, worldwide CFC usage fell. The U.S. Congress banned CFCs in aerosol cans in 1978, but continued to allow their use as refrigerants. DuPont publicly initiated a program to begin searching for CFC alternatives (while quietly spending millions of dollars to run full-page newspaper ads claiming that there was no evidence CFCs were associated with ozone depletion). Over the next several years, as concern for the ozone layer retreated from the front page of the newspaper (and out of the public eye), worldwide production and use of CFCs began to resume and then increase. The next breakthrough came in the analysis of some unusual data gathered by a British research team in Antarctica. They were measuring the thickness of the ozone layer over the continent and were shocked to discover a depletion cycle that showed seasonal fluctuation—**strangely, it appeared that there was a hole in the ozone layer that seemed to get worse through the summer months and recover partially during winter months.** The depletion was so significant and unexpected that they collected data for three years before finally publishing their discovery. Some of the confusion owed to the fact that NASA satellites had also been monitoring the thickness of the ozone layer, and had detected no depletion over the same period. The discrepancy was eventually traced to the satellite data having been "fitted" to remove outlying data points, meaning that particularly low measurements were systematically deemed inaccurate and were thrown out. When satellite data was reexamined with all data points intact, the measurements correlated with those taken by the British team. There was no longer any doubt: a large hole had opened up in the ozone layer over Antarctica (Image 10.12).

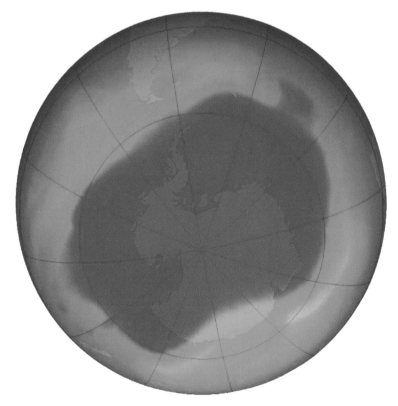

IMAGE 10.12. The ozone hole in 2006 as measured by NASA, where different colors correspond to different thicknesses in Dobson units (DU). Normal thickness is in the range of 300–500 DU. Blue areas, and particularly purple areas, are abnormally thin.

IMAGE 10.13. The thickness of the ozone hole (yearly minimum values) over Halley Bay, Antarctica. TOMS is the Total Ozone Mapping Spectrometer employed by NASA to monitor the layer. Ozone thickness is measured in Dobson units (DU), where the normal thickness is 300–500DU.

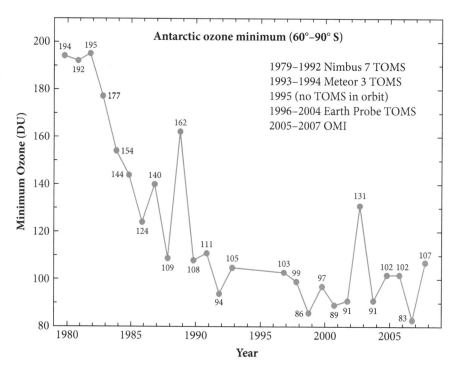

Image 10.12 shows 28 years of ozone layer thickness over Halley Bay, Antarctica. Notice that while year-to-year fluctuations in thickness occur, the trend has been toward a dramatically thinner layer.

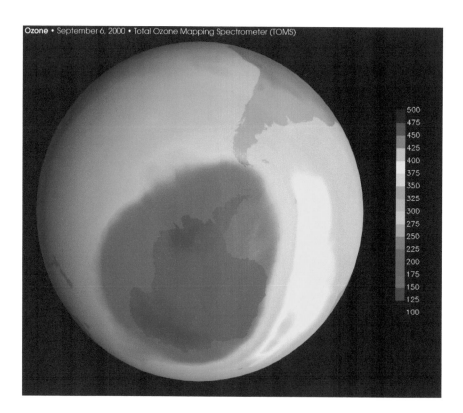

IMAGE 10.14. Polar stratospheric clouds are responsible for the location and seasonal fluctuation of the ozone hole.

While monitoring revealed the hole to be present year-round, it was particularly large in the austral summer months (winter in the Northern Hemisphere). It was eventually determined that both the seasonal variation and the polar location of the hole had to do with temperature. The Antarctic winters have the lowest average winter temperatures of any place on the planet. These low temperatures promote the formation of polar stratospheric clouds (Image 10.14).

Due to the low water content of the stratosphere, cloud formation at that elevation is rare. Only in the case of exceedingly low temperatures, such as in the midst of an Antarctic winter, can such clouds form, and when they do, they capture and provide a reaction surface for stratospheric ClO. Near the beginning of the austral summer (December), the increasingly intense sunlight causes rapid warming of the upper atmosphere, increasing air circulation. Ozone-depleted air moves away from the pole, and ozone-rich air from surrounding areas migrates southward. The result is a rapid catalytic destruction of ozone and an expanded area of depletion. As the ozone layer is Earth's protective shield from exposure to catastrophic amounts of UV light, areas of the planet no longer protected by adequate ozone thickness become susceptible to increased UV intensity. As the hole expands, larger portions of the planet are at risk for exposure to levels of UV light incompatible with life.

10.5   The Montreal Protocol

In September of 1987, the Montreal Protocol on Substances That Deplete the Ozone Layer became available for international ratification. **The protocol called for a phaseout of the production and use of CFCs and other** ozone-depleting substances (ODS). There have been many revisions to the original protocol that have generally served to increase the rate at which the phaseout will occur and the number of substances to be phased out; the most recent of these was signed in 2007 in Montreal. As this edition of the text goes to press, there are several new amendments to the protocol in the works, which will be considered for possible adoption late in 2015 at the Dubai Meeting of the Parties to the Montreal Protocol.

As of September 2009, the protocol became the first United Nations treaty to receive universal ratification. **That the Montreal Protocol has been able to engender global cooperation on an environmental problem of this magnitude—the solution to which has both developmental and economic ramifications—is beyond remarkable.** In formulating a global treaty of this sort, there are multiple obstacles that must be overcome. The first of these is the disparity between developed and developing nations. For the Montreal Protocol to be effective, it required a commitment to ODS phaseout the world over. Developing nations, however, initially expressed feeling that they were being put at an economic and developmental disadvantage in order to fix a problem they hadn't been a party to causing—the United States, Canada, the European nations, and so forth had benefited from ODS, and had enjoyed economic growth as a result of their implementation. These developed countries had also been directly responsible for the initial ozone depletion. Developing nations, while concerned about the environmental ramifications of continued ozone depletion, nevertheless expressed an understandable desire to pursue their own economic growth, including the requisite utilization of refrigerants in homes and in industrial applications. While it was understood under the Montreal Protocol that ODS refrigerants would be replaced by newer substances without the potential to deplete the ozone layer, these substances would require research and development and would also be more expensive to produce and obtain—at least in the short term—than the older CFCs. Developing nations would be at a huge disadvantage, from the standpoint of economic growth, if they had to put money into developing (or purchasing) newer refrigerants. The United Nations addressed these concerns by writing the Montreal Protocol to allow a grace period for developing nations: where developed countries were to phase CFCs out entirely by the end of 1995, the developing world was given an additional 15 years. Other ODS, including carbon tetrachloride and methyl bromide,[9] were also scheduled for phaseout by the mid-1990s in developed nations, with an additional 15–20 year grace period allowed for developing nations.

A second concern in authoring the protocol—and another whose solution can be credited with contributing to the global acceptance of the treaty—had to do with tabulating usage of the substances and sanctioning trade. Of concern, any country that did not sign the protocol would be in the powerful position of holding a monopoly on ODS production. In order to encourage universal compliance while fostering trade and economic development during the period of time in which developing nations were still allowed to utilize ODS, the protocol defined use as consumption plus production, minus export. This allowed developing countries continued access to ODS during their grace period—at least until such a time as replacement substances became available—without encouraging or inadvertently rewarding noncompliance. Additionally, a World Bank fund was established in 1990 to aid developing nations in their reduction of ODS use and adoption of alternative substances.

Taken together, the measures instated by the Montreal Protocol have had a significant positive environmental effect. Atmospheric data is beginning to show a positive response to the reduction in ODS production and use: effective atmospheric chlorine levels are falling (Image 10.15). Projections suggest that by 2050, mid-latitude ozone will have largely recovered, though it is expected that polar ozone will require an additional decade or two to regain its normal thickness. The reasons for the long delay between implementation of the Montreal Protocol and a recovery of the thickness of the ozone layer are twofold. First, CFCs have an extremely long atmospheric life; it is possible for CFCs to persist for decades in the atmosphere before reacting with UV light, so the effects of a phaseout are not seen right away. Additionally, CFC-containing devices manufactured and purchased before the phaseout continue to be in use and, once discarded, release their CFC contents into the atmosphere slowly over time. A CFC-era refrigerator or old automobile air conditioner, for instance, will leak its Freon into the atmosphere over

NOTE[9]

While CFCs represent a huge class of ODS, there are others. Carbon tetrachloride, for instance, was commonly used as an industrial solvent until its phaseout, while methyl bromide found application as a pesticide. The Montreal Protocol sought to address all possible ODS, regardless of application or chemical class.

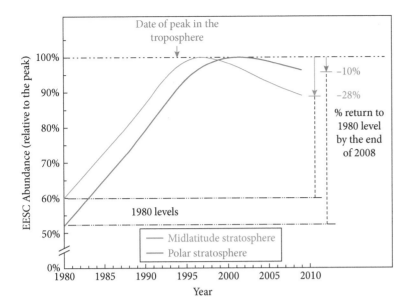

IMAGE 10.15. Effective stratospheric chlorine levels show the effects—and projected effects—of the Montreal Protocol (data from NOAA [National Oceanic and Atmospheric Administration]).

a period of many years as they sit at the dump or in the junkyard. As such, it took decades for stratospheric chlorine levels to begin falling following the initial period of ODS phaseouts, and it will be decades still before we see the ozone hole repair itself to its natural thickness.

Worldwide cooperation, regular reexamination and amendment, and evidence of the treaty's impact make the Montreal Protocol one of the most effective

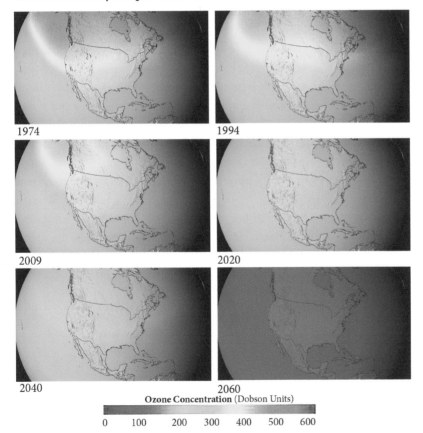

IMAGE 10.16. NASA data projecting the timeline of ozone depletion had there been no Montreal Protocol.

global environmental treaties ever developed—one need only glance at the NASA projection for the ozone destruction timetable had the treaty never been implemented (Image 10.16) to appreciate the magnitude of the accomplishment. In terms of the impact on humans alone, the EPA estimates that the Montreal Protocol has prevented more than 280 million cases of skin cancer (and almost 2 million skin cancer deaths) in the United States. Obviously, worldwide impact has been much greater than that.

10.6   Refrigerants—Moving Forward

While it is tempting to write off the Montreal Protocol as an unequivocal success that has heralded a new age of global cooperation and environmental security, the story doesn't end there. Once a technology has been implemented, it's nearly impossible to take it away again. As such, not one of the signatories expected that CFCs and other ODS would be phased out without replacements. DuPont and other companies quickly provided the first-generation substitute refrigerants—hydrochlorofluorocarbons (HCFCs) (Image 10.17). These compounds are more reactive than CFCs and have shorter lifetimes in the troposphere, meaning their ozone-depleting potential is somewhat less than that of the CFCs, though not nonexistent. The HCFCs were accepted as temporary substitutes and scheduled for phaseout by 2020 in developed nations (2040 in developing nations). Unfortunately, they are exceedingly potent greenhouse gases and contribute to global warming (see Chapter 13).

The second-generation replacement substances are hydrofluorocarbons (HFCs) (Image 10.17), whose use has been rising as developing nations near the end of their CFC and HCFC grace periods. These compounds lack any chlorine, and have no ozone-depleting potential. **Regardless, their astronomical potential to contribute to climate change (they are powerful greenhouse gases) has begun to cause concern in the scientific community.** The United States and several other countries have recommended adding HFCs to the list of banned compounds under the Montreal Protocol, not for their effect upon the ozone layer, but for their environmental impact elsewhere. As this edition of the textbook goes to press, there are formal proposals from North America, the European Union, India, and several Pacific

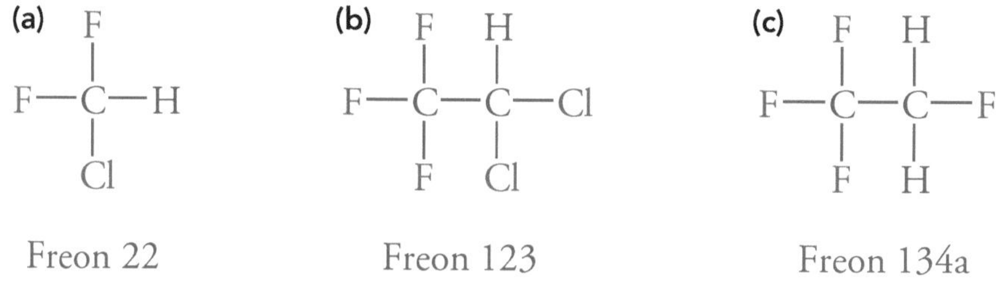

IMAGE 10.17. Two HCFCs (left and center) and an HFC (right).

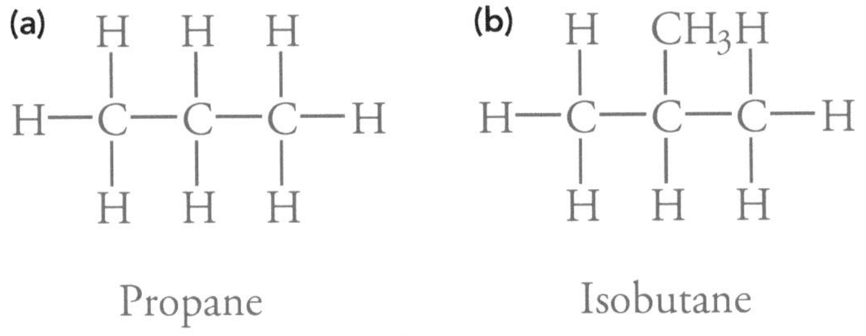

(a) Propane

(b) Isobutane

IMAGE 10.18. The components of Greenfreeze. Hydrocarbons have the potential to be non-ozone-depleting substances that refrigerate without contributing to global warming.

Island States—the latter group being among the nations most vulnerable to rising ocean level as a result of global warming—to amend the Montreal Protocol to include HFCs among substances to be phased out. These amendments will be considered for possible adoption in late 2015.

One promising refrigerant, called Greenfreeze, is a mixture of propane and isobutane (Image 10.18). Not only do these compounds have zero ozone-depleting potential and minimal global-warming potential, their energy efficiency in refrigerators and air conditioners is as good or greater than that of HFCs.

The politics of Greenfreeze have been interesting—and revealing. Originally developed in Germany, the technology captured the attention of the public: a consortium of fridge manufacturers immediately responded by denouncing Greenfreeze on the grounds that it was not a feasible refrigerant. Regardless, one company proceeded with the manufacture of a Greenfreeze-reliant refrigerator, which ended up winning an award from the German Ministry for the Environment. In 1993, the fridge manufacturers who had initially denounced the technology responded to public pressure and unveiled plans to produce fridges utilizing Greenfreeze. Since that time, most European nations and many other countries around the world have switched to Greenfreeze-based refrigeration; 40% of all refrigerators manufactured worldwide now utilize the hydrocarbon mixture. Hydrocarbon refrigeration technology has largely been ignored—and, until recently, was banned—in the United States, ostensibly because of the flammability of hydrocarbons (despite the empirical observation that widespread implementation of the technology elsewhere has proceeded without incident). It cannot be discounted that the ban may have been the result of pressure on the EPA from chemical companies in the United States. After all, without the ability to develop and sell alternative refrigerants under patent, chemical companies would stand to lose a source of revenue. However, in 2011, in response to pressure from a number of U.S. companies, including Ben and Jerry's and General Electric, the EPA ruled to make hydrocarbon refrigerants legal. It is still relatively rare to find a hydrocarbon refrigerator in the United States, but at least they are now legal, and hopefully numbers will climb in the coming years.

There are important lessons in our discussion of refrigeration. For instance, this topic highlights how difficult—even impossible—it is to anticipate all the effects of a technology before it leaves the lab and is released into the world. Even had he been a somewhat more cautious character than history shows him to be—even had he had a bit less faith in the inertness of his Freons—Midgley would never have been able to anticipate their effect on the ozone layer. Additionally, though, sometimes as we tackle one problem we create another. The development of CFCs allowed for economic growth and huge technological advancement, but the ozone layer suffered as a result. Rather than ban refrigeration—it's impossible to put the genie back in the bottle—the global community responded to the problem by replacing CFCs with alternatives, which themselves proved problematic, given their contribution to global warming. We now turn our attention to another global cycle in which we see many of these same themes: unanticipated effects, a problem that results from the solution to another problem, and global ramifications that leave developed and developing nations wondering how to compromise with one another to avert the coming storm.

10.7   The Nitrogen Cycle

The ozone–oxygen cycle is not the only one that has been perturbed during the last century. As we'll see in Chapter 13, the carbon cycle has also been badly disrupted. Perhaps no elemental cycle has been so altered, however, as that of nitrogen. The natural nitrogen cycle is maintained by a delicate balance of life and death, of assimilation and liberation. Abundant in the atmosphere, N_2 is nevertheless nonreactive—most living organisms lack the ability to incorporate it chemically. An essential elemental ingredient for many of the molecules of life, including proteins and DNA, nitrogen must first be *fixed*, or made into a reactive form, before it can be taken up and utilized by plants; animals then get their reactive nitrogen from the plants—or animals—that they consume. While a limited amount of nitrogen is fixed when lightning causes atmospheric oxygen to react with atmospheric nitrogen (generating reactive nitrogen species), **most reactive nitrogen has historically been produced by** nitrogen-fixing bacteria, **which are found both in soil and in the root nodules of leguminous plants (including alfalfa, soy, and beans).** The nitrogen-fixing bacteria produce NH_3, a basic compound that quickly reacts with water to form ammonium, NH_4^+ (Image 10.19). This compound is converted to nitrite (NO_2^-), and then to nitrate (NO_3^-), again by bacteria. Both ammonium and nitrate can be assimilated by plants, which use the compounds to produce the building blocks of DNA, amino acids, and other biomolecules.

The nitrogen cycle is actually somewhat more entangled and complex than presented above. Reactive nitrogen not assimilated by plants can be utilized and released into the atmosphere as N_2 by still other varieties of soil bacteria. In addition, plants (and the herbivores that feed on them, and the carnivores that feed on the herbivores) die and decay. When they do, decomposers including bacteria and fungi return the nitrogen to the soil as NH_4^+. Similarly, nitrogenous waste excreted by animals in urine and feces returns reactive nitrogen to the soil.

Regardless of the nitrogen cycle's complexity, two fundamental truths remain. First, under natural conditions, all nitrogen incorporated into any molecule in any multicellular organism can be traced back to nitrogen-fixing bacteria.[10] Second—again, under natural conditions—the cycle is in dynamic equilibrium, just like the Chapman Cycle that interconverts oxygen and ozone; while reactions occur

NOTE¹⁰

Less a very, very small amount produced by the effects of lightning.

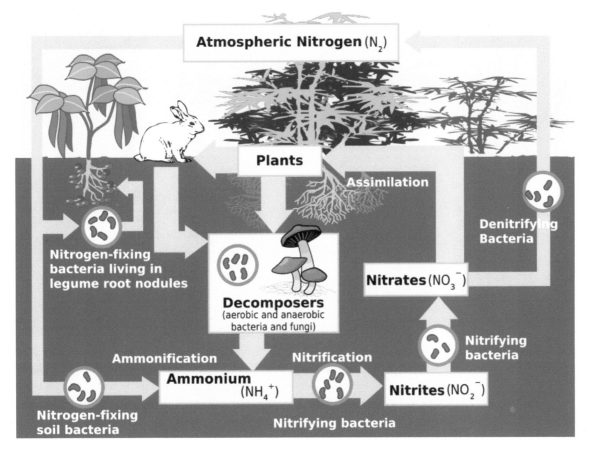

IMAGE 10.19. The nitrogen cycle, in all its complex glory.

constantly, they are in balance with one another, and total concentrations of each nitrogen species do not change over time.

Our ability to affect the nitrogen cycle was minimal until the twentieth century; we simply lacked the technology to reproduce the nitrogen fixation carried out by soil bacteria. Nevertheless, reactive nitrogen was an important commodity. Farmers, for instance, needed it to fertilize their crops, a feat generally accomplished by spreading manure on fields. It was also common to grow crops in cycles, rotating plants that depleted the soil of nitrogen, like corn, with leguminous plants that enriched the soil, like soy. While this limited the growing capacity of the land—a plant grows within the confines of the raw materials available to it—the system seemed to work well enough.

10.8   The Haber-Bosch Process

The real leap forward in nitrogen-fixing technology was inspired not by the needs of agriculture, but by those of war. In addition to providing for the needs of plants, reactive nitrogen is also a crucial ingredient in the production of explosives. Specifically, nitric acid (HNO$_3$), which can be made either from ammonia or nitrate,

is essential to the manufacture of TNT, nitroglycerine, and similar compounds. Natural, concentrated sources of nitrates are uncommon—in fact, one of the few known reservoirs of nitrate in the early twentieth century was the arid desert of Chile (Image 10.20), with its huge saltpeter ($NaNO_3$) deposits.

During the First World War, Germany found the British Navy firmly ensconced between them and Chile, leaving the Germans in desperate need of an alternate source of reactive nitrogen. **Fritz Haber, a German chemist, solved the problem by using high pressure and high temperature to accomplish what hitherto only soil bacteria had been able to do—he fixed nitrogen:**

$$N_{2(g)} + 3\,H_{2(g)} \rightarrow 2\,NH_{3(g)}$$

This reaction, initially called the Haber process but later renamed the Haber-Bosch process in honor of the chemist who industrialized it, had far-reaching ramifications. Through World War II, the United States was among the many countries manufacturing huge quantities of ammonium nitrate (NH_4NO_3) for explosives production.[11] At the end of the war, the United States found itself sitting on a large pile of reactive nitrogen, with a greatly reduced need for explosives. More importantly, a generation of young men coming home from the war needed jobs during peacetime. The problem of what to do with the ammonium nitrate—and how to create new jobs for the soldiers—was eventually solved by maintaining and expanding the ammonium nitrate factories, putting the young men to work in manufacturing, and distributing the product to farmers as fertilizer. This particular policy forever changed not only how food is grown, but from what it's grown.

NOTE[11]

The fixation process complete, transformation of ammonia to ammonium (and to nitrate) is a synthetically trivial process, easily accomplished in the lab.

Prior to the advent of chemical fertilization, plants were very nearly sources of "free" energy—they took up CO_2, water, and sunlight (a process that will be examined in greater detail in Chapter 11), they built sugars, starches, and the structural material cellulose, and they grew. While the farmer did need to spread manure from his cows or other stock on the field from time to time, and while he certainly expended personal energy in the pursuit of agriculture, no fuel was required to effect plant growth—all the necessary power was provided by the sun. Chemical fertilization changed the energetics of agriculture drastically. The Haber-Bosch process, for all its utility, is a tremendous consumer of energy. The requisite pressures (about 200 atm) and temperatures (400–500°C) to react elemental nitrogen with hydrogen are achieved only through the combustion of large quantities of fuel. In the same way that fuel is burned to run an automobile engine, fuel—generally petroleum—is now burned to fix nitrogen. **The production and utilization of chemical fertilizer shifted the energy balance of food production, for the first time, into a deficit: given that we burn fuel to make fertilizer, it takes more than a calorie of petroleum to produce a calorie of food.**

Industrial nitrogen fixation has had other effects as well. Cheap, abundant fertilizer encourages overfertilization of fields, leading to huge quantities of reactive nitrogen entering the soil. Some of this reacts to form NO_2, contributing to acid rain. Reactive nitrogen is also washed out of the soil through the process of field irrigation and can enter local municipal water supplies. Nitrite, an anion that forms from nitrate in aqueous solution, reacts with hemoglobin in the blood when nitrite-containing water is consumed; the reaction renders hemoglobin less able to carry oxygen to the tissues. Young babies who drink formula made with nitrite-rich water are particularly vulnerable and can suffer from potentially fatal "blue baby syndrome," a form of chemical suffocation. Also, reactive nitrogen washed off fields eventually works its way into lakes, rivers, and the ocean, where it acidifies aqueous environments and kills aquatic organisms. It also contributes to huge blooms of algae; the tremendous algal proliferation depletes oxygen in the water, suffocating other aquatic life. Finally, recall that many reactive nitrogen species are tropospheric pollutants (and greenhouse gases, as we'll learn in Chapter 13)—application of chemical fertilizer increases tropospheric concentrations of NO, NO_2, and N_2O.

The end result of the Haber-Bosch process is not to increase the number of atoms of nitrogen in the cycle—to do so is simply not possible, as atoms of elements can't be created through chemical processes. **The Haber-Bosch process accelerates the fixation of nitrogen without accelerating the denitrification reactions that restore N_2.** There is no concern that we will run out of N_2—it's tremendously abundant and serves merely as a "holding tank" for nitrogen atoms. Rather, the acceleration of nitrogen fixation increases the concentrations of reactive nitrogen species in the world, with myriad effects. The Haber-Bosch process has significantly accelerated global nitrogen fixation, and has become the major mechanism for formation of reactive nitrogen in the developed world. To illustrate, a century ago, every atom of nitrogen in an individual's body would have been fixed by soil or root-nodule bacteria. Today, assuming you are a resident of an industrialized nation and eat a typical diet, only about 2% of the nitrogen in your body was fixed by bacteria: the remaining 98% was fixed in an industrial lab. This is an incredible—an unimaginable—perturbation to the nitrogen cycle, even more so as it's happened in the course of only 100 years. In essence, industrialized nations have increased the rate at which nitrogen is being fixed into reactive forms by 5,000%.

Unfortunately, where science and society intersect on a global scale, complications abound. It would be simplistic to characterize the Haber-Bosch process as having a solely negative effect upon humanity. With the global population increasing rapidly, starvation and the availability of food are pressing issues. There is no doubt that chemical fertilization increases the growing capacity of the land—crop yields per acre are now as much as 15 times what they were 100 years ago—which increases the quantity and decreases the price of food. It's possible that without the process population would have stayed smaller. Perhaps more likely, given scientific advances that promote both fertility and longevity, starvation and disease would have become more common and widespread. Which of these scenarios—or a combination of the two—would have occurred in the absence of the Haber-Bosch process is anyone's guess, but it's a moot point: the genie is out of the bottle and can't be enticed back in again.

Speculation aside, there are lessons here that bear elaboration. First, as we saw with CFCs, it is difficult, and often impossible, to anticipate all the ramifications of a new technology. Second, global cycles are interconnected, and a disturbance in one arena tends to show its effects elsewhere: nitrogen fixation, for instance, is a thread connecting pollution and ozone, energy, and climate change. Finally, many of our global dilemmas were conceived as solutions to problems, and have since become the problems for which we are seeking solutions. CFCs allowed for population expansion and industrial growth, at the expense of the ozone layer. HFCs appeared to be promising replacement refrigerants until attention turned to global warming. The same nitrogen-fixing technology that has fed the world since the 1940s is now revealing itself to be responsible, in whole or in part, for a number of environmental and health-related problems. Nevertheless, solutions are not easy to come by or implement, and technological advancements become inextricably integrated into our way of life very quickly. **As much as we owe our global dilemmas—the ozone hole and climate change, acid rain and pollution, destruction of aquatic life and deforestation—to technology, we must remember that the innovations responsible for these dilemmas were developed and implemented for a reason.** Addressing the dilemmas, then, becomes very complicated indeed.

10.10   The Precautionary Principle Revisited

In our discussion of tetraethyl lead (Chapter 6), we saw an early attempt to use the precautionary principle as a means of forestalling an innovation for which there was sound scientific theory—but little existing evidence—of risk. A concerned physician suggested that the burden of proving tetraethyl lead's safety lay upon those wishing to manufacture and distribute it and that, in the absence of proof, caution should prevail. In the end, the substance was approved on the grounds that there was no hard evidence of risk, and history bears witness to the result: theory became reality as lead polluted the ground, air, and water.

Environmental advocates often point to this example to demonstrate the power of the precautionary principle: had it been heeded, lead pollution and lead-related deaths would have been avoided. This is a compelling argument. Looking back, few would suggest that tetraethyl lead should ever have made it out of the development phase and into the hands of the public. **Of course, in hindsight, such judgments are easily made: it's much harder to accurately assess risk when a technology is new.**

The precautionary principle's strength in hindsight is its greatest weakness when looking forward. As opposed to risk–benefit analysis, which takes into account the contribution an innovation will make, the precautionary principle focuses solely on risk. Further, it demands a logical impossibility: the proof that no risk exists. In reality, no activity or innovation is without risk, and had the precautionary principle been applied rigorously throughout the course of human civilization, we could not have progressed beyond building fires in caves—if we had ever allowed ourselves to build fires at all! Risk is unavoidable, inherent in day-to-day life, and certainly inextricable from technological advancement. **A risk–benefit analysis makes allowances for the inevitability of risk—in performing such an analysis, we compare an innovation's contribution to its potential or anticipated potential for harm, and we acknowledge and accept a certain degree of risk.** As discussed earlier in this chapter, a risk–benefit analysis is most powerful when it is most flexible: we do well to adapt our analysis as new evidence presents itself, and where new information shifts the risk-to-benefit ratio, we alter our actions accordingly.

Different paths appear sensible depending upon the type of analysis—precautionary principle or risk–benefit—used to determine a course of action. For instance, a 2001 United Nations convention in Stockholm addressed the use of the pesticide dichlorodiphenyltrichloroethane (DDT). Originally utilized during the second half of World War II, DDT was highly effective at preventing epidemics of the diseases typhus and malaria, both of which are spread by mosquito vectors and which threatened U.S. troops in the South Pacific (Image 10.21).

Later, DDT was used as an agricultural pesticide with remarkable effectiveness. Regulation was nonexistent, the chemical was inexpensive, and application of large quantities of the compound to huge areas of agriculture was logistically facile; these factors combined to result in U.S. farmers applying up to 25 times the necessary quantity to crops. Elsewhere—in the United Kingdom, for instance—DDT was used more judiciously. As a result, the negative environmental impact of DDT in the United States was avoided elsewhere. A *persistent organic pollutant*, DDT concentrates in the soil and degrades slowly. The compound is stored in fatty tissue and bioaccumulates, meaning that it concentrates up the food chain. Insects exposed to nonlethal doses of DDT store the chemical. The insects are eaten by small birds,

IMAGE 10.21. A U.S. soldier being sprayed prophylactically with DDT.

which end up storing the sum of all DDT from their insect prey. Large predator birds, then, concentrate the compound further as they feed on small birds, and so forth. The result of this bioaccumulation is that animals nearest the top of the food chain are exposed to the highest doses of the chemical. In 1962, the American biologist Rachel Carson published *Silent Spring*, an exposé of DDT's environmental impact. Carson revealed a variety of phenomena, including plummeting populations of a variety of birds—raptors, songbirds, and waterfowl among them—as the result of catastrophically thinned eggshells due to DDT's impact upon the birds' metabolisms. A U.S. ban on the insecticide followed shortly thereafter, and in the time since, the environment has largely recovered. The United Kingdom continued using DDT for many years after the U.S. ban—in the lower doses they'd utilized all along—and while wildlife testing revealed traces of the compound in large predators, no wildlife deaths or ill effects were ever recorded.

Unfortunately, many replacements for DDT are more toxic, more expensive, or less effective. This has not presented a problem in developed countries—which relied on the compound mainly as an agricultural pesticide and which have the economic means to utilize more expensive substitutes—but developing nations have struggled to curtail DDT use. The UN restricts using DDT for widespread agricultural spraying on the grounds that it disseminates large quantities of the compound, has the potential to adversely affect the environment, and can lead to insecticide resistance in mosquitoes. However, developing nations continue to find the insecticide highly effective in the prevention of malaria. It was this use of DDT—in which small amounts of the chemical are sprayed or painted on interior house walls—that was addressed in the 2001 Stockholm Convention.

Using the precautionary principle as a guide, one would be left with no choice but to categorically ban all uses of DDT: the risks are clear and unequivocal, the compound is pollutant and toxic, and the dangers to wildlife are known. While human health effects following exposure to low concentrations of DDT (such as through normal use) have never been observed, at high doses, DDT is known to cause a variety of ill effects in humans, including vomiting and seizures. In lab animals, DDT negatively impacts reproduction and liver function and, extrapolating from lab studies, it is considered a possible human carcinogen.

A risk–benefit analysis, however, leads us to a somewhat different conclusion. While the benefits of agricultural use of DDT are outweighed by the known risks, given the amount utilized, the risk of interior wall spraying or painting is much lower, since far less chemical is used. Further, the benefits are tremendous. With nearly 300 million cases of malaria per year, a million of which lead to death (mostly in Africa, and mostly among children), there is no doubt that malaria is a major public health concern. Mosquito-borne illness is also becoming more prevalent as global warming occurs (as we'll discuss further in Chapter 13), making mosquito-borne disease prevention a topic of global importance. Several countries including South Africa and Ecuador have reduced antimalarial DDT use in response to global pressure, only to watch the number of cases of the disease skyrocket. No other pesticide has the combination of effectiveness, relative safety, and low cost afforded by DDT. After careful consideration, the Stockholm Convention—to which there are currently 169 parties—determined to allow the use of DDT for disease vector control, while stipulating strict safety measures and routine reevaluation of the risk-to-benefit ratio. It is nearly impossible to estimate the number of lives that have been saved through this decision.

In facing global challenges, there is much we can learn from the Montreal Protocol, which serves not simply as an important and successful treaty, but as an example of the need for flexibility and adaptation. **In accepting that technology is critical to the advancement of our society, we must remember that risk is ubiquitous—no activity, technology, or chemical is without it—and that advancement comes only through conceding to acceptable risk.** We must constantly revise our risk–benefit analysis as risks, which can never be fully anticipated, present themselves, without forgetting that once a technology is introduced, our way of life is transformed, and it can be difficult or impossible to backtrack. We must be willing to add or subtract technologies from our "phaseout" list as new information becomes available, and we must be sensitive to the burden this places on the developing world.

10.11 The Last Word—Sunlight in the Troposphere

For all the importance of the ozone layer in protecting us from too much of the sun's ultraviolet rays, life on Earth would not be possible without our star and the full spectrum of its radiation. While UV rays can destroy tissue, causing sunburn and worse, they also allow for the synthesis of vitamin D from cholesterol in the skin. UV's ability to break bonds leads to chemical reactions—some to our detriment and some that we couldn't survive without.

UV is not the only radiation from the sun, however. Also emitted is infrared (IR) radiation, which plays an important role in maintaining the planet's livable temperature, as we'll see in Chapter 13. Perhaps most familiar to our everyday experience, though, is the visible light from the sun. This provides a mechanism for sensing our environment, is the fuel that powers plant growth, and translates into the rich palette of colors that paints our world. What we observe as "colorless" light is, in fact, a mix of the entire visible spectrum. Our eyes respond to wavelengths from approximately 380 nm (violet, highest in energy) to 750 nm (red, lowest in energy). If you're curious, it was Isaac Newton who determined that a rainbow contains seven colors: red, orange, yellow, green, blue, indigo, and violet. There are, of course, an infinite number of colors in the visible spectrum, but Newton thought seven had a nice symmetry to it; it is the same as the number of notes in a musical scale, and that pleased him … and because he was Isaac Newton, we all accept his decision. Our ability to see in color comes from the fact that some molecules absorb light, reflecting the portion of the visible spectrum they do not absorb. A molecule that strongly absorbs wavelengths other than those of red light will reflect the red,

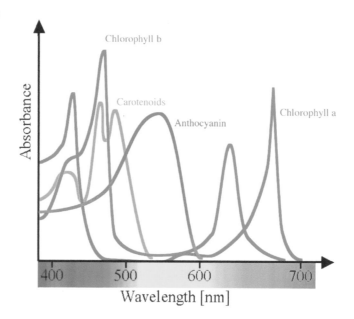

and we'll see the substance as being red in hue. Some molecules do not absorb visible light, and hence appear colorless (or white, if they reflect the entirety of the visible spectrum). For this reason, the air around us—mostly N_2 and O_2, neither of which absorbs in the visible portion of the spectrum—has no color. Substances that absorb the entire visible spectrum, reflecting none of it, appear black to us.

Tree leaves have a familiar green hue during spring and summer months. A major light-absorbing molecule in leaves is *chlorophyll*. Chlorophyll appears green not because it absorbs green light, but because it absorbs other colors of visible light, leaving green (and yellow) behind (Image 10.22) to reflect into the eye of the observer. The absorbed light, incidentally, is used by the plant as a source of energy to put carbon dioxide and water together, producing sugar, which plants can later burn for energy or use to build the structural material cellulose.

Chlorophyll, it turns out, is not the only pigment present in a leaf. A second class of pigments is the *carotenoids*, so named for their presence in carrots, and with a predictable color given the source. Carotenoids assist chlorophyll in collecting sunlight (this will be discussed more in Chapter 11), but their absorption spectrum is somewhat different than that of chlorophyll. Rather than absorbing orange and red light, as chlorophyll does, carotenoids absorb only violet, blue, and green. Generally, we don't see the color of carotenoids in leaves—the orange and red light they reflect is absorbed by chlorophyll, and neither molecule absorbs much yellow. Where carotenoids make their appearance is in the fall.

Dropping temperatures and decreasing light cause trees to slow their growth in fall and winter months. Chlorophyll, normally produced in large amounts and regularly recycled by the plant, is no longer synthesized and starts to degrade. As chlorophyll concentrations in the leaves decrease, the orange and red light normally absorbed by the molecule is instead reflected. Carotenoids persist longer than chlorophyll in the leaf, meaning that for a brief and dramatic period at the end of the

IMAGE 10.23. Fall color brings out brilliant yellows in aspen leaves and fiery reds in maple. (Aspens © Scott Lefler, 2007. Maple leaves © Scott Lefler, 2007.)

growing season and before the leaves are lost for the winter, the trees sport a fiery palette of yellow, orange, and (as the result of yet another class of pigments) red (Image 10.23).

The anthocyanins, though not present in all leaves, are undoubtedly the most dramatic-appearing of the leaf pigments. These are not generally produced during the growing season, nor are they manufactured by all species. Instead, synthesized from residual sugar in the leaf at the beginning of fall as a sort of arboreal sunscreen, the anthocyanins absorb the highest-energy visible light—violet through yellow—leaving behind the more benign orange and red wavelengths. The purpose seems to be to prevent excess light (which is no longer harvested to power growth as the tree downregulates its metabolism for the winter) from being absorbed by and damaging the tissues of the leaves. It is an eminently practical, and breathtakingly beautiful, phenomenon.

Our relationship with the sun is a bit like our relationship with technology: our star is at once essential and detrimental, exposing us to countless risks and providing us with incalculable benefits. It is creative and destructive, providing for life and taking it away again. Complicated and varied in its effects, our sun, like technology, is absolutely essential to our way and experience of life on our blue-green (violet and yellow, and red and orange) planet.

SUMMARY OF MAIN POINTS

- Refrigeration depends upon a cycle of vaporization (absorbs heat) and condensation (releases heat) of a refrigerant liquid.
- Good refrigerants must be not be toxic, reactive or explosive, and must have boiling points less than the melting point of water.
- The Freons (CFCs) were developed by Thomas Midgley in the late 1920s as a class of refrigerants; they are very chemically inert.
- The ozone layer is located in the stratosphere and shields us from UV light.
- UV is a form of EMR: all EMR has the same speed (written c in equations) and an inversely related wavelength and frequency.

- The wave equation $c = \lambda v$ relates measurable quantities for EMR.
- The energy of EMR is directly related to frequency by the equation $E = hv$, where $h = 6.626 \times 10^{-34}$ J×s.
- High-energy EMR (gamma rays, X-rays, and UV) are ionizing radiation and can break bonds in molecules (including DNA, leading to cancer).
- Low-energy EMR (visible light, IR, microwaves, and radio waves) is not ionizing radiation and can't break bonds.
- The sun emits UV, visible light, and IR; the UV is far too intense to be compatible with life in the absence of the ozone layer.
- The Chapman Cycle describes the dynamic equilibrium whereby ozone is converted to O_2, and O_2 back to ozone—both portions of the cycle require the absorption of UV light.
- Because of the Chapman Cycle, O_2 in the upper atmosphere absorbs much of the UV-C and UV-B incident upon the earth, while O_3 absorbs UV-B.
- No UV-C, very little UV-B, and a fair amount of UV-A (which is the least damaging of the three types of UV light) reach the surface of the earth.
- CFCs react with UV light, producing chlorine atoms which catalytically convert ozone to O_2.
- The idea that CFCs led to ozone destruction was theoretical at first, then verified by the discovery of the hole in the ozone layer.
- The ozone hole formed over Antarctica because it is the coldest place on Earth. The hole shows seasonal variation because polar stratospheric clouds form in winter, trapping chlorine and allowing for catalytic ozone destruction in the austral spring and summer.
- The Montreal Protocol and its amendments have put restrictions on CFCs and other ODS. Restrictions are more stringent for developed rather than developing nations.
- HCFCs were accepted as temporary substitute refrigerants, but they also deplete the ozone layer.
- HFCs do not deplete the ozone layer, but they are powerful greenhouse gases.
- Greenfreeze, a hydrocarbon refrigerant, has no potential to deplete the ozone layer or to affect the global climate and has been widely accepted in many countries (but not in the United States).
- The Montreal Protocol has begun to show a positive effect upon the ozone layer, with full recovery by 2065 (assuming full compliance with the Montreal Protocol).
- Nitrogen is required by all living organisms but must be fixed to a reactive form by soil bacteria before it can be assimilated.
- The Haber-Bosch process is an energy-dependent industrial process for fixing nitrogen that has greatly perturbed the global nitrogen cycle.
- Food production dependent upon chemical fertilizer represents a negative energy balance.

- The Haber-Bosch process has been responsible for greatly increasing global food production, which has had a positive effect upon starvation and disease worldwide.
- The precautionary principle is an examination of risk without consideration of benefit and fails to allow for the inherent risk associated with any technological advancement.
- Risk–benefit analysis can be used to weigh risks, which are unavoidable, against benefit in the attempt to produce the best possible solution to a global dilemma.

QUESTIONS AND TOPICS FOR DISCUSSION

1. How is the relationship between chemistry and society different on the global level than on the individual or community level? How is it the same?

2. What does it mean to say, "When a chemical, an invention, a discovery, or a technology emerges from the lab and debuts in the world, we are often greeted with unanticipated effects"? Is it possible to remove the element of surprise?

3. On a molecular or atomic level, why do you think vaporization is endothermic? What is happening? (Hint: think about intermolecular interactions.)

4. Why is the air behind a refrigerator warm?

5. Why is it important that a refrigerant gas not be explosive or toxic?

6. What are some of the major ways in which refrigeration changed the world for the better? What would the world be like without it?

7. How is EMR similar to an ocean wave? How is it different?

8. Based only upon our discussion of EMR so far, do you think exposure to microwaves can cause cancer? Why or why not?

9. Why is it helpful to think of a molecular bond as a spring? In what ways is this accurate?

10. What is a dynamic equilibrium, and why is it important to our discussion of ozone?

11. Why are CFCs so destructive to ozone?

12. What does it mean to say that chlorine is a catalyst for the conversion of ozone to oxygen?

13. A friend says that ozone destruction by CFCs is not actually a problem, on the grounds that the product of ozone destruction by chlorine is not dangerous—element oxygen is produced—and that sunlight can just convert the oxygen back into ozone. Evaluate these claims.

14. Why does more UV-A reach Earth's surface than UV-C?

15. Sometimes indoor tanning salons promote UV-A-only tanning beds as being safe because UV-A rays don't cause sunburn (this is because the body does not "recognize" the effect of UV-A, and a sunburn is the result of cellular inflammation and immune system activation in response to the recognized damaging effects of UV-B exposure). Evaluate the legitimacy of this claim.

16. Some indoor tanning salons claim that a certain amount of tanning is healthy, because UV light promotes vitamin D synthesis. Evaluate the legitimacy of this claim.

17. Why do you think that the initial public response to the *New York Times* article about ozone description didn't have a long-term effect?

18. Why did NASA satellite data miss spotting the ozone hole?

19. What elements of the Montreal Protocol made it successful? What lessons can we learn from this?

20. Why do you think that several nations are attempting to add HFCs to the list of chemicals banned by the Montreal Protocol? Are they ODS? If not, what might be the justification?

21. What are the special challenges associated with international environmental treaties? How can they be overcome?

22. What did the Montreal Protocol do to help developing nations cope with the hardships that they would face as a result of signing the Montreal Protocol? Do you think these actions were necessary? Not enough? Too much?

23. Why is it especially important to be sensitive to the needs of and hardships faced by developing nations when a substance or technology is removed from the marketplace? Try to think in terms beyond simple economic hardship and research limitations.

24. Has the Montreal Protocol been successful? Explain your reasoning.

25. Why are HCFCs less ozone depleting than CFCs? Why are HFCs not ozone depleting? Why are HFCs considered a "dangerous" substance?

26. What does the response to Greenfreeze in Germany and in the United States suggest about its potential as a refrigerant and about the politics of environmental science?

27. What is the nitrogen cycle, and why is it important?

28. Should the Haber-Bosch process be banned? Defend your position.

29. Apply the precautionary principle to an existing technology (microwave ovens, airplanes, vaccines). According to the principle, should the technology have been implemented?

30. Apply risk–benefit analysis to an existing technology (microwave ovens, airplanes, vaccines). According to the analysis, should the technology have been implemented?

31. Do you think that the precautionary principle has utility in global decision-making? Why or why not?

32. What does it mean to say, "The precautionary principle's strength in hindsight is its greatest weakness when looking forward"? Illustrate with an example.

33. How did the UN come to a decision regarding DDT at the Stockholm convention? What guided decision making, the precautionary principle, or risk–benefit analysis? If the other form of decision making had been used, how might the outcome have been different?

34. Do you agree with the UN's decision regarding acceptable and unacceptable uses of DDT? Why or why not?

PROBLEMS

1. Define: condensation, vaporization, CFC.

2. Name several ways in which CFCs changed the world (for the better).

3. What is an allotrope?

4. Why do we say ozone is protective in the stratosphere, but a pollutant in the troposphere?

5. Define: EMR; wavelength; and frequency.

6. What is the ramification of all EMR traveling at the same speed?

7. Place the following waves in order from highest frequency to lowest frequency: Wave 1, $v = 1.1 \times 10^{22}$ Hz; Wave 2, $v = 5.6 \times 10^{15}$ Hz; Wave 3, $v = 7.6 \times 10^{3}$ Hz.

8. Place the following waves in order from highest frequency to lowest frequency: Wave 1, $\lambda = 5.5 \times 10^{2}$ m; Wave 2, $\lambda = 3.1 \times 10^{-6}$ m; Wave 3, $\lambda = 7.5 \times 10^{0}$ m.

9. Place the following waves in order from longest wavelength to shortest wavelength: Wave 1, $\lambda = 1.2 \times 10^{-15}$ m; Wave 2, $\lambda = 9.1 \times 10^{-4}$ m; Wave 3, $\lambda = 8.2 \times 10^{-11}$ m.

10. Place the following waves in order from longest wavelength to shortest wavelength: Wave 1, $v = 8.4 \times 10^{0}$ Hz; Wave 2, $v = 5.1 \times 10^{20}$ Hz; Wave 3, $v = 2.2 \times 10^{12}$ Hz.

11. What is the frequency of an EMR wave with $\lambda = 5.4 \times 10^{-16}$ m?

12. What is the wavelength of an EMR wave with $v = 4.1 \times 10^{16}$ Hz?

13. Place the following waves in order from highest energy to lowest energy: Wave 1, $v = 3.6 \times 10^{12}$ Hz; Wave 2, $v = 1.1 \times 10^{8}$ Hz; Wave 3, $v = 2.4 \times 10^{20}$ Hz.

14. Place the following waves in order from highest energy to lowest energy: Wave 1, $\lambda = 7.0 \times 10^{-10}$ m; Wave 2, $\lambda = 6.3 \times 10^{-20}$ m; Wave 3, $\lambda = 7.1 \times 10^{8}$ m.

15. What is the energy of an EMR wave with $v = 8.4 \times 10^{12}$ Hz?

16. What is the energy of an EMR wave with $\lambda = 7.6 \times 10^{-14}$ m?

17. What is the frequency of an EMR wave with $E = 4.1 \times 10^{-15}$ J?

18. What is the wavelength of an EMR wave with $E = 8.6 \times 10^{-20}$ J?

19. Describe the effect of UV light on bonds in molecules.

20. Why does UV exposure lead to premature aging and cancer?

21. Why can gamma rays cause blood and deep tissue cancers, while UV rays can only cause skin cancers?

22. Why is the ozone layer important? Is ozone the only important component of the ozone layer?

23. How is the ozone layer maintained naturally? What is/are the overall reaction(s)?

24. What are the three types of UV, and which is most damaging? Which is highest in energy? Which has the longest wavelength? The lowest frequency?

25. Which type of UV light reaches the earth in greatest abundance? In least abundance?

26. Describe the reaction that a CFC undergoes in the stratosphere.

27. Why do CFCs cause catalytic destruction of ozone?

28. Why is the hole in the ozone layer located over Antarctica?

29. Why does the ozone layer get larger in the austral summer?

30. What alternatives were or are available to CFCs? What are the problems associated with each? Which is the most promising?

31. Why is it going to take so long for the ozone hole to repair itself?

32. What is the difference between reactive and nonreactive nitrogen?

33. Is elemental nitrogen reactive or nonreactive? Explain.

34. What is the Haber-Bosch process, and what does it do?

35. What's the most significant natural source of reactive nitrogen? The most significant *overall* source?

36. What are some of the negative effects of chemical fertilization?

37. What does it mean to say that the energy balance of food production is negative?

38. What is the difference between the precautionary principle and risk–benefit analysis?

1. Condensation is gas turning to liquid; vaporization is liquid turning to gas; CFCs are chlorofluorocarbons—they were introduced as refrigerants in the 1920s.

3. An allotrope is an alternate form of an element.

5. EMR is electromagnetic radiation, an electrical and magnetic disturbance in space; wavelength is a wave's distance from crest to crest; frequency is the number of waves past a given point in a given period of time.

7. Wave 1, $v = 1.1 \times 10^{22}$ Hz; Wave 2, $v = 5.6 \times 10^{15}$ Hz; Wave 3, $v = 7.6 \times 10^3$ Hz.

9. Wave 2, $\lambda = 9.1 \times 10^{-4}$ m; Wave 3, $\lambda = 8.2 \times 10^{-11}$ m; Wave 1, $\lambda = 1.2 \times 10^{-15}$ m.

11. $v = 5.6 \times 10^{23}$ Hz.

13. Wave 3, $v = 2.4 \times 10^{20}$ Hz; Wave 1, $v = 3.6 \times 10^{12}$ Hz; Wave 2, $v = 1.1 \times 10^8$ Hz.

15. $E = 5.6 \times 10^{-21}$ J.

17. $v = 6.2 \times 10^{18}$ Hz.

19. UV light breaks bonds in molecules by "overstretching" bonds or by ionization.

21. Gamma rays can penetrate much more deeply, and break bonds in deep tissue DNA.

23. $3 \, O_2 \rightarrow 2 \, O_3$ (high energy UV), $2 \, O_3 \rightarrow 3 \, O_2$ (lower energy UV).

25. UV-A reaches the earth in greatest abundance; UV-C reaches in least abundance.

27. Because Cl is released at the end of the reaction sequence, it can react with more ozone; one CFC can destroy hundreds of thousands of ozone molecules.

29. When the sun warms the atmosphere, ozone-depleted air circulates away from the pole and ozone-rich air from surrounding areas moves in, leading to more ozone destruction.

31. Stratospheric chlorine has a long lifetime, and it's not possible to remove it from the atmosphere. Additionally, old refrigerators and cooling systems will continue to release CFCs and other ODS even once they're discarded and have been replaced by newer systems.

33. Elemental nitrogen is nonreactive because of the very stable triple bond.

35. The most important natural source is nitrogen-fixing bacteria. The most significant overall source is the Haber-Bosch process.

37. When more than a calorie of energy is invested in the form of fuel to produce a calorie of food, the energy balance of food production is negative.

SOURCES AND FURTHER READING

Ozone Layer

http://ozonewatch.gsfc.nasa.gov/

Cicerone, R. J., Stolarski, R. S., and Walters, S. (1974). Stratospheric ozone destruction by man-made chlorofluoromethanes. *Science, 185*(4157), 1165–67.

Crutzen, P. (1974). Review of upper atmospheric photochemistry. *Canadian Journal of Chemistry, 52*(8), 1569–81.

Farman, J. C., Gardiner, B. J., Shanklin, J. D. (1985). Large losses of total ozone in Antarctica reveal seasonal ClOx/NOx interaction. *Nature, 315*(6016), 207–10.

Grabiel, D. F. (2007, September). Crucial crossroads. *Our Planet: The Magazine of the United Nations Environment Programme.*

James, R. W., and Missenden, J. F. (1992). The use of propane in domestic refrigerators. *International Journal of Refrigeration, 15*(2), 95–100.

Kowalok, M. E., (1993). Common threads—research lessons from acid rain, ozone depletion, and global warming. *Environment, 35*(6), 12–38.

Molina, M. J., and Rowland, F. S. (1974). Stratospheric sink for chlorofluoromethanes—chlorine atomic-catalyzed destruction of ozone. *Nature, 249*(5460), 810–12.

Stolarski, R. S., and Cicerone, R. J. (1974). Stratospheric chlorine—possible sink for ozone. *Canadian Journal of Chemistry, 52*(8), 1610–15.

Sullivan, W. (1974, September 26). Tests show aerosol gases may pose threat to Earth. *New York Times,* p. A1.

Nitrogen Fixation

Le Couteur, P., and Burreson, J. (2003). *Napoleon's buttons: 17 molecules that changed history.* New York, NY: Penguin Group.

Pollan, M. (2006). *The omnivore's dilemma.* New York, NY: Penguin Group.

Smil, V. (2001). *Enriching the earth: Fritz Haber, Carl Bosch, and the transformation of world food production.* Massachusetts Institute of Technology.

DDT and Malaria

www.who.int/topics/malaria/en/

Carson, R. (1962). *Silent spring.* New York, NY: Houghton Mifflin.

Timbrell, J. (2005). *The poison paradox.* Oxford: Oxford University Press.

United Nations Environment Programme. (2001). *Stockholm convention on persistent organic pollutants (POPs).* UNEP/Chemicals/2001/3. 50p.

"Current" Events—
Electrochemistry

11

Electricity is so commonplace that we rarely stop to think about what it is and how it's generated. Upon reflection, we know it's a form of energy, since it can be used to power machines. We can even surmise that electricity must therefore be transformed from another form of energy, because the first law of thermodynamics reminds us that energy can never be created or destroyed. Still, our everyday experiences with electricity tend to leave us a bit mystified as to its true nature—lightning strikes seem to come out of thin air and vanish without a trace. A mere flip of a switch turns a flashlight on and just as quickly turns it back off again. What's really going on in the clouds or inside the device to produce these events? Why do a few AA batteries make a flashlight work—and why can we recharge some batteries, but not others? Finally,

IMAGE 11.1. Lightning (a), power lines (b), and batteries (c) are just a few of the phenomena and objects associated with electrical energy.

how can sunlight be used to produce electricity, and is there a future for solar power as an alternative energy source? In order to answer these questions, we must first look at the nature of electrical energy.

11.1 Redox

We saw in Chapter 8 that a box held above the ground has gravitational potential energy: the box is at a higher energy state when separated from the earth, and the force of gravity will act upon the box to move it toward a lower energy state. Like any high-energy to low-energy transition, this happens spontaneously, with a release of energy. In much the same way, the electrostatic force between charged particles can result in particle motion. If separated from a negative charge, a positive charge has **electrical potential energy** due to the electrostatic attraction between particles. The charges will spontaneously rush toward each other, and as they start to move, their potential energy is transformed first into energy of motion (kinetic energy), and then into other forms of energy, depending upon the specific situation. We generally refer to moving charge simply as a **current**, which has **electrical energy**. The only difference between this scenario and that of the falling box is that electrostatic forces are responsible for both attraction and repulsion, depending upon the charges involved. For instance, two negatively charged particles near each other in space also have electrical potential energy—their proximity represents a high-energy state, and they will spontaneously move away from each other. The same is true for two positively charged particles. In the end, though, the result is the same: potential energy can be transformed into other forms of energy, and used to do work. Of course, electrical energy has to come from somewhere (all energy does!), and one common source is chemical energy.

Not all chemical reactions have the potential to create electrical energy. In fact, while all reactions involve electrons (in the sense that bonds are broken and formed), only some involve the actual transfer of electrons from one atom to another. **It is this transfer of electrons that can be used to generate electrical energy.** Reactions involving electron transfer have two components. One atom or molecule must lose electrons—this process is called **oxidation**. Another atom or molecule must gain electrons through a process called **reduction**.[1] Let's revisit an old reaction that we saw for the first time in Chapter 4—the formation of ions. We looked at the ionization of Na and F, which took place through an electron transfer:

It wouldn't be possible for Na to form the Na⁺ ion without F (or some other atom) present to accept the electron—electrons can't just disappear! Further, F couldn't ionize to F⁻ without the Na (or some other atom) present to donate an

NOTE[1]

There's a mnemonic to help you remember this—*LEO the lion says GER. LEO* stands for "Loss of Electrons—Oxidation," and *GER* stands for "Gain Electrons—Reduction." Apparently, aside from helping us remember which reaction is which, we also learn from this mnemonic that LEO the lion can't spell!

electron, since electrons can't come out of thin air. In fact, the reaction between Na and F is an oxidation/reduction reaction in which Na is oxidized and F is reduced. This same logic holds true for all electron transfer reactions: **oxidations and reductions must always occur in pairs**. For this reason, electron transfers are often called **redox** reactions, which is shorthand for reduction/oxidation.

We could write out a (theoretical)[2] redox reaction between Na and F as a complete chemical reaction:

$$Na + F \rightarrow Na^+ + F^-$$

Alternately, we could keep explicit track of the electron being transferred in the reaction by separating it into two parts:

$$Na \rightarrow Na^+ + e^-$$

$$F + e^- \rightarrow F^-$$

The above are called **half-reactions**, and they give us insight into the electron transfer taking place. We know sodium is losing an electron, because we understand the principles of ionization. However, even if we didn't understand the principles of valence that drive sodium's loss of its outermost electron, we could still surmise (using conservation of charge) that for Na to become Na^+, a single electron must be lost. Similarly, even without our understanding of ionization, conservation of charge tells us that F can only become F^- through gaining an electron. **Half-reactions can be written from redox reactions simply by analyzing the change in charge**. It's important to note that correctly written half-reactions always sum to the original redox equation. Further, the oxidation half-reaction (the one in which electrons are a product) must always contain the same number of electrons as the reduction half-reaction (the one in which electrons are a reactant) in order to sum to a correctly balanced chemical equation. This may result in our needing to do a bit of multiplication across one (or both) of the half-reactions in order that the electrons balance with one another. Let's revisit another familiar ionization reaction, this time finding the overall balanced equation by determining and summing the individual half-reactions. The reaction that takes place between Mg and F—and the half-reactions, and which species is oxidized, and which is reduced, and how many electrons are transferred—can all be determined by following a series of steps:

1. Determine the individual half-reactions involved:

We can use principles of ionization—an understanding of how many electrons each atom needs to gain or lose to have a full octet—to determine the relevant half-reactions.

$$\text{Oxidation: } Mg \rightarrow Mg^{2+} + 2\ e^-$$

$$\text{Reduction: } F + e^- \rightarrow F^-$$

CONCEPT CHECK

How do we know that sodium is losing one electron when it reacts with fluorine?

Answer: It has one electron in its valence; it forms a cation with a charge of +1.

NOTE[2]

I say theoretical because if we were to do this reaction in the lab, we'd use F_2 instead of atomic fluorine, which is very unstable. However, the mechanics of electron transfer are identical in both cases, and it's a little easier to focus on the important bit— the electron transfer—if we look at a simple case of atoms.

Note that these half-reactions do not contain equal numbers of electrons. When we looked at the magnesium and fluorine reaction in Chapter 4, we used Lewis structures to help us determine how many of each atom would be involved in a balanced reaction. Now, however, we can use the half-reactions: as long as the number of electrons lost in the oxidation half-reaction is equal to the number required in the reduction half-reaction, our half-reactions will add up to a correctly balanced overall equation.

2. Multiply one or both half-reactions by a whole number value to obtain equal numbers of electrons in the two half-reactions:

In this case, we multiply the fluorine reaction by 2; now both half-reactions involve two electrons.

$$2\,F + 2\,e^- \rightarrow 2\,F^-$$

3. Add the half-reactions to determine the overall reaction:

$$Mg \rightarrow Mg^{2+} + 2\,e^-$$

$$\underline{+\,2\,F + 2\,e^- \rightarrow 2\,F^-}$$

$$Mg + 2\,F + 2\,e^- \rightarrow Mg^{2+} + 2\,e^- + 2\,F^-$$

TRY THIS

Write a balanced equation for the reaction of Al with O, using the half-reaction method.

4. Just as in a regular addition problem, eliminate quantities appearing on both sides of the equation:

$$Mg + 2\,F + \cancel{2\,e^-} \rightarrow Mg^{2+} + \cancel{2\,e^-} + 2\,F^-$$

Note that this does not mean the electrons weren't involved; it just means they were transferred from Mg to F and aren't net reactants or products.

5. Rewrite the equation in its final form:

$$Mg + 2\,F \rightarrow Mg^{2+} + 2\,F^-$$

Answers:
$2\,Al + 3\,O \rightarrow 2\,Al^{3+} + 3\,O^{2-}$

Solution:
Oxidation: $Al \rightarrow Al^{3+} + 3\,e^-$
Reduction: $O + 2\,e^- \rightarrow O^{2-}$
Oxidation, multiplied through: $2\,Al \rightarrow 2\,Al^{3+} + 6\,e^-$
Reduction, multiplied through: $3\,O + 6\,e^- \rightarrow 3\,O^{2-}$; electrons in half-reactions are now equal.
Combined: $2\,Al + 3\,O + 6\,e^- \rightarrow 2\,Al^{3+} + 3\,O^{2-} + 6\,e^-$
Final: $2\,Al + 3\,O \rightarrow 2\,Al^{3+} + 3\,O^{2-}$

We can use these same principles to determine what overall reaction takes place between reactants in more complicated cases, as long as we are given the half-reactions. For instance, in nickel-cadmium batteries, electrons are transferred from cadmium to nickel:

$$\text{Oxidation: } Cd \rightarrow Cd^{2+} + 2\,e^-$$

$$\text{Reduction: } Ni^{3+} + e^- \rightarrow Ni^{2+}$$

These half-reactions look somewhat different from what we've seen so far; for one thing, both species are metallic. For another thing, the nickel reaction involves changing one ion into another ion, not into a neutral species. These curiosities, however, are incidental, and in no way affect our ability to build a balanced chemical reaction out of the two halves. Using our electron-balancing technique, we'd need to multiply the nickel reaction by two for the number of electrons to be equal. We'd then add our half-reactions and cancel out the electrons on both sides, and our full reaction would read:

$$Cd + 2\ Ni^{3+} \rightarrow Cd^{2+} + 2\ Ni^{2+}$$

Since these are both metallic elements, we might not have anticipated that they would react with one another. In fact, we know from principles of chemical bonding that they certainly can't form an ionic compound like the elements in the redox reactions we've discussed so far. However, it turns out that some metals are easier to oxidize than others—this has to do with a combination of many chemical properties—and can transfer electrons to other metals. The ease with which a chemical is oxidized is called its **activity**; more active chemicals lose electrons more easily than less active chemicals (Table 11.1).

Reactant species near the top of the table are more active, and therefore easier to oxidize than (less active) reactant species near the bottom. For instance, it's easier to oxidize Na to Na^+ than it is to oxidize Au to Au^{3+}.[3] Conversely, product species near the bottom of the table are easier to reduce than product species near the top (where a reduction would simply involve reversing the reaction shown in the table). From the table, we can therefore surmise that it is easier to reduce Au^{3+} back into Au than it is to reduce Na^+ back into Na. For this reason, we can use the table to predict which reactions can be paired up (one as an oxidation, the other as a reduction) to produce a spontaneous reaction. **As long as the species being oxidized is more active than the species that will be formed through reduction, the energetics of redox are favorable, and a spontaneous reaction will occur.** With regard to our nickel-cadmium battery, this theory bears out: since Cd is much more active than Ni^{2+} (which is what will be produced when Ni^{3+} is reduced), the reaction will proceed spontaneously. Why and how this generates electrical energy, we'll see shortly.

It's worth noting that not all redox reactions are **spontaneous**. A spontaneous reaction is one that happens on its own, does not require an input of energy, and in fact releases energy. If we try to react a less active species with one whose product is more active, no spontaneous reaction will occur. Nevertheless, we can write equations for **nonspontaneous** reactions, just as we can discuss a case in which a box is lifted into the air, which itself is nonspontaneous. How does a nonspontaneous chemical reaction take place? The same way that a box ends up in the air—through an input of energy! **Spontaneous processes release energy and nonspontaneous processes require it.** Put another way, spontaneous processes take a system from a higher-energy (less stable) state to a lower-energy (more stable) state; nonspontaneous processes do the opposite. Since spontaneous processes, including falling boxes

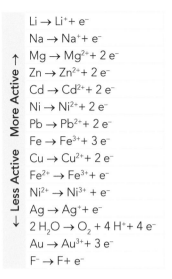

More Active →	
	$Li \rightarrow Li^+ + e^-$
	$Na \rightarrow Na^+ + e^-$
	$Mg \rightarrow Mg^{2+} + 2\ e^-$
	$Zn \rightarrow Zn^{2+} + 2\ e^-$
	$Cd \rightarrow Cd^{2+} + 2\ e^-$
	$Ni \rightarrow Ni^{2+} + 2\ e^-$
	$Pb \rightarrow Pb^{2+} + 2\ e^-$
	$Fe \rightarrow Fe^{3+} + 3\ e^-$
	$Cu \rightarrow Cu^{2+} + 2\ e^-$
	$Fe^{2+} \rightarrow Fe^{3+} + e^-$
	$Ni^{2+} \rightarrow Ni^{3+} + e^-$
	$Ag \rightarrow Ag^+ + e^-$
	$2\ H_2O \rightarrow O_2 + 4\ H^+ + 4\ e^-$
	$Au \rightarrow Au^{3+} + 3\ e^-$
← Less Active	$F^- \rightarrow F + e^-$

TABLE 11.1. Relative activity of common chemical species.

NOTE[3]

Incidentally, this is why we use the metals we do for jewelry making. Copper oxidizes a bit (but not as much as, say, aluminum). Silver is less active than copper, and gold even less so. When metals oxidize, they lose their luster, and some of the mass of the metal is lost in the form of a metal salt, which generally is rubbed or chipped off (imagine a nail rusting; that is an oxidation). In the case of copper and silver, some tarnishing (oxidation) does occur—and we respond by rubbing off the oxidized metal, or tarnish, and restoring the shine—but it's much less than would occur with a more active metal. Gold doesn't oxidize to any significant degree as we wear it, making it shiny, long lasting, and an ideal jewelry metal.

IMAGE 11.2. Copper is more active than water (the species produced
when oxygen is reduced). For this reason, green copper oxides
form on the surface of exposed metal; this is sometimes used to
aesthetic advantage in architecture (left). On the other hand, the
oxidation product of iron—rust—is not desirable. Regardless, it's
hard to avoid, given that iron is also more active than water.

and certain chemical reactions, release energy, it's actually possible to harness that
energy and transform it into a useful form. This is true whether we drop a box on a
plank to launch a rock in the air, as we discussed in Chapter 8, or burn molecules in
oxygen to release energy. In the specific case of spontaneous redox reactions, we can
transform chemical energy into electrical energy, as we'll see shortly.

11.2 Voltaic (Galvanic) Cells

From a historical perspective, much of early electrochemistry was discovered and
explored by two Italians, Alessandro Volta and Luigi Galvani, in the late eighteenth
century. It was Volta who built the first electrochemical cell (the building block
of the modern battery), and Galvani who noted the similarity between inorganic
electricity and neural conduction. The story goes that one day while Galvani was
working at his lab bench with a freshly dissected frog's leg, the sciatic nerve (a large
nerve innervating the leg) touched a piece of metal on Galvani's desk and received
a small shock of static electricity. The leg kicked as though the frog were still alive,
leading Galvani to begin an exploration into the nature of bioelectricity. In further
experiments, he set up a lightning rod and metal wire in his yard, and reportedly
took to hanging frog legs by their sciatic nerves from the wire during lightning
storms (Image 11.3). The resulting chorus line of kicking disembodied limbs must
have been quite a sight—and perhaps made Galvani a bit of an unpopular neighbor!

These two physicists are immortalized through having lent their names to
common electrical units and techniques—a *volt* is a unit of electrical current,
while *galvanized* metal has been electrically coated with another metal to prevent
corrosion. During their lives, they had a famously adversarial relationship; it is

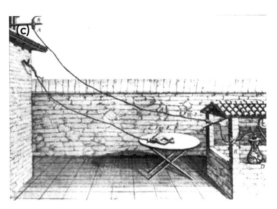

IMAGE 11.3. Alessandro Volta (left) and his contemporary Luigi Galvani (center). Both worked on the physics of electricity, and Galvani discovered the similarity between electrical and neural impulses through his frog experiments (right). Disappointingly, this image is relatively tame: a single frog on a table, connected to a lightning rod. Much more spectacular would have been a drawing immortalizing the anuran chorus line.

therefore a bit ironic they are immortalized jointly in the common name for the energy-generating electrochemical cell, which can be called either a galvanic cell or a voltaic cell. **The voltaic cell[4] converts chemical energy into electrical energy by harnessing energy released during spontaneous redox reactions.** Take, for instance, the reaction of solid zinc ($Zn_{(s)}$) and $Cu^{2+}_{(aq)}$. From Table 11.1, we can determine that these two species should react spontaneously:

$$\text{Oxidation: } Zn_{(s)} \rightarrow Zn^{2+}_{(aq)} + 2\ e^-$$

$$\text{Reduction: } Cu^{2+}_{(aq)} + 2\ e^- \rightarrow Cu_{(s)}$$

$$\text{Balanced Reaction: } Cu^{2+}_{(aq)} + Zn_{(s)} \rightarrow Cu_{(s)} + Zn^{2+}_{(aq)}$$

If we were to put a bar of solid zinc in a beaker of $Cu(NO_3)_{2(aq)}$, we'd expect a spontaneous reaction. Here, nitrate is a spectator ion and is left out of the chemical reaction. In fact, the only reason it's there at all is that it's not possible to find a sample of Cu^{2+} on its own: as we learned in Chapter 4, ionic compounds always occur in cation/anion combinations resulting in a neutral species. In fact, there are many compounds of Cu^{2+}, but the nitrate salt is commonly used because it's very water-soluble.

The beakers in Image 11.4 represent what our zinc bar and copper solution would look like with the passage of time. $Cu^{2+}_{(aq)}$ that came in contact with the zinc bar would be reduced by the zinc, and solid copper would begin to appear on the surface of the bar. The copper solution, originally blue, would slowly lose its color as the concentration of $Cu^{2+}_{(aq)}$ decreased. The zinc bar would shrink over time, because the oxidized zinc ($Zn^{2+}_{(aq)}$) would dissolve in the solution. Since the overall zinc-copper reaction is spontaneous, all of this would happen on its own, and would

NOTE[4]

In this text, we will use "voltaic cell" (rather than "galvanic cell") for a spontaneous electrochemical cell; there is no particular reason to use one over another, and while both are common in science texts, we'll select one and stick with it for the purposes of clarity.

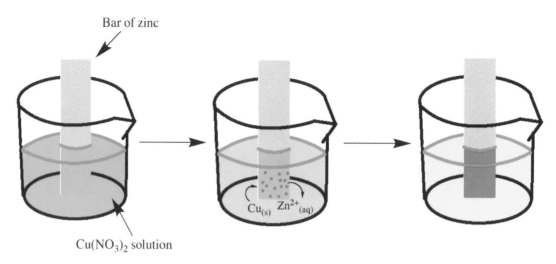

Bar of zinc

$Cu_{(s)}$ $Zn^{2+}_{(aq)}$

$Cu(NO_3)_2$ solution

IMAGE 11.4. A bar of zinc in aqueous $Cu(NO_3)_2$ solution (copper nitrate has a blue color) will react with the Cu^{2+}; zinc will be oxidized, and the copper cation will be reduced. Over time, solid copper starts to collect on the zinc bar, and the solution becomes enriched in Zn^{2+} (and loses its bluish hue as the concentration of Cu^{2+} decreases).

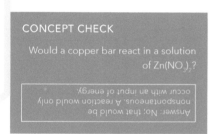

CONCEPT CHECK

Would a copper bar react in a solution of $Zn(NO_3)_2$?

Answer: No: that would be nonspontaneous. A reaction would only occur with an input of energy.

NOTE[5]

For clarity, not all redox reactions involve a change in formal charge—for instance, redox reactions are common between molecular compounds that lack formal charges—but a full analysis of that topic is beyond the scope of this text. Suffice it to say for our purposes that when charges change in the course of a reaction, electrons have changed hands, and when cations and anions swap partners (but do not change charge) in the course of a reaction, electrons have not changed hands.

proceed until either the zinc bar was completely covered with copper (rendering the solid zinc inaccessible for reaction) or the solution became depleted in Cu^{2+}.

Now, while this is an interesting phenomenon to observe, it's not necessarily useful unless we particularly want a zinc bar covered with copper. Even though the reaction is releasing energy (since it's spontaneous), we have no way to collect that energy—some of it is being slowly dissipated as heat, some of it is being dispersed as entropy, and so forth. Put simply, the energy released into the surroundings is being released in an unusable form. However, through a few small adjustments to the system, we can produce something much more useful.

In Image 11.5, the zinc bar has been separated from the solution of $Cu(NO_3)_2$. At first, this separation of reactants seems counterproductive—it's hard to imagine a reaction taking place without the reactants touching! HCl, for instance, could never react with NaOH if we put them in separate beakers; by necessity, the reactants must physically touch in order to rearrange:

$$HCl + NaOH \rightarrow H_2O + NaCl$$

It's worth taking a moment to notice that the reaction above is not a redox reaction. It's more like an "ion swap,"—technically called a *double displacement reaction*—in which each ion maintains its charge,[5] and therefore neither gains nor loses electrons. In order to see this, it can be helpful to keep explicit track of the ions, remembering that an acid acts ionic in solution:

$$H^+ + Cl^- + Na^+ + OH^- \rightarrow H^+ + OH^- + Na^+ + Cl^-$$

(Where H^+ and OH^- come together as a pair to form H_2O, and Na^+ and Cl^- come together as a pair to form NaCl.)

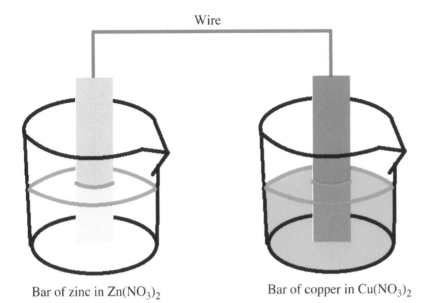

Wire

Bar of zinc in $Zn(NO_3)_2$ Bar of copper in $Cu(NO_3)_2$

IMAGE 11.5. While the zinc bar and copper solution are now physically separated from one another, a wire allows the free movement of electrons between beakers.

Since this reaction requires a recombination of ions, it would be impossible to achieve without a physical juxtaposition of the species. Even a wire connecting the beakers wouldn't allow the reaction to proceed, since ions can't travel through wires. Electrons, however, can, which is what differentiates our zinc-copper reaction from the neutralization above. **In the case of the zinc-copper reaction, there are only two requirements for the reaction to proceed: zinc must be in contact with something that will carry its electrons away, and the copper solution must be in contact with a source of electrons.** As we've seen, this can be accomplished by putting a zinc bar in a beaker of copper solution—the zinc is directly in contact with the copper ions, and they take electrons from the zinc. However, the same conditions are fulfilled if the reactants are in separate beakers, connected by a wire. The zinc metal bar (called an **electrode**) can lose electrons, which travel through the wire toward the copper metal bar. The copper bar itself doesn't react with them, since copper metal can't be reduced to an anionic species, so the electrons travel through the bar (also an electrode) to its surface, where they are collected by copper ions in solution. Those copper ions then plate out on the copper bar, just as they did in the single-beaker system. Since the only thing that must happen for the reaction to proceed is an electron transfer, it's perfectly possible to run a redox reaction this way.

The two necessary parts of this electrochemical cell are the zinc bar and the copper solution. The solution in the zinc beaker doesn't have to contain zinc ion—it could contain any unreactive aqueous solution. The bar in the copper beaker doesn't have to be made of copper—it can be anything that's conductive and doesn't react. However, by convention, the bars and solutions in each beaker are often (but not always) matched up by element. **The one problem with our system as shown in Image 11.5 is that the reaction will be prevented from proceeding so as to avoid charge imbalance in the beakers.** Imagine a theoretical scenario in which the reaction proceeds: as Zn^{2+} builds up in the zinc beaker, it becomes harder and harder for more zinc to oxidize, because forcing more positive charge into an already positively charged solution is quite difficult on account of the electrostatic repulsion between the charges. Similarly, as Cu^{2+} is reduced to solid copper and leaves the copper solution, an excess of negative charge from the nitrate left behind begins to accumulate. This makes it difficult for more copper ion to be reduced. This buildup of charge isn't merely unfavorable; it doesn't happen. To remedy this, we must allow for charges to remain balanced in both beakers. We accomplish this by adding a salt bridge between the cells, as in Image 11.6. **The bridge is made up of salts composed of spectator ions (common salts include KNO₃ and NaNO₃) that can flow freely into either beaker to balance charge; ions from the beakers can also flow into the salt bridge.** In Image 11.6, we see the utility of the salt bridge: NO_3^- from the bridge flows into the zinc beaker because that is the beaker in which positive charge is being added to solution. Similarly, Na^+ flows into the copper beaker, since the redox reaction is removing positive charge from that solution. Further, Zn^{2+} ions from the zinc beaker and NO_{3-} ions (left over from the $Cu(NO_3)_2$) from the copper beaker can flow into the bridge, further helping to maintain charge balance.

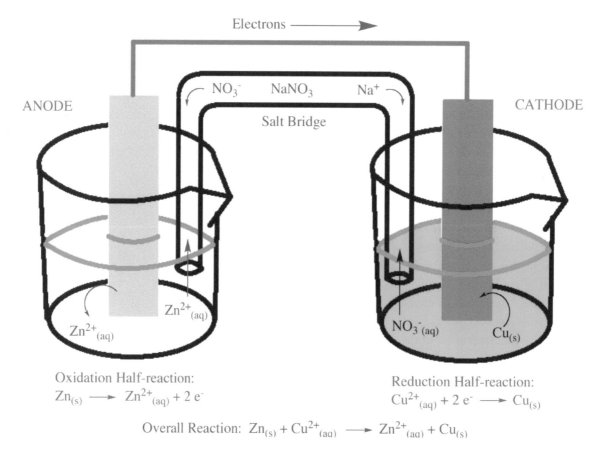

IMAGE 11.6. A copper-zinc cell with a NaNO₃-containing salt bridge in place to ensure that charges in solution remain balanced, allowing redox to proceed until either the zinc bar or the copper solution is depleted of reactant.

The two electrodes in the cell above are referred to as the anode and the cathode. **For any electrochemical cell, the anode is where oxidation takes place, while the cathode is the site of reduction.**[6]

So, certainly separating the zinc and copper from one another in beakers is interesting, but why is it chemically useful? **In separating the components of our reaction, we've forced the electrons to travel through a wire to get from the anode to the cathode, and in forcing electrons to move, we've generated an electrical current.** We can use that current to do work: for instance, we could attach a light bulb to the wire, and our zinc-copper battery (another word for a spontaneous electrochemical cell)[7] would light up the bulb. Why does an electrical current do work? An incandescent light bulb provides a fairly simple example. As electrons flow through a wire, they produce friction, which generates heat. As the light bulb filament (an extension of the wire, through which electrons flow) heats up, it starts to glow, producing light! This is precisely what happens when you put batteries in a flashlight, for instance.

11.3 Electrolytic Cells

Batteries are spontaneous electrochemical cells, but there are nonspontaneous cells as well. At first glance, the utility of this sort of cell appears questionable: Why would we be interested in setting up a redox reaction that requires an input of energy? There are actually several useful applications of nonspontaneous cells, also called electrolytic cells. First, they can be used to decompose compounds into their constituent elements. Like NaF, NaCl forms from its constituent elements (Na and Cl_2) through a highly spontaneous redox reaction. About the reverse reaction, then, we know two things: it is a redox reaction (if it took an electron transfer to go in the forward direction, it will take a transfer of electrons to go back the other way), and it requires an input of energy. By using an external source of electrical energy to force electrons out of chloride and into sodium, it's possible to reverse the reaction:

$$2 \, NaCl \rightarrow 2 \, Na + Cl_2$$

$$\text{Oxidation: } 2 \, Na^+ + 2 \, e^- \rightarrow 2 \, Na$$

$$\text{Reduction: } 2 \, Cl^- \rightarrow 2 \, e^- + Cl_2$$

A cell powered by an external source of energy in which the reaction above takes place is an example of an electrolytic cell. In fact, electrolysis means "splitting with electricity," which is exactly what's happening to NaCl in this case. Why would we want to do such a thing? Chlorine gas is very reactive, so it (like sodium metal) is not found in nature. However, it has a wide variety of industrial uses, including in the production of some plastics, so the ability to produce it through electrolysis is

TRY THIS

Using Table 11.1, determine how a voltaic cell could be made from lead and silver. Write out the half-reactions and overall reaction, indicate which way electrons would flow, and identify the cathode and anode.

Answers:
$Pb + 2 \, Ag^+ \rightarrow Pb^{2+} + 2 \, Ag$.
Oxidation: $Pb \rightarrow Pb^{2+} + 2 \, e^-$ (this would be the anode)
Reduction: $2 \, Ag^+ + 2 \, e^- \rightarrow 2 \, Ag$ (this would be the cathode)
Electrons would flow from the lead beaker to the silver beaker.

NOTE[6]

A common mnemonic is *An Ox and a Red Cat; An* (for anode) and *Ox* (for oxidation) go together, while *Red* (for reduction) and *Cat* (for cathode) go together.

NOTE[7]

Technically, our zinc-copper system is a cell. Two or more cells linked together in a series forms a battery. That said, in common usage, "cell" and "battery" are used interchangeably.

quite useful. Another common electrolytic reaction involves the decomposition of water into H_2 and O_2, also a nonspontaneous process. The hydrogen derived from this reaction can be used to produce power (more on this in Chapter 13) or used for chemical synthesis reactions.

Most relevant to our discussion of batteries, however, is the applicability of electrolysis to recharging a spent cell. We all know from experience that batteries have a limited lifetime; in light of their chemistry, this makes sense. At some point, either the anode will run out of material to be oxidized or the cathode will run out of material to be reduced and the reaction will cease, which will cease the flow of electricity. **In theory, however, it should be possible to recharge a battery by applying electricity to it, forcing the reaction to proceed in the reverse direction, and regenerating the starting materials.** For instance, with our zinc-copper cell, we'd use an outside source of electrical energy to force the electrons to flow from the copper side toward the zinc side, achieving the following nonspontaneous reaction:

$$Zn^{2+}_{(aq)} + Cu_{(s)} \rightarrow Zn_{(s)} + Cu^{2+}_{(aq)}$$

$$\text{Oxidation: } Cu_{(s)} \rightarrow Cu^{2+}_{(aq)} + 2\ e^-$$

$$\text{Reduction: } Zn^{2+}_{(aq)} + 2\ e^- \rightarrow Zn_{(s)}$$

This would regenerate the zinc bar and increase the concentration of Cu^{2+} in the copper solution. In theory, once we removed the outside source of electricity, the reaction would resume in the original (spontaneous) direction, and would once again generate a current.

While the theory is sound, in reality, not all batteries are rechargeable. Typical alkaline batteries, for instance, are powered by a fairly complex redox reaction involving zinc powder and manganese dioxide. Completely recharging these would require returning each atom to its original location in order to maximize surface area for the chemistry to take place. This simply isn't possible with existing technology. While rechargers for "nonrechargeable" batteries do exist, they confer limited additional lifetime upon the cells. Further (and more important), batteries not intended for recharging can heat up during an attempt to reverse the reaction, which leads to the potential for explosion. True rechargeable batteries are assembled in such a way that when external current is applied, the reaction is reversed quickly and efficiently. Car batteries, which use a lead-acid reaction, were among the first commonly used rechargeable cells. If a car's engine is off, the battery supplies power to the accessories, like the lights:

$$Pb_{(s)} + PbO_{2(s)} + 2\ H_2SO_{4(aq)} \rightarrow 2\ PbSO_{4(s)} + 2\ H_2O_{(l)}$$

$$\text{Oxidation: } Pb_{(s)} + H_2SO_{4(aq)} \rightarrow PbSO_{4(s)} + 2\ H^+_{(aq)} + 2\ e^-$$

$$\text{Reduction: } PbO_{2(s)} + H_2SO_{4(aq)} + 2\ H^+_{(aq)} + 2\ e^- \rightarrow PbSO_{4(s)} + 2\ H_2O_{(l)}$$

The half-reactions above are much more complicated that what we've seen so far in the chapter, and while we can see that electrons are in fact being transferred, it's not immediately obvious where they're coming from and where they're going. To fully dissect the half-reactions, we must look more closely at the individual species involved. In the oxidation half-reaction, neutral lead becomes Pb^{2+}—a particular that we can determine only by noting that the lead cation is paired with SO_4^{2-} on the product side. We can therefore represent the heart of the oxidation half-reaction as:

$$Pb_{(s)} \rightarrow Pb^{2+} + 2\ e^-$$

The reduction half-reaction reactants contain Pb^{4+}, which we determine by noting the pairing with 2 oxide anions (O^{2-}). In the products we find $PbSO_4$, which contains Pb^{2+}. The reduction half-reaction can thus be simplified to:

$$Pb^{4+} + 2\ e^- \rightarrow Pb^{2+}$$

The other species involved in the overall reaction (and complete half-reactions) do not change their charges. It's fairly easy to see this with regard to SO_4^{2-}, but the dispositions of H^+ and O^{2-} are a bit more difficult to track without an understanding of *oxidation numbers*,[8] a discussion of which is beyond the scope of this text. Suffice it to say, these species participate in the chemistry of a lead-acid battery, but are not involved in electron transfer.

Like any battery, a car battery will cease to provide power if its reactants become depleted. If you've ever parked your car with the lights on and returned to a dead battery, you're familiar with this phenomenon. Generally, however, a car is not left to run on battery power for long stretches of time, and the battery is designed so that it recharges while the motor is running. Since recharging a battery requires an input of energy, the car's engine is called upon to power the electrolytic reaction. While the motor is running, the battery is recharged by a complete reversal of the reactions observed during discharge:

$$2\ PbSO_{4(s)} + 2\ H_2O_{(l)} \rightarrow Pb_{(s)} + PbO_{2(s)} + 2\ H_2SO_{4(aq)}$$

Oxidation: $PbSO_{4(s)} + 2\ H_2O_{(l)} \rightarrow PbO_{2(s)} + H_2SO_{4(aq)} + 2\ H^+_{(aq)} + 2\ e^-$

Simplified oxidation: $Pb^{2+} \rightarrow Pb^{4+} + 2\ e^-$

Reduction: $PbSO_{4(s)} + 2\ H^+_{(aq)} + 2\ e^- \rightarrow Pb_{(s)} + H_2SO_{4(aq)}$

Simplified reduction: $Pb^{2+} + 2\ e^- \rightarrow Pb_{(s)}$

The relationship between car battery and engine is reciprocal: the battery powers the starter motor and ignition, which allow the engine to start. The engine, in turn, recharges the battery. In the case of a dead battery, the only way to start the motor

NOTE[8]

An oxidation number is a theoretical charge given to a neutral atom in a molecule. For instance, O is not anionic in the covalent molecule H_2O, but it has an oxidation number of -2, which serves as an indicator that, even though it's not anionic anymore, it hasn't actually lost electrons if water is formed from the oxide anion. A full discussion of the rules for determining oxidation numbers can be found in any general chemistry textbook.

is with an external source of electrical energy, as in a jump-start, which utilizes the power from another vehicle's battery. If a dead battery is jump-started, the car must be left running for a period of time to allow the engine to recharge the battery, so that it can start the motor the next time the vehicle is used.

11.4   Photovoltaic Cells

Like any voltaic cell, a **photovoltaic** cell produces electricity from another form of energy. Unlike typical voltaic cells, however, photovoltaic cells do not rely on chemical energy in order to accomplish this: instead, they harvest and transform radiant energy from the sun (Image 11.7).

IMAGE 11.7. A large array of solar panels.

The major reason solar power is so attractive as an energy source is that, aside from being completely renewable and nonpolluting, it is tremendously abundant. Currently, global energy consumption is estimated at about 12.5 terawatts (10^{12} watts; a watt is 1 J/s). That number is expected to approach 17 terawatts in the next two decades. Incident solar radiation upon accessible terrain[9] is approximated at nearly 600 terawatts—more than 30 times the projected demand. Further, proponents of solar power note that the global energy requirement estimates are based upon fossil fuel reliance. Coal and other combustion fuels are fairly inefficient sources of energy, meaning that the replacement of combustion-generated power with solar-generated electrical power would actually decrease global energy needs.[10] The same is true, incidentally, for other sources of sustainable electrical energy, including wind power.

NOTE[9]

Global incident solar energy is almost 11 times higher than accessible solar energy, but current technology precludes the harvesting of radiation from oceans and high mountain regions.

NOTE[10]

If gasoline is used to fuel an automobile, about 20% of the energy from the gas creates forward motion. If electricity is used to power the vehicle by way of a battery, up to 86% of the electrical energy creates forward motion.

Photovoltaic electrical generation relies upon the use of silicon semiconductors, which are somewhat less conductive than metals. While there are many different kinds of photovoltaic cells currently being utilized (with frequent new developments in technology), one type in particular is called the **silicon p-n junction solar cell**. In a p-n junction solar cell, silicon is separated into two distinct semiconducting regions, each of which is **doped** with small quantities of another element. **N-type silicon**, for instance, is purposely contaminated with group 5 elements like arsenic or phosphorus. Compared to silicon's four valence electrons, arsenic and other group 5 elements have a valence of five—this incorporates "extra" electrons into the silicon matrix. **P-type silicon**, on the other hand, incorporates small amounts of group 3 elements like gallium or boron. The "missing" valence electrons in the silicon matrix owing to the presence of the group 3 elements create **electron holes**. When sunlight strikes the solar panel, electrons absorb the radiant energy and become free to move through the silicon matrix—this is the nature of a semiconductor. Because the n-type silicon has a higher electron density than the p-type silicon, there is a separation of charge (or potential difference, or voltage) across the junction (Image 11.8). **Electrons can then flow through a wire (for example) toward the area of electrostatically attractive positive charge on the other side of the junction. One-way flow of electrons is an electrical current and a source of electrical**

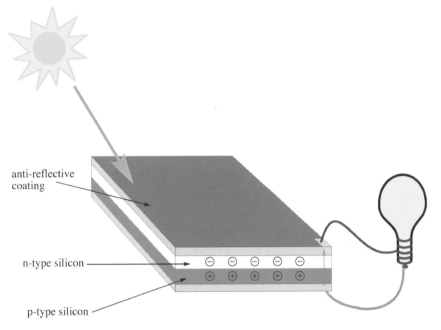

anti-reflective coating

n-type silicon

p-type silicon

IMAGE 11.8. A schematic of a photovoltaic cell. Note that the electrons will move from n-type to p-type silicon, but that by convention, the direction of current (black arrows) is said to be opposite the direction of electron flow (red arrows).

CONCEPT CHECK

Why do electrons flow from n-type silicon to p-type silicon?

Answer: Compared to p-type silicon, n-type is electron rich.

power. An electrically powered object, like a lightbulb, can be placed across that wire and powered by the flow of electrons.

A major challenge associated with expansion of solar power has historically been that photovoltaic cells can't be used to provide power at night. However, there is another method of harvesting solar power which offers a solution. **Parabolic trough solar-thermal power plants** concentrate the sun's heat onto a pipe of fluid using a long parabolic mirror (Image 11.9). **The hot fluid (often a molten salt)[11] can be used to generate renewable heat from noncombustion sources, or can be used to produce steam, which drives steam-powered electrical generators.**

 NOTE[11]

Why molten salt rather than water? Among the reasons are that it's possible to heat salt to a higher temperature than water. Also, salt holds the heat much longer than water, allowing for continued power generation through the night.

IMAGE 11.9. A parabolic trough solar-thermal power apparatus at Sandia National Laboratories in New Mexico.

There are several hurdles to overcome in the large-scale implementation of solar power. Some materials required for solar cells, for instance, are not abundant on the planet. However, the variety of different cell types (and the disparate selection of elements from which they are produced) offers opportunities to optimize production and efficiency of those cells for which materials are more available, and phase out or redesign cells for which materials are scarce. Also, cell reuse and recycling could play a valuable role in maximizing the lifetime of supplies. **Proponents of solar energy expansion suggest that, even using only existing technology, it would be economically and logistically feasible to end reliance upon fossil fuels by 2030.** Some solar advocates suggest that a combined approach, relying on wind- and water-generated power in addition to solar, would be most effective, while others argue that the sheer abundance of sunlight makes it the single most appropriate source of global energy.

Those who oppose further development of solar power sometimes cite the inefficiency of the technology with regard to energy transformation. Most photovoltaic panels convert radiant energy with an approximately 12–20% efficiency, with some newer panels approaching 30–40%. Regardless, even using the conservative estimate of 12% efficiency, global utilization of solar panels could provide about 72 terawatts of energy—more than four times the predicted global need in the coming decades. When one takes into account the reduced global energy requirements using an electricity-dependent (rather than fossil fuel–dependent) system, the availability of existing cells with far more than 12% conversion efficiency, and the likelihood of future technologies with increased efficiency, it becomes apparent that low efficiency is simply not a tenable argument against solar power. Not only is the "wasted" energy not necessary to meet global requirements, solar energy is also fully renewable and nonpolluting. In other words, a 40% efficient coal-fired power plant is distressing, since the 60% of energy lost as heat still results in the release of pollution and carbon emissions from combustion, as well as the depletion of nonrenewable fossil fuel resources. A loss of the majority of sunlight incident upon a solar panel, however, is neither polluting nor of any environmental consequence—the sunlight would hit Earth's surface whether we collected it or not.

How much land would be required—and how many solar plants would need to be built—in order to provide for global energy needs based upon current technologies? **Several estimates suggest that a square array approximately 100 miles on a side would be sufficient to power the United States—that's an area slightly smaller than the state of Massachusetts.**[12] It bears remembering, however, that continued technological advancement will reduce the necessary footprint. For comparison purposes, Hoover Dam generates electricity as water falls through the dam and spins generator turbines; Lake Mead, which was produced through building the dam, covers 250 square miles. A concentrated solar plant generating the same amount of annual electricity as the dam would cover only about 10 square miles of land. When one considers the footprint of a coal-fired power plant

NOTE[12]

Such an array would not actually be built; solar power has to be generated close to where it will be used, meaning that many smaller power-generating arrays scattered across the country—coupled with alternative energy-generation mechanisms for areas in which solar harvesting is impractical—would be necessary. Nevertheless, the size comparison serves the intended purpose, which is to illustrate that meeting national power requirements with solar would not require blanketing vast swaths of land in photovoltaic arrays.

(which must include the land required to mine the coal), solar-plant footprints compare very favorably indeed.

It's important to note that many of the arguments against development and implementation of solar technology come from fossil-fuel lobbies and companies invested in petroleum, coal, and the like. On one hand, it's understandable that a coal miner or oil field worker would want to protect his or her job, and one feels sympathy for the plight of individuals faced with losing jobs in an industry that global sentiment has begun to turn against. On the other hand, it's also important to remember that the global dependence upon fossil fuels has tremendous environmental costs, and that (despite the utilitarian sound of such a statement) in the end, the global community must make a decision as to whether the livelihood of a select group of workers, businessmen, and corporations warrants continued commitment to energy sources with known and significant negative environmental impact, the effects of which are felt by the global population as a whole: a group much larger than that which stands to be negatively financially impacted by a move toward a more sustainable energy economy.

IMAGE 11.10. The environmental impact of Hoover Dam (to say nothing of the 250-sq.-mi. artificial lake it creates) is tremendous compared to the small footprint and emission-free operation of a solar plant.

11.5 Plants—Natural Solar Power Generators

The ability to harness solar power and convert it to electrical energy is neither new nor is it a human invention. In fact, plants and a few other organisms have been doing it for billions of years. Unlike man-made photovoltaic cells, plants don't use doped silicon in their light-harvesting panels. Still, similarities exist: **photosynthetic organisms utilize a system designed to allow unidirectional electron movement in the presence of sunlight.**[13] The term photosynthesis comes from the ability of these organisms to collect light energy and store it as chemical energy.

Let's take a moment to focus on a major difference between plants and animals. Animals eat to fulfill their energy needs, while plants get their energy from the sun. This, however, fails to address the source of a plant's carbon. The molecules of life are carbon based; in order to grow, animals utilize the carbon building blocks from the nutrients they consume. Plants, on the other hand, obtain carbon in another way. The cellulose (structural material) and amylose (energy storage) that make up a plant can't be obtained through the roots—instead, they have to be synthesized in the plant itself, using CO_2 and H_2O as starting materials. As we saw in Chapter 9, glucose can be burned for energy by the reaction:

$$C_6H_{12}O_6 + 6\,O_2 \rightarrow 6\,CO_2 + 6\,H_2O$$

The reverse of this reaction, therefore, requires energy. **Since both cellulose and amylose are polymers of glucose, and the formation of glucose from starting materials requires energy, we see that a plant needs energy to build carbohydrate; this is where photosynthesis comes in.**

NOTE[13]

Those who feel man-made photovoltaic cells are inefficient would be horrified by plants—typical crops convert solar energy with somewhere in the range of 3–6% overall efficiency!

IMAGE 11.11. A general structure of chlorophyll. Differences at X and Y differentiate between forms, but all collect light energy.

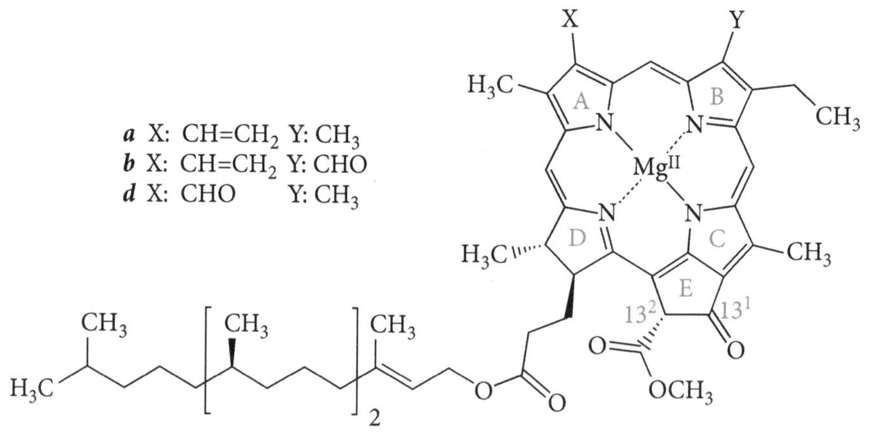

a X: CH=CH₂ Y: CH₃
b X: CH=CH₂ Y: CHO
d X: CHO   Y: CH₃

The process of photosynthesis starts with the action of a water-splitting enzyme, which oxidizes water by the reaction:

$$2\,H_2O \rightarrow 4\,H^+ + 4\,e^- + O_2$$

The electrons from water are transferred to a molecule of **chlorophyll** (Image 11.11). Chlorophyll is a light-absorbing molecule—it's responsible for the green color of leaves—and while it is the molecule most commonly associated with photosynthesis, it's by no means the only light-harvesting compound. As we saw in Chapter 10, carotenoids and anthocyanins can also absorb light.

A combination of many proteins, enzymes, and light-harvesting pigments make up a **photosystem**, of which the water-splitting enzyme is a part. Together, many photosystems and a number of additional enzymes and proteins comprise the cellular structure responsible for photosynthesis, called a **chloroplast**. The photosystems and most of the auxiliary enzymes are embedded in a membrane inside the chloroplast.

The schematic in Image 11.12 above represents the step-by-step processes of photosynthesis. The redox reaction whereby electrons are transferred from water to chlorophyll is nonspontaneous and requires energy. In order to accomplish this reaction, chlorophyll and other light-harvesting pigments in the photosystem absorb light energy from the sun,[14] which powers the electron transfer (step 1). The energy also excites the electrons, which loosens them within the molecular structure (just as light energy loosens electrons within the silicon matrix of a photovoltaic cell). The loose electrons flow from chlorophyll to another molecule (called pheophytin), and from there to another molecule (called a quinone), and so forth down a long line of molecules (steps 2–5). In each case, the molecule accepting the electrons forms a product that is less active than the molecule donating the electrons, meaning that each of these transitions is spontaneous and releases energy. Some steps release small amounts of energy, which are dispersed as heat or in other nonusable forms. Other steps release larger amounts of energy, which are represented in the schematic (steps 2 and 3). All these electron-passing components in the membrane combine to make up a unidirectional molecular "wire," which allows the flow of

NOTE[14]

Recall from Chapter 10 that all EMR, including visible light, has energy (which can be calculated as $E = h\nu$).

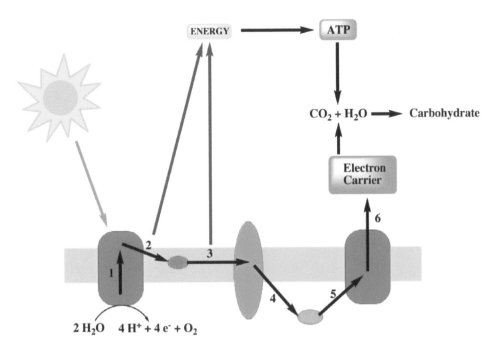

IMAGE 11.12. A photosystem (green, embedded in a tan membrane) is an important part of photosynthesis, shown here in a schematic representation. Other components include various proteins and molecules that take part in a series of electron transfer (redox) reactions, ultimately generating glucose.

electrons from the photosystem to its eventual destination via an electron carrier molecule (step 6). The energy released in steps 2 and 3 is then used to synthesize ATP, an energy-transfer molecule.

ATP (Image 11.13) is sometimes called "chemical currency" because of its ubiquity in the living world. A relatively small molecule, it nevertheless has very high-energy bonds owing to the tremendous electrostatic repulsion between negatively charged atoms, meaning that it takes quite a bit of energy to make it. Conversely, the breakdown of ATP via hydrolysis releases lots of energy, which makes it ideal for short-term energy storage and transfer of energy from one place to another in a cell. While animals use molecules like glycogen and fat to store large amounts of energy for later harvesting, it's not the glucose or fatty acids themselves—the breakdown products of glycogen and fat—that directly allow muscles to function, cells to grow, and so forth. Instead, glucose, fatty acids, and other nutrient molecules are metabolized, and the energy is used to make ATP. The ATP is then responsible for fueling processes in the cell. **Plants use the sun's energy to make ATP, which is then broken down to provide the energy to synthesize glucose, a building block for larger carbohydrates.**

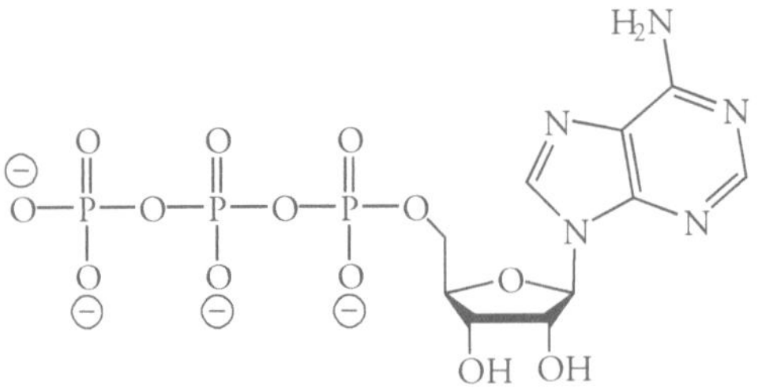

In the final portion of photosynthesis, carbohydrate is produced from water and CO_2. This requires both energy (from ATP) and electrons. The electrons are required to build the new bonds that will combine many smaller molecules into one larger one. **With the completion of this synthesis, the energy transformation and electron transfer are complete—energy from the sun has been stored as carbohydrate, and**

IMAGE 11.13. ATP—short for adenosine triphosphate—is a chemical currency molecule.

CONCEPT CHECK

What are the sources of energy and electrons that a plant uses to build carbohydrate?

Answer: Energy is from the sun, electrons are from water.

electrons from water have been transferred to carbohydrate through a long series of redox reactions. In the process, O_2 has been produced (as a waste product, incidentally), which is important to animals and other oxygen-dependent life forms. The stored solar energy, having been transformed into chemical energy, is now available for the plant itself (which stores some of the carbohydrate to fulfill its own energy needs) and for organisms that consume the plants. Further, the very important process of **carbon fixation** has taken place. This is the incorporation of atmospheric CO_2—which animals can't use to build the molecules of life—into macronutrients, which animals are able to utilize.

11.6   Human Batteries—Respiration

We know from our discussion of nutrition in Chapters 8 and 9 that carbohydrate consumption yields energy. In fact, we could literally burn glucose—just as we could burn methane in a Bunsen burner—to provide light and heat. In the cells, however, the process whereby glucose is "burned" to release energy is a bit more complex. Remember that simply because a reaction releases energy doesn't mean that energy is useful; burning glucose in a pot on the kitchen counter would make a little flame that we could use to heat water, or we could simply let the fire burn, and the energy would be dispersed as heat, to increase entropy, and so forth. **It's very important that our cellular process for "burning" macronutrients takes place in such a way that much of the energy released is stored in a usable form—for instance, by making ATP. This is accomplished through a process that shares many similarities with photosynthesis.**[15]

While a complete discussion of the chemical process by which glucose is broken down into CO_2 and H_2O is beyond the scope of this text, we can surmise a few things about it just from our discussion of photosynthesis. Because the breaking of glucose into CO_2 and H_2O is a reverse in many ways of what happens during photosynthesis, we know that it must yield energy, and it must release electrons. Just as electrons are required to produce carbohydrate from starting materials in photosynthesis, electrons are released during the reverse reaction, which is called **cellular respiration.**[16] Respiration liberates chemical energy from macronutrients (here, we are specifically looking at the breakdown of glucose) and stores that energy by using it to build molecules of ATP. The schematic in Image 11.14 provides a step-by-step representation of respiration, beginning with the transfer of electrons from glucose to an electron carrier, which in turn passes them to a respiratory enzyme (step 1).

There are many similarities between respiration and photosynthesis. First, as in photosynthesis, the enzymes of respiration are clustered together in the membrane of a cellular structure called the **mitochondrion** (the plural form is **mitochondria**). This membrane is shown in blue in Image 11.14, and the embedded enzymes are shown in various colors. Electrons are passed from one molecule to the next in a unidirectional fashion; this forms a molecular "wire." Also as in photosynthesis, for

NOTE[15]

Frankly, the similarity should come as no surprise. It's common in nature to see one successful process or structure duplicated (often in a modified fashion) in other organisms. For instance, despite significant differences in function, the form of the skeletal structures of human, bat, and whale forelimbs are remarkably similar!

NOTE[16]

We sometimes use the word *respiration* to refer to our process of breathing, but technically, it's a metabolic term. Oxygen-utilizing bacteria rely on respiration, despite the fact that they don't breathe.

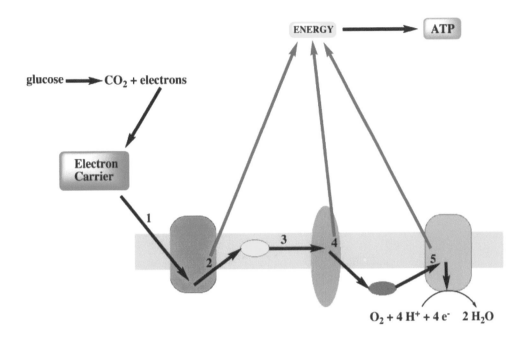

IMAGE 11.14. A schematic representation of respiration. Note the similarities to photosynthesis, despite the fact that one process is powered by glucose and the other produces it.

each electron transfer reaction, the product formed when a molecule accepts the electrons is less active than the molecule that donated the electrons, meaning each redox reaction is spontaneous, and yields energy. Finally, just as with photosynthesis, some of this energy is dispersed in an unusable form—heat, for instance—while some of it is collected and used to produce ATP. So far, there are two major differences between respiration and photosynthesis. First, the initial redox reaction in respiration is spontaneous and requires no energy input. The initial redox reaction in photosynthesis, on the other hand, requires energy from the sun. Additionally, while both processes share the ultimate goal of producing ATP, photosynthesis also uses the electrons collected from water to synthesize carbohydrate. In respiration, carbohydrate is broken down rather than produced. The only goal is to make ATP, and the electrons are essentially waste. Whether they're needed or not, however, the electrons have to go somewhere—if they weren't passed along to another molecule by the last respiratory enzyme (light green in the schematic), that last enzyme wouldn't be able to continue to accept electrons, and the entire process would stop. Ceasing the flow of electrons would cease ATP production, which would defeat the purpose of the reaction. As a result, electrons from the last enzyme must be disposed of. This is where oxygen comes into play. **In a reverse of the reaction that started photosynthesis, the last respiratory enzyme passes electrons to O_2, reducing it to water.** The water, then, is a metabolic waste product (which can be eliminated from the cell or used for other things), and as long as a continuous supply of O_2 is available, the energy production will continue. This is the reason that we, and other oxygen-dependent organisms, need O_2!

Incidentally, the processes whereby amino acids and fatty acids are metabolized for energy are somewhat analogous to that shown for carbohydrate: the breakdown of the nutrients yields electrons, which are transferred along the same molecular wire to oxygen, and energy from those transfers is collected to make ATP.

11.7 Anaerobic Metabolism—Fermentation

What happens, though, if there is not enough O_2 to allow the electron transfer reactions of respiration to continue to function? Conditions such as these occur in the case of intense exercise. As discussed in Chapter 9, power efforts require energy production—which we now know is the production of ATP—in excess of the body's ability to supply oxygen. The result is that cells must utilize an anaerobic strategy for metabolism of glucose. **Cells switch between aerobic and anaerobic metabolic strategies purely as a function of oxygen availability; as long as sufficient oxygen is available, respiration can proceed.** Of the two strategies, respiration is "preferred"; it is a far more efficient mechanism for production of ATP. In the absence of sufficient oxygen, though, respiration ceases to function. Without a final electron acceptor to reduce, the last respiratory enzyme cannot be reduced by the enzyme before it, which in turn can no longer accept electrons from its electron donor, and so on. As an alternative energy production strategy, fermentation occurs, as represented in Image 11.15. Unfortunately, the energy production of fermentation is far below that of respiration: about 15 times less energy is produced, meaning 15 times less ATP is generated per molecule of glucose. It is for this reason that fermentation is generally avoided by multicellular organisms as a metabolic strategy, except in cases of dire need.

Despite the fact that fermentation doesn't involve a long series of electron transfer reactions, the electron carriers must still participate—for oxidation to

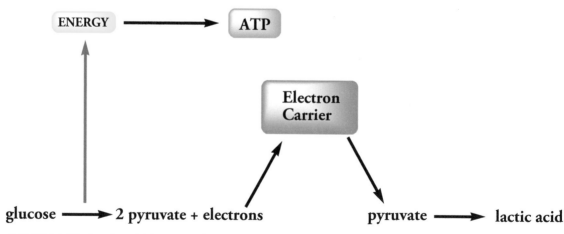

IMAGE 11.15. Lactic acid fermentation.

occur, something must be reduced. The oxidation of glucose yields a bit of energy, providing for production of a small amount of ATP. However, the process of fermentation doesn't end there. If electron carriers were continually reduced without ever dropping off their electrons, all available carriers would soon become full. This would be a disastrous state of affairs: without carriers to reduce, glucose oxidation, and therefore energy production, would cease. In order to solve this problem, something must accept electrons from the carriers. **Pyruvate**, the product of glucose oxidation, is able to accept electrons. The product (aside from empty electron carriers that can return to react with more glucose) is lactic acid.[17] Note that production of lactic acid from pyruvate does not increase the energy yield of fermentation, nor is it a particularly desirable outcome; it's a waste product, and must be disposed of. However, its formation is critical to continued energy production under anaerobic conditions and, as such, it's unavoidable.

While larger organisms simply can't produce sufficient energy to sustain function under anaerobic conditions—at least not for any length of time—there are many unicellular organisms capable of doing so, owing to their lower energy needs. Some bacteria and yeast utilize fermentation as a primary (or at least optional) energy-production strategy, a fact we use to our advantage in the cultivation of the interesting fermentation-based food products discussed in Chapter 5. Bacteria like *Lactobacillus acidophilus* ferment sugars to lactic acid, which reacts with milk to produce many of the fermented dairy products we enjoy. Yeast, on the other hand, ferment sugars to produce CO_2 and ethanol (Image 11.16), both of which are utilized in the production of various comestibles. The chemistry of ethanol-producing fermentation is quite similar to that of lactic acid fermentation. The difference lies only in the specific enzyme responsible for reduction of pyruvate; different enzymes are responsible for different fermentation products.

Regardless of the specifics, either type of fermentation accomplishes two important goals: production of energy and regeneration of empty electron carriers. In the end, also, it's just frankly intriguing to realize that the same metabolic strategy that

NOTE[17]

This is why muscles burn during hard efforts. Anaerobic work requires cells to produce energy by fermentation rather than respiration. Fermentation produces lactic acid, which is an irritant to the muscle cells.

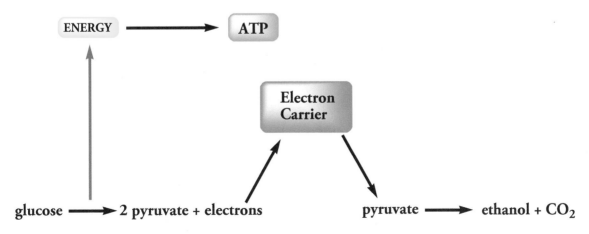

IMAGE 11.16. Ethanolic fermentation.

leaves us with burning muscles after a sprint or a session in the weight room is also responsible for the bubbles (and the alcohol) in a beer.

11.8 The Last Word—Lightning

Lightning is one of nature's most interesting phenomena, both beautiful and destructive, admired and feared, intensely studied, and still incompletely understood. While hypotheses abound regarding the specifics of the phenomenon, it has been known for some time that lightning is an electrical discharge—recall Galvani and his Franken-frogs. Within thunderclouds, falling drops of rain and ice become electrically polarized. One hypothesis holds that this is a consequence of the droplets passing through Earth's natural electrical field. Others suggest the involvement of forces including wind, humidity, and atmospheric pressure. Also debated is the mechanism whereby these ions are segregated within a cloud. Regardless of how it happens, charge does, indeed, separate. Large areas of positive and negative charge form within clouds, and this separation represents a tremendous potential for electrical current.

In the photovoltaic cells discussed earlier in the chapter, electrons were held within a silicon matrix, and energy from sunlight loosened them enough to allow their movement from slightly negative to slightly positive regions of doped silicon. In clouds, however, no such energy input is required; significant separation of charge results in the buildup of massive electrostatic forces. In the same way that plastic can be used to coat and insulate a wire, preventing it from transferring electricity if it's touched, the insulating medium of the air initially prevents the discharge of electrical potential energy from charged clouds. The areas of accumulating charge

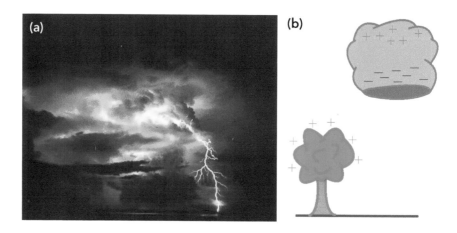

IMAGE 11.17. Lightning (a) results from the separation of charge within thunderhead clouds; this can result in either cloud-to-ground or cloud-to-cloud discharge (b).

begin to exert electrostatic forces upon molecules of air between the bottom of the cloud and the ground, with electrons pulled toward positively charged regions and pushed away from negatively charged regions. This creates dipoles in molecules that would otherwise have none, as in Image 11.18.

Eventually, the electrostatic forces upon air molecules produce such significant dipoles that ionization results, and a variety of interesting species not normally found in the atmosphere are formed. These include N^+, O^+, N_2^+, and O_2^+. Just as ions (electrolytes) in water allow the solution to conduct electricity, charged particles in the air allow electrical current to flow. This produces the massive electrical discharge that we see in a lightning strike. Further, the high-energy charged particles quickly combine with neutral molecules, initiating a sequence of reactions that concludes with the production of NO_2. As discussed in Chapter 6, NO_2 can react to form ozone, the sharp, metallic odor of which can sometimes be detected in the air shortly after a lightning strike. In fact, ozone gets its name from the Greek word *ozein*, meaning "to smell."

With lightning striking the earth 100 times a second (nearly three trillion times each year), and delivering an awesome electrical discharge with every strike, it's easy to understand why the phenomenon generates so much interest. At a time when many researchers are turning their attention to harnessing the energy of the earth, it's worthwhile to stop sometimes and simply admire the forms of tremendous power that exist all around us, and defy not only our ability to harness them, but even our ability to fully understand them.

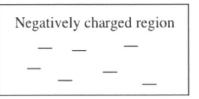

IMAGE 11.18. Separation of charge is induced in otherwise neutral molecules

SUMMARY OF MAIN POINTS

- Separation of opposite charge produces electrical potential energy, which can be used to do work.
- Moving charge is electrical energy and is called a current.
- Chemical reactions involving transfer of electrons can be used to generate electricity and are called redox reactions.
- Redox reactions must include both an oxidation (electron loss) and a reduction (electron gain); the two parts of the reaction are called half-reactions.
- Correctly written half-reactions can sum to give the overall chemical reaction.
- The spontaneity of a redox reaction can be determined using the oxidation potential, or activity, of the species involved: as long as the reactant being oxidized is more active than the product of the reactant being reduced, the reaction will be spontaneous.
- Nonspontaneous redox reactions will only occur if energy is added to the system. It's still possible to write out reactions for (and balance half-reactions for) nonspontaneous redox reactions, regardless of the energetics.
- A voltaic (galvanic) cell converts chemical energy into electrical energy by utilizing a spontaneous redox reaction.

- In a voltaic cell, the reactant being oxidized is isolated from the reactant being reduced—the only connections between them are a wire (to allow electron flow) and a salt bridge (to balance charge).
- As electrons flow from the material being oxidized (the anode) to the material being reduced (the cathode), they can be utilized to do work.
- Spectator ions are present in a reaction mixture but are not involved in the reaction.
- Voltaic cells have both anodes (where oxidation takes place) and cathodes (where reduction takes place); electrons always flow from anode to cathode.
- An electrolytic cell is a nonspontaneous cell that requires an input of energy to run a redox reaction—these are used to produce elements for industrial purposes.
- Recharging a battery involves turning a voltaic cell into an electrolytic cell by applying outside energy and running the spontaneous redox reaction backward.
- Batteries designed to be rechargeable are capable of reversing the redox reaction with greater efficiency than nonrechargeable batteries.
- Cars work through a reciprocal relationship between battery and motor. When the car is off, the battery powers the accessories and the starter motor, but while the engine is running, it recharges the battery.
- Photovoltaic cells transform radiant solar energy into electrical energy.
- Silicon p-n junction solar cells use group 5 element-doped silicon (n-type) layered with group 3 element-doped silicon (p-type) to create regions of electron density and electron paucity.
- Solar energy loosens electrons in silicon, allowing unidirectional flow from n-type to p-type silicon—this is an electrical current.
- Parabolic trough solar-thermal power plants concentrate solar energy onto a pipe of fluid that is heated. This can be used to provide heat or produce electricity through steam generation.
- Solar power plant land requirements are low compared to many other technologies.
- Photosynthesis is the conversion of solar energy into chemical energy (production of ATP, and then carbohydrate) through unidirectional electron flow; oxygen is a by-product.
- The electrons transferred during photosynthesis are eventually used to make carbohydrate, which provides structural material for the plant and energy for both the plant and plant consumers.
- Cellular respiration is the conversion of chemical energy from macronutrients into chemical energy in the form of ATP through unidirectional electron flow.
- The electron transport chain of respiration transfers electrons from carbohydrate to oxygen (forming water), and the energy released during electron transfer is used to make ATP.

- Fermentation is anaerobic energy production; energy yield is lower than for respiration.
- Electron carriers used in fermentation must react with the product of glucose oxidation (pyruvate) to be reoxidized to lactic acid or ethanol (both of which are waste products).

QUESTIONS AND TOPICS FOR DISCUSSION

1. What is electricity?

2. What is an electrical current?

3. Why must electron transfer reactions include both oxidations and reductions?

4. Why is our discussion of ionization from Chapter 4 so relevant to a discussion of redox?

5. Why is it useful to separate redox reactions into half-reactions?

6. Why do electrons show up in half-reactions, but not in overall reactions?

7. Why must the number of electrons produced in oxidation half-reaction be consumed in a reduction half-reaction?

8. What does it mean to say a species is very active (in an electrochemical sense)?

9. Why does silver become tarnished (react with air), but gold doesn't?

10. Platinum is prized as a jewelry metal because it doesn't react with air. What can you surmise about its activity?

11. When determining whether a redox reaction is spontaneous, why do we compare the activity of the species being oxidized to the activity of the species produced through the reduction? Why don't we just compare the activities of the reactants?

12. Why do we separate the species being reduced from the species being oxidized in a voltaic cell?

13. Why can we separate reactants in a voltaic cell, but we can't separate reactants in the reaction of HCl and NaOH?

14. What is the purpose of a salt bridge in a voltaic cell?

15. What is the purpose of the wire in a voltaic cell?

16. What determines the direction of electron flow in the wire that connects the halves of a voltaic cell?

17. What is the difference between a voltaic cell and an electrolytic cell?

18. When we say a reaction is spontaneous—that it releases energy—we can't assume that the reaction releases all (or even any) of that energy as heat. That is to say, a spontaneous reaction is not necessarily exothermic. Why is this so? Provide an example (try to think of an example you can describe on the basis of experience, rather than looking something up on the Internet) of a process that is spontaneous, but not exothermic.

19. When we say a reaction is spontaneous, we can't assume that it is also fast. A discussion of reaction rate is largely beyond the scope of the text, but try to think of an example on the basis of experience in which a process is spontaneous but happens slowly.

20. Why aren't most batteries rechargeable?

21. How can we be sure there is no electron transfer involving SO_4^{2-} in the lead-acid battery (car battery) reaction: $Pb_{(s)} + PbO_{2(s)} + 2\ H_2SO_{4(aq)} \rightarrow 2\ PbSO_{4(s)} + 2\ H_2O_{(l)}$?

22. How are photovoltaic cells similar to voltaic cells? How are they different?

23. What are the challenges to solar energy as a global alternative-energy strategy?

24. Do you think solar energy is a feasible global alternative-energy strategy?

25. What advantages do parabolic trough solar-thermal power plants have over photovoltaic plants?

26. Why do plants save/use the electrons harvested from water during photosynthesis, but we eliminate the electrons harvested from carbohydrate during respiration?

27. If plants use sunlight to make ATP, why do you think they store carbohydrate?

28. Why do plants produce both cellulose and amylose from the glucose they synthesize?

29. Would you expect a cave-dwelling organism to produce chlorophyll? Why or why not?

30. Why do you think plants use multiple different light-harvesting molecules (in addition to chlorophyll)?

31. Why do you think that photosystems and other enzymes and proteins involved in photosynthesis are all located in a chloroplast-contained membrane?

32. What about the structure of ATP suggests that it's a high-energy molecule?

33. How are photosynthesis and respiration similar? How are they different?

34. Why do we need oxygen?

35. Why is water considered a metabolic waste product in respiring organisms?

36. In common parlance (and in medicine), "respiration" refers to moving air in and out of the lungs. What does the word "respiration" mean in a chemical sense?

37. Why do our muscles burn when we engage in an intense physical effort? Use more detail than presented in Chapter 9.

38. Why do we end up with an alcoholic, carbonated beverage when yeast contaminates a solution of starch and water? Use more detail than presented in Chapter 5.

PROBLEMS

1. Define each of the following: electrical potential energy; current; redox; oxidation; reduction; half-reaction.

2. Label each of the following as an oxidation or reduction reaction:

$$Na \rightarrow Na^+ + e^-$$

$$Cl_2 + 2\,e^- \rightarrow 2\,Cl^-$$

$$Fe^{2+} \rightarrow Fe^{3+} + e^-$$

3. Label each of the following as an oxidation or reduction reaction:

$$O_2 + 4\,e^- \rightarrow 2\,O^{2-}$$

$$Ca \rightarrow Ca^{2+} + 2\,e^-$$

$$Ce^{4+} + e^- \rightarrow Ce^{3+}$$

4. For each of the following redox reactions, indicate which species is being oxidized and which is being reduced:

$$Cl_2 + 2\,Li \rightarrow 2\,Li^+ + 2\,Cl^-$$

$$Mg + Cu^{2+} \rightarrow Mg^{2+} + Cu$$

5. For each of the following redox reactions, indicate which species is being oxidized and which is being reduced:

$$Zn + 2\,H^+ \rightarrow Zn^{2+} + H_2$$

$$4\,K + O_2 \rightarrow 4\,K^+ + 2\,O^{2-}$$

6. Write out half-reactions for each of the following:

$$Cl_2 + 2\,Li \rightarrow 2\,Li^+ + 2\,Cl^-$$

$$Mg + Cu^{2+} \rightarrow Mg^{2+} + Cu$$

7. Write out half-reactions for each of the following:

$$Zn + 2\,H^+ \rightarrow Zn^{2+} + H_2$$

$$4\,K + O_2 \rightarrow 4\,K^+ + 2\,O^{2-}$$

8. Combine the following half-reactions to produce a balanced redox reaction:

$$Sn \rightarrow Sn^{2+} + 2\,e^-$$

$$Cu^{2+} + 2\,e^- \rightarrow Cu$$

9. Combine the following half-reactions to produce a balanced redox reaction:

$$Na \rightarrow Na^+ + e^-$$

$$I_2 + 2\,e^- \rightarrow 2\,I^-$$

10. Combine the following half-reactions to produce a balanced redox reaction:

$$S + 2\,e^- \rightarrow S^{2-}$$

$$Al \rightarrow Al^{3+} + 3\,e^-$$

11. Combine the following half-reactions to produce a balanced redox reaction:

$$Li \rightarrow Li^+ + e^-$$

$$2\,H^+ + 2\,e^- \rightarrow H_2$$

12. Indicate which (if any) of the following could react spontaneously with Li: Cu, Cu^{2+}, Cd, Cd^{2+}.

13. Indicate which (if any) of the following could react spontaneously with Fe: Na, Na^+, Ni^{2+}, Ni^{3+}.

14. Indicate which (if any) of the following could react spontaneously with Ag^+: Au, Au^{3+}, Zn, Zn^{2+}.

15. Indicate which (if any) of the following could react spontaneously with Na^+: F, F^-, Pb, Pb^{2+}.

16. How could the following half-reactions be combined to produce a spontaneous redox reaction? (Hint: use Table 11.1.)

$$Zn \rightarrow Zn^{2+} + 2\,e^-$$

$$2\,H_2O \rightarrow O_2 + 4\,H^+ + 4\,e^-$$

17. How could the following half-reactions be combined to produce a spontaneous redox reaction? (Hint: use Table 11.1.)

$$Ni^{2+} \rightarrow Ni^{3+} + e^-$$

$$Fe^{2+} \rightarrow Fe^{3+} + e^-$$

18. How could the following half-reactions be combined to produce a nonspontaneous redox reaction? (Hint: use Table 11.1.)

$$Mg \rightarrow Mg^{2+} + 2\ e^-$$

$$Pb \rightarrow Pb^{2+} + 2\ e^-$$

19. How could the following half-reactions be combined to produce a nonspontaneous redox reaction? (Hint: use Table 11.1.)

$$2\ H_2O \rightarrow O_2 + 4\ H^+ + 4\ e^-$$

$$Na \rightarrow Na^+ + e^-$$

20. If an iron bar were immersed in a solution of $NaNO_3$, would sodium metal plate out on the bar? What if the solution were $Au(NO_3)_3$? $Cd(NO_3)_2$?

21. If a magnesium bar were immersed in a solution of $NaNO_3$, would sodium metal plate out on the bar? What if the solution were $Au(NO_3)_3$? $Cd(NO_3)_2$?

22. Define each of the following: voltaic cell; cathode; anode; salt bridge; spectator ion.

23. For a voltaic cell consisting of magnesium and silver electrodes in beakers of soluble $Mg(NO_3)_2$ and $AgNO_3$ respectively, answer each of the following (Hint: use Table 11.1.):

    a. What is the overall reaction?
    b. What are the half-reactions?
    c. Which bar is the cathode and which is the anode?
    d. Which electrode dissolves over time?
    e. If the salt bridge contains KNO_3, in which direction does each salt ion flow?

24. Imagine a battery is used to recharge the voltaic cell in question 19 and answer each of the following:

    a.   What is the overall reaction?
    b.   What are the half-reactions?
    c.   Would the cathode and anode change or stay the same?
    d.   Which electrode dissolves over time?
    e.   If the salt bridge contains KNO_3, in which direction does each salt ion flow?

25. For a voltaic cell consisting of gold and lead, answer each of the following. (Hint: use Table 11.1.):

    a.   What is the overall reaction?
    b.   What are the half-reactions?
    c.   Which occurs at the cathode, and which at the anode?
    d.   Which electrode dissolves over time?
    e.   If the salt bridge contains KNO_3, in which direction does each salt ion flow?

26. Imagine a battery is used to recharge the voltaic cell in question 21 and answer each of the following:

    a.   What is the overall reaction?
    b.   What are the half-reactions?
    c.   Would the cathode and anode change or stay the same?
    d.   Which electrode dissolves over time?
    e.   If the salt bridge contains KNO_3, in which direction does each salt ion flow?

27. Write out the overall reaction for the electrolysis of water. Which species is oxidized, and which is reduced?

28. Write out the half-reactions and overall reaction for the electrolysis of KI.

For questions 29–32, use the lead-acid reactions given in the chapter.

29. In a lead-acid car battery, what half-reaction would take place at the anode when the car engine was off, but the lights were on (use half-reactions given in the chapter)?

30. In a lead-acid car battery, what reaction would take place at the cathode when the car engine was off, but the lights were on (spontaneous overall reaction is $Pb_{(s)} + PbO_{2(s)} + 2\,H_2SO_{4(aq)} \rightarrow 2\,PbSO_{4(s)} + 2\,H_2O_{(l)}$)?

31. In a lead-acid car battery, what reaction would take place at the anode when the car engine was on (spontaneous overall reaction is $Pb_{(s)} + PbO_{2(s)} + 2\ H_2SO_{4(aq)} \rightarrow 2\ PbSO_{4(s)} + 2\ H_2O_{(l)}$)?

32. In a lead-acid car battery, what reaction would take place at the cathode when the car engine was on (spontaneous overall reaction is $Pb_{(s)} + PbO_{2(s)} + 2\ H_2SO_{4(aq)} \rightarrow 2\ PbSO_{4(s)} + 2\ H_2O_{(l)}$)?

33. Define each of the following: doping; n-type silicon; p-type silicon; and silicon p-n junction.

34. Why does a p-n junction photovoltaic cell utilize alternating sheets of Si doped with group 3 and group 5 elements?

35. Could silicon in a p-n junction photovoltaic cell be doped with Ge? Why or why not?

36. Could silicon in a p-n junction photovoltaic cell that was doped with phosphorus be p-type or n-type? Why?

37. In photosynthesis, water is oxidized. What is reduced (initially)? What is the final electron acceptor?

38. Why is it important that each reduced enzyme in photosynthesis is lower in activity than the enzyme that reduced it?

39. What is/are the major goal(s) of photosynthesis?

40. Why does photosynthesis require energy, but respiration does not?

41. Glucose is oxidized during respiration. What is reduced (initially)? What is the final electron acceptor?

42. What is/are the major goal(s) of respiration?

43. Why must electron carriers be involved in fermentation?

44. What are the differences between ethanolic and lactic acid fermentation?

45. What is/are the major goal(s) of fermentation?

1. Electrical potential energy is the potential for electron movement/work from the separation of opposite charges or close positioning of like charges; current is the flow of electrons; redox is a chemical reaction involving electron transfer; oxidation is the loss of electrons; reduction is electron gain; half-reactions are the oxidation and reduction parts of a redox reaction.

3. Reduction; oxidation; reduction.

5. Zn is being oxidized, H^+ is being reduced; K is being oxidized, O_2 is being reduced.

7. Half-reactions shown under complete reaction below.

$$Zn + 2\,H^+ \rightarrow Zn^{2+} + H_2$$

Oxidation: $Zn \rightarrow Zn^{2+} + 2\,e^-$

Reduction: $2\,H^+ + 2\,e^- \rightarrow H_2$

$$4\,K + O_2 \rightarrow 4\,K^+ + 2\,O^{2-}$$

Oxidation: $4\,K \rightarrow 4\,K^+ + 4\,e^-$ (can also be written as 1:1:1)

Reduction: $O_2 + 4\,e^- \rightarrow 2\,O^{2-}$

9. $2\,Na + I_2 \rightarrow 2\,Na^+ + 2\,I^-$

11. $2\,Li + 2\,H^+ \rightarrow 2\,Li^+ + H_2$

13. Ni^{3+}.

15. None of these.

17. $Fe^{2+} + Ni^{3+} \rightarrow Ni^{2+} + Fe^{3+}$

19. $2\,H_2O + 4\,Na^+ \rightarrow 4\,Na + O_2 + 4\,H^+$

21. Sodium would not plate out; gold would; cadmium would.

23. a) $Mg_{(s)} + 2\,Ag^+_{(aq)} \rightarrow Mg^{2+}_{(aq)} + 2\,Ag_{(s)}$; b) the oxidation is $Mg_{(s)} \rightarrow Mg^{2+}_{(aq)} + 2\,e$, the reduction is $2\,Ag^+_{(aq)} + 2\,e^- \rightarrow 2\,Ag_{(s)}$; c) the magnesium bar is the anode, and the silver bar is the cathode; d) the magnesium bar dissolves; e) the K^+ would flow toward the silver side, and the NO_3^- would flow toward the magnesium side.

25. a) $3\,Pb_{(s)} + 2\,Au^{3+}_{(aq)} \rightarrow 3\,Pb^{2+}_{(aq)} + 3\,Ag_{(s)}$; b) the oxidation is $3\,Pb_{(s)} \rightarrow 3\,Pb^{2+}_{(aq)} + 6\,e^-$, the reduction is $2\,Au^{3+}_{(aq)} + 6\,e^- \rightarrow 2\,Ag_{(s)}$; c) the lead bar is the anode, and the gold bar is the cathode; d) the lead bar dissolves; e) the K^+ would flow toward the gold side, and the NO_3^- would flow toward the lead side.

27. $2\,H_2O \rightarrow 2\,H_2 + O_2$; water is both oxidized and reduced, since it's the only reactant.

29. $Pb_{(s)} + H_2SO_{4(aq)} \rightarrow PbSO_{4(s)} + 2\,H^+_{(aq)} + 2\,e^-$

31. $PbSO_{4(s)} + 2\,H_2O_{(l)} \rightarrow PbO_{2(s)} + H_2SO_{4(aq)} + 2\,H^+_{(aq)} + 2\,e^-$

33. Doping is adding a small amount of contaminant to a substance, as in silicon semiconductors; n-type silicon has extra electrons due to group 5 element doping; p-type silicon has electron holes due to group 3 element doping; a silicon p-n junction occurs when n-type silicon and p-type silicon are put in contact to allow electron flow.

35. It could not; Ge has four valence electrons, just like Si. The silicon would be neither p-type nor n-type.

37. The first chemical to be reduced by water is chlorophyll. The final electron acceptors at the end of the transfer process are CO_2 and H_2O, to make carbohydrate.

39. To store chemical energy (ATP) and provide electrons for carbohydrate synthesis.

41. The first chemical to be reduced is an electron carrier. The final electron acceptor at the end of the electron transport chain is O_2.

43. The oxidation of glucose requires a reduction (of electron carriers).

45. To produce some energy (ATP) and reoxidize electron carriers.

SOURCES AND FURTHER READING

Solar Power

Ashley, Steven. (2008, August). Sunny days for silicon. *Scientific American*, *299*(2), 32.

Delucchi, M. Z., and Jacobson, M. A. (2009, November). A path to sustainable energy by 2030. *Scientific American*, *301*(5), 58.

http://www.sandia.gov/csp/

http://www.eere.energy.gov/

Photosynthesis

Bugbee, B. G., and Salisbury, F. B. (1988). Exploring the limits of crop productivity. *Plant Physiology*, *88*(3), 869–78.

Lightning

Ferguson, E. E., and Libby, W. F. (1971). Mechanism for fixation of nitrogen by lightning. *Nature*, *229*(5279), 37.

CREDITS

From Alchemists to Oppenheimer and Beyond—Nuclear Chemistry

12

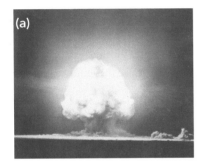

IMAGE 12.1. The word "nuclear" conjures many images, including those of a mushroom cloud from a nuclear blast (a) and a nuclear power plant (b). The symbol for radioactivity from a nuclear reaction (c) is internationally recognized.

The alchemists, early physicists, and chemists (including such notables as Sir Isaac Newton) spent their lives in pursuit of what are now recognized as mythical concepts. The Philosopher's Stone, for instance, was said to produce the Elixir of Life, a substance that would grant immortality. Further, the alchemists hoped to achieve *transmutation*, or the transformation of less valuable, or "base" metals (like lead), into gold. We learned in Chapter 1 that transmutation is chemically impossible—the Law of Conservation of Matter states that atoms of one substance cannot be transformed into atoms of another substance by any chemical means. In fact, it was John Dalton, an English chemist, who set forth the principles of atomic theory (including the Law of Conservation of Matter) in the early 1800s. As Dalton's atomic theory gained acceptance, the pursuit of alchemy declined.

12.1   Rise of Nuclear Chemistry

The field of chemistry enjoyed tremendous growth through the nineteenth and early twentieth centuries, as Dalton's atomic theory helped explain the empirical properties of substances and their reactions with one another. The idea that a chemical reaction is a rearrangement of atoms, involving the breaking and forming of bonds, stemmed directly from Dalton. Both inorganic and organic chemistry blossomed into major fields of study, giving rise to whole new industries and breathing life into old ones. The dye and pharmaceutical industries, explosives and plastics, were all developed or advanced during this time period. In 1896, a French physicist named Henri Becquerel noticed an interesting phenomenon during his investigation of uranium salts: they appeared to give off penetrating rays (much like X-rays, which had been discovered approximately 20 years prior).[1] This discovery intrigued a graduate student, Marie Curie, who was looking for a dissertation topic. She investigated uranium further, as well as looking for other elements that emitted similar penetrating rays. In the process, she and her husband, Pierre Curie, discovered the elements polonium and radium. The discovery of these strongly radioactive elements set the stage for later research by Ernest Rutherford, who categorized charged radioactivity into two classes (alpha and beta, as we'll see shortly), and later used alpha radiation to determine the structure of the atom and discover the nucleus.

In addition to allowing for the elucidation of atomic structure, nuclear chemistry and the study of radioactivity went on to turn science upside down. Newton's laws of physics were found to have limited applicability to subatomic particles, empirical evidence demonstrated that Dalton's theories about conservation of mass and matter didn't apply in nuclear reactions, and Einstein determined that mass and energy were more similar (and more interconvertible) than anyone had ever dreamed. It was truly the dawn of a new age.

NOTE[1]

While the medical applications of X-rays were recognized immediately, with the first radiology department already established by 1896 (in Scotland, incidentally), the dangers of routine exposure to radiation were not yet known. For some time after their invention, X-ray machines appeared at fairs and in exhibitions as a sort of sideshow entertainment.

IMAGE 12.3. Giants of physics and chemistry at the Solvay Conference, 1927. Among these great scientists, Marie Curie (seated third from left, front row), and Albert Einstein (seated fifth from left, front row).

12.2   Radioactive Decay

Unlike chemical reactions, which involve electrons from the electron cloud being shared or transferred between atoms, **nuclear reactions** take place within an atom's nucleus. A very simple decay reaction takes place when a **parent nucleus**[2] undergoes **decay**, meaning that it breaks down. A radioactive particle is lost from the parent nucleus, and the remainder of the parent is called a **daughter nucleus**:

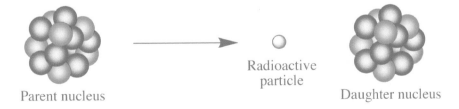

Parent nucleus   Radioactive particle   Daughter nucleus

 NOTE[2]

Of course, atoms are not devoid of electrons. The parent nucleus in question has an electron cloud, as all atoms do, but nuclear reactions don't involve the electrons surrounding the nucleus—they involve only the nuclear particles. For this reason, instead of referring to a "parent atom," we generally refer to a "parent nucleus," with the understanding that it is of course a complete atom.

The first decay reactions to be discovered involved the emission of charged particles from a nucleus. Charged particles can be deflected by an electromagnetic field, which is easily produced in the lab, and allows for detection of radiation. The direction in which particle rays are deflected provides information about the charge of the particles. Further, in an electromagnetic field, heavier particles are deflected to a lesser extent, meaning that the mass of the particle can also be established. One such charged particle is called the **alpha particle (a)**: it consists of two protons and two neutrons. The alpha particle is essentially a helium nucleus (not a helium atom—it has no electrons!), and is therefore sometimes written as $_1^2 H$. It's also quite common to see the particle written as $_2^4 \alpha$ in **nuclear equations**. A nuclear equation shares much in common with a chemical equation; just as in a chemical reaction, reactants are written on the left side of an arrow, and products are written on the right. Also as in a chemical reaction, mass is conserved in nuclear reactions. **Unlike a chemical reaction, however, matter is not conserved—in other words, if an atom of uranium appears on the reactant side of the equation, it may not appear on the product side.** In balancing chemical reactions, we don't explicitly check to be certain that mass is conserved; because every atom of a given element maintains its mass in the course of the chemical reaction, we can be confident that as long as the type and number of atoms are conserved, mass is conserved as well. In nuclear reactions, since mass is conserved but matter is not, we must explicitly check for conservation of mass in order to make sure that the equation is balanced. For this reason, it's very important to keep track not only of the identities of atoms involved in the reaction, but of their masses as well. Note that nuclear reactions are specific for particular isotopes of an element; that is, carbon-12 behaves differently than carbon-14 in a nuclear sense, despite their identical behavior in chemical reactions. As such, while in chemical reactions we simply indicate the presence of carbon (or whatever element), in nuclear reactions we must be explicit regarding which particular isotope of carbon (or whatever element) we mean. To do this, we indicate the mass number—and thereby identify the isotope—either by following up the name of an atom with its mass number (carbon-12 or C-12), or by providing atomic shorthand notation $(_6^{12}C)$, which was introduced in Chapter 2, and which includes not just the mass number but the atomic number as well. Take, for instance, the reaction below, which shows the alpha decay of uranium:

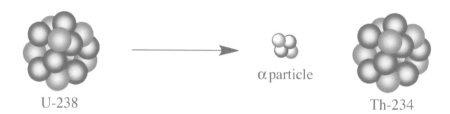

U-238   α particle   Th-234

We would read the above: *uranium-238 undergoes alpha decay to produce thorium-234*. Like a chemical equation, a nuclear equation is a symbolic representation of a reaction. It's a bit more common, though, to see the reaction written using atomic shorthand:

$$^{238}_{92}U \rightarrow\ ^{4}_{2}\alpha\ +\ ^{234}_{90}Th$$

Notice that this equation is balanced. The mass of the reactant is 238 amu. An alpha particle, weighing 4 amu, is lost, and the daughter nucleus has a mass of 234 amu. Thus, the total mass on either side of the equation is 238 amu. In addition to conservation of mass, nuclear equations (like chemical equations) must demonstrate conservation of charge. Since the nuclear reaction doesn't involve an atom's electrons at all, we ensure that charge is conserved by keeping track of protons. From the atomic number of uranium, we see that there are 92 protons in the parent nucleus. An alpha particle, containing two protons, is lost, and the daughter nucleus has 90 protons. In sum, there are 92 protons on each side of the equation. **In all nuclear equations, both mass and charge must be conserved.**

Note that on the periodic table, thorium does not have an atomic mass of 234 amu. In fact, thorium's atomic mass is 232. This does not mean that the daughter nucleus is not a "real" atom of thorium; rather, it's simply one of thorium's isotopes. The most stable form of thorium—Th-232—is the one indicated on the periodic table, but there are many isotopes of the element, each with different nuclear reactivity. What all the isotopes have in common, as we learned in Chapter 2, is their number of protons: all isotopes of thorium have 90 protons in their nucleus, or an atomic number of 90. As you begin to examine nuclear reactions, you will sometimes see the most common isotope of an element reacting (that would be the isotope whose mass number is equal or nearly equal to the atomic weight[3]), but you'll often see less-common isotopes participating in a reaction. This is nothing to worry about; all we have to remember is that an element is identified by its number of protons, so if we have a nucleus with 90 protons in it, it's thorium, no matter what the mass.

A second type of radioactive decay involves the loss of a **beta particle** (β) from the nucleus. A beta particle is an electron, and the idea that an electron could be coming from the nucleus understandably leaves us scratching our heads: our understanding of atomic structure tells us that the nucleus consists of protons and neutrons, but not electrons. In fact, it may be easier to think of the beta particle as a massless unit of negative charge (which, of course, makes it an electron) that comes from the nucleus. In order to conceptualize where negative charge might be found in the nucleus, let's mentally dissect a proton and a neutron. A proton consists of one unit of mass (1 amu) and one unit of positive charge:

NOTE³

Recall that in the case of an element with a fractional atomic mass, the most common isotope of the element can be identified by rounding the atomic mass to the nearest whole number, provided the atomic mass is already quite close to a whole number. For instance, with an atomic mass of 12.01, it's quite clear that carbon's most common isotope is C-12. With an atomic mass of 35.45, we can't determine the most common isotope of chlorine from the periodic table alone.

TRY THIS

Determine the number of protons and neutrons in each of the following: U-235, $^{15}_{8}O$, nitrogen-15, $^{210}_{84}Po$.

Answers:
92 protons, 143 neutrons; 8 protons, 7 neutrons; 7 protons, 8 neutrons; 84 protons, 126 neutrons.

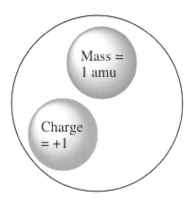

We've thus far been thinking of a neutron as consisting of one unit of mass (1 amu) and zero units of charge, but instead, let's imagine it slightly differently, as though it has one unit each of positive and negative charge:

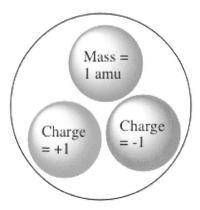

Conceptualized this way, the neutron still has a mass of 1 amu and a charge of zero, where this charge is the sum of a +1 unit and a -1 unit. When we think about a neutron in this way, it's much easier to imagine where in the nucleus a massless particle with a charge of -1 might come from: in beta decay, it's as though a neutron is simply losing its -1 charge unit. When this happens, the former neutron is now a particle with a mass of 1 amu and a charge of +1—it's a proton!

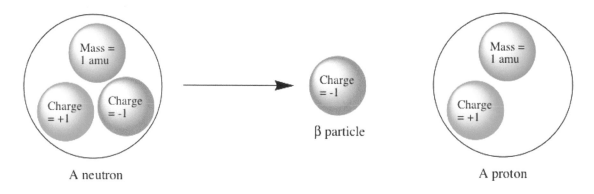

A neutron   β particle   A proton

Beta decay does not result in the loss of any mass from the parent nucleus; it's simply the loss of one unit of negative charge or, as we see from the above, the conversion of a neutron to a proton. Lead-214 reacts by beta decay to produce bismuth-214, as seen here:

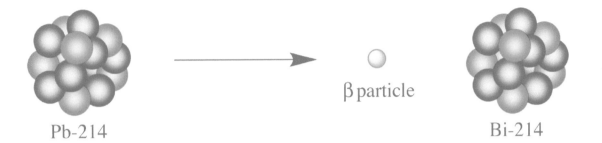

Pb-214   β particle   Bi-214

We can write this as a nuclear equation:

$$^{214}_{82}Pb \rightarrow \, ^{0}_{-1}\beta + \, ^{214}_{83}Bi$$

As an alternative to the symbol for beta decay in the equation above, it's quite common to see the beta particle written in such a way as to indicate that it is, in reality, an electron: $^{0}_{-1}e$. It's important to note the negative sign in front of the "atomic number" for the beta particle. Because this particle does not contain any protons, the number instead represents the particle's charge. As in all chemical and nuclear reactions, both mass and charge must be conserved in beta decay. We must remember that we are now working with both positive and negative charges when balancing the equation. On the reactant side, we have a total mass of 214 amu, and 82 protons. On the product side, we add the masses for a total of 214 amu. We then add the atomic numbers/charges, for a total of 82 protons (83 – 1 = 82). **A simple way to think about this is that within nuclear equations, the superscripts must balance, and the subscripts must balance.**

Because we always observe conservation of mass and charge in nuclear equations, it's possible to predict the product of a nuclear reaction as long as we know the identity of the parent nucleus (we have to know not only which element, but also which isotope) and of the decay particle. For instance, we can determine the daughter nucleus produced by the alpha decay of polonium-218:

1. Write out the equation including the known information, leaving a blank for the unknown information. Make sure to fill in the atomic number for the parent nucleus from the periodic table.

$$^{218}_{84}Po \rightarrow \, ^{4}_{2}\alpha + \underline{\hspace{2cm}}$$

Determine the daughter nucleus
produced when bismuth-214 decays
through emission of a beta particle.

Answers:

$^{214}_{83}Bi \rightarrow \,^{0}_{-1}\beta + \,^{214}_{84}Po$

Solution:
214 = 0 + (daughter mass number);
daughter mass number = 214
83 = −1 + (daughter atomic
number); daughter atomic number = 84

What type of decay produces N-14
from C-14?

Answers:

$^{14}_{6}C \rightarrow \,^{0}_{-1}\beta + \,^{14}_{7}N$

Solution:
14 = 0 + (daughter mass number);
daughter mass number = 14
6 = −1 + (daughter atomic
number); daughter atomic number = 7

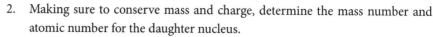

2. Making sure to conserve mass and charge, determine the mass number and atomic number for the daughter nucleus.

218 = 4 + (daughter mass number); daughter mass number = 214

84 = 2 + (daughter atomic number); daughter atomic number = 82

$$^{218}_{84}Po \rightarrow \,^{4}_{2}\alpha + \,^{214}_{82}\underline{\quad\quad}$$

3. Using the daughter nucleus's atomic number, determine the identity of the daughter. Make sure you use the atomic number, not the mass number—there are isotopes of many elements with a mass of 214 amu, but there's only one element with an atomic number of 82!

$$^{218}_{84}Po \rightarrow \,^{4}_{2}\alpha + \,^{214}_{82}Pb$$

Using this same technique, it's possible to determine the type of decay taking place in a nuclear reaction, or the identity of the parent nucleus, as long as the other two reactants or products are known.

A third decay particle, called a **gamma particle (γ)**, is both massless and charge-less. The gamma particle is written $^{0}_{0}\gamma$, and is pure electromagnetic energy. Recall from Chapter 10 that gamma rays (beams of gamma particles) are at the highest-energy end of the electromagnetic spectrum. Like all EMR, we can think of them as waves (with a measurable wavelength), or as particles (with a measurable energy). We'll think about them as particles for the time being, because this allows us to imagine their release from a nucleus. **Gamma particles are lost from a nucleus as it relaxes energetically, but since they are both massless and chargeless, they do not change either the mass number or atomic number.** Often, a high-energy nucleus is indicated with the use of an asterisk, but this is not critical:

Tc-99*   γ particle   Tc-99

The reaction above could be written as either of the following:

$$^{99}_{43}Tc * \rightarrow \,^{0}_{0}\gamma + \,^{99}_{43}Tc$$

$$\,^{99}_{43}Tc \rightarrow \,^{0}_{0}\gamma + \,^{99}_{43}Tc$$

If a gamma decay equation is written without an asterisk on the parent nucleus, it's nevertheless taken to mean that the parent was higher in energy, and that energy has been lost in the form of gamma radiation. Because neither mass nor atomic number changes from parent to daughter nucleus, gamma decay equations are easy to recognize.

12.3 Radioactive Decay Series

Recall from Chapter 2 that nucleons are subject to two different forces within a nucleus. The strong force holds any nucleon to any other nucleon, but is only effective over short distances. Additionally, there is electrostatic repulsion between protons. Therefore, larger nuclei must contain greater numbers of neutrons relative to protons. Despite the extra neutrons, very large nuclei (bismuth and larger) are always radioactive—they're so big, pieces of the nucleus (decay particles) tend to fall off. It seems to make sense, then, that the daughter nucleus of a decay reaction might be radioactive itself. U-238, for instance, decays by giving off an alpha particle to form Th-234, but thorium is also large, and therefore radioactive. **In fact, it's quite common for decay reactions to occur in a series (Image 12.4), with each daughter nucleus becoming the parent in a subsequent reaction, until a stable nucleus is reached.** This is called a **radioactive decay series** (Image 12.4).

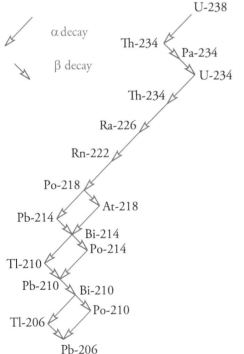

TRY THIS:

Determine the daughter nucleus produced when Ni-260 decays through emission of a gamma particle.

Answers:
$$\,^{60}_{28}Ni* \rightarrow \,^{0}_{0}\gamma + \,^{60}_{28}Ni \text{ or } \,^{60}_{28}Ni \rightarrow \,^{0}_{0}\gamma + \,^{60}_{28}Ni$$

Solution:
60 = 0 + (daughter mass number);
daughter mass number = 60
28 = 0 + (daughter atomic number);
daughter atomic number = 28

IMAGE 12.4. A radioactive decay series for U-238; Pb-206 is a stable nucleus. Note that there are several branch points from which more than one decay path is possible.

In general, alpha decay reduces nuclear bulk and is commonly emitted from large nuclei. Beta decay adjusts the ratio of protons to neutrons, and is common both in small nuclei (such as radioactive isotopes of carbon and nitrogen, which are not likely to produce alpha particles) and as an intermediate step in a decay series.

12.4 Nuclear Fission

An interesting development in nuclear chemistry arose in the early twentieth century as scientists began attempting to split the atom. Working under Rutherford, Hans Geiger (of the Geiger counter) and Ernest Marsden used alpha particles to determine the structure of the atom by firing alpha particles at them. They determined that the alpha particles often passed through the atom without being deflected (or with minimal deflection), but that they sometimes experienced significant deflection. From this, the scientists surmised that much of the atom was made up of a nearly massless cloud (made up of electrons). They further determined that at the very center of the atom was a very massive, positively charged core—the nucleus—that deflected the positively charged alpha particle. While a monumental discovery in its own right, this led Rutherford to further wonder what would happen if he purposely aimed the alpha particles at the nucleus. Would the nucleus continue to deflect all the radiation or would it be possible, given sufficient acceleration of alpha particles, to hit the nucleus and cause a rearrangement of its own subatomic makeup?[4] In 1919, he demonstrated that striking a nitrogen-14 nucleus with an alpha particle caused the incorporation of that particle, followed by a fragmentation of the new nuclear group:

$$^{14}_{7}N + {}^{4}_{2}\alpha \rightarrow {}^{18}_{9}F \rightarrow {}^{17}_{8}O + {}^{1}_{1}H$$

While this is a slightly more complex nuclear reaction than those we've seen so far, this equation is also balanced with regard to mass and charge: each reactant or product grouping has a mass of 18 amu and a charge of nine. Unlike the decay reactions from earlier in the chapter, the reaction above is induced—nitrogen-14 is not radioactive under ordinary circumstances. Rutherford wanted to continue to test his theory, and sought smaller, higher-energy particles with which to bombard a nucleus. The particle accelerator was invented shortly thereafter, allowing for the bombardment of stable nuclei with high-energy protons and neutrons, which also induced nuclear reactions. In 1932, Rutherford used the particle accelerator to bombard lithium with protons. The resulting nucleus split into two helium atoms:

$$^{7}_{3}Li + {}^{1}_{1}H \rightarrow {}^{4}_{2}He + {}^{4}_{2}He$$

NOTE[4]

Picture this as a game of "atomic billiards": Rutherford was trying to scatter the nucleus using an alpha particle as a "cue ball."

In addition to the intriguing split of a single larger nucleus into two equal parts (a departure from the previously known decay reactions in which a small particle was released from the nucleus), it was observed that lithium reaction also released a tremendous amount of energy, given that the reaction took place on such a small atomic scale. Further, and even more bizarre, a very small amount of mass was observed to have been lost in the reaction—about 0.02 amu. This lost quantity is called mass defect. **While the defect was too insignificant to show up in the balanced nuclear equation—the mass numbers appear to balance perfectly because the lost mass is so small compared to the total—it nevertheless puzzled researchers, as it seemed to violate the law of conservation of mass.** Unfortunately, the lithium-splitting reaction was quite difficult to achieve, with nuclear events taking place in only one out of billions of atoms in a sample. For this reason, neither the energy generated nor the mass defect initially attracted much attention.

In 1939, a German group, including the Jewish scientist Lise Meitner, achieved the first true heavy atom nuclear fission reaction. Induced to react by bombardment with neutrons, a uranium nucleus did not decay by releasing small particles but instead broke into two nearly equivalent pieces (while also releasing several neutrons):

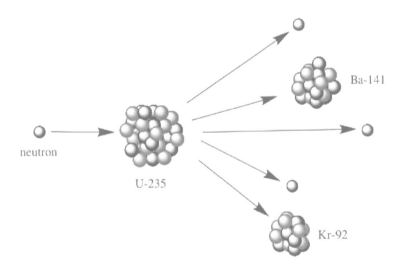

neutron

U-235

Ba-141

Kr-92

$$^{235}_{92}U + {}^{1}_{0}n \rightarrow {}^{236}_{92}U \rightarrow {}^{141}_{56}Ba + {}^{92}_{36}Kr + 3\ {}^{1}_{0}n$$

As with the earlier lithium experiments, both a tremendous energy release and mass defect were observed. Meitner refused to dismiss these observations; despite having to flee Germany and leave her research group for political reasons,[5] she continued to work on the problem of mass defect. The explanation, she eventually reasoned, lay in Albert Einstein's theory of relativity, which had been developed nearly 35 years prior but which had not yet been proven or applied. Based upon Einstein's observations that a moving body has slightly more mass than an identical

NOTE[5]

Making note of the year, Meitner's Jewish heritage, and the political environment at the time elucidates the reasons for her flight.

body at rest, he theorized that mass and energy were interchangeable quantities, with a relationship determined by the equation:

$$E = mc^2$$

Here, E is energy (J), m is mass (kg), and c is the speed of light (3.0×10^8 m/s). Because the speed of light is such a large quantity to begin with and is amplified by being squared in Einstein's equation, the energy yield from even a small amount of mass is enormous. **Meitner determined that in fact, the energy yielded in the fission experiments correlated exactly with the observed mass defect as predicted by Einstein's equation.**

As experiments in nuclear fission progressed, it was determined that a typical mass defect was in the range of 0.1% of the original fissionable mass: for 1.0 kg of reacted material, this would equate to approximately 1.0×10^{-3} kg (or 1.0 g) of defect. Per Einstein's equation, the energy yield from such a reaction would be 9.0×10^{13} J of energy, or the same amount released by the detonation of 22,000 tons (44,000,000 pounds) of TNT. In the historical context of these discoveries, it's no wonder that the first application of nuclear fission was as an instrument of war.

TRY THIS

What amount of energy (in Joules) would be released if a nuclear reaction resulted in a mass defect of 0.05 kg?

Answer:
4.5×10^{15}J.

Solution:
$E = mc^2$
$E = (0.05 \text{ kg})(3.0 \times 10^8 \text{ m/s})^2 = 4.5 \times 10^{15}$ J

12.5 The Atom Bomb

Developing a weapon from the nuclear fission of uranium posed some challenges. It was immediately apparent to scientists that the reaction had the potential to be self-sustaining, owing to the fact that a neutron was required to start the reaction, and three neutrons were yielded. In the event that each (or at least one) of these neutrons struck another nearby U-235 nucleus, another fission reaction would occur, and so on. It was also apparent that the reaction could grow exponentially (Image 12.5).

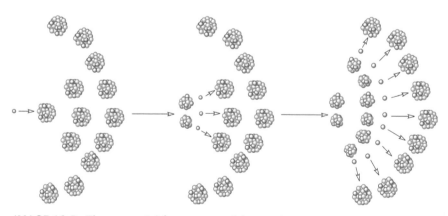

IMAGE 12.5. The potential for exponential growth of an induced uranium fission reaction.

The exponential growth of a nuclear reaction was a desirable outcome with regard to production of a weapon; it would result in the fastest possible release of energy and maximal explosive potential. It was therefore necessary to ensure that the released nuclei would contact other U-235 nuclei, rather than being lost in one way or another. As such, to produce a self-sustaining reaction, called a **nuclear chain reaction**, uranium samples had to contain high concentrations of U-235. Naturally occurring uranium is a mixture of isotopes, primarily U-235 and U-238. The heavier isotope is capable of absorbing neutrons, but generally does not fission in response, thus rendering absorbed neutrons "lost" with regard to the chain reaction. Furthermore, U-238 makes up approximately 99.3% of naturally occurring uranium (0.7% U-235); **researchers therefore determined that to sustain a reaction, the uranium used in nuclear reactions would need to be higher in U-235 than it was as found in nature.** Unfortunately for weapons chemists, isotopes of an element are chemically identical to each other, differing only in weight and nuclear properties. This makes the logistics of separating isotopes from one another exceedingly complex. In the end, the researchers invented a separation technique called **gaseous diffusion**. When gas phase uranium is passed through a filter, the lighter isotope filters faster than the heavier (though the separation is far from perfectly efficient), isolating a somewhat higher percentage of U-235 on one side of the filter (approximately 3.5% U-235), and leaving less of the reactive isotope (approximately 0.3% U-235) on the other side. The leavings—called **depleted uranium** or **uranium tails**—are radioactive waste that must be stored in high-level facilities. They are difficult and expensive to reprocess into usable uranium fuel.

The hurdle of uranium enrichment overcome, the final step in developing a weapon was to determine what quantity of nuclear fuel—called the **critical mass**—

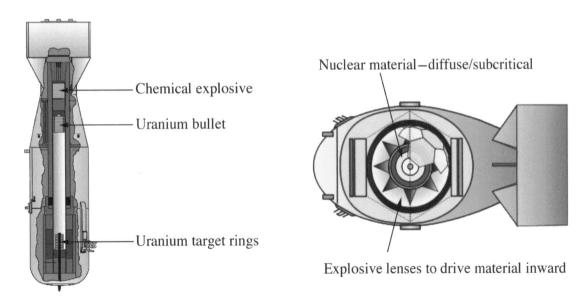

Chemical explosive

Uranium bullet

Uranium target rings

Nuclear material—diffuse/subcritical

Explosive lenses to drive material inward

IMAGE 12.6. Little Boy was a "gun-type" weapon, using a chemical detonation to drive a subcritical uranium bullet into subcritical uranium rings, creating a critical mass (a). Fat Man was an "implosion-type" weapon, using a chemical detonation to compact a diffuse core of plutonium into a critical mass.

would be required to ensure that released neutrons would strike reactive nuclei a sufficient proportion of the time, such that the reaction would grow in magnitude. It was determined that each nuclear fuel—Pu-239, which was discovered in the process of the uranium weapons studies, was also used in weapons development—displayed a characteristic mass at which a self-sustaining reaction could be produced. For U-235, the critical mass was approximately 15 kg (33 pounds). One major consideration in the building of a nuclear weapon was that a critical mass of either fuel, once assembled, could be counted upon to react immediately. While the fission reactions of both U-235 and Pu-239 are neutron-induced, it is not necessary to introduce neutrons into a critical mass in order to generate a reaction—stray neutrons from sources such as cosmic rays abound, meaning that weapons would have to be designed so as to ensure that fuel masses were kept subcritical until detonation. Two types of nuclear weapons were developed through the **Manhattan Project**, an undertaking of the U.S. Army, led by physicist Robert Oppenheimer and staffed by a team of chemists and physicists. **Both weapons functioned by bringing together subcritical masses of nuclear fuel through chemical detonation, resulting in the assembly of a critical mass that would then undergo nuclear fission.**

An "implosion-type" bomb was tested at the Trinity Site near Alamogordo, New Mexico, on July 16, 1945. The "gun type" was never tested prior to its use later that summer. The results were impressive: reports of the sound of an explosion came from as many as 200 miles away, with the mushroom cloud rising to a height of more than 7 miles. Oppenheimer, upon observing the blast, reportedly simply said, "It worked." Years later, in a 1965 television interview, he recounted his reaction to the weapons test:

> We knew the world would not be the same. A few people laughed, a few people cried. Most people were silent. I remembered the line from the Hindu scripture, the Bhagavad Gita … "Now I am become Death, the destroyer of worlds." I suppose we all thought that, one way or another.

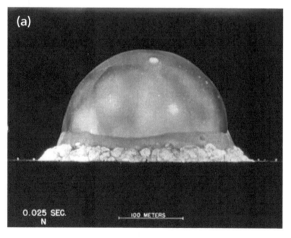 

IMAGE 12.7. An original film shot from nuclear testing at the Trinity Site (a), and the commemorative sign that marks the area along the highway today (b). In the film still, the blast diameter is nearly 500 m, with a height of approximately 250 m, only 0.025 seconds after detonation.

On August 6, 1945, the B-29 bomber *Enola Gay* dropped "Little Boy" on Hiroshima, Japan, which had been spared conventional bombing so as to serve as an ideal experimental site for testing an atomic weapon in war. The explosive force has been estimated at around 15,000 tons of TNT. Three days later, "Fat Man" was dropped on Nagasaki. While reports of the death toll from the two explosions and their aftereffects vary—numbers appear to be well over 300,000—it has been difficult to determine exactly how many lives were lost in the attack and its aftermath. The reason for this is that, in addition to the effects of pressure, fire, and radiation resulting in immediate death, lower doses of radiation from the bomb and its fallout (unreacted radioactive nuclear material dispersed into the atmosphere, which later rained down over a large portion of Japan) resulted in increased incidences of various cancers among survivors. The appropriateness of the nuclear attack upon Japan has been debated globally, and never since has a nuclear weapon been used in war.

CONCEPT CHECK

Why is the concept of mass defect important to the development of a nuclear weapon?

Answer: The mass defect is the source of a nuclear weapon's explosive energy.

12.6   Nuclear Fusion

In a reaction nearly opposite that of fission, it's possible for smaller nuclei to combine in a process called **nuclear fusion**. The energetic yield of fusion is tremendous—it is this type of reaction that accounts for the heat and energy produced by the sun and other stars.[6] The bulk of the sun's energy stems from the fusion of hydrogen into helium, through a process that can be simplified as follows:

$$_1^1H + {}_1^1H \rightarrow {}_1^2H + {}_1^0e + neutrino$$
$$_1^1H + {}_1^2H \rightarrow {}_2^3He + {}_0^0\gamma$$
$$_2^3He + {}_2^3He \rightarrow {}_2^4He + 2\,{}_1^1H$$

Not to be confused with a beta particle, in the fusion series above, $_{-1}^0e$ is the symbol for a **positron**, which is a massless, positively charged particle. Like the similar (but negatively charged) beta particle, positrons must be considered when balancing a nuclear reaction. A *neutrino* is a massless, neutral particle, and as such, doesn't affect balancing of the reaction. While fusion releases an enormous amount of energy (owing, as in fission, to the conversion of a small amount of mass to a large amount of energy), it also takes an enormous amount of energy to initiate fusion. Because all nuclei are positively charged and repel each other, energy is required to accelerate nuclei into one another at phenomenal speed. In a reactor such as the sun, fusion becomes self-sustaining—the energy released from fusion reactions supplies the energy required to initiate further fusion reactions. Man-made fusion, however, presents a logistical challenge. With an eye toward utility in power-generation and weapons, physicists and chemists applied themselves to the production of fusion as

NOTE[6]

This process is called *nucleosynthesis*.

a side arm of the Manhattan Project. It was determined that, appropriately fueled, deuterium and tritium (both isotopes of hydrogen) could be fused:

$$_1^2H + {}_1^3H \rightarrow {}_2^4He + {}_0^1n$$

A fission reaction was identified as one potential source of energy powerful enough to fuel fusion, due to its phenomenal energy release. **As a result, where the original nuclear weapons ("atom bombs") were fission weapons with chemical explosions as triggers (driving the critical mass of fuel together), the "hydrogen bombs" initially created and tested in the early 1950s were fusion weapons triggered, essentially, by atom bombs.**

As shown in Image 12.8, a hydrogen bomb likely consists of two distinct areas (one fission, one fusion), suspended in polystyrene foam. The primary area is made up of a subcritical plutonium core, much like that at the heart of "Fat Man," which is compressed into a critical mass through chemical detonation. The fission reaction releases tremendous heat and radiation, and it compresses the polystyrene foam. These conditions compress the fission spark plug within the secondary area, initiating a second fission reaction—this time much closer in proximity to the fusion

material, which consists of $_3^6Li$ and lithium deuteride, a source of $_1^0H$.

Upon bombardment with neutrons from the secondary fission reaction, lithium-6 fissions, producing tritium, which then fuses with deuterium:

$$_3^6Li + {}_0^1n \rightarrow {}_2^4He + {}_1^3H$$
$$_1^3H + {}_1^2H \rightarrow {}_2^4He + {}_0^1n$$

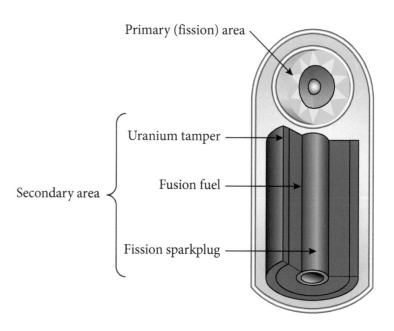

Primary (fission) area

Uranium tamper

Secondary area

Fusion fuel

Fission sparkplug

IMAGE 12.8. A likely schematic of the hydrogen bomb.

While hydrogen bombs have been tested (generally underwater, so great is their power), none has ever been used in wartime. Given the limited testing, it's difficult to accurately approximate the potential destructive power of such a weapon, but most estimates hold that a hydrogen bomb would have an explosive power of up to 10 megatons (equivalent to 10 million tons of TNT). By comparison, "Little Boy" had a 13,000-ton equivalency. The combined explosive power of every single bomb dropped during the course of World War II equals only 2 megatons.

12.7   Nuclear Energy

While the atomic bombs required that nuclear fission reactions "go critical" (become exponential chain reactions), it was noted by those individuals working on the Manhattan Project that a similar energetic yield, without the explosion, might be achieved through a controlled fission reaction. In order to produce such an outcome, masses of fuel would need to be kept subcritical, and neutron-absorbing materials would be necessary to ensure that while a chain reaction was maintained, exponential reaction growth could not result. The development of these safety measures heralded the age of nuclear power—the first nuclear reactor to be used for generation of electricity was built in the USSR in 1954, and 1957 saw the first nuclear reactor built in the United States. Nuclear power generation potential grew rapidly, both owing to expansion of existing reactors and the building of new ones. **Worldwide, from the early 1960s to the late 1980s, the amount of nuclear power generated on a yearly basis increased by a factor of 300.**

Nuclear reactor fuel is essentially identical to weapons fuel, though the purity may be significantly lower, as over-purified fuel increases the likelihood of a critical reaction. The general design of a nuclear plant involves a controlled reaction taking place within a **nuclear reactor**, which is an isolated reaction vessel. **Fuel assemblies**, made up of pellets composed of oxides of U-235 or Pu-239, are interspersed with **control rods**, made of elements that absorb neutrons without fissioning, such as boron and cobalt (Image 12.9).

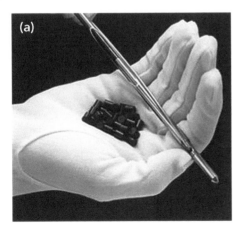

IMAGE 12.9. Fuel pellets (a) are stacked into rods, and then grouped into fuel assemblies (b). Control rods are interspersed to absorb neutrons and control the rate of reaction.

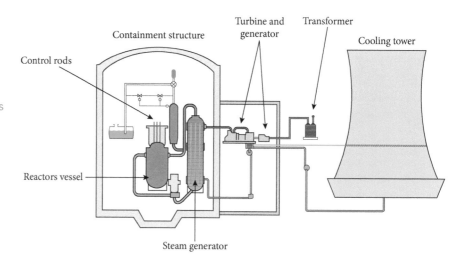

IMAGE 12.10. A schematic of a nuclear power plant. Source of heat aside, a nuclear plant is nearly identical to a coal-fired power plant—heat vaporizes water, which forms steam, which turns a turbine and generates electricity.

In addition to fuel assemblies and control rods, the nuclear reactor is filled with **primary coolant**, a liquid (often metal or molten salt) that heats up as it absorbs the energy released by the fission reaction. The heat is transferred from one medium to another, eventually being used to generate steam.

It should be noted that neither the water entering a nuclear power plant (which will eventually be vaporized into steam) nor the steam exiting the cooling tower is ever in contact with radioactive material. As Image 12.11 illustrates, primary coolant—which is in direct contact with radioactive material—is circulated through a closed loop by means of a pump. As the primary coolant passes through a heat exchanger, its heat is transferred to a secondary coolant (also a metal or molten

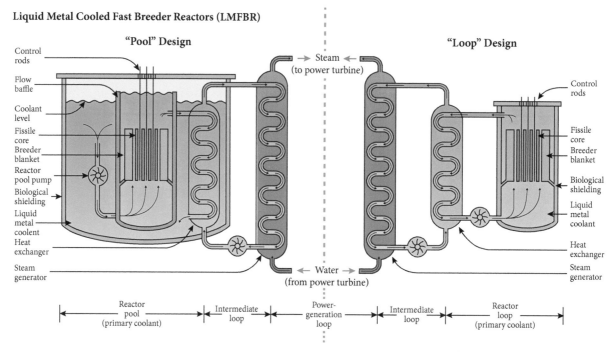

IMAGE 12.11. Water that enters from (or steam that exits to) the environment outside a nuclear power plant is separated from the nuclear reaction by two completely closed loops.

(a)

(b)

IMAGE 12.12. Steam exits the cooling tower, after having moved upward past a turbine in a steam-powered electrical generator—resulting in production of electricity—in both coal-fired power plants (a) and nuclear power plants (b).

salt). This coolant is also circulated within a completely closed loop. Secondary coolant, contained in its pipes, passes into a steam generator, where it transfers heat to water pumped in from the outside.

The water is vaporized, and the steam exits by way of a cooling tower. **In the end, while the original source of heat is different in coal-fired and nuclear power plants, the result is the same: water is heated to steam, which, as it rises, rotates a turbine in a steam-powered electrical generator—this produces electricity** (Image 12.10).

While the primary coolant can take on temporary radioactivity from the fuel, it is not possible to transfer that radioactivity to the secondary coolant,[7] meaning that the closed loop system prevents water exiting the cooling tower from ever having been exposed to radiation. **In fact, individuals living within 50 miles of a nuclear power plant are exposed to an average of 3 times less radiation than those living within the same proximity of a coal-fired power plant:** the naturally occurring radioactive isotopes of carbon and other elements present in coal (which are vaporized and released as particulate matter upon combustion) account for significantly more radiation exposure than any released by a nuclear generator.

 NOTE[7]

Some types of radiation can cause nonradioactive substances to become radioactive—neutrons, for instance, can induce radioactivity (hence the induced radioactivity of the primary coolant). However, exposure to other types of radioactivity, such as gamma radiation, can't induce radioactivity—a gamma ray is nothing more than energy, and it therefore can't change the nucleus of an atom or induce it to fission. For this reason, neutron-releasing nuclear fuel can induce radioactivity in other substances (like primary coolant), but primary coolant (which doesn't release neutrons) can't cause secondary coolant or anything else to become radioactive.

12.8   Half-Life

An important concept in a discussion of radioactivity is that of half-life. **All radioisotopes (meaning radioactive nuclei) decay at predictable rates. Half-life is the amount of time required for half the nuclei in a sample to decay into another nuclear form.** Carbon-12 is a stable isotope, and does not have a half-life. Carbon-14, however, is radioactive, and has a half-life of 5,730 years. This means that given a sample of 100 grams of carbon-14, half the sample will have reacted to become something else—and the other 50 grams of C-14 will remain unreacted—after 5,730 years. Should another 5,730 years be allowed to pass, half of the 50 grams remaining of C-14 will react to become something else, leaving just 25 grams of C-14. Note that

Radioisotope	Half-Life
U-238	4.47×10^9 years
U-235	7.03×10^8 years
Pu-239	24065 years
C-14	5730 years
H-3	12.35 years
Ra-228	5.75 years
Ra-225	14.8 days
Rn-222	3.82 days
C-11	20.4 minutes
N-13	9.97 minutes
O-15	2.04 minutes
N-16	7.13 seconds
Rn-219	3.96 seconds
Po-212	3.05×10^{-7} seconds

TABLE 12.1 Half-lives of some radioisotopes.

since half the remaining sample reacts with each subsequent half-life, the decay curve is asymptotic: mathematically, the mass of isotope approaches but does not actually reach zero, as in Image 12.13.

The half-lives of radioisotopes vary tremendously; some decay in a matter of microseconds, while others have half-lives in the billions of years (Table 12.1).

It's noteworthy that while we tend to keep track of the mass remaining of the original radioisotope, the rest of the mass doesn't disappear. Rather, it is present in a different nuclear form. Whether this new form goes on to react is entirely dependent upon what kind of decay is taking place, and which new nuclear form is produced. For instance, C-14 generally decays by beta emission:

$$^{14}_{6}C \rightarrow \,^{0}_{-1}\beta + \,^{14}_{7}N$$

Because N-14 is a stable nucleus, no further decay takes place. One may therefore assume that if 100 grams of C-14 has been allowed to decay for 5,730 years (1 half-life), 50 g of C-14 remain, and the remaining 50 g have been converted to N-14. After a second half-life, 25 g of C-14 would remain, and a total of 75 g of N-14 would have been formed. In the case of U-238 (half-life of 4.47×10^9 years), determining the quantity of the original radioisotope remaining is merely a matter of counting half-lives, but figuring out the mass (and identity) of each decay product is much more complex. U-238 decays to form Th-234, with a half-life of 24 days, which decays to form Pa-234, with a half-life of 1.2 minutes, and so on. Thus, given 100 grams of U-238, we'd be assured that after 4.47×10^9 years,

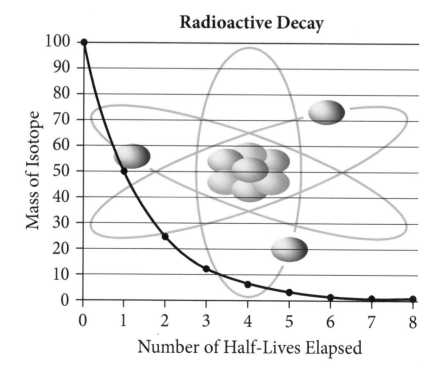

IMAGE 12.13. Mass of an original 100 g radioactive sample remaining after a given period of time. Note that the decay is asymptotic.

Radioactive Decay

(Graph: x-axis "Number of Half-Lives Elapsed" from 0 to 8, y-axis "Mass of Isotope" from 0 to 100)

only 50 grams of U-238 would remain. The remaining 50 g, however, would be working its way through a chain of nuclear reactions in an asynchronous manner, making it very difficult to determine exactly how much of each radioisotope would be present.

It's very important to realize that while half-lives are discrete "checkpoints" in time, radioactive decay itself is a continuous process. **That is, while we can be assured that half of an initial sample will have decayed after one half-life has passed, the actual decay takes place throughout the half-life.** As such, if less than one half-life has passed, there are fewer than 100 g (but more than 50 g) of an original 100 g sample remaining.

There is a mathematical equation we can use to solve half-life problems:

$$N_t = N_0 \left(\frac{1}{2}\right)^{\frac{t}{t_{0.5}}}$$

Here, N_t is the quantity of radioisotope remaining after a given period of time (t), N_0 is the original amount of radioisotope, and $t_{0.5}$ is the length of the half-life. Note that time and half-life may be in any units—seconds, minutes, years, and so forth—as long as they match. For instance, how much of a 55 g sample of Ra-228 remains after 17.25 years (see Table 12.1 for half-life)?

$$N_t = N_0 \left(\frac{1}{2}\right)^{\frac{t}{t_{0.5}}}$$

$$N_t = 55g \left(\frac{1}{2}\right)^{\frac{17.25}{5.75}}$$

$$N_t = 6.875g$$

While the half-life equation can be quite useful, many students will find that an intuitive approach to solving problems is somewhat simpler. Using an intuitive approach, the same problem could be addressed as follows:

1. Determine how many half-lives have elapsed.

$$17.25 \; years \; total \times \frac{5.75 \; years}{1 \; half-life} = 3 \; half-lives$$

 TRY THIS

How much of a 45.6 g sample of N-13 remains after 4 half-lives?

Answer:
2.85 g.

Solution:
$45.6 \, g \, initially \times 0.5 = 22.8 \, g \, after \, half-life \, 1$
$22.8 \, g \times 0.5 = 11.4 \, g \, after \, half-life \, 2$
$11.4 \, g \times 0.5 = 5.7 \, g \, after \, half-life \, 3$
$5.7 \, g \times 0.5 = 2.85 \, g \, after \, half-life \, 4$

How much time does it take to reduce 200 g of U-235 to 12.5 g?

Answer:
2.81×10^9 years.

Solution:

$200\ g\ initially \times 0.5 = 100\ g\ after\ half-life\ 1$
$100\ g \times 0.5 = 50\ g\ after\ half-life\ 2$
$50\ g \times 0.5 = 25\ g\ after\ half-life\ 3$
$25\ g \times 0.5 = 12.5\ g\ after\ half-life\ 4$

$4\ half-lives \times \dfrac{7.03*10^8\ years}{1\ half-life} = 2.81*10^9\ years$

2. Determine how much of the original sample remains after each half-life.

$$55\ g\ initially \times 0.5 = 27.5\ g\ after\ half-life\ 1$$
$$27.5\ g \times 0.5 = 13.75\ g\ after\ half-life\ 2$$
$$13.75\ g \times 0.5 = 6.875\ g\ after\ half-life\ 3$$

12.9 Storing Nuclear Waste

A major logistical problem associated with nuclear power generation is the production of **high-level radioactive waste**, which includes spent fuel and the waste from reprocessed spent fuel. **While the rate of fission in this material has slowed to the point that it can no longer be used for energy generation, the spent fuel is still hot (thermally) and highly radioactive:** recall that heavy atoms are all radioactive, and that even if nearly all the U-235 or Pu-239 in a sample has reacted, the daughter nuclei are still reactive.[8] To date, there is no permanent high-level waste storage facility in the United States, meaning that nuclear plants are responsible for storing their own spent fuel. Additionally, reprocessing of spent fuel does not currently take place in the United States, due to concern that the enriched fuel could be used in weapons, though scientists are trying to find a way to recycle spent fuel so as to increase the amount of energy that can be extracted from a given mass of fuel and reduce the environmental impact of nuclear power generation.[9] As such, spent fuel is treated as waste without any current attempt to reprocess. For an average power plant, that comes out to about 20 tons of spent fuel per year, all of which must be stored. **Currently, nuclear plants in the United States store their spent fuel in pools initially built to house waste temporarily.** Pools ensure that spent rods are under at least 20 feet of water, which is a sufficient barrier to shield nearby workers from the worst of the radiation. Many once-operational plants have been decommissioned because they've run out of storage. Most operational U.S. nuclear plants are likely to run out of space before 2020. In the meantime, some plants have taken to an alternative method of storing, called **dry-casking**, in which spent fuel is kept above ground in metal containers within a concrete structure. This practice is of questionable safety.

Yucca Mountain, a site in southern Nevada (Image 12.14), has been proposed as a permanent **deep geological repository** for radioactive waste. Targeted for investigation in 1978, the site was originally slated to begin accepting waste in 1998. This date has been delayed many times, and currently the status of Yucca Mountain is in a state of flux; the Obama administration and supporters in Congress have been working since 2009 to permanently close the site, to no effect thus far, other than that federal funding is blocked. The idea of geological repositories has been the subject of global debate. Proponents argue that nuclear technology offers an

NOTE[8]

The daughter nuclei are generally not fissionable, making them unusable as fuel.

NOTE[9]

Japan, Russia, and Europe do routinely engage in fuel reprocessing.

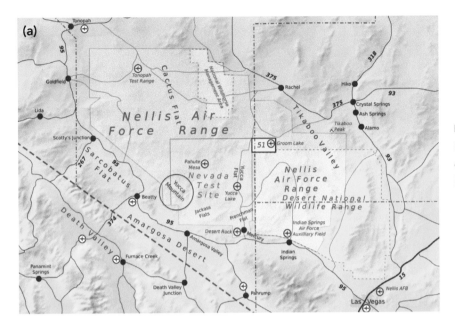

IMAGE 12.14. Yucca Mountain (a) is located on the western end of the Nevada Test Site (b).

abundance of emissions-free power, and that with many facilities out or nearly out of storage space for spent fuel, an alternative must be made available. Opposition to the plan comes from scientists and environmental groups who cite the potential for radiation exposure. Because the Yucca Mountain site lies on and near several major geological faults, there is concern that existing (and future) fractures in the rock structure could allow radioactive gases to escape as decay processes produce gaseous elements. Further, storage containers will inevitably fail sometime in the next several hundred years; the regulations currently in place require only that storage containers be assured 300–1,000 years of integrity. When containers do fail, liquid waste will seep easily through the fractured rock structure into the local water supply. While the land in Nevada was originally targeted because it was largely unsettled and *nonarable* (did not support agriculture), the water table under Yucca Mountain serves populated (and agricultural) areas. It is worth noting that the fault lines in the Yucca Mountain area cannot be assumed to be inactive; seismic activity is quite common, and an earthquake would accelerate the dissemination of radioactive material into the air and water. Finally, residents of Nevada (a state with no nuclear power plants of its own), generally object to the idea of a waste repository in their home state.

Even under ideal circumstances, the idea of a geological repository is fraught with debate. Some radioisotopes in the decay chains of U-235 and Pu-239 have short half-lives and would decay quite quickly. Others have half-lives in the tens of thousands of years. As a result, it is quite difficult to predict to what extent farmers or citizens of nearby populated areas, sometime in the future, would be exposed to radiation. While regulation is in place to keep potential exposure levels low for the next several hundred years, it becomes impossible to predict how the factors of time, seismic disturbance, container failure, inevitable leakage, and unforeseen events may combine to produce a hazard.

CONCEPT CHECK

Why is the proposed repository in Nevada?

Answer: The proposed land is nonarable and uninhabited.

Despite the surge in nuclear power production from the 1960s through the 1980s, this period of growth was followed by a rapid and dramatic decline. Plans to build new plants were canceled, with about two thirds of those commissioned after the early 1970s never coming to fruition. This was the result of a number of factors. Oil prices fell dramatically during that period of time, which made the promise of vast amounts of nuclear power less attractive in the face of the rising costs of building new reactors. In 1979 the Three Mile Island Nuclear Generating Station in Pennsylvania suffered a partial meltdown—meaning containment failure—due to a cooling system malfunction. The U.S. Nuclear Regulatory Commission reports no deaths or injuries from the incident (though it can be difficult to separate the delayed effects of mild to moderate radiation exposure from those due to routine aging and environmental factors). Regardless, the meltdown captured global attention, and significantly dampened national support for expanding the nuclear program. A number of systemic upgrades and changes to prevent similar events were implemented in the wake of the incident.

In 1986 the city of Chernobyl, Ukraine (formerly part of the USSR), suffered the most dramatic reactor disaster in history during an experimental procedure designed to test safety feature responses to a potential power outage. A number of factors combined to create the catastrophe. The experiment was timed to take place during a shift change, rather than being seen through by a single group of workers. Inexperienced workers shut off safety features in excess of what was necessary to perform the experiment. An operator inserted control rods too far, bringing the reactor to a near standstill, following which control rods were extracted too much in order to restart the reaction. This initiated a chain of events that culminated in heat and pressure buildup, control system failure, and eventual explosion of the reactor casing. Radiation exposure was intense—many workers received lethal doses within minutes, with death following in the next few weeks. Residents of Chernobyl and the nearby city Pripyat were evacuated, and both remain unpopulated today, as they are within an exclusion zone of continued contamination surrounding the old reactor.

IMAGE 12.15. The Chernobyl reactor (a). An abandoned house within the exclusion zone around the site of the nuclear disaster (b).

The death toll of the Chernobyl disaster is difficult to quantify accurately. According to the World Health Organization, fewer than 50 people died as a direct result of radiation exposure from the incident. Regardless, a team of scientists working under the auspices of the United Nations has suggested that the number could eventually rise as high as 4,000 owing to delayed effects. Some environmental organizations argue that even this estimate fails to fully represent the extent of the disaster.

The Chernobyl event significantly impacted global attitudes regarding the safety of nuclear power generation. Several countries began phasing out nuclear reactor construction and operation in response. It is difficult to gauge the effect of the Chernobyl event in the United States, as attitudes had already been impacted by the Three Mile Island incident seven years prior. In any event, no new power plant has become active in the United States since 1996.

12.11   The Future of Nuclear Power

While there are proponents of expanded nuclear power generation in the United States and globally, the logistics of waste disposal, expense of building new reactor sites, and public fear surrounding nuclear technology have prevented any significant growth. Particularly with attention turning to emission- and pollution-free energy technologies, it is difficult to justify further development of a technology that produces high-level waste that must be stored permanently (for all intents and purposes) and that carries with it the risk of radiation exposure in surrounding areas. Finally, it is problematic to suggest that additional funding should be shunted into fission-based nuclear technology, given our ever-expanding global energy requirements. If funding for energy technology goes toward building nuclear power plants and a waste repository, then that money is not available for development of newer technologies that don't produce radioactive waste. While some might like to see a global nuclear economy (waste-containment issues aside), in order to fully address world energy needs with current nuclear technologies, the number of operational power plants would need to increase tremendously. Calculations by one Cal Tech research group suggest that to meet current global energy needs, a power plant would need to be built every other day for the next 50 years (discounting any increase in population or power requirements during that time). Note also that fissionable fuel is nonrenewable and limited in quantity. **On a global scale, it appears that current nuclear technologies present a bit of an impasse—it is prohibitively expensive to maintain and expand nuclear power as a partial energy strategy, and problematic to the point of impossible on many levels to switch to a fully nuclear global energy economy.**

Some researchers are exploring the possibility of a fusion-based nuclear power generator, hoping that the astronomical energy potential can be used to justify years of research and development costs until the technology is fully fledged. The idea under consideration involves producing and trapping *plasma* (an electrically charged gas) within a reactor cell. The hot plasma would act as a source of energy for the fusion reaction. The hydrogen bomb also uses plasma to effect fusion: the plasma in that case is created through extreme heat and pressure from a fission reaction applied to polystyrene foam. Of course, a nuclear reactor could not rely on a fission bomb in order to provide energy for fusion; as a result, the production, trapping, and safe maintenance of plasma through other means are under investigation. *Magnetic confinement* involves using electromagnetic fields and radiation to confine hydrogen gas, heat it, and pressurize it sufficiently to induce the plasma state, whereupon fusion would take place. Alternately, *inertial confinement* relies upon lasers to accomplish the same task. In any case, the fusion reaction most frequently targeted for investigation is that of deuterium with tritium, owing to the abundance of deuterium and lithium, which is the nuclear precursor to tritium (see Nuclear Fusion, above). This technology is still highly experimental, however, and no plans are currently in place for a fusion-based nuclear power facility.

12.12   Health Effects of Irradiation

As the radiation victims of the atomic bomb and Chernobyl make all too clear, exposure to nuclear radiation is detrimental to health. The types of radiation vary greatly in their abilities to penetrate the human body and the severities of their effects. **The three types of decay particles addressed in this chapter—and X-rays— are ionizing radiation, capable of breaking chemical bonds and damaging DNA, proteins, and other molecules**, as discussed in Chapter 10. In the case of a severe acute (one-time, massive-dose) exposure, extensive cellular damage can result in death. Chronic exposures (smaller doses, repeated over time) can accumulate and lead to mutations and cancer as well as diseases and symptoms associated with aging. While alpha radiation can cause damage if taken internally or inhaled, the particles themselves are of insufficient energy to penetrate the skin. Radon, an alpha-emitter ubiquitous in Earth's crust, is a major source of alpha particle exposure; it's also the second-leading cause of lung cancer, behind smoking. Beta radiation is somewhat higher in energy and more penetrating than alpha, but exposure is less common. Carbon-14 is a beta-emitter, but is dispersed in the environment and generally not encountered in concentrations sufficient to cause acute radiation effects. Of the major decay particles, gamma rays are the most energetic and damaging. Acute radiation sickness—a set of symptoms resulting from radiation exposure—is almost always due to gamma radiation. Aside from nuclear accidents and use of gamma rays in medical applications (see below), most exposure is through contact with naturally occurring radioisotopes. In fact, we are bombarded with radiation on a routine basis—radon in the ground is a significant natural source,[10] as are isotopes of nitrogen and carbon in the ground and air. Cosmic rays, which are more highly concentrated at higher elevations

NOTE[10]

Radon is also a product of uranium decay, which further underscores the need for extreme care in storing high-level radioactive waste so as to minimize the potential for exposure.

(including on airplanes), also contribute to background radiation, a term for unavoidable radiation from the natural world.

Radiation exposure is measured in units of *rem*, which stands for *roentgen equivalent man*. This unit multiplies a radiation dose, measured in *rads*, or Joules of radiation absorbed per mass of tissue, by a biological effectiveness factor. The idea here is that some types of radiation are more penetrating than others, and specifics of absorption (where on the body, internal versus external, and so forth) partially determine effectiveness of the dose. While it's difficult to define a "safe" dose of radiation (particularly as effects are cumulative), the Environmental Protection Agency (EPA) correlates doses to predictable symptoms (see Table 12.2).

While the thought that we are constantly absorbing radiation—from the ground, the air, the cosmos, and from within our own bodies—is somewhat disconcerting, remember that we're exposed to background radiation in incredibly low dosages (Table 12.3). **The EPA estimates that the average U.S. resident absorbs on the order of 300–400 mrem/year (0.3–0.4 rem/year), far below the threshold necessary to cause even the mildest radiation symptoms.** Further, while radiation effects are cumulative, they also dissipate somewhat over time as a result of cellular repair mechanisms: provided that exposures don't occur in rapid succession, the whole is less than the sum of the parts. Finally, it may be of interest to those who use consumer products and engage in certain habits that some voluntary activities result in radiation exposure far more significant than that from background sources. For instance, one coast-to-coast flight across the United States results in as much radiation exposure as an entire year's worth of fallout from nuclear energy production, weapons testing, and coal burning—about 3 mrem. Smoking contributes 50–280 mrem of radiation per year, depending upon the type of cigarette and quantity smoked. This means a heavy habit can nearly double annual exposure to radiation. In general, while some people are subject to special circumstances (including cancer treatment, nuclear medicine, and significant nuclear fallout), most individuals will not accumulate enough radiation over a lifetime to experience symptoms of radiation sickness.

Exposure (rem)	Health Effect
5–10	Changes in blood chemistry
50	Nausea within hours
55	Fatigue
70	Vomiting
75	Hair loss within a few weeks
90	Diarrhea
100	Hemorrhage
400	Possible death within months
1,000	Destruction of intestinal lining, bleeding, death within weeks
2,000	Damage to central nervous system, unconsciousness within minutes, death within days

TABLE 12.2. Physical effects of acute radiation exposure. Chronic effects vary somewhat.

Source of Exposure	Annual Dose
Radon (inhaled, from ground—varies with region)	200 mrem
Background (cosmic rays—varies with elevation)	55 mrem
Medical (X-rays, CAT scans, nuclear medicine)	53 mrem
Internal body (food, water, air, incorporated radioisotopes)	39 mrem
Consumer products (TVs, computers, smoke detectors)	10 mrem
Other (nuclear fallout, weapons testing)	3 mrem

TABLE 12.3. Average radiation absorbed by United States residents from naturally occurring and common man-made sources. Note: 1 rem = 1,000 mrem.

CONCEPT CHECK

Under normal circumstances, what is the source of the majority of radiation to which we are exposed?

Answer: Background (natural) sources such as radon, cosmic rays, etc.

Chemotherapeutic pharmaceuticals also specifically target rapidly dividing cells. The common side effects of both chemotherapy and radiation treatment for cancer occur because not all rapidly dividing body cells are cancerous—hair follicles and epithelial cells (in skin and the mucous membranes) also divide rapidly and can easily be damaged or killed by cancer treatments. For this reason, hair loss and mouth sores are common in cancer patients undergoing treatment.

Despite a conflicted global philosophy with regard to nuclear energy production, nuclear technology is nevertheless commonly utilized (and very important) to other applications, including medicine, archaeology, and food science. The X-ray machine was the first of the diagnostic techniques to utilize radiation, and afforded physicians the ability to see inside the human body. **Computed tomography (CT)**, a three-dimensional image created by computer from a series of X-ray images, soon followed. X-rays do not technically fall under the auspices of nuclear medicine, as they are not generated by nuclear phenomena—rather, they are the result of acceleration of electrons within an electromagnetic tube—but they nevertheless result in significant exposure to ionizing radiation, for which reason they are generally avoided in a medical setting without indication. Single X-rays typically result in dosages less than 100 mrem per exposure, while large-area CT scans may provide 1 rem or more. The utilization of tissue-penetrating radiation has since advanced far beyond the X-ray; beta- and gamma-emitting sources now find application in both diagnosis and treatment of certain conditions.

Many cancers are susceptible to radiation treatment. A tumor is composed of cells proliferating at an abnormally high rate, meaning that they copy their DNA more frequently than nearby cells. **Since DNA is particularly sensitive to ionizing radiation, rapidly dividing cells are most likely to be affected (and killed) by radiation exposure.**[11] For this reason, beta and gamma rays (and even high-intensity X-rays) have been used as cancer treatments. The Gamma Knife, for instance, is used to focus several beams of gamma radiation from cobalt-60

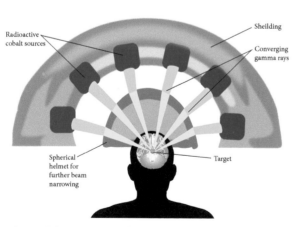

Stereotaxic Radiosurgery - Gamma Knife Concept
Multiple radiation beams converge on target tumor, delivering high-dose radiation to the tumor, but little to surrounding tissues. It is a single treatment and to ensure proper patient positioning and immobility, a positioning frame is secured to the patient's skull, then attached to the radiation source. Treatment lasts 45 to 60 minutes.

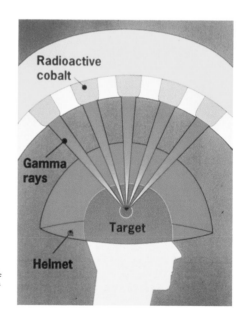

IMAGE 12.16. The Gamma Knife functions by focusing several gamma rays onto a small target.

upon a brain tumor. Where the beams meet, the radiation intensifies to the point that a tumor is typically destroyed in a single day. Since beta radiation is not nearly so penetrating, external beams of beta particles can only be used in the irradiation of surface tumors.

Frequently, radioisotopes are incorporated directly into the body. **Positron emission tomography (PET)** scans produce many images over a period of time, so that metabolic activity can be gauged. A radioisotope is introduced as part of a larger molecule normally utilized by the body (C-11, for instance, may be incorporated into glucose).[12] The location of the isotope-containing molecules is then monitored by tracking the source of the radiation. These scans allow for imaging of tumors, which are highly metabolically active, and tend to take up glucose out of proportion with the cells around them. They can also be used to diagnose some brain diseases, and map heart and brain function.

The thyroid, a gland responsible for production of several hormones and regulation of metabolism, takes up iodine as part of its normal function. In the case of hyperactivity of thyroid cells (such as in Graves' disease) or thyroid cancer, I-131, a beta- and gamma-emitter, can be introduced into the body. This radioisotope has a half-life of eight days, and is concentrated by the target gland. For this reason, systemic effects of radiation are mitigated by the localization of the radioactive substance. I-131 can be used to monitor thyroid activity if given in small doses or, in larger doses, is effective at killing thyroid tumor cells. Other radioisotopes show site-specific activity as well, based upon which body organ tends to concentrate these elements. Strontium-89 and samarium-153, for instance, are both beta-emitters and are particularly effective against bone tumors. Phosphorus-32, also a beta-emitter, is often injected into hollow brain tumors. Bone and brain tumors are far too deep to access using externally administered beta radiation, but internal administration of a radioisotope not only brings the beta-emitter into close proximity with the tumor; it also allows for localization of radiation, preventing unnecessary exposure of healthy tissue.

NOTE[12]

Remember that all isotopes of an element are identical in their chemical properties. An atom of C-11 can bond with other atoms of C, H, and O to form glucose just as well as an atom of C-12 can. The difference, though, is that the C-11 is radioactive, meaning that its location in the body can be tracked by monitoring radioactive decay.

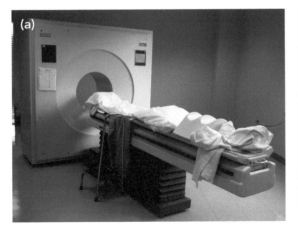

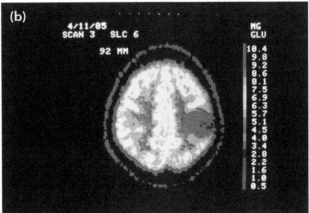

IMAGE 12.17. A PET scan (a) can reveal metabolically active tissues, such as the tumor in the right side of this brain (b).

Nuclear technology can also be applied to the carbon dating of fossils, textiles, and other objects of historical interest or significance. A naturally occurring isotope of carbon, C-14 is produced in the upper atmosphere when N-14 is bombarded with cosmic radiation:

$$^{14}_{7}N + {}^{1}_{0}n \rightarrow {}^{14}_{6}C + {}^{1}_{1}H$$

The majority of carbon on Earth is C-12 (about 99% of total carbon), with C-13, a stable isotope, representing about 1% of total carbon. C-14 is present in concentrations on the order of one part per trillion.

There are two important principles to carbon dating. The first is that carbon is taken up by plants during the process of photosynthesis (see Chapter 11). Since the three naturally occurring isotopes of carbon are chemically indistinguishable from one another, plants take up these isotopes in direct proportion to their representation in the atmosphere. During a plant's life, it would be expected to contain approximately 99% C-12, 1% C-13, and one part per trillion C-14. Because herbivorous animals get their carbon from plants (and carnivorous animals get their carbon from other animals), living animals would also be expected to have carbon isotope ratios identical to those of the atmosphere. The second major principle of carbon dating is that C-14 is radioactive, and therefore decays predictably, with a half-life of 5,730 years (Image 12.18). **For this reason, it is possible to use the proportion of C-14 remaining in a bit of dead organic matter to determine its year**

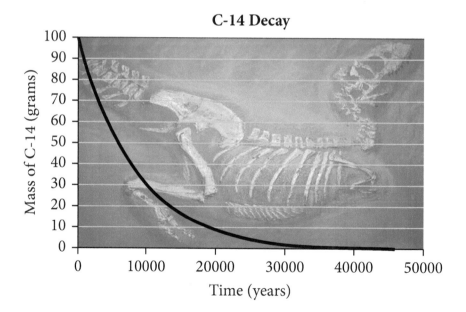

C-14 Decay

IMAGE 12.18. Carbon-14 decays predictably with a half-life of 5,730 years.

of death, since at the time of death, the organism stopped equilibrating its carbon ratio with that of the atmosphere.

If, for instance, a mummy was discovered in which C-14 was present at only .5 parts per trillion (half the normal concentration), one would determine that the mummified individual perished one C-14 half-life ago—that is, 5,730 years ago. Recent advancements producing highly accurate analytical techniques allow the different isotopes of carbon within a specimen to be sorted and counted, which makes carbon dating quite accurate. A limitation of the procedure, of course, is that it is only relevant to organic (once living) material. A further limitation is that, with the passage of long periods of time, C-14 levels fall so far that it is no longer possible to accurately gauge the time of an organism's death. Carbon dating can only be used to date samples from up to about 50,000 years ago, though researchers are working on techniques that will increase sensitivity and hopefully allow for dating of older samples.

12.15 Irradiation of Food

A final application of nuclear technology worth discussing—particularly in light of its generally negative perception by the public—is that of food irradiation. Typically accomplished using a cobalt-60 source of gamma rays, food—often produce or meat—is irradiated for the purpose of killing viruses, bacteria, and insects that may be present. It is important to recall that gamma radiation consists of high-energy photons. **While direct exposure can be harmful or fatal (hence the efficacy of this technique against microbes and insects), gamma radiation cannot induce radioactivity in other nuclei. For this reason, it is simply impossible for irradiated food to become radioactive.** Early in the investigation of irradiation in the food industry, there was concern that exposure to gamma rays might decrease the vitamin (or other nutrient) content of foods. A number of studies have examined this issue, and in general, irradiation does not appear to decrease vitamin or nutrient content to any significant degree. Regardless, there are several reasons irradiation of comestibles is globally debated. Proponents argue that reducing microbial and insect loads keeps food fresh longer, meaning less spoilage during transport. This is particularly important in rural and undeveloped countries, where produce may need to be brought in from quite a distance away over a period of days or weeks. Opponents, however, are concerned that irradiation provides a veneer of safety with regard to bacterial proliferation in foods; simply because bacteria have been killed doesn't mean the toxins they produced while alive have been removed from the food (this is of particular concern with meat, which can frequently host bacteria that produce toxins). Additionally, the limited shelf life of perishable items has historically prevented them from being transported over very long distances (which occurs at a tremendous energy expense, and can be detrimental to the local economy). Irradiation, opponents argue, would simplify the logistics of extremely long-haul food transport, with ramifications for sustainability in terms of increased fuel combustion and decreased support of local agriculture.

Nuclear science, which lies at the intersection of chemistry and physics, is a rapidly evolving field. Even if the promise of an unimaginable quantity of cheap nuclear power, anticipated since the 1950s, is never realized—whether for logistical reasons or by consensus—many other applications of the technology are and will remain important to our way of life. The lessons of the past remind us that any time a technology holds great potential for effecting positive change, it also has the potential to cause harm, either through intentional misuse or unanticipated consequences. Irrespective of how we as individuals or nations feel about nuclear weapons, energy, and technology, the fact remains that no society can possibly make decisions without affecting others. We share a global responsibility for determining in which direction we will take nuclear science, and in which direction we will allow it take us.

12.16 The Last Word—Microwaves

Since gaining household popularity in the 1980s, the microwave oven has become an integral part of the American kitchen. The technology is completely unrelated to nuclear phenomena: microwaves are not generated through nuclear reactions, nor are they a form of ionizing radiation. Still, the public perception of microwaves, what they are, and how they cook is frequently confused with nuclear technology, as apparent in common slang whereby we refer to microwaving food as "nuking" it.

Microwaves are a form of EMR (Chapter 10). Also in this category are gamma and X-rays (which probably serve to strengthen the misperception that microwaves are a nuclear phenomenon). However, the wavelength of microwave radiation is much longer than that of ionization radiation, meaning that the frequency and energy are much lower. As a result, microwaves, like visible light, are far too low

Microwave radiation

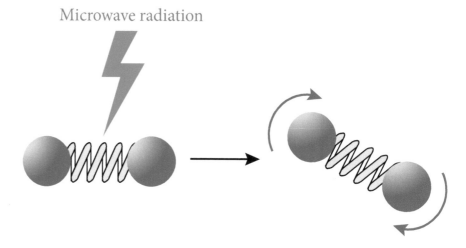

IMAGE 12.19. Microwave radiation causes molecular rotation, which generates heat through friction.

in energy to break molecular bonds—they're actually lower in energy than visible light. Instead, microwaves cause molecules to rotate,[13] which produces heat through friction (just as rubbing your hands together leaves them feeling warm).

This heat is distributed throughout the food, defrosting, warming, or cooking it. Really, a microwave oven cooks in a manner absolutely analogous to your stovetop. The only difference is that on the stovetop, the heat comes from the stove element, while in a microwave, the heat comes from water molecule friction. In no way are the vibrating water molecules chemically altered—there is simply no truth to the common misconception that microwaving renders food radioactive, unfit to eat, or devoid of nutrients (unless, of course, you overcook it, which is always detrimental to vitamin content, even on the stovetop). In fact, because cooking times are generally shorter in a microwave than on the stovetop—and because vitamins deteriorate with exposure to heat—studies of properly cooked (not overcooked) steamed veggies done in the microwave versus on the stovetop show, at worst, analogous vitamin content and, at best, better retention of vitamins in the microwaved sample. As for whether microwaves can leak radiation and cause injury, the answer is no, not really. First, the grid pattern on all microwave doors is there to stop the escape of waves; the rest of the oven is similarly insulated against leakage. Second, ovens are designed to stop functioning if the door is opened. Finally, though, the most important consideration is that, in the event that microwaves did escape the oven in any significant quantity, the only thing they'd be capable of in a human is what they're capable of in food—production of heat. That is to say, in the same way you'd burn your finger if you touched a hot stove element, you could burn your finger if you found a way to stick it in a functioning microwave. Under no circumstances, though, can microwaves cause cancer—they're simply the wrong type of radiation!

NOTE[13]

Specifically, microwave ovens cause water molecules to rotate, which is why it's not possible to heat completely dry substances in a microwave and why some dishes (specifically ceramics, which can absorb water) are not microwave safe and get very hot if used in an oven.

SUMMARY OF MAIN POINTS

- Unlike chemical reactions, nuclear reactions can change the identity of atoms by changing the number of protons in the nucleus.
- Radioactive decay is the loss of a small decay particle from a parent nucleus to form a new daughter nucleus.
- Common decay types include alpha $\left(_{2}^{4}He\right)$, beta $\left(_{-1}^{0}e\right)$ and gamma $\left(_{0}^{0}\gamma\right)$, which is pure energy.
- Nuclear equations are balanced with regard to mass (superscript) and number of protons (subscript), but NOT with regard to atoms or atom identities.
- Atoms of the same element with different masses are called isotopes. The mass of an element on the periodic table reflects the mass of the most common isotope.
- All large elements (bismuth and larger) are radioactive: if one decay event does not produce a stable nucleus, a decay series takes place until a stable nucleus results.
- Nuclear fission is the splitting of a larger nucleus into two smaller nuclei.

- Fission takes place with a mass defect—the lost mass is too inconsequential to show up in the balanced nuclear equation, but is converted to energy by the equation $E = mc^2$.
- Nuclei can be induced to fission by bombardment with neutrons and other high-energy particles.
- U-235 and Pu-239 are both fissionable fuels and are used in nuclear weapons and fuels.
- The quantity of fissionable fuel necessary to self-sustain a chain reaction is called the critical mass.
- U-235 must be enriched out of the more common (and nonfissionable) U-238 by gaseous diffusion, a process that leaves radioactive waste in the form of depleted uranium.
- The first nuclear weapons used chemical detonations to bring critical masses of fissionable fuel together; the fuel would then undergo a nuclear reaction.
- Nuclear fusion is the combination of two smaller nuclei to form a larger one—like fission, it results in a mass defect and yields tremendous energy.
- Fusion reactions fuel the sun and stars and are responsible for producing heavier elements from hydrogen.
- Fusion reactions require tremendous initial energy input; for this reason, the hydrogen bomb uses a fission reaction to initiate a fusion reaction.
- If a nuclear reaction is controlled, the energy can be harnessed and used to produce electricity. This is accomplished in nuclear power plants.
- Nuclear plants use fissionable fuel (U-235 or Pu-239) in fuel assemblies that are interspersed with neutron-absorbing control rods to keep the reaction from growing exponentially.
- Radioactivity is sequestered from the outside of the plant through two separate closed loops.
- Radioisotopes decay predictably with a given half-life: one half-life is the time for half of a sample to decay.
- Nuclear waste from power plants is radioactive and must be stored. Most plants in the United States are out of or are running out of storage and are dry-casking waste.
- A geological repository has been proposed in Nevada but has met with opposition and is decades behind schedule.
- Nuclear disasters like Three Mile Island and Chernobyl have significantly impacted public support for nuclear power plant proliferation.
- While researchers are exploring the possibility of fusion-generated power, this technology is far from fully developed.
- Radioactive decay produces ionizing radiation—radiation with the ability to break chemical bonds.
- Exposure to ionizing radiation can have severe health ramifications. Some exposure from natural sources is avoidable, while other sources of exposure can be avoided.

- Nuclear technology contributes significantly to medicine through diagnostic and treatment techniques including imaging, targeted destruction of tumors, and radiotherapy.
- The predictable decay of radioisotopes allows for carbon dating, a technique that determines the age of an organic sample based upon its remaining C-14.

QUESTIONS AND TOPICS FOR DISCUSSION

1. How is nuclear chemistry similar to the concept of reactions held by the ancient alchemists? By Dalton? How is it different?

2. How are nuclear reactions similar to chemical reactions? How are they different?

3. Why do we think about nuclei, but not electrons, in nuclear chemistry? When we refer to a "parent nucleus" or "daughter nucleus," do we mean that the particle reacting has no electron cloud?

4. How is alpha radiation like an atom of helium? How is it different?

5. Given that nuclei don't contain electrons, where do beta particles come from?

6. Why doesn't gamma radiation change the identity of the parent nucleus?

7. Why are all the large nuclei radioactive? Explain in terms of nuclear forces.

8. Why do nuclear equations balance with regard to mass, despite the mass defect?

9. How did the burgeoning field of nuclear chemistry help to advance our understanding of the structure of the atom?

10. Why is the fission of U-235 self-sustaining under the right circumstances?

11. Explain the basics of the atomic bomb. What is the purpose of the chemical detonation?

12. Describe the process by which fuel-grade uranium is produced from naturally occurring uranium.

13. Why must uranium be enriched for weapons or fuel purposes?

14. Why are uranium tails environmentally hazardous?

15. How is nuclear fission similar to nuclear fusion? How are they different?

16. Why does the hydrogen bomb require a primary fission reaction?

17. A friend is considering buying a new house but is concerned that the lot is only half a mile from a nuclear power plant. He worries that the steam from the plant, visible overhead, will rain radioactive mist down on his family. What might you say to him?

18. How does a nuclear power plant generate energy? How is it similar to (and how is it different from) the way a coal-fired power plant generates energy?

19. Are you more likely to be exposed to radiation from a nuclear power plant or coal-fired power plant? Explain.

20. How is nuclear fission used to generate power?

21. Why is a discussion of half-life relevant to the issue of high-level radioactive waste?

22. What are the risks associated with building a deep geological repository? What are the benefits?

23. Does risk–benefit analysis support the idea of building a deep geological repository in Nevada? Why or why not? What about the precautionary principle? (See Chapters 6 and 10 to review this concept.)

24. The nuclear disasters presented in the chapter significantly affected public opinion of nuclear power. Do you think this was warranted? Why or why not?

25. Do you think that nuclear power (fission, fusion, or both) should continue to be evaluated as potential energy sources? Why or why not?

26. Many people fear nuclear technology and radiation but willingly submit to X-rays, which are also a source of ionizing radiation. Why do you think this is so?

27. A person tried to disprove the accuracy of carbon dating by trying it out on his titanium wristwatch. He showed that the watch was 60,000 years old. Does this convince you that carbon dating is fallible? Why or why not?

28. Can we use carbon dating to assess the age of dinosaur fossils? Why or why not?

29. At the grocery store, you overhear a woman comment to her friend that she won't buy conventional produce because it's been irradiated and she is worried that her family will get cancer. Are her concerns legitimate? Why or why not?

PROBLEMS

1. What is conserved in a chemical reaction? In a nuclear reaction?

2. What is not conserved in a chemical reaction? In a nuclear reaction?

3. Of the common decay particles, which have mass? Which have charge?

4. Why is it legitimate to refer to a beta particle as an electron?

5. How does losing an alpha particle change a nucleus?

6. How does losing a beta particle change a nucleus?

7. How does losing a gamma particle change a nucleus?

8. How many protons and neutrons are in: ^{55}Mn, ^{201}Hg, potassium-39, Ba-141?

9. How many protons and neutrons are in: Bi-214, ^{11}C, Np-239, radium-225?

10. Complete each of the following nuclear equations with the missing particle or nucleus:

$$^{210}_{82}Pb \rightarrow \underline{\quad} + ^{210}_{83}Bi$$

$$^{230}_{90}Th \rightarrow \underline{\quad} + ^{226}_{88}Ra$$

$$^{210}_{84}Po \rightarrow ^{4}_{2}He + \underline{\quad}$$

$$\underline{\quad} \rightarrow ^{228}_{89}Ac + ^{0}_{-1}e$$

11. Complete each of the following nuclear equations with the missing particle or nucleus:

$$^{226}_{88}Ra \rightarrow \underline{\quad} + ^{222}_{86}Rn$$

$$\underline{\quad} \rightarrow ^{229}_{90}Th + ^{0}_{0}\gamma$$

$$^{59}_{26}Fe \rightarrow ^{0}_{-1}e + \underline{\quad}$$

$$\underline{\quad} \rightarrow ^{210}_{81}Tl + ^{4}_{2}\alpha$$

12. Write balanced equations for each of the following: Bismuth-214 undergoes beta decay; Pu-242 emits alpha radiation; ^{81}Se emits beta radiation; U-235 undergoes gamma decay.

13. Write balanced equations for each of the following nuclear transformations: I-131 to Xe-131; Ra-226 to Rn-222; ^{201}Hg to Tl-201; rubidium-81 to strontium-81; Th-231 to Ra-227.

14. Which type of decay helps to reduce the size and mass of a large nucleus?

15. What is a radioactive decay series, and when does it occur?

16. Define: nuclear fission; mass defect; critical mass.

17. Mass defect is typically about what percent of the original mass of fissionable fuel?

18. A reaction takes place with a mass defect of 0.030 kg. What is the energy yield?

19. The nuclear fission reaction of 2 kg of Pu-239 occurs with a 0.1% mass defect. What is the energy yield?

20. What mass defect is required to generate 5.5×10^{16} J of energy?

21. What percent mass defect would be necessary to generate 2.5×10^{14} J of energy from 35 kg of fissionable fuel?

22. The half-life of tritium is 12.3 years. What mass of tritium from a 48.0 g sample remains after 24.6 years? (Use intuitive approach.)

23. Approximately how long does it take for the sample of tritium in Problem 19 to be reduced to 2.0 g of tritium? (Use intuitive approach.)

24. A radioactive sample with an initial mass of 100.0 g has been reduced to 1.56 g of the initial radioisotope in 85 years. How long is the radioisotope's half-life? (Use intuitive approach.)

25. Using Table 12.1, how long does it take 75% of a sample of N-16 to have reacted? (Use intuitive approach.)

26. Using Table 12.1, how much U-235 remains from an original sample of 54.3 g after 9.9×10^9 years? What has happened to the rest of the uranium? (Use equation.)

27. Using Table 12.1, how much C-11 remains of an original 35.0 g sample after 146 minutes? (Use equation.)

28. What are the primary sources of (unavoidable) radiation to which we are routinely exposed?

29. What are some major sources of radiation to which we are exposed that are avoidable?

30. What is ionizing radiation, and what does it do in the body?

31. What are the symptoms of radiation sickness?

32. Why can radiation be used to treat certain cancers? What are the side effects?

33. Why is carbon dating only useful to date organic materials?

34. An archeologist carbon dates a sample of fabric found in an ancient gravesite. C-14 is present at .4 parts per trillion. Approximately how old is the fabric?

35. Why does carbon dating look at the concentration of C-14 rather than C-13 or C-12?

ANSWERS TO ODD-NUMBERED PROBLEMS

1. In a chemical reaction mass, matter (identity of atoms), and charge are conserved. In a nuclear reaction, mass (except for defect) can be said to be conserved, and charge is conserved.
3. Alpha particles have mass, both alpha and beta have charge; gamma particles have neither.
5. Alpha decay results in the loss of 4 amu (2 protons and 2 neutrons).
7. Gamma decay causes stabilization of a nucleus (loss of energy).
9. Bi-214: 83 protons, 131 neutrons; ^{11}C: 6 protons, 5 neutrons; Np-239: 93 protons, 146 neutrons; radium-225: 88 protons; 137 neutrons.
11. Complete each of the following nuclear equations with the missing particle or nucleus:

$$^{226}_{88}Ra \rightarrow \, ^{4}_{2}He + \, ^{222}_{86}Rn$$
$$^{229}_{90}Th* \rightarrow \, ^{229}_{90}Th + \, ^{0}_{0}\gamma$$
$$^{59}_{26}Fe \rightarrow \, ^{0}_{-1}e + \, ^{59}_{27}Co$$
$$^{214}_{83}Bi \rightarrow \, ^{210}_{81}Tl + \, ^{4}_{2}\alpha$$

13.

$$_{53}^{131}I \rightarrow _{54}^{131}Xe + _{-1}^{0}e$$

$$_{88}^{226}Ra \rightarrow _{86}^{222}Rn \; _{2}^{4}He$$

$$_{37}^{81}Rb \rightarrow _{38}^{81}Sr + _{-1}^{0}e$$

$$_{80}^{201}Hg \rightarrow _{-1}^{0}e + _{81}^{201}Tl$$

$$_{90}^{231}Th \rightarrow _{2}^{4}He + _{88}^{227}Ra$$

15. A decay series occurs when the daughter nucleus of a decay reaction is also radioactive, and decays again. These are particularly common with large nuclei.

17. 0.1%.

19. 1.8×10^{14} J.

21. 0.008%.

23. Between 4 and 5 half-lives; between 49.2 and 61.5 years.

25. 14.26 seconds.

27. 0.24 g; the rest of the material has decayed to become daughter nuclei.

29. Medical sources, consumer products including cigarettes, airplane travel, nuclear fallout/weapons testing.

31. Symptoms include (from mild to severe): changes in blood chemistry, nausea, fatigue, systemic symptoms (various), damage to nervous system, and death.

33. Carbon dating compares remaining C-14 to original percentage of C-14 in a sample. While an organism is alive, its C-14 is at the same percentage as that in the environment. Upon death, the C-14 is no longer replaced. Nonorganic material does not equilibrate its C-14 with the environment, so the remainder can't be used to date the material.

35. C-14 is radioactive, while neither C-13 nor C-12 are.

SOURCES AND FURTHER READING

Early Nuclear Science

Heilbron, J. L. (2003). *Ernest Rutherford and the explosion of atoms.* New York, NY: Oxford University Press.

Preston, D. (2006). *Before the fallout: From Marie Curie to Hiroshima.* New York, NY: Penguin.

Willis, K. (1995). The origins of British nuclear culture, 1895–1939. *The Journal of British Studies, 34*(1), 59–89.

Nuclear Bombs

Hughes, J. (2002). *The Manhattan Project: Big science and the atomic bomb.* New York, NY: Columbia University Press.

Nuclear Power and Waste

www.nrc.gov

Lemons, J., and Malone, C. (1989). Siting America's geological repository for high-level nuclear waste: Implications for environmental policy. *Environmental Management, 13*(4), 435–41.

Lewis, N. (2007). Powering the planet. *MRS Bulletin, 32*, 808–20.

Mould, R. F. (2000). *Chernobyl record: The definitive history of the Chernobyl disaster.* London, UK: Institute of Physics Publishing.

http://www.ocrwm.doe.gov/

U.S. Nuclear Regulatory Commission. (2002). *Radioactive waste: Production, storage, disposal* (NUREG/BR-0216, Rev.2) [Brochure]. Washington, DC: Office of Public Affairs.

Health Effects

www.epa.gov

http://www.ocrwm.doe.gov

Applications

Duodu, K. G., Minnaar, A., Taylor, J. R. N. (1999). Effect of cooking and irradiation on the labile vitamins and antinutrient content of a traditional African sorghum porridge and spinach relish. *Food Chemistry, 66*(1), 21–27.

Early, P. J., Landa, E. R. (1995). Use of therapeutic radionuclides in medicine. *Health Physics, 69*, 677–94.

Graham, W. D., Stevenson, M. H. (1999). Effect of irradiation on vitamin C content of strawberries and potatoes in combination with storage and with further cooking in potatoes. *Journal of the Science of Food and Agriculture, 17*(3), 371–77.

Graham, W. D., Stevenson, M. H., and Stewart, E. M. (1998). Effect of irradiation dose and irradiation temperature on the thiamin content of raw and cooked chicken breast meat. *Journal of the Science of Food and Agriculture, 78*(4), 559–64.

Orton, C. G., (1995). Uses of therapeutic X-rays in medicine. *Health Physics, 69*, 662–76.

www.cancer.org

CREDITS

Energy, Chemistry, and Society

13

As a global community, the last several decades have seen us forced to confront many challenges. Our expanding population and increasing reliance on technology will require us to face many more such challenges in the coming years, some of which we foresee and some of which will surely catch us by surprise. At present, however, none loom so pressingly and dauntingly large as climate change[1] and the related global energy crisis. We must power the world, but our current sources of energy are also sources of pollution that perturb global cycles. In previous chapters, we've discussed how combustion reactions produce tropospheric pollutants and nuclear reactions produce high-level radioactive waste. We've examined our use of chemical fertilizer—which we rely upon to provide for the nutritional energy needs of a hungry population—and its effect upon the nitrogen cycle. Now, however, we turn our attention to the big picture, and look at how each of these perturbations (and many others) affects the atmosphere on a much larger scale than previously discussed. In this chapter, we will examine the relationship between human energy requirements and the planet's carbon cycle. We'll use a scientific and skeptical analysis of available data and evidence to determine whether it suggests that human activity is perturbing the cycle and whether that perturbation is warming the planet, resulting in climate change. Finally, we'll discuss how global, local, and individual actions affect energy balance.

NOTE[1]

Some deniers of global warming and climate change suggest that "they" (whomever they are) changed "global warming" to "climate change" in some sort of tacit admission that global warming does not exist. In fact, as we will see in this chapter, the planet is warming (global warming), which is in turn having a host of effects on precipitation patterns, weather severity, etc. (climate change). In popular parlance, global warming and climate change are used interchangeably. In this text, to reflect that the effects go far beyond simply a warming planet, we will refer to climate change, with the understanding that global warming is an important facet—and precipitator—thereof.

13.1   An Important Introductory Note from the Author

The topics we've discussed up to this point in the text are not matters of major debate, either in the scientific world or within society as a whole. There is a global, scientific, and popular consensus on the environmental effects of pollution. The reality of ozone depletion and the steps necessary to reverse it are not questioned. The central topic of this chapter—climate change—is also widely accepted as truth, within both scientific and the vast majority of global communities. **In the US, however, politics and science often mix; as such, and fueled by the manner in which topics of science and politics are presented by the media, it may be easy to fall under the impression that climate change is still hotly debated, or that there's a lack of global scientific consensus on the issue of climate change.** In order to separate ourselves from the politics and the rhetoric, the central topic of this chapter will be presented as something of a hypothesis. We will be the critical thinkers and will examine the evidence to determine whether that evidence supports or refutes the hypothesis. As we begin our discussion, I ask you to be a skeptic, but I also ask you to remember what that means. **A skeptic believes nothing he or she is told, but instead asks to see the evidence, and personally evaluates it on its own merits.** A skeptic neither believes in climate change nor categorically dismisses it as a hoax until the evidence has been reviewed. A skeptic is not swayed by politicians or journalists, friends or Internet bloggers, popular media, or even isolated scientific sources. Rather, a skeptic insists on seeing the data firsthand, and insists that it be scientifically gathered, reproducible, and corroborated. To accept climate change without evidence is foolhardy and, similarly, to dismiss it without examination of the evidence—or to dismiss the evidence because it does not support a preexisting belief—is not skepticism, but simply denial. I encourage you to read this chapter as you would any other source—with a skeptical and discerning eye—and I further encourage you to seek out additional information that corroborates or refutes what is presented here.

13.2   Our Global Energy Dependence and Fossil Fuels

As a global community, we consume a tremendous (and ever increasing) amount of energy. In 2008, estimated global energy consumption was nearly 5×10^{20} J. While this is a staggeringly large quantity, it is much less than the amount of energy the earth receives from the sun each year, which is estimated at 2×10^{23} J—nearly 500 times our annual global energy consumption. Given that planetary energy input so greatly outweighs that which we consume, why do we so often hear reference to a global energy "crisis"? The reason is that most of our energy requirements are not fulfilled by sunlight—at least, not by the sunlight we're receiving now. **Instead, the majority of fuels we burn—called** fossil fuels—**were created millions of years ago when prehistoric organisms used photosynthesis to capture and store energy from the sun.** The fossil fuels, which include coal, petroleum, and natural gas, collectively provide for approximately 85% of energy currently consumed worldwide (Image 13.1).

In order to understand fossil fuels and our reliance upon them, let's take a few minutes to discuss where they come from and why they are such an important source of energy. From some of our discussions in earlier chapters, we know that energy can neither be created nor destroyed; instead, it must be transformed

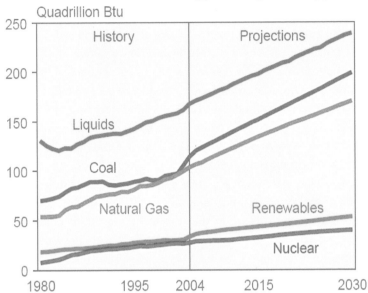

World Marketed Energy Use by Fuel Type

Quadrillion Btu

History | Projections

Liquids

Coal

Natural Gas | Renewables

Nuclear

250
200
150
100
50
0

1980   1995   2004   2015   2030

IMAGE 13.1. Global energy use history and projections from the U.S. Energy Information Administration (EIA). A BTU is 1.06 kJ, or 252 calories. The units on the y-axis are as multiples of 10^{15} BTU. "Liquids" refers to petroleum.

from one form into another. Incident sunlight, for instance, can be transformed into chemical energy (such as glucose) by plants. If the sunlight falls upon a man-made photovoltaic panel, it can be transformed into electrical energy. The simplest energy transformations, however, rely purely upon the capture and transfer of heat energy—a coal-fired power plant, for instance, uses the heat released when coal is burned to boil water and produce steam, which can drive a piston (as in the early steam engines) or turn a generator turbine. The same principle is applied in the cases of a parabolic solar power trough, which uses sunlight to heat a liquid, and nuclear power, in which the energy released by fission is used to heat liquid. In both of these cases, the hot liquid is then used to heat water, generating steam.

Preindustrialization, wood was the most accessible combustion fuel. However, wood fires don't burn particularly hot, which limits their ability to provide power. Further, wood is not a very dense source of energy: it takes a large volume—and a large mass—of wood to provide a relatively small amount of heat. This means that extracting energy from wood is very labor intensive, given the amount of fuel that must be gathered, and also, that it is an impractical source of energy for transportation, given the weight and quantity of fuel that must be carried. Hence, the ability to mine **coal** was a necessary prerequisite to the Industrial Revolution, as discussed in Chapter 6. Compared to wood, coal is a very dense source of energy—it contains more than twice as much energy per unit mass. Like wood, coal contains energy in the form of molecules that were made by storing the sun's energy; coal is the fossilized remains of ancient plant life that died, partially de-cayed, and was then buried beneath layers of earth. The compression of the organic material stalled its decomposition, meaning that while coal contains much less water than wood (hence its increased energy yield per gram burned, since water represents noncombustible mass), the energy-containing carbon-based molecules are still partially intact. **Coal is not a pure substance, but is instead a mixture of many organic molecules that are composed largely of carbon and hydrogen, with some nitrogen and oxygen also figuring into the mix. Trace sulfur—the exact percentage varies significantly from sample to sample—is also present.** As discussed in Chapter 6, the presence of nitrogen and sulfur make coal combustion a very polluting process. The major products of coal burning, however, are CO_2 and H_2O—the compounds produced when any hydrocarbon or hydrocarbon derivative burns in oxygen. We'll soon see that CO_2, while not toxic, has tremendous ramifications as an environmental pollutant.

Like coal, petroleum and natural gas contain stored solar energy from millions of years ago. Generally found together, these two fossil fuels are the partially decomposed remains of sea organisms. As tiny oceangoing plants and animals died, they fell to the sea floor where they mixed with and were buried in sediment. Over time, the shifting layers of sediment created covered, pressurized pockets of organic material that, like coal, had many of the original energy-containing molecules still partially intact. When these pockets of fossilized organic material are accessed today, they are found to contain a mixture of gas and liquid. The liquid, made up of larger hydrocarbons and other organic material, is petroleum. **Due to its low sulfur content, petroleum combustion produces fewer SOx emissions than coal. The lighter hydrocarbons are not found as liquid components of petroleum, but instead make up natural gas.** With minimal contamination by nonhydrocarbon organics (and composed primarily of methane, which is too small to produce combustion particulates), natural gas is the cleanest burning of the fossil fuels. Table 13.1 compares the elemental breakdown of coal and petroleum.

The challenges associated with mining petroleum and natural gas made coal the first popular fossil fuel. Petroleum followed, gaining popularity in the early to mid-1900s. Natural gas, however, initially presented a logistical dilemma. It couldn't be carried like coal, or piped like petroleum. It was difficult to collect, store, and transport. As a result, for many years it was common practice to simply light the gas on fire as it escaped from a newly drilled oil well. With the problematic gas out of the way, removal of petroleum could commence. Since the mid-1900s, however, natural gas retrieval and processing technologies have progressed to the point that it is now used in proportions very similar to those of coal. In the United States, petroleum is still the major source of energy, however (Image 13.2).

One problem associated with an energy economy that depends largely upon fossil fuels is that they are nonrenewable: the partial decomposition and pressurization process that transformed living plants and animals into dense sources of energy took millions of years. **Since fossil fuels can't be produced on demand, there's no way to replenish the supply in response to increasing need. Our tremendous global energy demand threatens to outstrip the availability of these fuels.** This is one reason that the search for alternative energy sources, including solar, has been gaining momentum. While estimates for when oil will be depleted vary

% by Mass	Coal	Petroleum
Carbon	75–90%	83–87%
Hydrogen*	4.5–5.5%	10–14%
Oxygen	5–20%	0.5–1%
Nitrogen	1–1.5%	0.1–0.5%
Sulfur	1–7%	0.5%

TABLE 13.1. Comparison of the percent composition by mass of coal and petroleum. *Recall that hydrogen is much lighter than the other organic elements, meaning that even significant quantities will represent relatively low percentages by mass.

NOTE²

It's difficult to predict exactly what our future energy dependence and available energy sources will be, particularly given that there are new petroleum deposits discovered from time to time. At the same time, existing but currently inaccessible deposits may or may not become accessible in the future, depending upon regulations and advancements in technology.

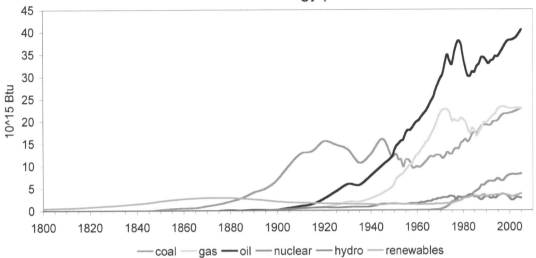

sources of U.S. energy production

IMAGE 13.2. Sources of energy production in the United States per the Energy Information Administration (EIA). On this graph, "renewables" refers to wood through the 1940s, and thereafter refers to wood, biofuel, wind, and solar. Oil is petroleum.

tremendously,[2] there is little doubt that the supply of petroleum will be exhausted in the not-too-distant future. In the likely event that oil availability dwindles dramatically in the next few decades (and in the absence of alternative energy technologies becoming more predominant than they are), coal is oil's most likely successor—there is about 40 times as much retrievable coal as there is petroleum. Even so, the nonrenewable nature of fossil fuels and their ever-diminishing quantity is not the biggest problem they present. Rather, their greatest threat to our welfare comes from one of their combustion products—the hitherto innocuous-seeming CO_2.

13.3 The Greenhouse Effect

The term "greenhouse effect" is often used synonymously with "global warming" or "climate change." In fact, nothing could be further from the truth. **The greenhouse effect is essential to the habitability of our planet—without it, Earth would be an average of about 33°C (60°F) cooler than it is and incapable of supporting life.** Much in the same way that a literal greenhouse (a glass building meant to trap the sun's energy) maintains plants through cool nights or seasons by keeping them warm, some of the molecules comprising our atmosphere naturally trap the sun's heat and warm the earth. This process is outlined in Image 13.3. When sunlight hits the planet, some is reflected by the earth's atmosphere (1), and some by the earth itself (2). The remaining radiation is absorbed either by the earth (3) or by atmospheric gases (4). The radiation absorbed by the earth is

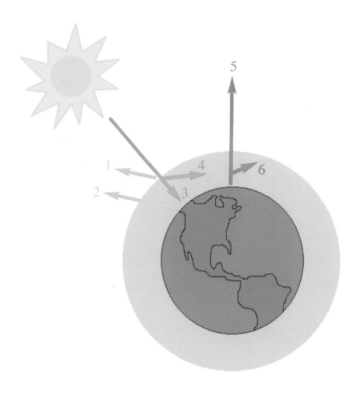

IMAGE 13.3. The greenhouse effect.

released as heat energy, some of which escapes into space (5), and some of which is trapped by atmospheric gases (6).

The collective action of atmospheric gas traps about 81% of the energy radiated by the earth, which would otherwise be lost to space. The gases then reradiate that heat upon the earth, maintaining the planet's warmth. Understanding the mechanisms whereby gases are able to absorb and reradiate heat requires that we spend a little time discussing yet another type of EMR, because what we feel as heat is actually **infrared radiation (IR)**.

IR has a longer wavelength (and is therefore lower in frequency and energy) than the UV light we discussed in Chapter 10. It is not ionizing radiation and is incapable of breaking bonds. However, here again, our bond-as-spring model serves us well, because IR is capable of compressing and stretching bonds, causing a molecule to vibrate (Image 13.4)

IMAGE 13.4. The effect of IR on a bond. Note that IR doesn't simply stretch the bond; instead, it initiates a vibration along that bond where atoms move closer together and farther apart as the bond compresses and expands.

IR Radiation

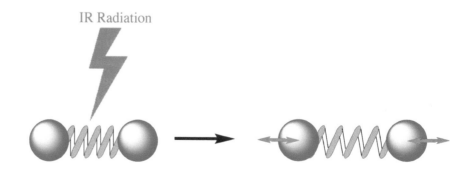

Absorption of IR is not permanent—shortly after a bond begins to vibrate, the molecule relaxes. This not only stops the vibration, but also results in the release of IR energy. **This absorption and subsequent release of IR is the key to understanding the greenhouse effect, because when a molecule traps IR, it's collecting heat, which is released some time later upon molecular relaxation, and reradiated upon the earth.**

Actually, not all molecular vibrations (and not all molecules) absorb IR. Table 13.2 shows many of the vibrations of CO_2, some of which absorb IR and others of which do not. It turns out that only vibrations producing a change in net dipole are capable of absorbing IR. **As such, molecules that have at least one IR-absorbing vibration can trap heat, and are therefore greenhouse gases.**

The first vibration listed in Table 13.2 shows CO_2 bending symmetrically around the central carbon. As this vibration takes place, the oxygen atoms will move together (as the molecule is drawn, we could call this an "up and down" motion), while the carbon atom will move in a direction opposite that of the oxygens. During the vibration, the molecule's shape is constantly moving between the extremes of an inverted "V" (as shown in the table) and a "V" (the shape created when carbon is at the bottom of its trajectory, and the oxygen atoms are at the top of theirs—Image 13.5). Since the C-O bonds are polar, the molecule's net dipole will also vary throughout this vibration process. When the molecule is bent as shown in the table, the net dipole points downward.[3] At the other extreme of the vibration, the dipole points upward (having

NOTE[3]

Remember, of course, that CO_2—like all molecules—is a three-dimensional object that can be turned in space. As such, references to "upward" and "downward" refer only to the perspective shown here and are for the sake of simplicity of language. Regardless of which way the molecule is oriented—and which way the atoms are moving—however, the net dipole will change during this particular vibration.

Vibration	Effect on Net Dipole	IR Absorbing?
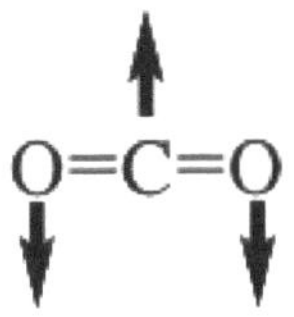 O=C=O (This can also take place in and out of the plane of the page, to the same effect)	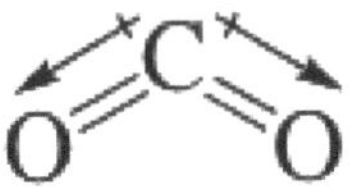 Change in net dipole	Yes
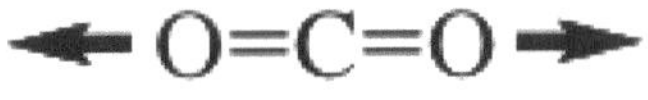 O=C=O	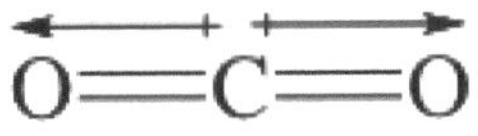 None	No
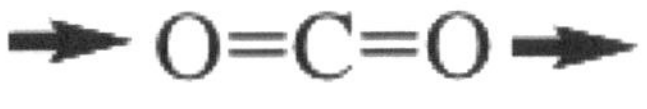 O=C=O	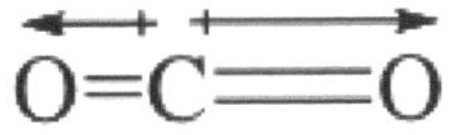 Change in net dipole	Yes

TABLE 13.2. Vibrations of CO_2 and IR absorption.

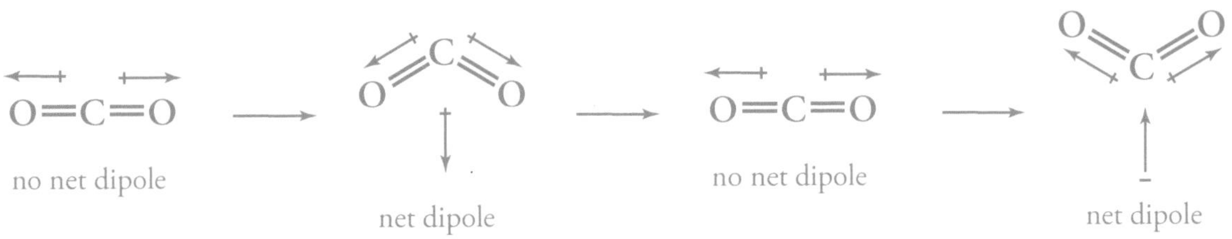

O=C=O → O=C=O → O=C=O → O=C=O

no net dipole   net dipole   no net dipole   net dipole

IMAGE 13.5. The variation of a carbon dioxide molecule's net dipole as it vibrates in one possible mode.

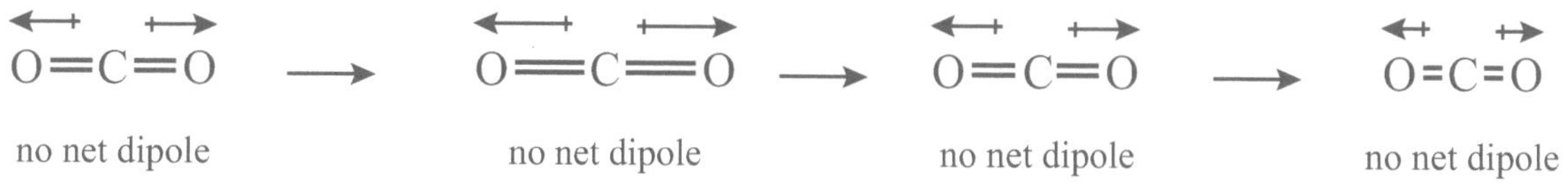

O=C=O →   no net dipole

O=C=O →   no net dipole

O=C=O →   no net dipole

O=C=O   no net dipole

IMAGE 13.6. A vibration of carbon dioxide in which the net dipole does not change, on account of symmetry. Regardless, carbon dioxide does have other vibrations that render it capable of IR absorption; this is simply not one of them.

moved through the neutral position in which CO_2 has no net dipole). Vibrations that change a molecule's dipole are high-energy vibrations—IR supplies the energy to produce this movement, and when the IR is released, the movement ceases.

In the second vibration on Table 13.2, in which carbon stays in place and the oxygens move together (away from carbon, then closer to carbon), the bonds change length, but the net dipole is not affected (Image 13.6). Since the C=O bonds always stretch or compress at the same time—and since there's no change in geometry around the central atom—the dipoles are always equal and opposite. As a result, the net dipole is always zero. This particular vibration, then, does not absorb IR. Regardless, CO_2 has other IR-absorbing vibrations, and is therefore a greenhouse gas.

There are greenhouse gases besides CO_2. SO_2, for instance, was mentioned in Chapter 6 as being both a tropospheric pollutant and a powerful greenhouse gas. Making this determination requires that we analyze SO_2's potential for IR-absorption, which requires us to identify at least one dipole-changing vibration. A good technique for going about this process is as follows:

1. Draw a Lewis structure of the molecule in question, including bond and net dipoles.

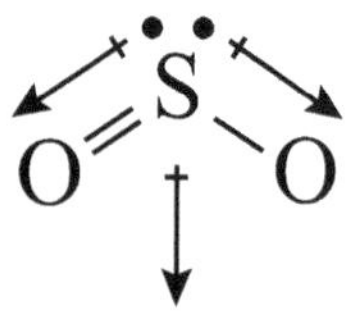

Note that lone pairs on outer atoms do not affect the geometry of the molecule, so they've been omitted from this structure in the interest of simplicity. SO_2 has a bent structure with a net dipole.

2. Try to determine a way in which the molecule could bend or bonds could stretch that would change the net dipole. It may be helpful to sketch this, even though it won't look like a normal or correct Lewis structure.

If SO₂ vibrated with the oxygens moving together (up and down) while the sulfur moved in the opposite direction, the molecule would pass through a linear geometry (no net dipole) on its way to a "V" shape (net dipole in the opposite direction of the original). This represents an IR-absorbing vibration. A given molecule will often have many IR-absorbing vibrations—determining that it's a greenhouse gas requires identifying only one such vibration. SO₂ is therefore a greenhouse gas.

In the end, whether a relaxed molecule has or does not have a net dipole is not the determining factor in whether it's a greenhouse gas—SO_2 has a net dipole, while CO_2 has none. Both, however, are greenhouse gases. **A shortcut to identifying a greenhouse gas without sketching dipole-altering vibrations is to look for bond dipoles: a greenhouse gas will always have at least one bond dipole.** Any molecule with at least one bond dipole can be bent or stretched to produce a change in the net dipole, and that bending or stretching represents an IR-absorbing vibration. By extension, any molecule with no bond dipoles is not a greenhouse gas. N_2, the most prevalent of the atmospheric gases, is not a greenhouse gas. With no bond dipoles, there are no bending or stretching motions capable of IR-absorption:

13.4   The Carbon Cycle

Atoms of carbon, like atoms of nitrogen, cycle through many different forms on Earth. Inorganic carbon, as found in carbonate rocks or CO_2, can be taken up by living organisms and transformed into forms of organic carbon. Oceans, rocks, organisms, and the atmosphere itself are all important **sinks** (meaning natural repositories of) and sources of carbon in the world.

The carbon cycle (Image 13.7) includes many transitions and transformations through which an atom of carbon can pass. Atmospheric carbon (CO_2) can dissolve in the ocean (1), where it may be used by photosynthetic organisms, incorporated into carbonate rocks, be taken up into the shells of aquatic organisms and skeletons of coral, or simply remain in solution. CO_2 is also released from the surface of the ocean (2). Plants participate in the carbon cycle in several ways. They take up CO_2 and store it in the form of organic molecules during photosynthesis (3). They also release CO_2 when they break down stored glucose for energy through the process of plant respiration (4). When plants are eaten, the consumer assimilates the plant's organic carbon. Plants and animals die and decay, a process that releases carbon back into the atmosphere (5). Under the right circumstances, carbon from decaying

TRY THIS

Using the technique outlined above, determine whether HCl is a greenhouse gas.

Answer:
It is a greenhouse gas.

Solution:
Stretching HCl along the bond lengthens (changes) the dipole. This is IR-absorbing.

TRY THIS

Using bond dipoles as your guide, determine whether each of the following is a greenhouse gas: O_2; CH_4; and CH_2Cl_2.

Answers:
O_2 is not a greenhouse gas (no bond dipoles), CH_4 is a greenhouse gas,[4] and CH_2Cl_2 is a greenhouse gas.

NOTE[4]

Note that while we considered the C-H bond *essentially* nonpolar for purposes of determining solubility in the earlier chapters of this text, there is a very small dipole since C is slightly more electronegative than H. The dipole, in this application, is very important—CH_4 is a potent greenhouse gas!

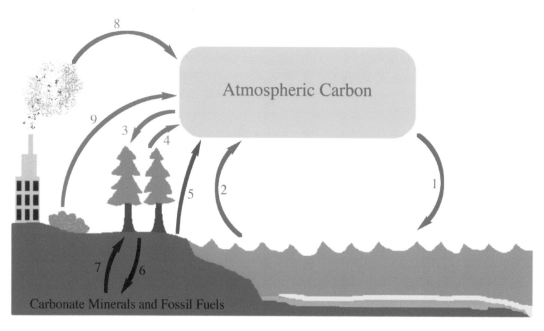

IMAGE 13.7. Some important components of the carbon cycle.

organisms can be fossilized, which adds it to the carbon sink of the fossil fuels; what's more, carbon can be taken up to form terrestrial carbonate minerals (6), from which it can later be rereleased to the atmosphere (7). Anthropogenic inputs into the carbon cycle include the combustion of biomass and fossil fuels for direct energy (8) and carbon-intensive land use (9), including the production of chemical fertilizing agents. Note that the complexity of the cycle means many actions affect several different processes—deforestation, for instance, reduces the tree population, which reduces the amount of carbon taken up through photosynthesis. The decaying tree releases its carbon to the atmosphere at a relatively slow rate, but if the removed tree is burned, its carbon is released much more quickly. Even subtle perturbations, such as temperature changes, can significantly affect the cycle. Warmer oceans, for instance, don't dissolve as much CO_2 (proportionally) as cooler ones. This disrupts the CO_2 exchange at the ocean surface. To be clear, rising atmospheric levels of CO_2 and temperature, the proof of which will be presented in this chapter, do not result in a net decrease in the amount of CO_2 in the ocean—quite the opposite. While a warmer ocean dissolves a smaller proportion of atmospheric CO_2 than a cooler ocean, when atmospheric CO_2 is high, the total *amount* dissolved by a warmer ocean is still greater than the total amount dissolved by a cooler ocean. As we've seen, carbon dioxide combines with water to form carbonic acid, so oceans with large amounts of dissolved CO_2 are more acidic. This has ramifications for aquatic life: coral reef bleaching, for example, is a direct effect of warming, acidifying oceans. As coral reefs support an incredible amount of biodiversity, this is hugely problematic. In fact, though they cover a mere 1% of the ocean floor, coral reefs support a staggering 25% of all ocean life, which highlights the importance of this ecosystem and the potential for catastrophe if it is lost (Image 13.8).

Looking at some numbers, our perturbation to the carbon cycle seems insignificant—human activities add about 8 gigatons (Gt, 10^9 tons) per year of carbon

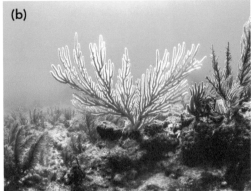

IMAGE 13.8. A colorful coral reef, together with the huge mass and variety of life it supports (a), stands in sharp contrast to the bleached, lifeless expanse of a reef fallen victim to ocean warming and acidification (b).

to the atmosphere, compared with an ocean-atmosphere exchange of about 115 Gt/year. Plants take up around 120 Gt/year through photosynthesis, and release about 50 Gt/year through the process of respiration. **Regardless of our comparatively small contribution to the atmospheric carbon, however, the imbalance compounds over time.** As with the perturbation of any other chemical cycle which would otherwise be in a state of dynamic equilibrium, our addition of carbon to the atmosphere amounts to speeding up one set of reactions (carbon entering atmosphere) without affecting the rate of the others (carbon exiting atmosphere). It's worth noting that while some factions doubt the connection between increasing atmospheric carbon and climate change, the statement that increasing carbon emissions are being released into the atmosphere is a scientific absolute and not a topic of debate. As Image 13.9 shows,

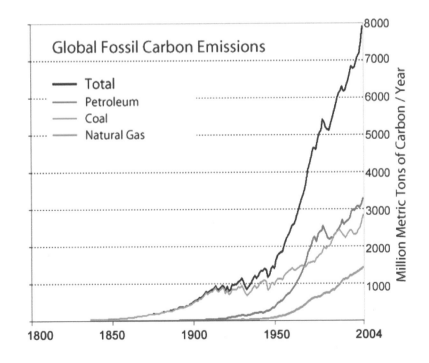

IMAGE 13.9. Global carbon emissions from various sources. Prior to the mid-1800s, the primary source of anthropogenic carbon emissions was combustion of wood. The quantity of these preindustrial emissions was insignificant.

CONCEPT CHECK

How is our perturbation to the carbon cycle similar to our perturbation to the nitrogen cycle?

Answer: In both cases, we're accelerating the movement of an element through one part of a cycle but not the other part (increasing reactive nitrogen, increasing atmospheric carbon).

our global carbon emissions have been steadily increasing ever since the Industrial Revolution.

Armed with an understanding of the carbon cycle and the science behind the greenhouse effect, we can now begin to address two very important issues: whether the planet is getting warmer, and, if it is, whether it's our fault or part of a natural cycle. Historically, the first indication that certain gases had the ability to affect the global temperature came about in the mid-1800s, when two scientists named John Tyndall and Jean Baptiste Fourier did some lab experiments that proved CO_2 (and some other molecules) had the ability to absorb IR radiation. While the idea didn't attract much notice at the time, it intrigued Svante Arrhenius, a Swedish chemist, many years later. In the late 1800s, Arrhenius was attempting to solve the riddle of how the planet had cooled enough to produce the historical ice ages. He ended up reasoning that decreased concentrations of heat-trapping gases might have precipitated the cooling trends. Through a labor-intensive series of calculations that dazzles modern climate-modeling scientists (who have the benefit of computers to accomplish the same task), Arrhenius managed to correlate atmospheric CO_2 concentrations with predicted temperature changes. He went on—purely out of academic curiosity—to calculate the effect increasing CO_2 concentrations would have on global temperature. **In this way, Arrhenius essentially invented the theory of global warming in the 1800s by showing that higher concentrations of greenhouse gases increase the extent to which heat is trapped.** This was the first time that a link was firmly established between the concentration of greenhouse gases and the strength of the greenhouse effect. The reality of anthropogenic climate change didn't particularly concern him, however—it was unimaginable at that time that we would ever produce enough carbon emissions to significantly alter atmospheric CO_2 concentration. As an aside, it is important to note the difference between CO_2 emissions and CO_2 concentration. Emissions are what we add to the atmosphere each year, while concentration is the amount of CO_2 in the atmosphere. If we stopped emitting CO_2 tomorrow (an impossible task in any case!), the atmospheric CO_2 concentration would still be much higher than it was prior to the Industrial Revolution—and would remain so for quite some time. Since Arrhenius's day, we've gone on to develop technologies that have proved him right in several respects: historical temperatures and corresponding atmospheric CO_2 concentrations have been determined, revealing that his calculations accurately predicted the relationship between the two. We have also, however, proven him wrong in a very disturbing way—we've increased the atmospheric CO_2 concentration more profoundly than he anticipated was humanly possible.

13.5 Measuring Atmospheric Carbon

While historians in Arrhenius's time knew that ice ages had been a part of the geological record, there was no specific data available on global temperature patterns and atmospheric CO_2 during either the ice ages or the interglacial periods. The collection of historical temperature and CO_2 data came with the development of technology to remove and analyze deep ice cores from various locations around the globe. The Vostok station on Antarctica is an important site for obtaining ice-core data.

As snow falls through the winter, air pockets composed of trapped atmospheric gases form within the snowpack. At the earth's poles, where summer temperatures never climb high enough to allow a thaw, each year's snow is simply covered over by the next, eventually compressing snowfall from earlier years into ice. Air pockets in the snow remain trapped in the ice matrix as millions of tiny bubbles. The result is that a core drilled deep into the ice reveals a series of rings (Image 13.10)—not unlike the rings of growth in a tree, except arranged from bottom to top rather than inside out—each of which represents a different period of time. Because the ice at the polar regions represents thousands and millions of years' worth of snow, with the earliest lying farthest beneath the surface, a drilled core provides a historical record of snow from hundreds of thousands of years ago up to the present, stacked in neat layers. **Because the bubbles in the ice are composed of the same gases (in the same concentrations) that were present in the air when the snow fell, they can be used to piece together a record of atmospheric CO_2 concentrations.** A deep Vostok core drilled to a layer more than 400,000 years old provides interesting information about historical CO_2 (Image 13.11). According to the Vostok core, atmospheric carbon has fluctuated on an approximately 100,000-year cycle.

There is further information to be obtained from the layers in an ice core. The atomic composition of H_2O in the ice itself allows for determination of global

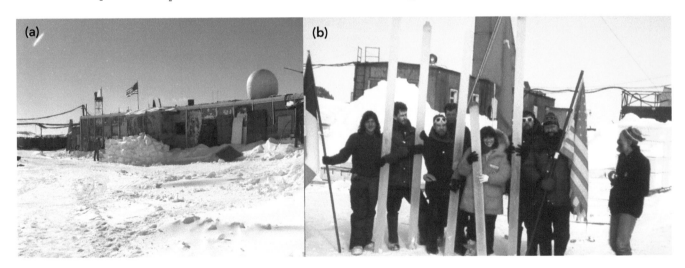

IMAGE 13.10. Vostok Station (a), and the 1990 Vostok research team (b). A deep ice core is composed of visible layers made up of snow that has long since been compressed into ice. Historical atmospheric composition and temperature record data can be extracted from the cores.

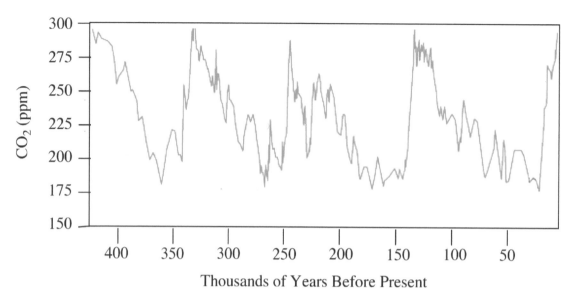

IMAGE 13.11. The historical atmospheric CO_2 record.

temperature patterns. Approximately 0.015% of hydrogen atoms are actually the

naturally occurring, stable isotope deuterium (^2_1H, sometimes just written ^2H). Because deuterium is heavier than hydrogen, water molecules containing deuterium (about 1 in every 6,700 water molecules) are slightly heavier than normal. This makes them incrementally slower to evaporate from the surface of the ice, an effect that is amplified in warmer weather. As a result, **the percentage of ^{2}H in an ice layer correlates with the average seasonal temperature when the ice was produced via snowfall; higher deuterium percentages indicate that the snow fell during warmer**

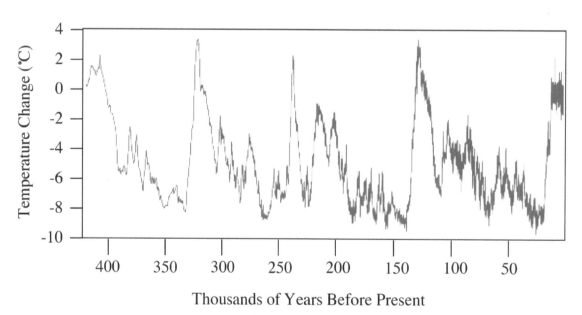

IMAGE 13.12. The historical temperature record from Vostok ice-core data. Temperatures are recorded in degrees above or below average 1990s Antarctic temperatures (the decade during which the deep core was analyzed).

weather. The same Vostok core that provided a historical CO_2 record was used to produce a 400,000-year temperature record (Image 13.12).

A superficial glance at the temperature data reveals that it, like atmospheric carbon, fluctuates on an approximately 100,000-year time scale. Particularly interesting is that when the two graphs are superimposed upon one another, it becomes clear that the fluctuations actually occur in tandem (Image 13.13).

A shrewd skeptic will note that while the CO_2 concentration and temperature data is very neatly matched, it is not perfectly synchronized: in some cases increasing CO_2 appears to slightly precede increasing temperature, while in other cases, the opposite seems to be true. As it turns out, it is logistically more difficult to date air bubbles than snow. While ^2H concentrations can be extracted neatly and precisely from a layer of known age (once snow falls, it neither shifts to a lower layer nor rises to a more superficial one), gas exchange between layers can occur for anywhere from 5,000 to 7,000 years after a layer becomes buried by new snow. This is because the compacting process that seals bubbles permanently takes quite some time; during that time, gas exchange between layers is commonplace. For this reason, it is possible to approximate the date of air bubbles to within several thousand years, but nearly impossible to pinpoint their exact age. A skeptic further notes that the Vostok data, while interesting, is correlational—it is not

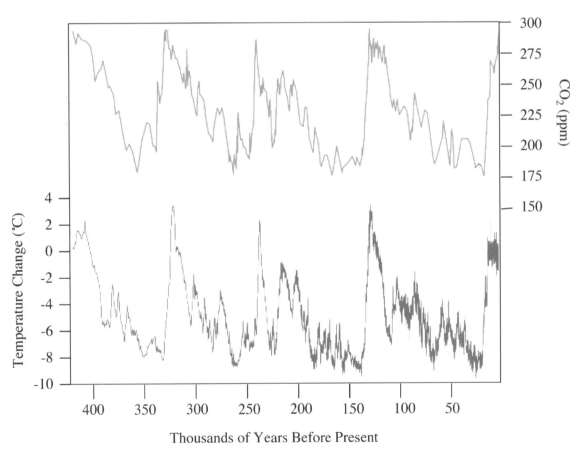

IMAGE 13.13. Combined CO_2 and temperature records from the Vostok data.

possible to determine from this data whether increasing atmospheric CO_2 caused global temperatures to increase, or vice versa. It is even possible that a third as-yet-unidentified factor affected both atmospheric carbon and global temperature. In any event, drawing no further conclusions as yet, the strong correlation between the two variables is interesting.

Useful as they are in yielding information about our geological past, ice cores cannot be used to generate data for more recent history. Thankfully, however, there are several hundred years' worth of global temperature data available, most of which was recorded for purposes with no connection whatsoever to the study of global warming, but which can still be put to good use in this capacity. Further, starting in the early 1900s, a few curious individuals began measuring atmospheric CO_2 from time to time. While a mild increase in carbon concentration was first noticed in 1938 (up 6% from the value recorded in 1900), no real attention was paid to the finding. Charles Keeling, a scientist and outdoor enthusiast, was the first researcher to track atmospheric CO_2 with any real dedication. From 1958 until his death (and thereafter taken over by other members of his research team), he monitored CO_2 concentrations from the top of Mauna Loa in Hawaii every four hours, with such dedication that he reportedly missed the birth of his first child. The remote placement (and elevation) made the mountain a perfect spot for accurate measurements,

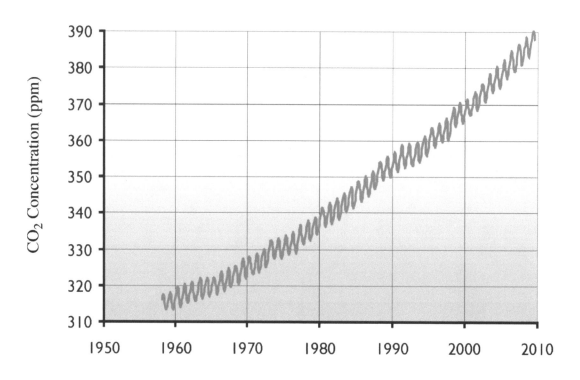

IMAGE 13.14. The Keeling Curve is a record of atmospheric CO_2 from 1958 to the present. The fluctuations are seasonal: photosynthesis in the spring results in a minor and temporary reduction in atmospheric CO_2. Despite these, however, a clear upward trend is noted.

as it was unaffected by local pollution and weather patterns. The record of his findings is called the **Keeling Curve** (Image 13.14).

One important point worth noting is that **while the Vostok data shows that atmospheric CO_2 fluctuates regularly, there is no evidence of it having been higher than approximately 280 ppm at any point in the last 400,000 years—a period that encompasses many ice ages and interglacial periods.** In 1958, when Keeling began his Mauna Loa measurements, CO_2 was already over 310 ppm and was steadily climbing. Today, the atmospheric CO_2 is just under 400 ppm, and concentrations continue to increase by around 2 ppm annually (though it has been noted recently that the rate of change is itself increasing). Superimposed upon the Vostok data, the Keeling Curve appears as a disconcerting, nearly straight line (shown in red in Image 13.15).

It's clear from gathered data that increasing carbon emissions (recorded by individual countries and compiled into global emissions records) have resulted in increasing atmospheric CO_2 concentrations in the last hundred or so years. Without an indication of climate change over the same period, however, there's no reason to examine the issue any further. Such indications fall into two different categories: **empirical evidence** of climate change, based upon observation and measurement, and **climate modeling**, which is used to predict future climate patterns.

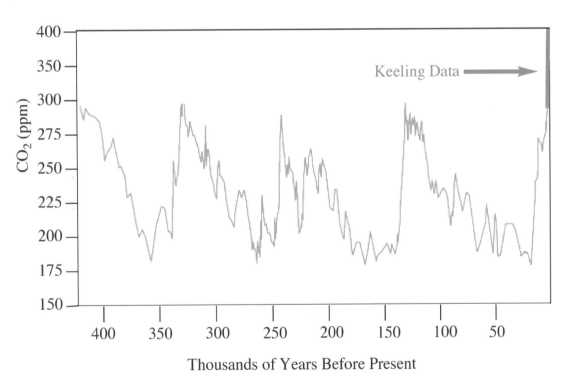

IMAGE 13.15. What appears on the Keeling curve to be a gradual increase in CO_2 concentration is more dramatic alongside the Vostok data.

There is a significant body of empirical evidence to suggest a recent (and accelerating) warming trend. Global mean temperatures have been rising steadily for the past 150 years, with the rate of change increasing through incrementally smaller sampling ranges, meaning that the fastest warming has taken place most recently (Image 13.16).

Certain weather pattern changes are associated with warmer average global temperatures—namely increasing frequency and severity of weather events, coupled with shifting distribution of precipitation. Many of these predicted changes have been observed over the last hundred years. The National Oceanographic and Atmospheric Administration (NOAA) has collected data indicating that parts of the world (including portions of eastern North and South America, northern Europe, and north-central Asia) are significantly wetter than they were a century ago, while southern Africa, the Mediterranean, and southern Asia are much drier. More high-latitude precipitation falls as rain rather than snow, and heavy or severe precipitation events are far more common than they were in the early 1900s. An analysis by the National Climatic Data Center found a 14% increase in heavy rain events (defined as more than 2 inches in a day) and a 20% increase in very heavy rain events (more than 4 inches in a day) during the 20th century.

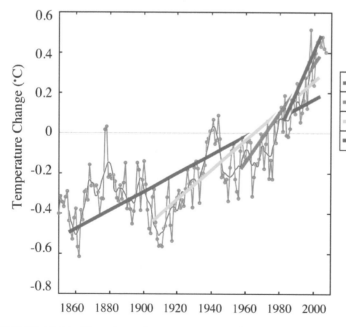

Period (Years)	Rate (°C per decade)
25	0.177 +/- 0.052
50	0.128 +/- 0.026
100	0.074 +/- 0.018
150	0.045 +/- 0.012

IMAGE 13.16. This chart shows 150 years of global mean temperatures (annual averages in red, five-year means in blue). Regression lines showing average rates of change are superimposed.

Extreme weather events have become not only more common but also more severe, particularly in the last two decades. NOAA uses a measurement called the U.S. Climate Extremes Index to track severe weather and changing patterns. The index quantifies extreme weather event severity and frequency by year, including conditions such as abnormally high and/or low temperatures, abnormally high and/or low precipitation, and high-intensity rainstorms. Higher index values indicate more extreme weather patterns—the average value through the past century has been 20. Values in the high 30s and 40s have become increasingly frequent since 1980, with particularly high values in 1998 (index value 44) and 2005 (index value 41). Hurricanes, a very specific and destructive extreme weather event, are known to develop over warm ocean surfaces. Recent studies show that sea-surface temperatures have been increasing through the 20th century, with the warmest temperatures noted most recently. Further, the number and severity of extreme hurricanes (class 4 and 5) has increased significantly since 1970—so much so that hurricane scientists have considered introducing a sixth class for storms.

For this data to be a meaningful indicator of climate change seems to require that the last several decades of increased extreme weather be unique in recorded history. In fact, however, they are not. Hurricane intensity was also high around the 1940s. This period was preceded (and followed) by periods of reduced hurricane severity and frequency. Sea-surface temperatures closely parallel hurricane activity for all periods, with warmer temperatures during more active periods (Image 13.17).

However, climactic events capable of producing each warming and cooling trend prior to the most recent are matters of record. Six volcanic eruptions between 1875 (Iceland) and 1912 (Alaska), including the massive eruption of Krakatoa in 1883, spewed tremendous amounts of ash and other aerosols into the lower atmosphere. These aerosols are known to have cooling effects, as they block incident solar radiation from reaching the earth's surface. In 1963, a volcanic eruption in Indonesia produced another massive cloud of dust and aerosols—this is likely responsible for declines in global (and sea-surface) temperature noted through the early 1980s. In the periods between and following these forcings—environmental or anthropogenic factors that affect climate—a steady warming trend was observed.

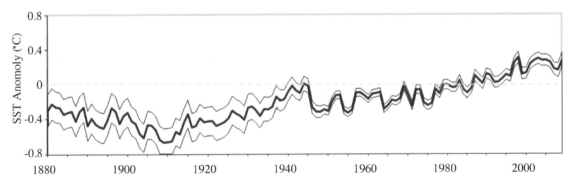

IMAGE 13.17. Sea-surface temperature (SST) data from NOAA, 1880–2009 (dark line). The warming trend is much larger than the uncertainty (lighter lines bracketing data) in recent years.

Additional global observations include that most mountain glaciers are diminishing in size. Sea ice in the Arctic is reduced year-round, and the latest estimates suggest that it could disappear entirely as early as the summer of 2030. The 2002 collapse of the Larsen B ice shelf in Antarctica—a shelf that is estimated to have formed at the end of the last ice age and which disintegrated in just weeks—resulted in the loss of a huge mass of ice with a surface larger than the state of Delaware. In earlier editions of this text, the collapse of the Larsen B shelf was referred to as "shocking." Since the previous edition of the book went to press, a new trend has arisen, with two additional monstrous ice sheets beginning their collapse. Perhaps the appropriate phrase to describe this melting is no longer "shocking," as it no longer comes as any surprise, but rather "deeply alarming." The West Antarctic Ice Sheet, a segment of the continental ice sheet covering Western Antarctica, has begun its collapse. While it is impossible to predict precisely the degree to—or the rate at—which the sheet will melt, relatively conservative estimates suggest that a complete collapse of the sheet would result in a 10–15 foot (3–5 meter) rise in sea level over the 21st century. The Totten Glacier in East Antarctica has similarly begun its collapse, with estimates suggesting that a complete loss of the ice mass would effect an additional 10–15 feet of sea-level rise. **Melting sea ice has resulted in a measurable—and accelerating—rise in sea level over the past century, which is only expected to continue.** Of course, the earth's history is filled with ice ages and interglacial periods, so huge swaths of ice coming and going is nothing new. The issue here is not the melting itself but the rate at which it is occurring; in general, the space between ice ages is about 100,000 years, with glaciers slowly disintegrating over a period of about half that time and then slowly forming again. The collapse of a huge mass of ice in a period of weeks—or months, or even decades—is unprecedented in the natural world and cannot be explained by natural cycles of cooling and warming.

In the living world, a number of studies have noted that insect populations are beginning to alter their reproductive cycles,[5] with implications for several insect-borne diseases, including malaria and dengue. With warmer temperatures extending farther from the equator and an extended breeding season afforded by longer summers, mosquitoes and other disease-vector insects appear to be flourishing. With the mosquitoes' increased prevalence, rates of and areas prone to mosquito-borne illness are increasing.

Of course, even if the data indicate a steady land- and ocean-warming trend over the last hundred years, the warming could be the result of increased energy input. One popular hypothesis to explain warming suggests that increased solar radiation (either due to a change in Earth's proximity to the sun or increased solar activity) is responsible for the temperature change. If this is true, then the amount of energy radiated by the Earth back into space should also increase—since a normal greenhouse effect maintains 81% of Earth's radiated heat and releases the remaining 19% into space, increased incident radiation would both increase the amount maintained and the amount released. Even if increased incident radiation somehow shifted the percentages, it stands to reason that the amount of energy radiated into

NOTE[5]

Insects, with exponentially shorter life spans than other animals, respond to changes in the environment before larger organisms. Animals that feed upon insects generally follow close behind, as already noticed in several bird studies that indicate changes in migration patterns.

space would nevertheless increase: all other factors held constant, more energy in must mean more energy out. **That is, unless something was also happening to change the ability of the atmosphere to maintain heat, a larger amount of incident energy should result in a larger amount of energy released to space.** Several studies, however, have shown that, in fact, the earth is releasing less heat with each successive year. Further, while global temperatures are increasing, the average temperature of the stratosphere—recall that the stratosphere occurs at elevations above about 15 km—has been falling over the same period (Image 13.18). This cannot be explained by increased solar activity: more radiation incident upon the planet would increase the temperature of all atmospheric levels. **Instead, a falling stratospheric temperature can be explained only by an increase in heat trapping at lower atmospheric levels.** In other words, it can only be explained by an enhanced greenhouse effect: the trapping of more than 81% of Earth's heat. Earth's decreasing

Lower Troposphere

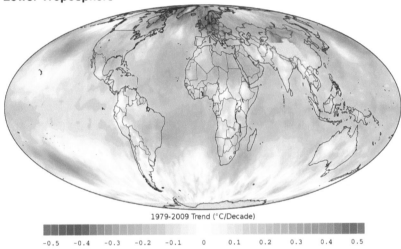

1979-2009 Trend (°C/Decade)

-0.5   -0.4   -0.3   -0.2   -0.1   0   0.1   0.2   0.3   0.4   0.5

Lower Stratosphere

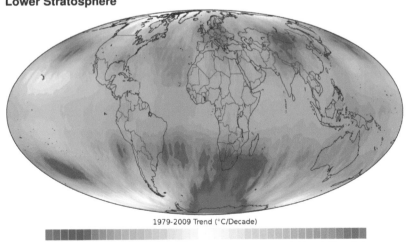

1979-2009 Trend (°C/Decade)

-0.6   -0.48   -0.36   -0.24   -0.12   0   0.12   0.24   0.36   0.48   0.6

IMAGE 13.18. The troposphere has experienced a warming trend during the same 30-year period (1979–2009) that the stratosphere has cooled. Surface warming (not included on image) is even more pronounced than lower tropospheric warming. Temperature change is recorded as a change in °C/Decade.

CONCEPT CHECK

Why is the information in Image 13.18 considered a "smoking gun" for the enhanced greenhouse effect?

Answer: Because if the earth were still radiating 81% of heat into space, warming trends on the surface would be reflected in the stratosphere.

radiation of energy into space has been called the "smoking gun" of climate change: it doesn't prove anthropogenic causes for global warming, but it certainly proves that global warming is due to an enhanced greenhouse effect.

13.7 Climate Modeling

With its roots in the calculations undertaken by Arrhenius in the late 19th century, climate modeling has the potential to be a powerful tool. Unfortunately, it's also incredibly complex, as climate may be affected by any number of factors. Volcanoes, solar activity, Earth's place in its orbit, the percentage of the earth's surface covered by ice, composition of the atmosphere, and ocean currents are just a few of the myriad factors that have the potential to affect climate, and that must be considered by modelers. Further, forcings are often dynamic, and their effects can compound in a positive-feedback loop, as we'll shortly see. In the end, it's simply not possible to include every possible forcing and its predicted effect in any given climate model. This, coupled with the fact that many climate models do not include the feedback effects of forcings, means that variability in predicted climate outcomes is significant. Some models predict major climate change over the next decades and century, while others suggest that the effect will be more minor. **Regardless, all scientifically based models agree that both increased global temperature and sea level rise will characterize the coming century.** The Intergovernmental Panel on Climate Change (IPCC), a United Nations panel with the mission to assess climate change research as well as its social and economic ramifications, released a report in 2014 that included the latest projections for the 21st century. Global mean temperatures are predicted to increase by at least 1.5°C, with most models agreeing on an increase of at least 2°C. Bearing in mind that an average global temperature change of only 4–6°C is the difference between an ice age and an interglacial period, this is significant. Models also agree that sea-level rise will occur and will likely be significant, though with both the West Antarctic Ice Sheet and Totten glacier only recently (and unexpectedly) beginning their collapse, projections are difficult to make accurately. **It's worth noting, though, that many climate scientists feel that the IPCC models may underestimate coming changes, as they do not include feedback effects, and the historical record shows that climate tends to respond to forcings in a way that amplifies, rather than mitigates, them.**

These amplifications, or feedback effects, can be quite significant, and are already recognized in the climate change literature. Gases are less soluble in warm liquids than in cool ones (the reason a soda goes flat sooner if it's warm). As a result, increasing global temperature—and the concomitant increasing ocean temperature—would decrease CO_2 solubility and result in release of carbon dioxide from the ocean (despite a net increase of CO_2 *in* the ocean, simply because of the increased amount of CO_2 in the system as a whole). This would increase atmospheric concentration of CO_2, further enhancing the greenhouse effect. Soils behave similarly; they take up less CO_2 (and release more) when warm. Net loss of gas in the soil has been recorded in England and elsewhere. Tundra is yet another source of global warming positive feedback effects. It is essentially frozen mulch—organic photosynthetic material that does not decompose and release its carbon because of the exceedingly low temperature at which it is found. As tundra melts in response to increased temperature, the stored carbon is released into the atmosphere. Because of the vast quantity of carbon dioxide currently locked in tundra (around 3.6 trillion tons), defrosting of the land represents a significant potential for increased CO_2 release. Frozen peat bogs, too, are a liability on a warming planet. The sub-Arctic (Siberia) is a massive shelf of frozen peat, which contains methane ice (called *clathrate*) that was generated when the bogs were liquid, and that has since become frozen beneath the surface of the ice. The methane (estimated at 70 billion tons) is released as bogs melt, a process that is already occurring. Like CO_2, methane is a powerful greenhouse gas—in fact, it has a global warming potential (GWP) 12 times that of carbon dioxide. GWP is a standard measurement established by the IPCC that compares a molecule's ability to store and reradiate IR to that of CO_2 (Table 13.3).

It's also worth noting that even without massive releases of greenhouse gases from melting tundra and peat, loss of ice alone amplifies warming. Because ice reflects sunlight to a greater extent than does water or land (this is referred to as *albedo*), a loss of ice leads to more heat energy being absorbed by the Earth, further increasing global temperatures.

Compound	GWP
CO_2	1
CH_4	21
N_2O	310
CH_2FCF_3	1300

TABLE 13.3. GWP for a few of the many greenhouse gases. Note that N_2O is in equilibrium with NO_2 and other NOx in the atmosphere, but is the nitrogen oxide for which GWP is most commonly reported. CH_2FCF_3 is an HFC (see Chapter 10).

13.8 Causality

At this point, we've examined large bodies of data establishing carbon dioxide and other molecules as greenhouse gases. They are important to keeping our planet warm and livable, but in high concentration, they maintain too much of Earth's heat. We've also reviewed the myriad studies, observations, and models that indicate our planet is in the midst of a clear warming trend, with significant and distressing ramifications. A skeptical analysis of these data, however, still demands proof of a causal relationship between the gases and the warming. After all, given the positive feedback discussed above, it appears possible that warming could cause a release of greenhouse gases rather than the other way around. In a sense, this argument is circular—we've established that certain gases are responsible for the greenhouse effect, so debating which comes first (the increase in gas or the increase in temperature) is tantamount to arguing about chickens and eggs. Still, there are many factors that affect climate, and we must examine each one in turn in an attempt to tease out the cause(s) of global warming.

One forcing commonly suggested as a nonanthropogenic cause of global warming is solar activity. However, while an increase in solar flares was recorded through the 1950s and 1960s, since the 1980s solar activity has been on the decline while

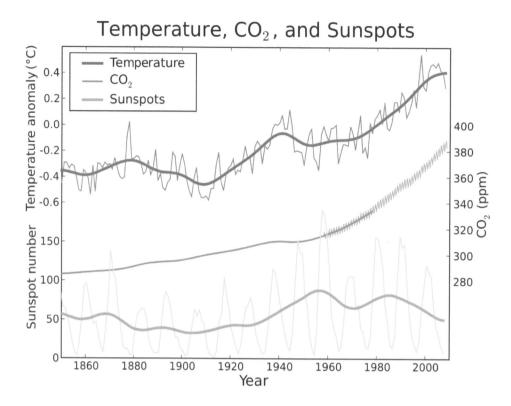

Temperature, CO₂, and Sunspots

IMAGE 13.19. Solar activity over the last 150 years (data from Solar Influences Data Center).

temperatures have continued to climb more steeply than before (Image 13.19). Additionally, a portion of the mid-20th-century solar activity peak does coincide with a mid-century increase in temperature, but the temperature change actually preceded the sunspot activity by more than a decade. If sunspots do affect global temperature (and they are certainly considered to be a valid forcing in their own right), they aren't solely responsible for climate change—particularly not lately.

A weakness of climate modeling in the traditional sense is that it's a "wait and see" science—once a model has been created, there's no way to know whether it is accurate until many years, or decades, in the future. However, it's possible to test a model (and to determine which factors are important in producing an accurate model) by simulating a past-to-present modeling scenario. That is, by programming known information about the last century's climate forcings—solar activity, volcanic activity, atmospheric composition, and the like—and allowing the computer to produce a temperature trend prediction for the period in question. A model produced in this way has the advantage of predicting temperature effects for years that have already passed, meaning that the data can be compared to known values and checked for accuracy. For instance, a model using only volcano-related forcings (no other effects on climate were considered) predicted temperature fluctuations quite out of harmony with actual observations (Image 13.20).

Similar models were produced for solar activity, ozone concentrations, aerosols, and other natural forcings, each of which predicted that temperature would

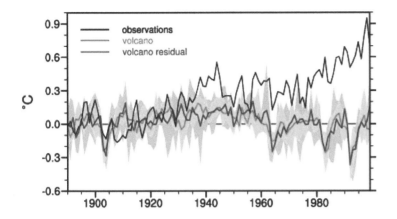

IMAGE 13.20. Model including volcanic forcings as compared to observed data. © American Meteorological Society. Reprinted with permission.

fluctuate throughout the 20th century, but not increase. Some forcings, taken alone, actually predicted a temperature decrease as the century progressed. One interesting model, however, showed a close parallel between predicted and observed temperature. The forcings used to produce the model were the collective greenhouse gases (Image 13.21).

The same study also produced two models that combined forcings. One of these took all significant natural factors into account, while the second included all significant natural factors and anthropogenic factors. The data speak for themselves (Image 13.22).

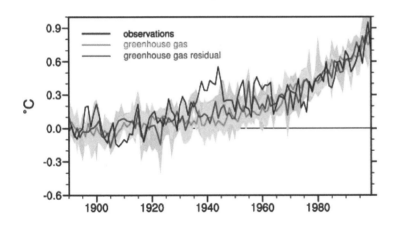

IMAGE 13.21. Model including greenhouse gas forcings as compared to observed data. © American Meteorological Society. Reprinted with permission.

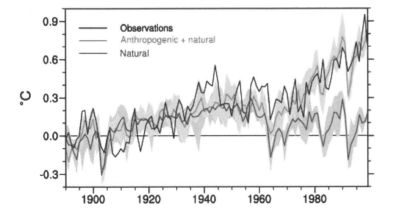

IMAGE 13.22. Model including all significant natural forcings (purple) and model including all significant natural and anthropogenic forcings (red) compared to observed data. © American Meteorological Society. Reprinted with permission.

CONCEPT CHECK

Why do retrospective models prove
that we are responsible for climate
change?

Answer: Because accurate models of past
temperatures can only be predicted when
models include anthropogenic factors.

Clearly, the best prediction of temperature change through the last century is produced only through inclusion of anthropogenic factors; in the absence of anthropogenic factors, models predict temperatures below those currently observed—and falling. With forcings other than greenhouse gases failing to produce a reasonable projection of temperature change—and greenhouse gases alone producing a model that closely fits observed temperature—the causal relationship between gas and temperature is established. **Natural forcings taken alone—even including natural sources of greenhouse gases—are insufficient to model climate accurately. The inclusion of human-generated emissions is critical to accurate climate modeling.** If Earth's decreased heat radiation into space is the smoking gun for the enhanced greenhouse effect, then the climate model in Image 13.22 is the smoking gun of its principle cause—and it has our fingerprints on it.

13.9   Why Focus on CO_2?

Given that there are many greenhouse gases (including water!), why is it that all colloquial discussions of global warming center on carbon dioxide? In reality, this is nothing more than a simplification that takes place in the public sector. Climate scientists and global organizations recognize that CO_2 is not the only greenhouse gas, nor is it the only one that's increasing in atmospheric concentration. That said, it is the most major anthropogenic greenhouse emission, making it an important compound to track for monitoring atmospheric health. In fact, though, methane (emitted anthropogenically as the result of fossil fuel production, burning of biomass, and large-scale livestock operations), N_2O, HCFCs, and HFCs are among the many greenhouse gases. Water falls into a somewhat different category: while it is a greenhouse gas, its concentration in the atmosphere varies greatly with temperature, so a GWP for water is generally neither calculated nor independently meaningful. Furthermore, the oceans regulate water concentration in the atmosphere; as concentration increases, oceans increase water uptake, meaning that it's not useful to track water concentration in the atmosphere for the purpose of predicting climate effects. **The fact is that all the major greenhouse gases are increasing in concentration through human activity.** Global warming reduction measures, therefore, take into account the effects of all greenhouse gases, and discussions of emissions reduction measures include all compounds of anthropogenic origin with global warming potential.

13.10   Making Sense of Rhetoric

Those who deny the legitimacy of global warming science sometimes point to differences in model forecasts, suggesting that this represents dissention in the scientific world (and illegitimacy of the science). This is tantamount to suggesting that there's dissention in the ranks of a panel of Olympic gymnastics judges when a performer receives a range of scores from 9.7 to 10. Did the judges all award the same number? No. Why? They were looking at different things. Judging involves subjective analysis. They were focusing on different aspects of the performance. Does this mean the judges didn't agree? Not at all; a score of 9.7 isn't quite as high as a score of 10, but they're both top-level performances. Similarly, when climate scientists are asked to model projections for the future, they include different forcings (and feedback effects) to different degrees, which impacts the predictions. Some models show that effects will be significant but moderate; others predict catastrophic temperature increase and sea-level rise. Regardless, all scientifically based models agree that temperature *will* increase and sea level *will* rise over the coming century. Models differ by degrees, not by consensus.

Again, modeling is a complex field: while it's possible to use models such as those discussed above to tease out key forcings, retrospective models are far easier to build—and build accurately—than prospective models. The reason is twofold: first, it's impossible to know what the future holds with regard to our continued global emissions. Scenarios modeled on the assumption of ever-increasing emissions will necessarily be more alarming than scenarios modeled on reduced global carbon emissions, with the latter optimistically anticipating a global response to the climate crisis. Second, the positive feedback effects upon climate are well established, but incompletely understood. For instance, it is difficult to know exactly how much polar ice melt will accelerate energy absorption. It's known that melting peat bogs will release methane, but tough to determine how much of that methane will be released unaltered, and how much will react with oxygen to form CO_2—the two gases have different GWPs, and therefore different magnitudes of impact. Prospective climate modelers must work at the intersection of science and art, combining theory with instinct and known values with anticipated human responses to produce their best scientific hypothesis for what the future holds. **In reality, what is sometimes seen outside the scientific world as a disparity of outcomes is seen by scientists as a consensus of trajectories. In 2001 Donald Kennedy, the editor of *Science* (a peer-reviewed scientific journal) noted, "Consensus as strong as the one that has developed around [global warming] is rare in science."**

Further, *Science* published a review article in 2004 that collated and analyzed almost 1,000 climate studies, drawing the conclusion, "Politicians, economists, journalists, and others may have the impression of confusion, disagreement, or discord among climate scientists, but that impression is incorrect." 97% of climate scientists agree that the planet is warming and that the cause is anthropogenic. Only 1% of climate scientists deny this (with the other 2% undecided). In addition to consensus regarding climate change and causality, climate scientists are also in agreement upon the degree to which humans—as opposed to natural forcings—are responsible for climate change. Image 13.23 shows the results of a group of studies that used various methods (modeling, etc.) for tabulating human versus natural contributions to climate change. In all cases, scientists agreed that humans were at least fully responsible; in many cases, scientists agreed that natural forcings alone

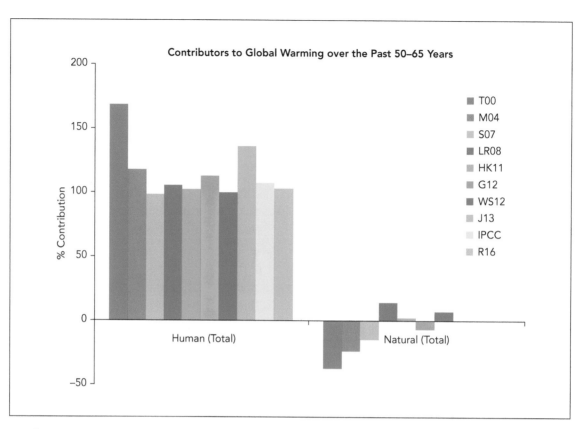

IMAGE 13.23. A selection of peer-reviewed studies that allocate contribution of human actions and contribution of natural forcings to climate change. Note that, in all cases, human contribution is ranked at or above 100% responsible for climate. In a majority of the studies listed here, natural forcings—left to their own devices—would have resulted in a cooling, rather than a warming, planet.

would have resulted in a cooling climate, meaning that human actions are not only overriding a natural cooling trend but creating a warming trend instead. Truly, consensus such as this is nearly unprecedented in science.

One last word about model disparity: given that all scientifically derived models agree that temperatures will increase and seas will rise (with predictions differing only in degree), model disparity is discomforting rather than reassuring. Earth's climate is changing, and in a very immoderate way. How hot it will get, how wild the weather will become, and how high the sea will rise can't be known for sure. As a global community, we have several legitimate options to evaluate through risk–benefit analysis. Hoping, however, that all the models and signs are wrong, and that the evidence is being misinterpreted, is not a justifiable scientific response. Unlike risk–benefit analysis, which acknowledges and accepts some risk to achieve a desired outcome (and unlike the precautionary principle, which accepts no risk), blind hope considers only the benefits of inaction, without acknowledging or evaluating the risks. In a way, our response to climate crisis is analogous to a driver who has started to lose control of his vehicle at high speed, and has to decide quickly how to respond. Some would turn the wheel to steer into the skid, while others might decide to hit the brakes. Some would worry more about trajectory while others might be more concerned with speed. But how many would simply take their feet off the pedals, their hands off the wheel, and hope for the best?

13.11 The Kyoto Protocol

In 1997, the Kyoto Protocol was signed by many countries around the world in an effort to reduce total greenhouse emissions. Developed countries were required to cut emissions first—the protocol required a 5% cut from 1990 emissions levels by 2012. While there were no restrictions put on developing countries for the first phase of the protocol, these countries nevertheless stood to benefit from reduced carbon emission: reducing emissions would afford them carbon credits, which they could sell to developed nations that were overemitting. The reasons for the disparity in treatment of developed versus developing nations were outlined by the United Nations. First, the developed nations were most responsible for carbon dioxide and other greenhouse gases currently (as of 1997) in the atmosphere—they had benefited economically from the same industrialization and development processes that had resulted in emissions. As such, they were expected to respond first to the crisis. Additionally, they were expected to use some of the wealth accrued at the expense of atmospheric CO_2 concentrations to assist developing countries in establishing carbon-reducing technology. Second, the emissions from developing countries were (as of 1997) still low. Even now, while China's total emissions have eclipsed those from any other single country (10,330 million metric tons of CO_2 in 2013, with the next largest emitter being the United States at 5,300 million metric tons), their per capita emissions are much lower than those of developing nations (Table 13.4).

Country	Total Emissions (in millions of metric tons)	Per Capita Emissions (in tons/capita)
1. China	10,330	7.4
2. United States	5,300	16.6
3. India	2,070	1.7
4. Russia	1,800	12.6
5. Japan	1,360	10.7
6. Germany	840	10.2
7. South Korea	630	12.7
8. Canada	550	15.7
9. Indonesia	510	2.6
10. Saudi Arabia	490	16.6
16. Australia	390	16.9

TABLE 13.4. The top ten carbon-emitting countries by total emissions (2013 data). Note that, due to population, countries can have very high total emissions but low per capita emissions. Industrialized countries may, by virtue of lower population, have lower total emissions but high per capita emissions.

In 2012, a second period was added to the Kyoto Protocol which established new emissions commitments for the period from 2013–2020. As of 2015, there has been limited acceptance of this second period, with only 36 countries agreeing to binding emissions targets (countries shown in green in Image 13.24). Global negotiations regarding emissions capping are ongoing, but in the meantime, global emissions continue to rise.

Compared to the Montreal Protocol, the Kyoto Protocol hasn't enjoyed the same degree of success and global popularity. The United States, for instance, has categorically refused to sign the protocol, ostensibly on the grounds that it doesn't do enough to curb emissions. In particular, the United States has objected to the extent to which China's and India's emissions have been allowed to grow. **It's important to note that while developing nations are now becoming large emitters, it will take decades of such emissions before they have contributed as much carbon to the atmosphere as the United States and other developed nations already have.** It's also worth noting that China and India have large *total* emissions but lower *per capita* emissions than industrialized countries; this is because they are populous countries. In other words, the US is not only the second-greatest emitter of carbon in the world, it's also among the top emitters per capita; if China had per capita emissions equivalent to those of the US, their total emissions would be monumentally higher than they currently are. In many ways, per capita emissions are a more appropriate measure of environmental impact by country than total emissions, as population is not distributed evenly across the globe. With regard to the United States' refusal

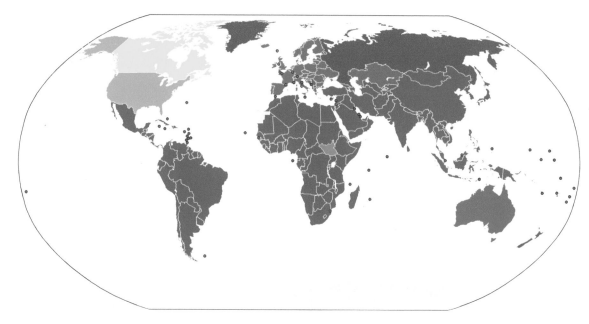

IMAGE 13.24. Signatories to the Kyoto Protocol as of 2014. Green countries are signatories with binding emission targets in the second period of the protocol. Purple countries had binding targets in the first period but not the second. Blue countries have no binding targets. Canada (yellow) had a binding target in the first period but withdrew from the protocol. The United States (orange) is a protocol signatory but has not ratified it. Countries in red are not parties to the protocol.

to ratify the protocol, developed nation signatories suggest that leading by example and assisting where necessary may be the best route to encouraging developing nations to reduce emissions. Reluctance on the part of the United States to ratify the protocol may be a sign of distaste for its economic ramifications. Quite simply, to an even greater extent than the Montreal Protocol, the Kyoto Protocol potentially threatens national economic security and forces a changeover to alternative energy sources—requiring a huge investment in research, development, and placement of infrastructure—and that has been difficult for many developed nations to accept.

13.12   Global and National Strategies

The failure of the Kyoto Protocol to effect significant global cooperation leaves us in a troubling place. Clearly, developing a unified global strategy that addresses climate change and greenhouse emissions is a matter of political will as much as it is a matter of developing and implementing new technologies. There are many interesting discussions of global politics and climate change (see Sources and Further Reading), but as this is a chemistry text, we'll focus on technologies with chemical significance. Strategies to reduce carbon dioxide concentration in the atmosphere fall into two groups: carbon sequestration and fossil fuel alternatives.

Carbon sequestration is the capture of CO_2 released during combustion processes, with subsequent condensation and injection into deep geological repositories. While the technology in no way addresses the fossil fuel shortage, it attempts to curb emissions and slow the pace of global warming. Since petroleum reserves are in far shorter supply than coal, most carbon sequestration technologies being developed are for implementation in coal-fired power plants. Unfortunately, carbon emissions from traditional smokestacks are very diluted by the air, and it's nearly impossible to capture and sequester CO_2. As a mechanism for addressing this, a new technology in development seeks to combust coal in the presence of both oxygen and water, generating H_2 (the combustion of coal provides energy to produce hydrogen from water) and CO as the major products. The CO would then be converted to CO_2, which would be sequestered, while the H_2 would be utilized for energy through a mechanism addressed in a later section of the chapter. The problem with this is that it's very energetically expensive—25% of energy extracted from the coal must be used simply to power the plant, representing a large decrease in efficiency. Another 20% of the energy generated from burning a given quantity of coal is required to liquefy the CO_2, and a variable (but also significant) quantity of the energy is necessary to transport and pipe the liquid underground. The volume of liquid CO_2 is also staggering—an estimated 50 km³ would need to be stored on a daily basis to completely eliminate global carbon emissions. Of course, the earth's crust is a finite space—it's not difficult to imagine that at some point, its capacity for accepting CO_2 would begin to diminish. Finally, the storage containers, like nuclear waste storage containers, run the risk of developing leaks. If liquid CO_2 began to leak from its container, it would quickly turn to gas (high pressure is required to liquefy carbon dioxide at livable temperatures), and could flood the atmosphere with years' or decades' worth of carbon emissions in a matter of hours to days. **While sequestration as a limited or auxiliary strategy for reducing emissions may have merits, it nevertheless fails to address an important issue—continued reliance on fossil fuels—and creates several new problems.**

The fossil fuel alternatives encompass a wide variety of technologies and materials, but all alternative-energy strategies generate energy through means other than combustion of fossil fuel. Solar and nuclear power have been discussed elsewhere (Chapters 11 and 12, respectively), and while wind and hydroelectric power have intriguing potential, they are not strictly chemistry related and so will not be discussed in this text. There are several alternative fuels worth discussing, however; the most viable of these are nonpolluting and carbon neutral, meaning that they do not result in the net release of carbon dioxide into the atmosphere. One alternative fuel class with the potential to reduce both fossil fuel dependence and net carbon emissions is biofuel, which is combustible material from renewable sources. Biofuels, like fossil fuels, release solar energy that was originally captured and stored through the process of photosynthesis. The major difference is that biofuels rely upon organisms capturing the sun's energy *now*, rather than organisms that captured energy millions of years ago. While either form of fuel yields its stored energy through a CO_2-releasing combustion reaction, when fossil fuels are burned, the carbon released was taken up from the atmosphere at the same time solar energy was stored—eons ago. As a result, adding that carbon back into the atmosphere represents an increase in current atmospheric carbon dioxide concentration. Biofuels, on the other hand, can be cultivated in fields (or in water, in the case of algae). They engage in photosynthesis, storing the sun's energy in the form of cellulose, starch, and sugar and removing carbon dioxide from the atmosphere. When the organisms are harvested and burned (either as whole biomass or in a processed form), the carbon is rereleased, meaning that CO_2 emission is in proportion to CO_2 uptake and the carbon cycle is balanced. Additionally, some biofuels have low environmental impacts compared to fossil fuels: they don't require mining or drilling, and their combustion is associated with much lower tropospheric pollutant emissions than fossil fuels. Owing to these considerations, some organisms are far more appropriate to biofuel production than others—tall trees, for instance, are nowhere near as renewable nor as carbon-neutral as rapid-growth grasses because of their long lifetimes. Switchgrass (a rapid-growth tall grass), corn, soy, sugarcane, and some species of algae have been identified as the most plausible cultivated biofuels.

We must note, however, that simply because a biofuel is carbon neutral and environmentally low impact in theory doesn't make it carbon neutral in practice. For instance, the cellulose and sugar in corn can be used to produce ethanol, which is incorporated into the fuel mixture burned in many vehicles worldwide (including the United States) and can be used alone as a biofuel. Corn in the United States, however, is farmed using chemical fertilizers, which have an environmental impact all their own (see Chapter 10), and which are produced through the combustion

IMAGE 13.25. Switchgrass is a tall, rapid-growth plant with biofuel potential.

of fossil fuels. Additionally, there is an energy cost associated with refining ethanol from corn. In fact, many researchers who have undertaken calculations to determine the energy efficiency and biofuel potential of cultivated corn note that there is lower environmental impact and fewer carbon emissions if petroleum is burned directly, as opposed to being used to produce chemical fertilizers, fuel corn growth, and generate biofuel. In the end, it takes more calories of energy to produce ethanol from corn than are contained in the ethanol itself, because of the energy dissipated to entropy during each step of the process. Recent estimates suggest that 131,000 BTU of energy are required to produce a gallon of corn ethanol containing 77,000 BTU of energy, meaning that 70% more energy is required to produce corn ethanol than is yielded.

Another biofuel concern has to do with the amount of land necessary for cultivation. For instance, planting switchgrass on otherwise "useless" (nonagricultural, nonforested) land can produce sustainable, carbon-neutral biofuel. However, destroying (generally by burning) thousands upon thousands of acres of rainforest in Brazil to plant sugarcane for biofuel preempts any carbon neutrality or low environmental impact. Since forests and rainforests are carbon sinks containing hundreds of years' worth of sequestered carbon, removing the forests not only releases all that CO_2 into the atmosphere (and all at once), but also removes plants that would have otherwise continued to take up and store atmospheric CO_2. This is to say nothing of the loss of biodiversity and impact to soil that are effects of deforestation.

Biofuels have the potential to be important, low-impact energy sources, but care must be taken when evaluating their environmental effects. In general, if they are cultivated, they must be grown upon land with ample water resources (to reduce unnecessary water dependency in areas that may lack sufficient precipitation). The land should not have been deforested to provide for the fuel cultivation, nor should the organisms require chemical fertilization. Processing of raw material to cellulose or ethanol should be minimal and as energetically conservative as possible in order to minimize usable energy loss and unnecessary energy input.

Another energy option worth discussing is hydrogen, which combines with oxygen in a combustion reaction that produces water:

$$2\,H_2 + O_2 \rightarrow 2\,H_2O$$

Hydrogen fuel cells (voltaic cells) running this highly spontaneous reaction can be used to provide energy without any carbon emissions, making them attractive alternative fuels. **The problem, however, is that unlike the carbon compounds synthesized by biofuel organisms, there are no renewable, natural sources of hydrogen gas—at least not in any great quantity.** Instead, H_2 must be produced through the electrolysis of water. This reaction is endothermic and requires an energy source. The question, then, becomes one of what is supplying the energy to effect the production of H_2. Currently, most hydrogen produced in the United States and elsewhere in the world is fueled by petroleum, nullifying the carbon neutrality and low impact of the hydrogen fuel. Proponents of hydrogen fuel suggest that

CONCEPT CHECK

Why is carbon dioxide from burning grass "better" than carbon dioxide from burning petroleum?

Answer: The grass took up the carbon recently (removed it from the atmosphere), so we're not producing a net increase in atmospheric carbon dioxide by rereleasing it.

it will be more environmentally friendly when energy to produce the reaction is provided by alternative sources, such as solar or wind. Opponents, however, argue that the problem with that logic is twofold: first, it is predicated upon further development of solar, wind, or other technologies; second, once those technologies are developed, it would be far more energy-efficient (because of the second law of thermodynamics) to use them directly as sources of electricity. The reason for this becomes clear through an examination of one proposed application of hydrogen fuel cells—as energy sources for transportation. A regular internal combustion engine burns a hydrocarbon fuel, the result of which is the production of a large amount of heat (essentially a small explosion), which is used to physically drive a piston. Hydrogen fuel cells, however, do not literally burn hydrogen—instead, hydrogen and oxygen are the anode and cathode (respectively) of a voltaic cell, and the electricity generated by the cell is used to power an electric motor. As a result, a hydrogen fuel cell is energetically inefficient compared to both traditional fuels and alternatives. Rather than burn petroleum to produce hydrogen, which is then used to generate electricity that is used to power a motor (two energy transformations), it makes more sense to burn petroleum directly in a traditional engine. Rather than use alternative energy sources to generate electricity to produce hydrogen, which is then converted back into electricity that is used to power a motor (three energy transformations), it makes more sense to use the alternative energy directly to charge an electrical battery that is used to power a motor. **Hydrogen fuel cells are sleek sounding and have consumer appeal, but until the technology advances to the point that hydrogen generation doesn't require many cycles of energy transformation, the energetics won't add up.**

13.13   Community and Individual Strategies

Politics, technologies, and global strategies aside, there are ways for communities and individuals to reduce their global impact and the emissions for which they are responsible. **Generation of electricity consumes the largest percentage of energy used in the United States (Image 13.26).** Further, 66% of all electricity generated in the US is produced through combustion of coal and natural gas, meaning that while electricity can "feel" like an emissions-free form of energy, such is often not the case. Electric cars, for example, are commonly advertised as nonpolluting, but if the electricity used to charge the car's battery is generated through combustion of fossil fuels, the impact remains the same as if the fossil fuels were burned directly in the vehicle's engine. Conserving electricity, then, is an important strategy to reduce carbon emissions.

Reduced consumption of electricity to heat and cool homes and buildings (even a few degrees makes a difference) significantly reduces daily electrical demand. New construction should incorporate energy-efficient and insulating building materials, while older construction can be updated in small ways that make

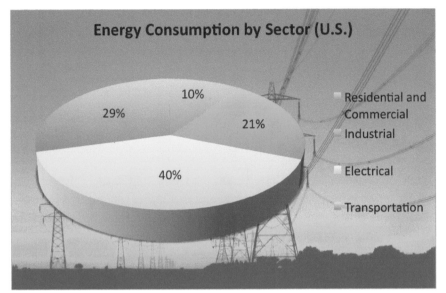

a big difference: replacing single-pane windows with dual glass and exchanging curtains or vertical blinds for closed-cell blinds both help prevent heat loss (or gain) through windows. Most appliances (including televisions and computers) draw electricity even when switched off; unplugging these can reduce unnecessary electrical consumption. Chargers for cell phones, MP3 players, and other small electrical devices draw power when plugged in, even if the device is not attached. Old refrigerators, dishwashers, washing machines, and water heaters are less energy efficient than newer models.

Transportation is another major source of energy consumption and therefore offers many opportunities for reducing environmental impact. Using public transportation (which operates, and therefore produces emissions, whether it's used or not) is a great way to reduce individual impact, as are biking and walking. Air travel produces significant emissions under even the best of circumstances, but direct flights are less energy-intensive than flights with multiple stops. Purchasing local food not only supports the local economy, but also reduces global emissions—locally grown food need not be shipped or trucked from farms outside the state or country. The carbon cost of food transportation worldwide is tremendous. On a similar topic, consuming food which is grown without chemical fertilizers also reduces carbon dioxide and other greenhouse emissions because the energy input is purely solar: no fossil fuels are burned to produce fertilizer.

Finally, because all commercially processed foods and goods require energy input, decreased reliance upon processed foods and decreased consumption of manufactured goods positively impacts the environment. In addition, it's far more energy efficient to reprocess plastic (or paper, or metal, or any other commonly recycled material) into a new form than it is to refine or produce the material from scratch. In general, the more energy transformations a particular action or material requires, the more energy it consumes, and the more carbon emissions it's responsible for producing. Bottled water, as discussed in Chapter 5, represents a tremendous energy expenditure as compared to what's available (nearly free of charge!) from the tap. A variety of sources with many more suggestions for reducing personal and community impact on the environment are listed in Sources and Further Reading.

13.14   A Parting Thought

While modern technologies have provided us with conveniences that are now hard to imagine doing without, we must remember that as with many of the topics discussed in this text, our solutions to problems have become the problems for which we now require solutions. It's easy, sometimes, to allow ourselves to become convinced that global problems require global solutions—and that we as individuals aren't able to effect change. However, a global community is nothing more than the sum of national communities, which are merely the sum of local communities, which themselves are the sum of individuals. We are all a part of the global organism, and we all affect its well-being.

13.15   The Last Word—Stomping or Tiptoeing?

In the last section of the chapter, we discussed some strategies for reducing individual impact upon the environment. Carbon emissions–related environmental impact is commonly called a **carbon footprint**. While we may want to make a difference, most of us have a tendency to feel very small and insignificant in a very big world. It's easy to start wondering whether our attempts at conservation really matter, particularly when the people around us—and some governments—don't appear to be making much of an effort. It's easy, too, to wonder whether the sacrifices and lifestyle changes make any sort of a dent at all in total carbon emissions; is making a personal lifestyle change really worth it?

We'll start with some national data: the average American is personally responsible for 19.8 tons of CO_2 emissions per year; if every human on Earth lived like the average American, we'd need 6.7 Earths to sustain our consumption of resources and production of emissions, and that would be for humans alone—the 6.7 Earth tally doesn't leave room for animal habitat.[6] The average citizen of the United Kingdom has a somewhat lower impact, responsible for about 9.7 tons of CO_2 per year (3.3 Earths, if everyone lived as they do). The average Indian is responsible for about 1.2 tons of CO_2 per year. India is so often cited as a "major polluter" which must bring its emissions under control, but the fact is that the country's large total emission footprint is the result of a population over 1 billion. Ironically, given India's reputation as a major polluter, if everyone lived like the average Indian, we would have enough resources to sustain ourselves on a single planet—and could even increase our population somewhat. Obviously, then, the average citizen of an industrialized

NOTE[6]

If you're curious about your own carbon footprint, there are several web-based calculators listed in the Sources and Further Reading at the end of the chapter.

country is disproportionately emitting relative to other cultures, and as such—and separate and apart from national or global action on climate change—we have a responsibility to do what we can to curb our impact. To that end, let's look at how much a conscientious individual can reduce the carbon emissions for which they are personally responsible in a year. Each of the calculations below is based upon EPA and other data, and is for a *single individual* (averaged over whether they live alone or in a group), with carbon emissions savings given in tons of CO_2 per year:

- Efficient heating and cooling of a home = **1.5–2.5 tons**, depending upon your home state. Extremely hot or cold states have greater discrepancy between emissions for efficient versus inefficient heating and cooling.
- Installing energy-efficient lighting = **0.5 ton**.
- Using energy-efficient appliances = **1.2 tons**.
- Reducing hot water usage by turning down the thermostat on the water heater and switching the heater to vacation mode when out of town = **0.7 ton**.
- Walking, biking, or using public transportation one day a week = **1 ton or more**, depending upon your vehicle and commute.
- Reducing meat consumption from every meal to once a day = **1.8 tons**.
- Eating mostly organic food = **1.8 tons**.
- Recycling and composting whenever possible = **0.8 ton**.

None of the measures above are extreme—certainly there are greater lengths to which an individual can go in order to reduce the size of their carbon footprint. Using alternative transportation multiple days a week, for instance, multiplies the reduction in emissions. **However, by focusing on just a few key points, it's possible *for a single individual* to prevent the release of nearly 10 tons (that's 20,000 pounds) of CO_2 annually.** Note that's also more than half of the yearly carbon emissions of an average American: a significant effect, to be sure. Furthermore, a person who engaged in all of the above would enjoy the added benefits of a lower electric bill, a little extra exercise, and a healthier diet. Frankly, those are all worthy goals in their own right, to say nothing of environmental impact!

SUMMARY OF MAIN POINTS

- Our global energy needs are increasing annually. Most are fulfilled through the burning of fossil fuels.
- Fossil fuels include coal, petroleum, and natural gas, all of which are partially decomposed organisms that captured and stored solar energy millions of years ago.
- Fossil fuels are nonrenewable and emit CO_2 when combusted.
- The greenhouse effect helps maintain the earth's temperature: certain gases trap heat radiated by the earth and prevent it from being lost to space.
- Infrared radiation (IR) is absorbed by molecules and induces bond bending or stretching that changes the molecule's net dipole.

- When IR is released, the molecule relaxes; in the atmosphere, this released IR (heat) is reradiated upon the planet.
- Molecules with at least one bond dipole are greenhouse gases.
- The carbon cycle is a dynamic equilibrium whereby carbon is released to the atmosphere from various sources and taken up by various sources; human emissions perturb the carbon cycle.
- Global carbon emissions from the combustion of fossil fuels are increasing dramatically.
- Historical atmospheric and global temperature data is extracted from ice cores.
- Bubbles in ice core layers provide a record of historical atmospheric CO_2 content.
- ^2H concentration in ice core layers provides a record of historical global temperature.
- Both the global temperature and atmospheric carbon records show fluctuations through the last 400,000 years.
- Recorded data from temperature and CO_2 monitoring shows that for the last 100 years, global temperature and atmospheric CO_2 have been increasing dramatically, with the most significant increase in the last 25 years.
- Sea-surface temperatures and extreme weather events have been increasing.
- Empirical evidence—weather, insect life cycles, bird migrations, glacial disintegrations—all suggest significant warming.
- Earth's radiation of heat into space has been decreasing over this same period. This is the "smoking gun" for an enhanced greenhouse effect.
- Due to complications associated with model building, climate models disagree as to how much the temperature will increase over the next century.
- Many factors affect models, some of which (ice melting, tundra melting, soil CO_2 solubility) show positive feedback effects.
- CO_2 is not the only important greenhouse gas—several other gases have global warming potential.
- Sunspots, volcanoes, and other natural forcings can't explain recent climate change.
- Only models that include anthropogenic factors can accurately predict observed global temperatures over the past century.
- The Kyoto Protocol put restrictions on developed nations, and encouraged emissions capping from developing nations. It has had limited success.
- Global emission reduction strategies include carbon sequestration (which is logistically difficult and energetically expensive) and alternative energies.
- Alternative fuels may be low impact and carbon neutral, but their ability to reduce atmospheric carbon while providing efficient energy depends on how they are produced.
- Energy transformations result in loss of usable energy, making hydrogen fuel cells likely to be an inefficient alternative energy strategy for the time being, unless technology changes.

- Individuals and communities can help reduce carbon emissions by conserving electricity and reducing personal and commodity transportation.
- Reducing dependence upon commodities, reusing, and recycling commodities results in reduced greenhouse emissions.

QUESTIONS AND TOPICS FOR DISCUSSION

1. People who believe climate change to be a hoax are sometimes called "climate skeptics" by the media. Is this an accurate label? Why or why not?

2. In the US, the perception that there's a global/scientific debate regarding climate change is relatively common. Is this perception accurate? What produces it?

3. What is the difference between "global warming" and "climate change"?

4. What does it mean to say that the fossil fuels are nonrenewable? In light of global energy requirements, what are the ramifications?

5. *The greenhouse effect* is often used to mean the same as *climate change*. Are the concepts synonymous? Explain.

6. Why is the term "greenhouse effect" different in meaning than "climate change"?

7. Why is the greenhouse effect important to our planet?

8. What is meant by a molecular vibration? Why do some vibrations absorb IR?

9. Why is our perturbation to the carbon cycle significant, even though it's relatively small compared to other flux quantities?

10. In earlier chapters, we considered the C-H bond to be essentially nonpolar. Is this completely accurate? Why or why not? Why do we make this approximation?

11. Why did Arrhenius begin examining the relationship between carbon dioxide and global temperature? Why was he unconcerned by his theory of global warming?

12. What are the limitations of ice-core data? How do you think these may have fueled climate change debate (in the public)?

13. Is it legitimate to say that our current temperature and CO_2 concentrations are part of a normal global fluctuation based upon ice-core data? Why or why not?

14. Why can ice-core data not be used to prove that CO_2 concentrations cause global climate change?

15. Why can't we use the Vostok data to prove that increasing CO_2 concentrations cause temperature change?

16. What challenges do climate-modeling scientists face in trying to predict future temperatures?

17. Does the disagreement among climate predictions undermine the statement that climate is changing and that there will be global ramifications? Why or why not?

18. Does the empirical evidence support climate change? Why or why not?

19. Why is extreme weather considered evidence of climate change?

20. Why is stratospheric cooling evidence of the enhanced greenhouse effect?

21. What positive feedback systems amplify climate change, and how?

22. A friend says that he doesn't believe in climate change because where he lives, it's been getting colder and colder in recent years (and raining more). What might you say in response?

23. A friend says that he doesn't believe in climate change because this last year was a record snow year in his state. What might you say in response?

24. Given that there are many greenhouse gases, why do we hear so much about carbon dioxide?

25. A friend says that it's silly to pay attention to greenhouse gases like CO_2, which are in relatively low concentration in the atmosphere, when there's so much water (also a greenhouse gas) in the atmosphere, over which we have no control, meaning that we cannot possibly be responsible for climate change. What might you say in response?

26. Can sunspot activity be used to explain climate change? Why or why not?

27. Why are retrospective models powerful mechanisms for determining which forcings are contributing most to climate change?

28. What were the major goals of the Kyoto Protocol? Was it successful? Why or why not?

29. What do you think needs to happen for the global community to come to agreement on climate action?

30. The US refused to sign the Kyoto Protocol on the grounds that it "didn't do enough to curb emissions." Do you think this was a legitimate reason? Do you think it accurately represented the reason for reticence? Explain your thinking.

31. Several industrialized nations currently refuse to commit to reducing emissions until India and China commit to reducing theirs. Is this fair, or unfair, and why?

32. What provisions do you think a new global emissions treaty should contain?

33. Is it completely accurate to say that China is the world's largest carbon emitter today? What about in a historical context? Do you think China should be required to reduce emissions?

34. Why is the distinction between atmospheric carbon and carbon emissions important when it comes to placing restrictions on nations? Do you think nations (like China and India) that are large emitters today, but have contributed minimally to atmospheric carbon, should have to cut emissions as much as industrialized nations?

35. What are some of the challenges associated with using carbon sequestration? Do you think the technology should be pursued? What does a risk–benefit analysis reveal?

36. How might global economics and politics be negatively affected by biofuels? Think about developing agricultural nations.

37. What are the considerations that must be taken into account in determining the environmental impact of biofuels?

38. The U.S. government heavily subsidizes corn, some of which is converted into ethanol. Is this a good sustainable-energy strategy? Explain.

39. What are the considerations that must be taken into account in determining the energy efficiency of hydrogen fuel cells?

40. Can you imagine a system that would make hydrogen fuel cells a good alternative energy? What technologies would need to be developed?

41. What are ways that an individual or community can reduce the carbon emissions for which they are responsible?

42. A common environmental slogan is "reduce, reuse, recycle." How does each of these help reduce emissions?

PROBLEMS

1. What are the three fossil fuels and what are their sources?

2. Where does the energy in fossil fuels come from (originally)?

3. Which fossil fuel contains the most sulfur and nitrogen? Why is this a problem?

4. Which fossil fuel contains the least sulfur and nitrogen?

5. Natural gas is sometimes called the "cleanest-burning" fossil fuel. How is this accurate? How is it inaccurate?

6. What are the possible dispositions for energy that is radiated toward Earth by the sun? Which of these are considered part of the greenhouse effect?

7. What is the difference, numerically, between the greenhouse effect and an enhanced greenhouse effect?

8. What kind of radiation is absorbed by greenhouse gases?

9. What kinds of molecules are greenhouse gases?

10. What's the simplest (shortcut) method of identifying a greenhouse gas?

11. Without drawing Lewis structures, identify the greenhouse gases among the following: H_2; CH_2Cl_2; and CHF_3.

12. Without drawing Lewis structures, identify the greenhouse gases among the following: NO_2; N_2; and CO_2.

13. Using Lewis structures, explain why H_2O is a greenhouse gas but O_2 is not.

14. Using Lewis structures, explain why CO_2 and SO_2 are both greenhouse gases, despite having different shapes.

15. Must molecules have bond dipoles to be greenhouse gases? Must they have net dipoles?

16. Why are oceans an important part of the carbon cycle?

17. What are the three major sources of U.S. energy?

18. What are the three major sources of global carbon emissions (fuels)?

19. What is likely to be the major source of global energy in coming decades in the absence of a new technology?

20. How is temperature record data extracted from ice cores?

21. How is atmospheric CO_2 data extracted from ice cores?

22. What data is used to provide information about trends in the last century with regard to temperature and CO_2?

23. What is the Keeling Curve and why is it important to a discussion of climate change?

24. How has global temperature changed over the last 150 years? Comment not just about trend, but upon the rate of change.

25. How is sea-surface temperature related to hurricane frequency and severity?

26. In terms of weather, what sorts of patterns are associated with climate change?

27. What is the "smoking gun" of the enhanced greenhouse effect?

28. What is a climate forcing?

29. Why do the IPCC predictions for climate change contain so much uncertainty?

30. What is a GWP and what does it mean quantitatively?

31. How might volcanoes have contributed to temperature patterns in the last century?

32. What do retrospective models show us about the cause of climate change?

33. What is carbon sequestration, and why is it being developed mainly for coal burning?

34. What is a biofuel?

35. Why are CO_2 emissions from fossil fuels "worse" than CO_2 emissions from biofuels?

36. Based on current technology, why are hydrogen fuel cells responsible for carbon emissions?

37. A friend is proud of a new car, an electric vehicle that he claims is "emissions free." The car is charged at home with electricity supplied municipally. Do you think it's fair to say that the car avoids carbon emissions entirely? What adjustments to the friend's strategy would make the car truly emissions free?

ANSWERS TO ODD-NUMBERED PROBLEMS

1. Coal, petroleum, and natural gas are the fossil fuels. All three are partially decomposed organic matter: coal is composed mostly of terrestrial organisms, where petroleum and natural gas are mostly ocean organisms.

3. Coal contains the most sulfur and nitrogen by percent mass. This makes it a more polluting combustion fuel than petroleum or natural gas.

5. Natural gas produces the fewest SOx and NOx, but it still generates carbon dioxide.

7. The greenhouse effect is re-reflection of 81% of Earth's radiated energy. The enhanced greenhouse effect is re-reflection of more than 81% of the energy.

9. Molecules that can vibrate in a way that changes their net dipole are greenhouse gases.

11. H_2 has no bond dipole, so it's not a greenhouse gas. CH_2Cl_2 and CHF_3 have bond dipoles, and therefore are greenhouse gases.

13. H_2O can vibrate in a way that changes the net dipole, but O_2 can't.

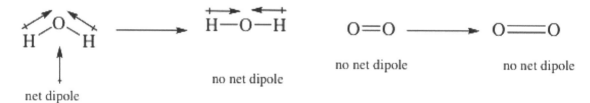

15. Molecules must have at least one bond dipole, but needn't have a net dipole to be a greenhouse gas.

17. Petroleum, natural gas, and coal.

19. Coal will likely be the next major source of global energy.

21. Bubbles in the ice layers contain historical atmospheric gas samples.

23. The Keeling Curve is a record of atmospheric carbon dioxide since 1958. It shows a huge increase in atmospheric CO_2 over any values in the last 400,000 years.

25. Hurricanes are more frequent and severe when sea-surface temperature is higher.

27. The "smoking gun" is the data showing that Earth is radiating less heat into space.

29. Climate modeling is a difficult science because there are many factors involved, some of which are unpredictable (including positive feedback effects and human behavior/response to climate change).

31. Volcanoes release clouds of particulates and dust, which can cause cooling. Volcano-related cooling was observed early in the 20th century and again in the 1960s.

33. Carbon sequestration is the capture of carbon dioxide, whereupon it is liquefied and injected into geological repositories. It is mainly being developed for coal because there is much more recoverable coal than petroleum on Earth.

35. Growing a biofuel results in removal of carbon dioxide from the atmosphere so that when the fuel is burned, there is no net increase in atmospheric CO_2. Fossil fuel carbon was removed from the atmosphere millions of years ago, so releasing it represents a net increase in atmospheric CO_2 in today's terms.

37. Coal and natural gas, taken together, are combusted to supply 66% of electricity in the United States. A car charged from the electrical grid is ultimately being powered by the burning of fossil fuels. If the car could be charged from a solar panel, however, the vehicle would be truly emissions free.

SOURCES AND FURTHER READING

Climate Change

http://www.ipcc.ch

www.realclimate.org

www.skepticalscience.com

Bellamy, P. H., Loveland, P. J., Bradley, R. I., *et al.* (2005). Carbon losses from all soils across England and Wales. *Nature, 437*(7056), 245–48.

Emanuel, K. (2005). Increasing destructiveness of tropical cyclones over the past 30 years. *Nature, 436*(7051), 686–88.

Groisman, P. Y., Knight, R. W., Karl, T. R., *et al.* (2004). Contemporary changes of the hydrological cycle over the contiguous United States: Trends derived from in situ observations. *Journal of Hydrometeorology, 5*(1), 64–85.

Hansen, J., Nazarenko, L., Ruedy, R., *et al.* (2005). Earth's energy imbalance: Confirmation and implications. *Science, 308*(5727), 1431–35.

International Panel on Climate Change. (2007). *Climate change 2007 synthesis report*. Geneva, Switzerland: Pachauri, R. K., and Reisinger, A. (eds.)

Kennedy, D. (2001). An unfortunate U-turn on carbon. *Science, 291*(5513), 2515.

Oreskes, N. (2004). Beyond the ivory tower—the scientific consensus on climate change. *Science, 306*(5702), 1686.

Paaijmans, K. P., Read, A. F., and Thomas, M. B. (2009). Understanding the link between malaria risk and climate. *Proceedings of the National Academy of Sciences in the United States of America, 106*(33), 13844–49.

Romm, J. (2007). *Hell and high water.* New York, NY: HarperCollins.

Slenning, B. D. (2010). Global climate change and implications for disease emergence. *Veterinary Pathology, 47*(1), SI 28–33.

Vegvari, Z., Bokony, Barta, Z., *et al.* (2010). Life history predicts advancement of avian spring migration in response to climate change. *Global Change Biology, 16*(1), 1–11.

Webster, P. J., Holland, G. J., Curry, J. A., *et al.* (2005). Changes in tropical cyclone number, duration, and intensity in a warming environment. *Science, 309*(5742), 1844–46.

Wirawan, I. M. A. (2010). Public health responses to climate change health impacts in Indonesia. *Asia-Pacific Journal of Public Health, 22*(1), 25–31.

Zimov, S. A., Schuur, E. A. G., and Chapin, F. S. (2006). Permafrost and the global carbon budget. *Science, 312*(5780), 1612–13.

Global and National Strategies

Fargione, J., Hill, J., Tilman, D., *et al.* (2008). Land clearing and the biofuel carbon debt. *Science, 319*(5867), 1235–38.

Nelson, G. C., Robertson, R. D. (2008). Green gold or green wash: Environmental consequences of biofuels in the developing world. *Review of Agricultural Economics, 30*(3), 517–29.

Romm, J. (2004). *The hype about hydrogen.* Washington, DC: Island.

Scharlemann, J. P. W., and Laurance, W. F. (2008). Environmental science—how green are biofuels? *Science, 319*(5859), 43–44.

Individual and Community Strategies

http://www.nature.org/ourinitiatives/urgentissues/global-warming-climate-change/index.htm

http://www.epa.gov/climatechange/wycd/index.html

http://www.350.org/

www.populationconnection.org

Yarrow, J. (2008). *Ways to reduce your carbon footprint.* London, UK: Duncan Baird.

Carbon Footprint Calculators

http://www.nature.org/greenliving/carboncalculator

http://myfootprint.org

CREDITS

Source: http://commons.wikimedia.org/wiki/File:Sources_of_US_energy_
production_%28historic%29.gif. Copyright in the Public Domain.

Jim E. Maragos, "Colorful underwater landscape of a coral reef," https://commons.wikimedia.org/wiki/
File%3AColorful_underwater_landscape_of_a_coral_reef.jpg. Copyright in the Public Domain.

Copyright © U.S. Geological Survey (CC by 2.0) at https://commons.wikimedia.org/wiki/
File%3ABent_Sea_Rod_Bleaching_(15011207807).jpg.

Copyright © Mak Thorpe (CC BY-SA 3.0) at http://commons.wikimedia.org/wiki/File:Global_
Carbon_Emission_by_Type_to_Y2004.png.

Source: http://photolibrary.usap.gov/Portscripts/PortWeb.dll?query&field1=Filename&op1=matches&
value=vostokmain.jpg&catalog=Antarctica&template=USAPgovMidThumbs. Copyright in the Public
Domain.

Source: http://www.ncdc.noaa.gov/paleo/slides/slideset/15/15_305_slide.html. Copyright in the Public
Domain.

Copyright in the Public Domain.

Copyright © William M. Connolley (CC BY-SA 3.0) at http://commons.wikimedia.org/wiki/
File:Vostok-ice-core-petit.png.

Copyright © William M. Connolley (CC BY-SA 3.0) at http://commons.wikimedia.org/wiki/
File:Vostok-ice-core-petit.png.

Copyright © William M. Connolley (CC BY-SA 3.0) at http://commons.wikimedia.org/wiki/
File:Vostok-ice-core-petit.png.

Source: http://commons.wikimedia.org/wiki/File:CO2-Mauna-Loa.png. Copyright in the Public
Domain.

Copyright © Giorgiogp2 (CC BY-SA 3.0) at http://commons.wikimedia.org/wiki/File:RSS_tropo-
sphere_stratosphere_trend.png.

Copyright © Leland McInnes (CC BY-SA 3.0) at http://commons.wikimedia.org/wiki/File:Temp-
sunspot-co2.svg.

Gerald A. Meehl, Warren M. Washington, Caspar M. Ammann, Julie M. Arblaster, T. M. L. Wigley, and
Claudia Tebaldi, Journal of Climate, vol. 17, issue 19. Copyright © 2004 by American Meteorological
Society. Reprinted with permission.

Gerald A. Meehl, Warren M. Washington, Caspar M. Ammann, Julie M. Arblaster, T. M. L. Wigley, and
Claudia Tebaldi, Journal of Climate, vol. 17, issue 19. Copyright © 2004 by American Meteorological
Society. Reprinted with permission.

Gerald A. Meehl, Warren M. Washington, Caspar M. Ammann, Julie M. Arblaster, T. M. L. Wigley,
and Claudia Tebaldi, Journal of Climate, vol. 17, issue 19, pp. 3722. Copyright © 2004 by American
Meteorological Society. Reprinted with permission.

Danlaycock, "Kyoto Protocol Parties," https://commons.wikimedia.org/wiki/File:Kyoto_Protocol_par-
ties.svg. Copyright in the Public Domain.

Source: http://www.ars.usda.gov/is/graphics/photos/may04/k11202-1.htm. Copyright in the Public
Domain.

http://www.tonyboon.co.uk/. Copyright in the Public Domain.

achiral: Superimposeable upon its mirror image; a symmetric molecule.

acid: A chemical that ionizes to produce H^+ in aqueous solution.

acid rain: Rain with pH < 5; produced from SOx and NOx.

acidic: A solution of pH < 7.

activity: A chemical's ability to be oxidized (highly active substances are easily oxidized).

aerobic exercise (aerobic zone): Moderate-intensity efforts in which sufficient oxygen for fuel combustion is available; the heart rate zone in which such exercise occurs.

aerosolized: To be made airborne.

alcohol: A compound containing the –OH functional group. Also refers specifically to the compound *ethanol*.

allotrope: Alternate form of an element (i.e., O_3 is an allotrope of oxygen, where O_2 is the most common form).

alpha particle (α) (decay): A helium nucleus released in nuclear decay; a decay event in which alpha particles are released.

alternative energy: Ways to generate energy through means other than combustion.

alternative fuels: Energy-production strategies that do not involve the combustion of fossil fuels.

amine: A nitrogen-containing functional group (these are basic).

amino acid: A small molecule with an amine group, carboxylic acid group, and variable side chain. The structural unit of proteins.

ammonium: A reactive nitrogen species that can be assimilated by plants.

amphiphilic: Possessing both water- and fat-soluble regions.

amylose: A polysaccharide. Plant starch.

anaerobic exercise (anaerobic zone): High-intensity efforts in which sufficient oxygen is not available for fuel consumption; the heart rate zone in which such exercise occurs.

anion: A negatively charged atom or group of atoms (polyatomic anion).

anode: Site of oxidation in an electrochemical cell.

anthropogenic: Of human origin.

antioxidant: A chemical that ameliorates damage from free radical formation, ionizing radiation, and toxins.

AQI (Air Quality Index): An EPA schema for quantifying air quality.

aqueous solution: A solution in which the solvent is water.

aquifers: Underground pools of water; groundwater.

aromatic ring: A highly fragranced hydrocarbon and hydrocarbon-derivative functional group. Includes benzene.

arthritis: Degeneration of the joints; can occur from overfat.

atherosclerosis: Hardening of the arteries.

atom: The smallest particle of an element that maintains the chemical identity of the element.

atomic mass: The average mass of an atom of an element (in amu). The atomic mass is a weighted average mass, and depends upon the masses (and prevalences) of isotopes of that element.

atomic mass units: The units used to measure the masses of atoms and molecules, 1 amu = 1.67×10^{-24} g.

atomic number: The number of protons in the nucleus of an atom of an element.

atomic symbol: The letter or letters representing an element on the periodic table.

ATP: An energy molecule; chemical energy currency.

Avogadro's number: The number of atoms/molecules/etc. in a mole (6.02×10^{23}).

background radiation: Unavoidable radiation from sources in the natural world.

baking powder: Baking soda in combination with a source of acid such as cream of tartar, and sometimes starch. Used to leaven baked goods that don't contain an acidic wet ingredient.

baking soda: The ionic compound $NaHCO_3$, used as an antacid or in combination with an acidic wet ingredient to leaven baked goods.

balanced equation: A chemical equation in which the identity and number of atoms is conserved from products to reactants.

barometric pressure: Ambient atmospheric temperature at a particular location at a particular time.

base: A chemical that can take H^+ from water or from an acid.

basic: A solution of pH > 7.

battery: A spontaneous electrochemical cell (in common parlance); a series of such cells (technical).

benzene ring: A ring of six carbon atoms (and requisite hydrogens) connected with alternating single and double bonds. Also called an aromatic ring.

beta particle (β) (decay): An electron released from the nucleus during nuclear decay; a decay event in which beta particles are released.

binary acid: An acid composed of hydrogen and one other element.

binary compound: A compound consisting of only two elements.

bioaccumulate The ability of select compounds to concentrate in organisms near the top of a food chain.

biofuel: Combustible material from renewable, carbon-neutral sources.

blood glucose (blood sugar): The concentration of glucose in the blood stream; it is carefully regulated.

BMI (body mass index): A metric for assessing body composition; a ratio of weight to height.

body composition: The amount of lean and fat tissues on the body. Measured to assess risk for weight-related disease.

boiling (boiling point): The transition from solid to liquid; the temperature at which the transition occurs.

bond: See **chemical bond, covalent bond, ionic bond, polar covalent bond.**

bond dipole — The polarization of a bond, indicated using an arrow (points toward more electronegative atom).

bond dissociation: The breaking of a chemical bond.

bran: The fibrous husk around a grain.

calcification: The precipitation of calcium and magnesium salts onto fixtures and surfaces. Results from hard water.

calorie/Calorie: A measure of energy in food. If "calorie," then the amount of heat to increase 1 mL of water by 1 °C. If "Calorie," then 1 kcal.

carbohydrate: A compound of C, H, and O consisting of one or more sugar (monosaccharide) units; sugars and starches.

carbon credits: Credits awarded under the Kyoto Protocol to nations emitting less than target values.

carbon dating: The use of C-14 concentration to determine the age of a sample of organic material.

carbon fixation: Reaction of carbon dioxide to produce organic molecules.

carbon footprint: An individual's impact upon the environment; the amount of carbon emissions for which an individual is personally responsible.

carbon monoxide: An environmental pollutant from hydrocarbon combustion, it reduces oxygen-carrying capacity of the blood.

carbon-neutral: Practices that do not result in the net release of carbon dioxide into the atmosphere.

carbon sequestration: The capture of CO_2 released during combustion processes, with subsequent condensation and injection into deep geological repositories.

carboxylic acid: A compound containing the $-COOH$ arrangement of atoms.

carcinogen: A substance that causes cancer.

catalyst: A chemical that speeds up the rate of a reaction without being consumed itself. Enzymes are biological catalysts.

cathode: Site of reduction in an electrochemical cell.

cation: A positively charged atom or group of atoms (polyatomic cation).

cellular respiration: See **respiration.**

cellulose: A polysaccharide. Plant fiber.

CFCs (chlorofluorocarbons): Substituted methane and ethane molecules, also called Freons; refrigerant compounds that deplete the ozone layer.

Chapman Cycle: The cycle whereby UV light produces ozone from O_2 and vice versa. A dynamic equilibrium.

chemical bond: An attraction that holds two atoms together, forming a molecule. See **covalent bond, ionic bond.**

chemical combination: A combination in which two or more atoms become chemically bonded together, acquiring a new and distinct chemical identity.

chemical equation: A symbolic representation of a chemical reaction.

chemical reaction: The breaking and reforming of chemical bonds to yield new combinations of atoms in molecules.

chiral: Nonsuperimposable upon its mirror image; an asymmetrical molecule.

chlorophyll: A light-harvesting pigment involved in photosynthesis.

chloroplast: The cellular machinery of photosynthesis.

cholesterol: A biomolecule required for synthesis of vitamin D and maintenance of appropriate cell membrane rigidity.

***cis* double bond**: A double bond that kinks a straight hydrocarbon chain.

Clean Air Acts: Regulatory measures passed by the United States in the 1960s and 1970s (amended in 1990) to reduce environmental pollution.

climate modeling: A science that uses computational and predictive techniques to predict the effects of forcings upon the climate.

coal: Fossilized remains of ancient plant life that died, partially decayed, and were then buried beneath layers of earth. A fossil fuel.

coal scrubbing: Chemical washing/filtering technique to reduce SOx emissions from coal combustion.

coenzyme: A molecule that works with an enzyme to produce a reaction.

cohesive: The ability or tendency of a substance to stick to itself.

collagen: An elastic protein in the skin. Damaged by UV light.

combustion: Reaction of a compound with oxygen; yields heat.

compound: A chemical combination of two or more elements in fixed proportions.

computed tomography (CT): A three-dimensional X-ray image.

concentration: The proportion of solute to a given volume of solution.

condensation: Transition from a gas to a liquid.

conductive: Can conduct electricity.

control rods: Rods of a neutron-absorbing material designed to prevent nuclear reactions in power plants from becoming critical (growing exponentially).

correlational: Data that shows a relationship but for which causality has not been established.

covalent bond: A shared pair (or shared pairs) of electrons that count toward the valence of both atoms involved in bonding. A physical connection between atoms.

critical mass: The mass of fissionable fuel that is capable of producing a sustained or growing nuclear reaction.

current: The movement of electrons; electricity.

daughter nucleus: A product in a nuclear reaction.

DDT (dichlorodiphenyltrichloroethane): A pesticide used in developing countries for malaria control. Banned for agricultural use because it is a persistent organic pollutant.

decay: See **nuclear decay**.

deep geological repository: A proposed underground storage facility for high-level radioactive waste.

dehydration synthesis: Linking two molecules together through removal of the atomic constituents of water.

density: The mass per unit volume of a substance.

depleted uranium: The waste from uranium enrichment; uranium low in U-235.

diatomic molecule: A molecule made of exactly two atoms.

diet: Properly used to mean food and drink that is taken habitually to support life and health. Improperly (but commonly) used to mean a departure from normal behavior for the purpose of weight loss.

Dietary Supplement Health and Education Act (1994): Considered herbals to be dietary supplements. A 2007 amendment stipulated that herbals and supplements must be free from contamination and correctly labeled.

dipole: A separation of charge.

direct effect (of exercise): The calories burned during an exercise activity.

disaccharide: Two sugar units linked together. A sugar.

dissociation: The separation of anions and cations in aqueous solution.

DNA (deoxyribonucleic acid): Genetic material.

doping (of silicon): To add nonsilicon elements to silicon in order to affect electron density.

double bond: A covalent bond made up of two shared pairs of electrons.

dry-casking: Method of storage in which spent nuclear fuel is kept above ground in metal containers within a concrete structure.

ductile: Can be drawn into wires.

dynamic equilibrium: A process in which no net changes to a system occur over time. A reaction system in which concentrations of chemical species remain constant, despite individual reactions taking place.

electrical energy: The movement of electrons; electrical current; electricity.

electrical potential energy: The potential for charged particles to move toward or away from one another on the basis of electrostatic forces.

electricity: The movement of electrons; electrical current; electrical energy.

electrochemical cell: A cell in which a redox reaction takes place.

electrode: The contact point at which oxidation or reduction occurs in an electrochemical cell.

electrolysis: The use of electrical energy to power a nonspontaneous chemical reaction (generally one that reduces a compound to elements).

electrolytic cell: A nonspontaneous electrochemical cell.

electrolyte: A compound that dissociates in aqueous solution; an ionic compound, a salt. An aqueous solution of an electrolyte conducts electricity.

electronegativity: A quantitative measure of how tightly an atom holds its electrons. A measure of the extent to which an atom will pull electrons in a bond. This is a periodic trend.

electron holes: The electron paucity in p-type as compared to n-type silicon.

electrons: Subatomic particles found in a nearly massless cloud surrounding the nucleus with mass = 0 and charge = -1. The components of covalent bonds.

electron shells: Location of electrons in an atom; lower numbered shells are closer to the nucleus.

electrostatic forces: The attractions between oppositely charged particles and repulsions between like-charged particles.

elements: The simplest type of matter; the substances listed on the periodic table.

elevated post-exercise metabolic rate: The higher RMR observed for up to 24 hours following exercise.

empirical evidence: Taken from observation.

empty calories: Calories containing little nutritional benefit apart from energy.

enantiomers: A pair of nonsuperimposable mirror image molecules.

endothermic: A reaction that requires energy.

enhanced greenhouse effect: A greenhouse effect phenomenon of greater than 81 percent.

entropy: Disordered energy.

enzyme: A protein that makes a chemical reaction happen faster. A biological catalyst.

ephedrine: A chemical compound used in diet pills; a sympathomimetic.

EMR (electromagnetic radiation): Electrical and magnetic disturbance that propagates through space; radiant energy.

essential body fat: The amount of body fat required to maintain normal function. Around 2–4 percent in men, 10–12 percent in women.

ester: A compound functional group made from a carboxylic acid and an alcohol. Often highly fragranced.

ether: A functional group consisting of two hydrocarbons connected by a common oxygen. Refers specifically to the compound *diethyl ether*.

exothermic: A reaction that releases energy.

exposure: A measure of how much of a pollutant substance is likely to be encountered.

fatty acid: A long-chain carboxylic acid.

fermentation: Anaerobic glucose metabolism.

first law of thermodynamics: Energy of the universe is constant; energy can neither be created nor destroyed.

fixation (of nitrogen): See "nitrogen fixation."

fixed proportions: Relative quantities of two or more components must be constant in a particular substance; changes in relative quantities will change the identity of the substance. Part of the definition of a compound.

Food and Drug Act (1906): Required that substances transported across state lines were free from contaminants and correctly labeled.

Food, Drug, and Cosmetic Act (1938): Added toxicity study requirements to the above.

Food, Drug, and Cosmetic Act Amendment (1962): Increased safety requirements and added proof of efficacy to the above.

forcings: Environmental or anthropogenic factors that affect climate.

fossil fuels: Fossilized terrestrial or sea organisms that contain solar energy captured through photosynthesis; coal, petroleum, and natural gas.

free radical: Highly reactive chemical with unpaired electrons.

frequency (n): The number of waves per unit time that pass a given point in space, measured in Hz (or s^{-1}).

fructose: A monosaccharide commonly found in fruit and as part of the disaccharide sucrose.

fuel assemblies: Pellets of fissionable fuel assembled for use in a nuclear reactor.

functional group: A characteristic arrangement of atoms in a molecule that lends a predictable set of properties to the molecule.

galvanic cell: A spontaneous electrochemical cell; a voltaic cell.

gamma radiation (gamma decay) (γ): Photons released from the nucleus during nuclear decay; a decay event in which gamma-wavelength photons are released.

gamma particle (γ): A particle of gamma radiation.

gamma rays: Very high energy EMR; ionizing radiation. A form of nuclear decay releasing a photon with no mass or charge.

gas: A physical state of matter. Gases are compressible and take the shape of their container.

gaseous diffusion: A separation process for uranium enrichment in which lighter isotopes pass through a filter faster than heavier ones and are concentrated.

GI (glycemic index): A measure of the rate at which a carbohydrate enters the bloodstream.

glucagon: A hormone that increases blood sugar.

glucose: A common monosaccharide.

glycerol: A trifunctional alcohol to which fatty acids are bonded to make triglyceride.

glycogen: A short-term energy-storage polysaccharide, found in the liver and muscles. Is broken down into glucose when required. The liver secretes resulting glucose into the bloodstream for general distribution, while the muscles generally utilize it for their own purposes.

gravitational potential energy: The energy an object has when above the ground (the potential to fall).

Greenfreeze: A substitute refrigerant that is not ozone-depleting and not a greenhouse gas.

greenhouse gases: Gases that contribute to the greenhouse effect; gases with at least one bond dipole.

greenhouse effect: The phenomenon whereby atmospheric gases trap 81 percent of Earth's radiant energy (heat) and re-radiate it back toward Earth.

groundwater: Water found underground; aquifer water.

GWP (global warming potential): A standard measurement established by the IPCC that compares a molecule's ability to store and re-radiate IR to that of CO_2.

Haber-Bosch process: Industrial nitrogen fixation; the industrial production of NH_3 from H_2 and N_2.

half-life: The time it takes for one half of a sample of radioactive isotope to decay.

half-reactions: Individual oxidation and reduction components of a redox reaction.

hard water: Water containing dissolved calcium and magnesium; generally ground water.

HCFCs (hydrochlorofluorocarbons): First-generation CFC substitutes; refrigerants. These are less ozone depleting but are greenhouse gases.

HDL (high-density lipoprotein): So-called "good" cholesterol; a transporter that carries cholesterol from the cells to the liver for excretion.

HFCs (hydrofluorocarbons): Second-generation CFC substitutes; refrigerants. These are not ozone depleting but are greenhouse gases.

hemoglobin: The protein that carries oxygen from the lungs to the tissues in the blood.

high-level radioactive waste: Highly radioactive spent fuel and the waste from reprocessed spent fuel that must be stored indefinitely to prevent contamination.

homeostasis: Maintenance of a constant environment.

hormone: A chemical used for cell-to-cell communication in the body; may be peptide or steroid.

HRmax (maximum heart rate): The highest heart rate an individual can sustain.

hydrocarbon: A compound composed of hydrogen and carbon. They combust in O_2 to produce carbon dioxide and water (idealized).

hydrogen bond: The electrostatic attraction between molecules containing FH, OH, or NH bonds.

hydrogen fuel cells: Voltaic cells that generate electricity from the reaction of hydrogen with oxygen to produce water.

hydrolysis: Breaking a bond by adding the constituent atoms of water across the bond.

hydrophilic: Water loving. A descriptive term for polar compounds in terms of their solubility behavior.

hydrophobic: Water fearing. A descriptive term for nonpolar compounds in terms of their solubility behavior.

hyperglycemia: High blood sugar.

hypoglycemia: Low blood sugar.

ice core: A sample of ice drilled for temperature, atmospheric, and other historical data. Generally taken from polar regions.

immersion test: A metric for measuring body composition; based upon weight underwater.

incompressible: Can't be packed into a smaller volume.

Industrial Revolution: Historic period during the eighteenth and nineteenth centuries in which machine-generated power replaced human- and animal-generated power. Inception of major dependence on fossil fuel.

inert: Chemically unreactive.

insulin: A hormone that decreases blood sugar; a hormone that causes cells to take up glucose.

intermediate: A chemical formed in one step of a reaction series and used in a later step.

ion: A nonneutral atom or group of atoms (polyatomic ion).

ion exchange: The process whereby hard water is softened by exchanging calcium and magnesium ions for sodium ions.

ionic bond: The electrostatic attraction between two charged atoms or groups of atoms.

ionizing radiation: Radiation capable of breaking chemical bonds and knocking electrons out of molecules.

IPCC (International Panel on Climate Change): United Nations panel with the mission to assess climate change research as well as its social and economic ramifications.

IR (infrared radiation): Nonionizing EMR; heat. Trapped by greenhouse gases, absorbed by some molecular vibrations.

irradiation (of food): Use of gamma rays to kill microbes and insects in food.

isotope: Different types of atoms of the same element. All have the same atomic number, but have varying numbers of neutrons.

Keeling Curve: A record of CO_2 concentration in the atmosphere (since 1958) vs. time. The curve shows that not only is CO_2 concentration increasing year by year, it's doing so at an unprecedented rate in geological history.

ketogenic diet: A low-carbohydrate diet that results in ketosis.

ketone bodies: Chemicals including acetone that are produced via fatty acid metabolism and can enter the bloodstream in times of starvation.

ketosis: The systemic circulation of ketone bodies.

kinetic energy: The energy of motion.

lactic acid: A metabolic waste product of anaerobic metabolism.

lactose: A disaccharide. Milk sugar.

Law of Conservation of Matter: Atoms cannot appear, disappear, or change into atoms of other elements by any chemical means.

LDL (low-density lipoprotein): So-called "bad" cholesterol; a transporter that carries cholesterol from the liver to the cells.

lead: A heavy metal pollutant. A potent neurotoxin.

leaven (leavening): To cause dough to rise; the agent that causes dough to rise.

Lewis dot structure: A representation of the valence electrons in an atom or a representation of bonds and lone pairs in a molecule.

line-angle formula: A simplified representation of a molecule (usually used for large molecules) in which hydrogen atoms attached to carbon are implicit and the atomic symbol for carbon is omitted.

lipid: A hydrophobic molecule.

liquid: A physical state of matter. Liquids are incompressible and take the shape of their container.

lone pair: A nonbonding pair of electrons.

macronutrient: Carbohydrate, protein, and fat. Nutrients required in large amounts that provide energy.

malleable: Can be hammered into sheets.

Manhattan Project: The U.S. Army project to develop nuclear weapons.

mass defect: The mass "lost" (converted to energy) during a nuclear reaction.

mass number: The mass of an atom of an element (in amu). Mass number is always a whole number, and is the sum of the number of protons and neutrons in the nucleus.

matter: Anything that occupies space and has mass.

meltdown (nuclear): A containment failure at a nuclear facility.

melting (melting point): The transition from solid to liquid; the temperature at which the transition occurs.

mercury: A heavy metal pollutant. A potent neurotoxin.

metabolism: The sum of all chemical reactions in the body.

metalloids: Elements with physical properties in between those of metals and nonmetals. Located on a diagonal line near the left side of the periodic table. Semiconductors.

metals: Elements that are shiny, ductile, conductive, and malleable. The majority of the periodic table.

micelle: A spherical arrangement of fatty acids in water, with grease sequestered at the inside.

micronutrient: Vitamins and minerals. Important chemicals required in small amounts.

Midgley, Thomas: An inventor and mechanical engineer. Developer of tetraethyl lead and Freons.

minerals: Metal and halide ions required in various amounts for normal function.

mitochondrion: The machinery of respiration.

mixture: Physical combinations of two or more substances in variable proportions.

molar mass: The mass of a mole of a substance.

molarity: Quantitative measure of solution concentration, moles solute per liter of solution.

mole: The name for a specific number of objects; 1 mole = 6.02×10^{23} objects.

molecular compound: A compound that does not dissociate in aqueous solution; a nonelectrolyte. Composed of covalent bonds.

molecular mass: The mass of a molecule (in amu).

molecule: A pure substance made up of a chemical combination of two or more atoms.

monosaccharide: A single sugar unit; the constituent unit carbohydrates.

Montreal Protocol (1987): International treaty to phase out ODS.

mutagen: A substance that causes mutation.

myoglobin: A chemical related to hemoglobin, it stores oxygen in the muscle (primarily type I fibers).

n-type silicon: Silicon that has been doped with group 5 elements.

natural gas: Partially decomposed remains of sea organisms, found with petroleum; a fossil fuel.

net dipole: The sum of the bond dipoles within a molecule.

neurochemical: A compound with activity in the brain or nervous system.

neurotoxin: A compound toxic to or damaging to brain or nerve cells.

neurotransmitter: A chemical used for cellular communication within the nervous system.

neutral: Having no charge; neither acidic nor basic (for solutions).

neutralize: To react an acid with a base.

neutrons: Subatomic particles found in the nucleus with mass = 1 amu and no charge.

nitrate: A reactive nitrogen species that can be assimilated by plants.

nitrogen dioxide: A NOx species and environmental pollutant. A respiratory irritant.

nitrogen fixation: The process of making atmospheric nitrogen reactive, accomplished by soil or root-nodule bacteria.

nitrogen-fixing bacteria: Bacteria in the soil and root nodules of leguminous plants, responsible for the process of nitrogen fixation.

nitrogen oxides: Binary compounds of nitrogen and oxygen (including NOx).

nomenclature: A system of naming.

nonelectrolyte: A molecular compound; solutions of nonelectrolytes do not conduct electricity.

nonmetals: Elements that are dull, brittle, and nonconductive. Located on the left-hand side of the periodic table.

nonnutritive sweeteners: Sweet compounds that provide no calories because they are indigestible.

nonpolar: A molecule with no net dipole.

nonrenewable (fuel): A resource that is nonrenewable; fossil fuels.

nonspontaneous: A reaction that requires the input of energy; a reaction that does not occur spontaneously.

normal body fat: The amount appropriate to normal function. Around 6–26 percent in men, 14–31 percent in women.

NOx: Mononitrogen oxides. Important species include NO_2 and NO. Produced by the reaction of atmospheric nitrogen with oxygen at high temperature.

nuclear chain reaction: A self-sustaining nuclear reaction in which decay particles initiate reactions in other nuclei.

nuclear decay: A nuclear reaction in which a parent nucleus loses a small particle to become a daughter nucleus.

nuclear equation: A symbolic representation of a nuclear reaction.

nuclear fission: The splitting of a large nucleus into two or more pieces.

nuclear fusion: The combination of two nuclei to form a larger nuclei, often with subsequent decay.

nuclear reaction: A reaction in which the nuclei of atoms gain or lose subatomic particles and/or rearrange.

nuclear reactor: An isolated reaction vessel in a nuclear power plant.

nucleons: Subatomic particles found in the nucleus (protons and neutrons).

nucleus: The small-volume, massive center of an atom. Contains protons and neutrons.

obesity: Too much body weight relative to body size, defined in several ways (by percentage body fat or BMI).

octet rule: Atoms are most stable when they have a full (generally 8, except in the case of row 1 elements) outer shell of electrons.

ODS (ozone-depleting substances): Substances like CFCs that deplete the ozone layer.

omega-3 fatty acid: A fatty acid with a *cis* double bond three carbons from the omega (tail) end.

omega-6 fatty acid: A fatty acid with a *cis* double bond six carbons from the omega (tail) end.

organic chemistry: The study of carbon-containing compounds/the molecules of life.

organic compound/organic molecule: A carbon-based molecule.

overweight: Too much body weight relative to body size, defined in several ways (by percentage body fat or BMI). Less severe than obesity.

oxidation: The loss of electrons.

oxyacid: An acid containing hydrogen, oxygen, and one other nonmetal.

ozone (the molecule): An allotrope of oxygen; a reactive molecule.

ozone (stratospheric): Forms a protective layer that protects Earth from UV light.

ozone (tropospheric): An environmental pollutant (secondary to combustion products). Formed from the reaction of NOx with sunlight. A respiratory irritant.

p-type silicon: Silicon that has been doped with group 3 elements.

PAHs (polycyclic aromatic hydrocarbons): Large organic molecules. Sources include combustion of petroleum and coal. Known carcinogens.

pancreas: An organ that releases insulin and glucagon.

parabolic trough solar-thermal power plants: Plants that concentrate the sun's energy on a pipe containing molten salt to store solar energy.

parent nucleus: A reactant in a nuclear reaction.

partial charge (negative or positive): The slight charge imbalance that results from the formation of a polar covalent bond.

partial hydrogenation (partially hydrogenated oils): A chemical process in which some double bonds are removed from fatty acids and others are changed from *cis* to *trans*. A liquid oil that has been converted to solid through the process of partial hydrogenation.

particulate matter: Includes PM_{10} and $PM_{2.5}$.

peptide bond: The bond that links amino acids to form a protein.

Percent Daily Value: The amount of each nutrient recommended by the FDA and USDA for daily consumption.

periodic table: An organizational scheme that groups elements by chemical properties.

periodic trend: A physical or chemical trend that can be predicted using the periodic table. A trend that follows a pattern in groups or rows.

perturbation: A disruption.

petroleum: Partially decomposed remains of sea organisms, found with natural gas; a fossil fuel.

pH — A measure of solution acidity or basicity. Related to concentration of hydrogen ion.

pharmacophore: A core functional area of a drug or group of drugs that allows binding to the receptor.

phenol: A compound containing an –*OH* functional group attached to a benzene ring. Refers specifically to the compound *phenol*.

photosynthesis: Storage of the sun's energy as chemical energy (carbohydrate) through electron transfer.

photosystem: Light-harvesting pigments and enzymes involved in photosynthesis.

photovoltaic cell: A cell that transforms the sun's energy into electrical energy.

physical activity energy expenditure (PAEE): The calories burned through daily activity.

physical combination: A combination in which each substance maintains its individual identity and properties.

Planck's constant: A multiplier used in the wave equation relating energy to frequency. Given the symbol h. Equal to $6.626 * 10^{-34}$ Js.

plaques: Accumulations of cholesterol in the arteries.

PM2.5: Fine particles of diameter < 2.5 μm.

PM10: Inhalable coarse particles of diameter 2.5–10 μm.

polar: A molecule with a net dipole.

polar covalent bond: A polarized bond; the bond that exists between two atoms of unequal electronegativity.

polar stratospheric clouds: High clouds that form over Antarctica in winter and capture chlorine; associated with rapid ozone depletion.

polarization: The pulling of electrons within a bond; the forming of a polarized bond.

polyatomic ion: An ion composed of two or more atoms, covalently bonded together.

polyatomic molecule: A molecule made up of two or more atoms.

polymer: A large molecule composed of repeating units.

polypeptide: A large molecule composed of amino acids; a protein.

polysaccharide: A polymer of sugar units. Starch or fiber.

positron: A positively charged, massless particle; the antiparticle of an electron.

positron emission tomography (PET): Scans tracking the location of radioactive particles (in organic molecules used by the body) over time.

potable: Safe to drink; used with regard to water.

ppm (parts per million): The number of particles of a substance of interest per million particles of sample. A measure of concentration.

precautionary principle: A formalization of maxims such as *better safe than sorry*. The idea that it is not reasonable to take risks in cases where risk may exist.

precipitation (precipitate): The process whereby a salt falls out of solution and forms a solid; the solid itself.

primary coolant: The coolant in a nuclear reactor that dissipates heat directly from fissionable fuel.

products: Substances produced in a chemical reaction.

protein: Large molecules formed as polymers of amino acids. A macronutrient; a structural and functional component of cells.

protons: Subatomic particles found in the nucleus with mass = 1 amu and charge = +1.

pure substances: Nonmixtures. Elements and compounds.

pyruvate: The molecule produced from glucose that accepts electrons in fermentation.

racemic mixture: A 50:50 mixture of two enantiomers.

radiation: EMR or nuclear wave/particle rays.

radiation sickness: A set of symptoms resulting from radiation exposure.

radical: Unpaired electron.

radioactive decay series: A series of decay reactions that continue until a stable daughter nucleus is produced.

radioactive particle: A small decay particle given off in a nuclear reaction.

radioisotopes: Radioactive nuclei.

reactants: Substances that react in a chemical reaction.

receptor: A large molecule (generally on a cell membrane) that binds a substrate and responds in some physiological way.

redox: An electron transfer reaction.

reduction: The gaining of electrons.

refined grains: Grains which have had bran (fiber) removed.

respiration: The process by which energy is extracted from nutrients and stored as ATP based upon electron flow.

RMR (resting metabolic rate): The number of calories burned daily through maintenance of cellular function alone.

salt bridge: A bridge containing spectator salt (ions) that connects the halves of an electrochemical cell to balance charge.

salts: Ionic compounds, electrolytes.

satiety signal: Something that communicates a sense of fullness and satisfaction.

saturated fat: Triglyceride containing fatty acids without double bonds.

second law of thermodynamics: For all processes, entropy of the universe increases.

secondary coolant: The coolant in a nuclear reactor that dissipates heat from primary coolant and uses the heat to vaporize water.

shells: See **electron shells**.

silicon p-n junction solar cell: A photovoltaic cell in which p-type and n-type silicon are juxtaposed for unidirectional electron flow.

single bond: A covalent bond made up of one shared pair of electrons.

sink: A repository of a substance. Generally used with regard to carbon.

skin fold tests: A metric for assessing body composition; based upon thickness of the subcutaneous fat in several places on the body.

sleep apnea: Periodic temporary cessation of breathing during sleep. Can result from overfat.

smog: Combination of smoke and fog.

soft water: Water that does not contain high concentrations of calcium and magnesium ions. Chemically softened water or rain water.

solid: A physical state of matter. Solids are incompressible and do not take the shape of their container.

solute: A chemical that is dissolved to form a solution.

solution: A mixture made up of one or more substances dissolved in another substance. If water is the solvent, the solution is aqueous.

solvent: A chemical that dissolves other chemicals to form a solution.

SOx: Monosulfur oxides. Important species include SO_2 and SO_3. Primary anthropogenic source is coal combustion.

spectator ion: An ionic species that is present in a reaction mixture, but not involved in the chemical reaction.

spontaneous: A reaction that occurs without the input of energy; a reaction that yields energy.

stoichiometric equivalents: The ratios of substances involved in a chemical reaction.

stratosphere: The region of the atmosphere between 15 and 50 km above sea level; the location of the ozone layer.

strong force: The strong attraction between nucleons over very short distances.

substituents: Atoms or groups bound to an atom or molecule of interest.

substrate: A compound that binds to a receptor.

sucrose: A disaccharide of glucose and fructose. Table sugar.

sulfur dioxide: An environmental pollutant, a major source of which is coal combustion. A respiratory irritant.

superimposable: Two objects that can be placed next on top of each other so that all parts match up.

surface water: Water on Earth's surface; lakes, rivers, streams (fresh water).

sympathomimetic: A chemical that mimics the effects of adrenaline or the sympathetic nervous system ("fight or flight").

teratogen: A compound known to cause developmental defects in an embryo or fetus.

tetraethyl lead: A gasoline additive invented to prevent engine knocking.

tetrahedral: The shape of a four-sided pyramid. This shape is formed when four electron groups separate as much as possible around a central atom.

thermic effect of food: The calories expended to digest and metabolize food.

thermodynamics: The study of the energetics of reactions and processes.

thought experiment: A way of finding the answer to a question that is impossible or impractical to test in real life through theoretical means.

toxicity: A measure of the inherent harm-causing potential of a pollutant.

trans **double bond**: A double bond that does not kink a straight hydrocarbon chain.

trans fat: Triglyceride containing fatty acids with *trans* double bonds.

triglyceride: Glycerol and three fatty acids chemically combined; storage fat.

trigonal planar: The shape of an equilateral triangle. This shape is formed when three electron groups separate as much as possible around a central atom.

triple bond: A covalent bond made up of three shared pairs of electrons.

troposphere: The region of the atmosphere from sea level to 15 km above sea level; the region of the atmosphere we inhabit.

type 2 diabetes: Insensitivity of body cells to insulin; often results from overfat.

type I muscle fibers: Fibers specializing in aerobic efforts; slow twitch fibers.

type II muscle fibers: Fibers specializing in anaerobic efforts; fast twitch fibers.

unsaturated fat: Triglyceride containing fatty acids with *cis* double bonds.

uranium enrichment: The process of concentrating U-235 above what is found in nature.

uranium tails: The waste from uranium enrichment; uranium low in U-235.

UV (ultraviolet radiation): High energy radiation; ionizing radiation. Filtered by the ozone layer.

UV-A: Lowest-energy UV radiation.

UV-B: Moderate-energy UV radiation.

UV-C: Highest-energy UV radiation.

valence: An atom's number of valence electrons.

valence electrons: The electrons in an atom's outermost shell.

van der Waals forces: The attractions between nonpolar molecules. Forces are stronger between larger molecules.

vaporization: Transition from a liquid to a gas; boiling.

variable proportions: Relative quantities of two or more components can vary without affecting the identity of a substance. Part of the definition of a mixture.

vitamins: Organic molecules required in small amounts for normal function.

VOCs (volatile organic compounds): Small organic molecules that vaporize easily and enter the atmosphere. Sources include solvents and building materials.

voltaic cell: A spontaneous electrochemical cell; a Galvanic cell.

wavelength (l): A wave's distance crest to crest, measured in meters.

Western Paradox: The idea that Western societies are simultaneously consuming a terrible diet and putting a premium on physical perfection.

work: The potential to move an object against a force.

wrist, neck, and waist circumference: A metric for assessing body composition based upon circumferences.

X-rays: Very high energy radiation; ionizing radiation. Can pass through soft tissue, but are reflected by bone (hence their utility in diagnostic medicine).

yeast: Fungal microorganisms that consume sugar and generate CO_2 as a by-product. Used to ferment beverages (to generate alcohol) or leaven bread.

INDEX

— Note that terms commonly appearing as abbreviations in the text are listed
alphabetically by abbreviation.

CPSIA information can be obtained
at www.ICGtesting.com
Printed in the USA
LVHW072326231121
704263LV00002B/3